Energy Technology Roadmaps of Japan

Yukitaka Kato • Michihisa Koyama •
Yasuhiro Fukushima • Takao Nakagaki
Editors

Energy Technology Roadmaps of Japan

Future Energy Systems Based on Feasible Technologies Beyond 2030

Springer

Editors
Yukitaka Kato
Tokyo Institute of Technology
Tokyo, Japan

Michihisa Koyama
Kyushu University
Fukuoka, Japan

Yasuhiro Fukushima
Tohoku University
Sendai, Japan

Takao Nakagaki
Waseda University
Tokyo, Japan

ISBN 978-4-431-56735-6 ISBN 978-4-431-55951-1 (eBook)
DOI 10.1007/978-4-431-55951-1

Printed on acid-free paper

This Springer imprint is published by Springer Nature
The registered company is Springer Japan KK

Foreword

The Society of Chemical Engineers, Japan (SCEJ) will celebrate its 80th anniversary in 2016. The SCEJ is devoted to the development of Japanese industry and society through technological developments based on chemical engineering. Recent human societies have experienced drastic changes in lifestyle, social structure, and the global environment. With such change, it is expected that qualitative transfiguration of chemical engineering technologies will be required to support future industrial production and material and energy consumption in society. The SCEJ has continued to foster new developments based on the VISION2023 proposal. Following VISION2023, the Committee on Future Energy and Social Systems (FUENSS), SCEJ Center for Strategic Planning, planned a project toward publication of a book envisioning future energy systems for Japan, based on chemical engineering knowledge. This book resulted from that plan, with academic and industrial researchers given full support by the SCEJ.

The energy system is a basic infrastructure for social activities. Research subjects for energy systems range from primary energy supplies to energy consumption to waste management. Innovative technologies are required not only in energy and material conversion processes but also in carbon dioxide emission mitigation and radioactive waste management associated with the Fukushima Daiichi nuclear power plant accident of 2011.

Development of innovative technologies is needed to improve conventional methods. These technologies should also be optimized for local consumers, with minimized risk and technological robustness. In future energy systems, renewable energy resources should be fully utilized and consumption of high-density conventional energy resources minimized. Public acceptance is mandatory for a choice of technology options.

Chemical engineering can overview relational energy systems, find the proper control step in the system, and adapt optimized technologies to it. Chemical engineering has great potential for solving problems during the establishment of future energy systems.

It is difficult to solve such problems solely through personal effort, so cooperation between researchers in different fields will be required. This book proposes potential technological options derived from discussions based on chemical engineering standpoints, with strong cooperation between diverse researchers in broad fields. The associated technologies are largely treated in relation to Japan but may be applicable worldwide. I hope that the technologies proposed in this book are useful contributions toward the development of future energy societies.

President, The Society of Chemical Engineers, Kazuhiro Mae
Tokyo, Japan

Preface

Japan has the world's third largest GDP and, fortunately, has experienced economic prosperity in recent decades. Because the country has few fossil resources, industries have efficiently imported raw materials and offered industrial products with high value and low cost. Japan has been successful through such industrial activities. However, stable resource imports present a risk, and the country has attempted to reduce this risk. Very efficient use of these imports and saving of resource consumption have been achieved, such that Japan has the most efficient energy utilization in the world.

Outlooks and discussion on Japan's future energy system are important for establishment of a bright future society, because the country needs to overcome risks to the supply of energy. Energy technology developments are keys to this energy system. These technologies have great diversity and numerous aspects, and each citizen has an opinion on future energy systems. Chemical engineering has the potential to quantitatively give an overview and analyze and optimize complex energy systems. A group in the Division of Energy Engineering, Society of Chemical Engineers, Japan (SCEJ) has reviewed future energy systems based on chemical engineering perspectives. Books discussing the potential of energy technologies and proposals for a future Japanese energy society based on this review were published as a first edition in 2005 [1, 2] and second edition in 2010 [3].

Following publication of the second edition, Japan suffered the Great East Japan Earthquake on 11 March 2011. This caused a drastic change in its energy system. The "Urgent Proposal for Energy Crisis in East Japan After the Great Earthquake" [4] was submitted by second-edition group members in March 2011. This was the first proposal on sustainable and feasible energy supply and demand for overcoming the crisis. This was followed by several other proposals from researchers and academic societies. The Temporary Committee for Urgent Proposals for Energy Systems After the Great East Japan Earthquake, SCEJ, organized in April 2011, continued with proposal improvement and publicity. The Committee on Future Energy and Social Systems (FUENSS), Center for Strategic Planning, SCEJ, was formally established for taking over the Temporary Committee and subsequent discussions on future energy systems in 2012.

This book was planned as an activity of FUENSS and represents the third edition of the books. People were able to overcome difficult social situations after the earthquake if they had hope. This book was designed to review the potential of Japanese energy technologies beyond 2030, toward establishment of a bright future society in the country. Reducing energy and material consumption and CO_2 emission mitigation are key global topics for future societies. High-quality industrial technologies and methods cultivated from Japanese industrial developments have great potential to solve current problems.

Nevertheless, the digitization of products has promoted their commoditization, and it has become easy for new companies to enter the global market. Thus, Japan is losing its global market for commodities such as home electric appliances. However, the country has a unique manufacturing method called *Suriawase* or *Gijyutu-no-Suriawase* (optimized integration between related technologies), and its industries retain a large market share throughout the world, especially among material industries [5, 6]. *Suriawase* is harmonized matching between production processes, from raw materials, components, and equipment to final products, through strong mutual communication and cooperation between researchers, engineers, and organizations. The value of final industrial products can be improved to world-leader status through cost reduction, applying harmonized adjustments between processes based on *Suriawase*. *Suriawase* efforts by researchers and engineers have achieved high-quality industrial products, as typified by robust and low-cost automobiles, high-quality and stable electricity supply infrastructure, punctual train operation, and others.

Suriawase shows mutual cooperation and improvement achieved by people related to production processes, and have great potential for social contribution, although the ability of individuals is limited. In this book, the subject is confined to Japan for clear evaluation of the quantitative potential of energy technologies. However, these technologies can be used globally. All the authors hope that these technologies contribute globally, via mutual cooperation based on the *Suriawase* approach.

This work was supported by the Society of Chemical Engineers, Japan (SCEJ), MEXT KAKENHI Grant Number 26285080, and the Environment Research and Technology Development Fund of 2011, No. RFe-11T1, Department of the Environment, Japan. On behalf of the editors and authors, I am greatly indebted to Prof. Ian S. Metcalfe, Newcastle University, UK, for his comprehensive review and effective comments on the book, and to all supporting individuals for their kind assistance.

Tokyo Institute of Technology, Tokyo, Japan Yukitaka Kato

References

1. Kameyama H, Kato Y (eds) (2005) HONEBUTO energy road map (Honebutono energy road map) (in Japanese). Kagaku-Kogyo-Sha, Tokyo
2. Kato Y (2007) Innovative energy materials from HONEBUTO energy road map. J Chem Eng Japan 40:1141–1149
3. Kato Y, Yasunaga T, Kashiwagi (eds) (2010) HONEBUTO energy road map, 2nd edn (Honebutono energy road map 2) (in Japanese). Kagaku-Kogyo-Sha, Tokyo
4. Fukushima Y, Kikuchi Y, Kajikawa Y, Kubota M, Nakagaki T, Matsukata M, Kato Y, Koyama M (2011) Tackling power outages in Japan: the earthquake compels a swift transformation of the power supply. J Chem Eng Jpn 44:365–369
5. Nakagaki T (2014) Manufacturing technology for indispensable materials (Nakutewa-naranai-monozukuri). Chem Eng Jpn (Kagaku Kogaku) 78:512–513
6. Manufacturing white paper in 2013 (Monozukui hakusyo in 2013). Ministry of Eco., Trade and Ind., Tokyo, Japan. http://www.meti.go.jp/report/whitepaper/mono/2013/. Accessed 23 Sept 201.

Contents

Part I
Introduction to Future Energy Systems Based on Feasible Technologies Beyond 2030

Yukitaka Kato and Michihisa Koyama

This book aims to describe the current status of Japan's energy system and roadmaps of energy technologies. An exteriorized vision for future energy systems in Japan beyond 2030 based on feasible technologies discussed in each chapter in this book is presented in a figure. Parts II–VIII of this book cover widely categories related with Japan's energy system. Executive summaries for the parts are introduced. The background and concepts of roadmaps are elaborated, and the function, potential risks, and usage of a technology roadmap are also discussed.

Y. Kato (✉)
Tokyo Institute of Technology, Tokyo, Japan
e-mail: yukitaka@nr.titech.ac.jp

M. Koyama
Kyushu University, Fukuoka, Japan
e-mail: koyama@ifrc.kyushu-u.ac.jp

Future Energy System and Executive Summaries of the Parts

Yukitaka Kato and Michihisa Koyama

Abstract This book aims to foster visualization of bright future energy societies by describing energy technology options. An exteriorized vision for future energy systems in Japan beyond 2030 based on feasible technologies discussed in each chapter is presented in a figure. The system consists of domestic primary, secondary, and ternary industrial sectors plus energy supply infrastructures, electric power generation, and transportation, with consideration of international relationships. Parts II–VIII of this book cover these categories. Executive summaries for the parts that contain chapters for technological terms described in the figure are introduced in this chapter.

Keywords Future energy system • Energy technology option • Japan

1 Energy System Beyond 2030

This book aims to provide a basis for envisioning a bright future society that uses energy sustainably, by providing future perspectives on major technology options. Figure 1 illustrates an exteriorized vision of a future energy system in Japan beyond 2030, based on feasible technologies discussed in the chapters of this book. Japan's energy system will be developed in consonance with its unique geopolitical, economic, and social structures. The figure consists of categories of domestic primary, secondary, and ternary industrial sectors, together with the residential and transportation sectors and energy supply infrastructures such as large-scale electric power generation systems. Parts II–VIII cover topics in each of those

---

The online version of this chapter (doi:10.1007/978-4-431-55951-1_1) contains supplementary material, which is available to authorized users.

Y. Kato (✉)
Tokyo Institute of Technology, Tokyo, Japan
e-mail: yukitaka@nr.titech.ac.jp

M. Koyama
Kyushu University, Fukuoka, Japan
e-mail: koyama@ifrc.kyushu-u.ac.jp

Y. Kato et al. (eds.), *Energy Technology Roadmaps of Japan*,
DOI 10.1007/978-4-431-55951-1_1

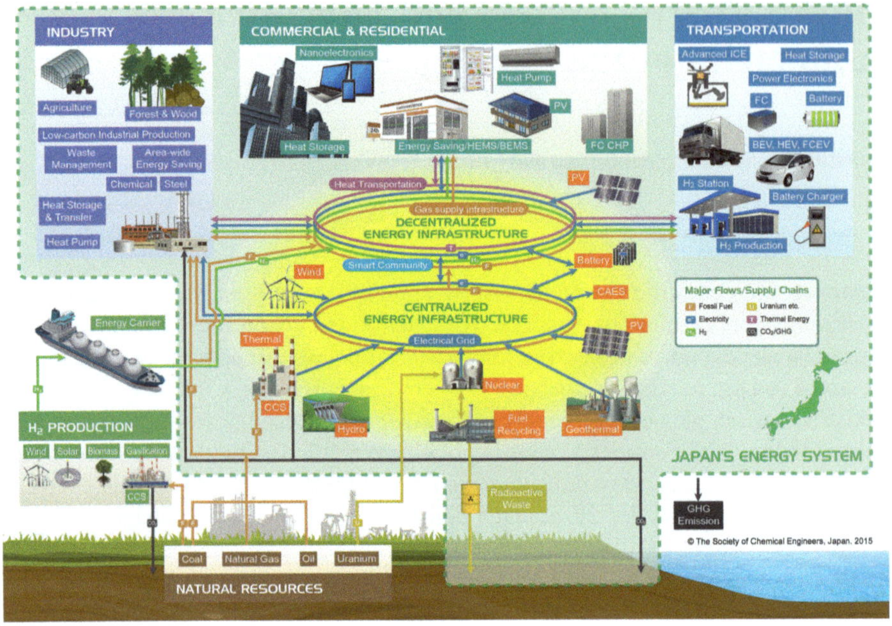

Fig. 1 Future energy system in Japan beyond 2030 based on feasible technologies (*BEMS* building energy management system, *BEV* battery electric vehicle, *CAES* compressed air energy storage, *CCS* carbon dioxide capture and storage, CO_2 carbon dioxide, *FC* fuel cell, *FC CHP* FC combined heat and power, *FCEV* FC electric vehicle, *GHG* greenhouse gas, *HEMS* home energy management system, *HEV* hybrid electric vehicle, *ICE* internal combustion engine, *PV* photovoltaic solar cell) (Published with kind permission of © the Society of Chemical Engineers, Japan, 2015. All Rights Reserved)

categories. Technological terms in the figure are representative of the chapters. There are many items not shown in the figure that are important in the future energy system. Therefore, we recommend that readers first review the figure to grasp the big picture of future energy systems and the relationships of each technology option and then read chapters of interest to discover detailed perspectives in the field.

The future energy system shown in Fig. 1 is based on two main networks, i.e., centralized and decentralized energy infrastructures. Presently, Japan mostly depends on imported resources of coal, natural gas, crude oil, and uranium. Depending on the period of focus, these will remain major resources, but drastic increases in domestic energy production will raise the self-sufficiency ratio, so the contribution of such conventional resources could become minor. As a possible energy resource shift, overseas hydrogen (H_2) produced from wind, solar, and biomass energies, plus coal gasification implemented with carbon capture and storage, is illustrated in the figure. Establishing feasible H_2 carrier systems is necessary to support the transport of H_2 from overseas production sites to domestic consumption sites. The centralized energy supply system will maintain its present role, with both conventional and nontraditional technologies. In addition to the efficient use of fossil fuels or the increased role of renewable energy sources, backend measures such as carbon capture and storage and radioactive waste treatment should be considered for mitigating environmental loads. Decentralized energy infrastructures will be developed for electricity, H_2, fossil fuels, and thermal energy, with integration with the smart community. Advancement of electricity and gas supply infrastructure as well as heat transportation systems is inevitable. In the industrial sector, energy-consuming industries such as chemical and steelmaking are targeting drastic reductions of carbon dioxide emissions by introducing innovative processes or radical reformation of current processes. Another important approach is integration, such as area-wide energy saving. In addition to energy consumption, one must consider waste management measures. Also, primary industry such as agriculture, forest, and wood should be maintained sustainably, considering their various aspects or roles in the local community. Many advanced technologies such as highly efficient fuel cells, heat pumps, and ultralow energy-consuming electric appliances will be implemented in the commercial and residential sectors. Their integration with energy management systems in a smart community will be key to strong penetration of intermittent renewable energy systems such as photovoltaic cells and wind power. In the transportation sector, dramatic advancement of the power train is expected together with the progress of electronic devices and heat management systems. For nonconventional power train systems, there will be establishment of infrastructure such as battery chargers and H_2 refueling stations.

We expect a low-carbon, sustainable, and bright future for society as a result of harmonized contributions of advanced technologies and concepts described in this book.

2 Executive Summaries of Book Parts

This book consists of eight parts. Parts II–VIII contain chapters describing the details of items shown in Fig. 1. Executive summaries for the parts are provided in the following.

2.1 Part II: Multiple Aspects of Energy Systems in Japan: Present and Future Perspectives

This part addresses the background, present status, and future directions of Japan's energy system.

Energy systems and policy are influenced by geopolitical, economic, social, and technological factors and regimes. Because Japan has few fossil fuel resources, its energy system is heavily dependent on imports. Thus, it is considered important to diversify energy resources, replacing fossil fuel with other energy resources such as nuclear power generation and renewable energy technologies. It is also important to increase efficiency of energy use in electricity generation and the industrial, transportation, commercial, and residential sectors.

The accident at Fukushima Daiichi Nuclear Power Plant caused by the earthquake and tsunami of 2011 changed public attitudes toward nuclear power. The short-term challenge in supplying peak demand during the summer was overcome by collective measures, including urgent enhancement of power supplies, reduction of peak demand through lifestyle and behavioral changes, and others. However, the future energy system is uncertain and under debate.

The design of future energy systems and energy policy involves diverse stakeholders with varying priorities toward energy security, the environment, economic efficiency, safety, and industrial and economic development. Alternative guiding principles for the energy system and policy design are necessary and discussed in this part.

2.2 *Part III: Advanced Use of Secondary Energy Media*

This part discusses advanced use of secondary energy media. Large-scale electrical energy storage systems with electrochemical batteries offer the promise of better utilization of electricity, with load leveling and massive introduction of solar and wind power renewable energies. Sodium-sulfur and redox-flow batteries are candidate technologies. Greater energy efficiency and cost reduction are technical challenges.

Hydrogen production involves the consideration of energy conversion efficiency, carbon dioxide (CO_2) emission, and cost. Even when hydrogen is produced conventionally from fossil fuel, a fuel-cell vehicle's well-to-wheel CO_2 emission can be less than that of conventional gasoline engine vehicles. Hydrogen production methods using biomass, renewable energy-based electrolysis, thermochemical methods, and photoelectrochemical water splitting are important to reduce CO_2 emission.

The importance of the energy carrier for hydrogen produced by renewable energy use has been recognized. Long-term and stable storage of hydrogen becomes subject to the complete hydrogen supply chain in the form of liquid hydrogen or chemical hydride, such as ammonia or organic hydride.

Waste heat recovery has great potential in Japan. Efficient heat use requires optimized technologies, with a combination of heat storage, transportation, and transfer. Thermochemical energy storage has potential for high-temperature storage at temperatures >200 °C. Heat transportation at <200 °C by latent heat storage has practical possibilities for waste heat utilization. Technology for heat exchange in the gas phase will become increasingly important. Consistent design of material, reactor, heat exchanger, and heat utilization systems at low cost is an ideal technology goal for heat recovery.

2.3 Part IV: Energy Supply Infrastructure

Part IV discusses the infrastructure of energy supply for people and economic activities in Japan, focusing on the infrastructures for gas, electricity, and automobiles. After imported natural resources such as primary energy, heat amount, component, and quality adjustments are made at refineries and power plants, those energy media supply consumers. Pipelines are used for gas and transmission grids for electricity. Those infrastructures must change per current trends of electricity and gas market reformation, discussions of infrastructure robustness after the Great East Japan Earthquake, and requirements of low-carbon energy use.

For gas supply infrastructure, stable and effective energy supply is expected through diversification of resources and extensive geographical and smart gas transmission networks. In city gas production, the calorific value of liquefied natural gas is likely to be less by using unconventional sources such as shale gas and coalbed methane, which have not yet been used extensively. Facilities and equipment must respond to this change, and a robust energy supply with the use of such diversified sources is expected. In gas transmission, it is also expected that wide-area gas pipelines will be constructed through cooperation of the government and gas companies. New technologies including underwater gas pipelines are also anticipated. The introduction of smart gas meters has progressed for consumer convenience, followed by comprehensive energy supply services using telecommunication networks. Challenges in gas supply businesses include energy infrastructure replacement needed for electricity and gas market reformation, transportation, and mixing of hydrogen for current city gas, plus more complex fuel adjustment owing to the use of hydrogen. Although there are some technical issues, gas supply infrastructure is likely to have a central role in Japan's future energy supply.

In the electricity infrastructure, it is anticipated that nationwide power interchange will increase and regional-level demand-side management (DSM) will expand in the long term. This trend is expected to be strengthened with nationwide and regional efforts. A robust supply chain and reductions in energy consumption and CO_2 emissions will be realized by the expansion of power interchange over wide areas, demand response, and smart grids in micro areas, plus increased use of renewable energy. Power generation efficiency improvement, the introduction of CCS technology, and expansion of renewable energy are likely after the 2030s, in response to greater demand for CO_2 emission reductions. Japan is tackling issues such as uncertainties of fluctuating energy prices and nuclear power plant rebuilding or replacement. However, the aim for its electricity infrastructure remains a resilient power supply and greater reduction of CO_2 emissions.

In the automobile fuel supply infrastructure, the number of gasoline refueling stations for internal combustion engine vehicles (ICEVs) and hybrid electric vehicles (HEVs) will continue to decline in the long term, while deployment of electric car charging and hydrogen refueling stations will progress with active government support to increase battery electric vehicles (BEVs), plug-in hybrid electric vehicles

(PHEVs), and fuel-cell electric vehicles (FCEVs). To realize a new low-carbon energy supply chain, it is anticipated that charging and hydrogen refueling stations will have more attractive features. Examples include a demand response system for BEV and PHEV chargers that works as energy storage by absorbing fluctuating supply energy from renewables. Furthermore, stable power supply is expected by integrating these into smart grid networks. In addition to their role as fuel supply facilities for vehicles, hydrogen refueling stations are also anticipated to have different functions, e.g., as energy storage and conversion stations from electricity to hydrogen. As part of energy storage and supply networks, they supply natural gas-reforming hydrogen to local areas similar to community gas businesses. It may take a while for hydrogen to become widespread, because it is currently controlled under the High Pressure Gas Safety Act, which has strict regulations. Regarding negative aspects of BEVs and PHEVs, if such vehicles expand, power demand is likely to rapidly increase. However, successful charging infrastructure deployment for BEVs, PHEVs, and FCEVs will boost market penetration of these vehicles, and consequently they are anticipated to contribute to greater CO_2 emission reductions.

2.4 Part V: Electric Power Generation and Its Backend Technologies

Part V deals with electric power generation and its backend technologies. Electricity is the most important secondary energy in the modern world. It is converted from various primary energy sources, such as fossil fuels, uranium, and renewable sources. This part covers principal technologies for large-scale electric power generation systems connected directly to power grid systems, including the back end of the nuclear fuel cycle. Reviewing Japan's energy flow, 7.44 EJ primary energy from several primary energy sources was fed into the electric power generation system and converted into 3.97 EJ of electricity and nearly the same amount of rejected energy in FY 2010. In particular, the share of nuclear power generation was ~30 % before the Great East Japan Earthquake, contributing to securing a stable supply of energy and low CO_2 emission. Thermal power generation, which evenly consumes coal, natural gas, and petroleum as fuel because of energy supply security, has been the main electric power source since the 1960s (the strong growth period of the Japanese economy). These increased in importance after the earthquake. Owing to the country's mountainous land and warm oceanic climate, hydropower can generate electricity year round. In recent decades, pumped storage hydropower has increased for load leveling, especially in the summer peak season.

The share of solar, wind, and geothermal power generation is <1 %, whereas solar power generation has been greatly increasing after enforcement of the Japanese "feed-in tariff." The main renewable source is still hydropower. From the viewpoint of industrial structure, the country is unique in that many heavy industry companies, including contractors for nuclear power plants, compete with each other and thereby produce the most efficient electric power in the world.

In the fourth strategic energy plan, the national government described in their policy that "nuclear power is an important baseload power source." However, no disposal site for high-level radioactive nuclear waste has been identified. Also, Japan had ~47 tons of plutonium within and outside the country at the end of 2013. In this difficult situation, reprocessing technologies of used nuclear fuel for recycling fast breeder reactors may be a solution, in addition to recycling back into metal oxide nuclear fuel for thermal reactors. To avoid CO_2 emission in tail gas, imported hydrogen is expected in the future, even for large-scale power generation and carbon capture and storage technology.

2.5 Part VI: Primary and Secondary Sectors of Industry

Part VI describes primary and secondary sectors of industry. There are three major dimensions that each industry can explore to plan their contribution toward a sustainable society: (1) enhancement of efficiency, (2) increasing availability and the use of renewable raw materials and energy, and (3) cascade utilization of materials and energy beyond the borders of business entities, sectors, and locations. It is generally said that for (1), Japan is already a top performer in the world in many of its industries. For (2) and (3), there remains much room for exploration. In reality, the situation varies by industry. Therefore, which of the three dimensions should have more emphasis will vary by sector.

Chapters in this part describe road maps to transform current industrial sectors other than electricity (Part V) and residential and commercial (Part VII), toward a sustainable future. Among industries in the primary sectors, the focus is on forestry, wood, and agriculture. Historical, cultural, and technological aspects are explained, to better guide readers through issues confronting Japan within these industries. From the secondary sector, chemical industries are discussed, with various emphases on the aforementioned dimensions. Finally, because energy recovery from waste has emerged as an increasingly important topic in the coming decades, a chapter is devoted to this topic. The authors share their vision on how different sectors in Japan may engage in endeavors toward the formation of sustainable industrial systems through management of technologies and resources. Sustainable systems in primary and secondary industrial sectors based on renewable resources require coprosperity of urban and rural communities. The chapters also highlight how new technologies can help in this regard and where the opportunities lie in the envisioned societal circumstances. In addition, many interesting individual projects are presented to deepen understanding, based on actual examples.

Although greenhouse gas emission from Japanese industries currently contributes substantially to the global total, its quantitative importance will decline because of slowing population growth. Projects of the country's industrial systems can play key roles in creating and innovating advanced technologies that can first be applied in Japan and then in other parts of the world that are rapidly developing. Because Japan is facing sustainability challenges in advance of most countries and regions in Asia, its vision for adaptation to changing socioeconomic and environmental circumstances will provide a reference to many regions that will face similar challenges in the near future.

2.6 Part VII: Commercial and Residential Energy Utilization

Part VII addresses technologies for commercial and residential energy use. It deals with energy technology road maps in commercial and residential sectors. These sectors, which include homes and buildings, account for about a third of primary energy consumption in Japan, and their energy consumption has been continuously increasing in recent decades; this consumption in 2012 was 2.4 times greater than that in 1973. Hence, there is a great demand for reduction of energy consumption in the two sectors. As shown in the chapter by Kikuchi et al. in Part II, the residential sector consumes electricity (46 %), natural gas (26 %), and oil (27 %). Electricity, natural gas, oil, and coal account for 44 %, 27 %, 28 %, and 1 % of energy consumption in the commercial sector, respectively. In the residential sector, electricity is used for lighting, power supply, space cooling, and others. However, ~6 % of electricity use is dissipated by standby power; this energy loss should be reduced by a novel device using nanoelectronics technology.

Fossil fuels such as natural gas and oil are converted into heat for space heating, hot water supply, and cooking. Consequently, nearly half of consumed energy in both sectors is supplied as electricity, and the other half used as heat. Therefore, it is important to use electricity and heat efficiently and reduce losses of both energies in the commercial and residential sectors. From this point of view, combined heat and power (CHP) systems with fuel cells (FCs) were introduced in the residential market in 2009. At the end of 2014, a FCEV was launched in the Japanese market. As a result, the establishment of a hydrogen economy has been pursued in Japan. Moreover, after the Great East Japan Earthquake in 2011, distributed power generation such as gas-fueled CHP systems has gained attention for reinforcing energy resiliency and for improving energy utilization efficiency. These systems can supply electricity via islanded operation, even during outages. Therefore, introduction of CHP systems in Japan will be further promoted. In addition to improvement of energy efficiency of facilities, smart energy systems, in which distributed generators are connected to form a network with user equipment and power storage as nodes, have also attracted attention in the country. Examples are smart houses, smart communities, and smart cities. With this background, this part contains five chapters and two topics on energy technologies using electricity.

2.7 *Part VIII: Transportation*

Part VIII portrays technologies for the transportation sector. As next-generation automobiles, BEVs and FCEVs have been developed because they have high energy efficiency and low CO_2 emissions. However, as alternatives to gasoline cars, they have major problems such as mileage, price, and infrastructure.

Thermal management of automobiles improves their total energy efficiency, not only for internal combustion engines but also for BEVs and HEVs. Chemical heat pump systems can be driven by both engine waste heat and various waste heats in HEVs or electric power from charging equipment, saving on energy use with expensive batteries for heating and cooling.

Power electronics that facilitate DC-AC or AC-DC conversion of electric power are a critical technology for energy saving. Electrification of vehicles contributes to CO_2 emission reduction. Wide-bandgap semiconductors such as SiC or GaN, which can be driven at high frequency and high temperature with downsized converters, will be important in future power electronics.

The transportation section must achieve further reductions in fossil oil consumption and CO_2 emissions. ICEVs (including HEVs) are expected to remain dominant in 2050. The thermal efficiency of HEVs with next-generation ICEs will reach 50 % in the future, via ICE technology developments for high compression/expansion ratios, super lean combustion, and higher boost pressure.

Roadmap of Energy Technologies for Envisioning Future Energy Systems

Michihisa Koyama, Takuya Hasegawa, and Yuya Kajikawa

Abstract Roadmaps, visions, and scenarios are different representations of certain aspects of thinking or perceptions of the future. Technology roadmaps have been used as a planning and management tool in industry and organizations. This book describes the current status of Japan's energy system and roadmaps of energy technologies. In this chapter, we elaborate on the background and concepts of roadmaps. We also discuss the function, potential risks, and usage of a technology roadmap.

Keywords Technology roadmap • Uncertainty • Risk • Future energy system

1 Aim of This Book

Energy functions a basis for modern human life. It supports daily living and industrial, economic, and social development. Energy is a matter of exploration across broad disciplines. New discoveries and inventions in energy and related technologies, such as turbines, nuclear power, and photovoltaic cells, have changed our ways of living. Exploration with intense curiosity has pioneered frontiers of science and established disciplines like thermodynamics and nuclear science and technology. The history of energy is also one of resource exploitation. Fossil fuels, which remain and will likely be a major energy resource, are regenerated little within human history and thus are defined as a nonrenewable resource. Such fuels

M. Koyama (✉)
Kyushu University, Fukuoka, Japan
e-mail: koyama@ifrc.kyushu-u.ac.jp

T. Hasegawa
Nissan Motor, Ltd., Yokohama, Japan

Ritsumeikan University, Kyoto, Japan
e-mail: takuya-hasegawa@mail.nissan.co.jp

Y. Kajikawa
Tokyo Institute of Technology, Tokyo, Japan
e-mail: kajikawa@mot.titech.ac.jp

Y. Kato et al. (eds.), *Energy Technology Roadmaps of Japan*,
DOI 10.1007/978-4-431-55951-1_2

have been exploited at increasing speeds during the last two centuries. Exploitation of energy resources has degraded local environments and led to global climatic change. International conflicts over limited access to resources have sometimes led to war. The importance of transformation into renewable energy systems and sustainable society is currently well recognized, but appears to have limited influence on social change through motivating stakeholders with a sense of urgency.

It is known that a single technology or social system will not solve current issues or bring about sustainable society; thus, it is mandatory to implement a number of technologies in society in an integrative manner, possibly with suitable support by policies or realization of new societal systems. Someone intending to draw a pathway to a future sustainable society may face difficulties in identifying feasible enabling technologies, which play key roles in realizing that society, from among a variety of options. Nowadays, we typically face difficulties in developing a comprehensive view of an issue while keeping the details reliable and neutral. Information infrastructure enables efficient access to information. However, this poses challenges, even for researchers who purport to have deeper knowledge of technology futures relative to other types of stakeholders, because disciplines are continually divided into many branches or subdisciplines because of the continuous accumulation of knowledge and development of academism.

This book is edited so as to help stakeholders envision a possible future society on the basis of feasible technology options, as well as to appropriately identify related technology in the big picture of a possible future society. By comparatively reading the thematic chapters, readers will find the mutual relationships of technology options, such as competitive, coexisting, and tightly coupled. This will facilitate fair discussion, ideally avoiding one biased toward a certain technology option or an a priori presumed conclusion. In this book, a future perspective is described in each chapter to the author's best knowledge. This will support comparison of technology options not limited to the present or near future, but also considering the temporal development of technology maturity or change of social systems. Such comparison will help identify related technology in the future society in a more objective manner, or help to envision that society in a more integrative and robust way.

The book is organized considering the present situation of Japan. Although editors are aware that the country's contribution to fossil resource savings and greenhouse gas (GHG) emission reduction is small in the global context, we believe the contents will be of great use in other countries. As pointed out by Dr. Komiyama, Japan is facing problems in advance of other countries. That is, many of its problems have emerged first or early relative to the rest of the world, whose countries will eventually face them. Energy presents a typical problem for Japan because of the country's scarce domestic resources. To conquer or counter problems that have emerged, the country has put a number of state-of-the-art technologies into practice, including light-emitting diode lighting, low-fuel-consuming vehicles with ultra-low harmful exhaust emission, residential fuel cell systems, hydrogen-fueled vehicles, and ultra-low GHG emission landfills for garbage treatment. We thus believe the present status and future prospects of technologies described in this book provide a useful showcase for many other countries.

In the subsequent parts of this book, the reader will find chapters describing the present status and future perspectives, as well as risks associated with technology penetration in the sector addressed. The authors of each chapter have an engineering background in their respective fields. Most have strong expertise in the technologies that will be important in the focused-on sector. This makes the book a compilation of state-of-the-art knowledge, covering a variety of sectors related to society. The editors acknowledge that the granularity and development phase of the described content depends on the chapter, i.e., sector. Some sectors adopt long-lived conventional technologies with continuous retrofitting of improvements, while others include alternatives to current technologies and unique and emerging directions to follow.

The editors also acknowledge that each author is not necessarily an expert in other technologies or sectors, even if the author is closely tied to the relevant sector. In other words, an author has expertise in a specific technology described in the chapter, but does not necessarily have comprehensive perspectives of competing technologies. Thus, the reader is encouraged to read chapters of interest followed by related chapters, irrespective of chapter order, to discover various ways of thinking, unexpected relationships, trade-offs, and potential discrepancies. More importantly, such a comparison may stimulate the reader to consider possible integration of technologies that are conventionally considered only for certain sectors.

For example, one will find as a current serious issue that intermittent outputs of photovoltaic and wind power generation systems are detrimental to the electrical grid, which must maintain a stable frequency. This situation may hinder the large-scale penetration of such renewable technologies by 2030 or beyond. One may also notice that a decrease in nuclear and increase in fossil-fuel-based thermal power plants is used to legitimate the need for large-scale penetration of photovoltaic and wind power generation systems in Japan, as a transition from a fossil-fuel-based to a low-carbon society. The chapters on nuclear systems guide the reader to appropriately understand the issue of nuclear waste, even if the country chooses a nuclear-free society in the future. The proposed concept of power generation from innovative waste treatment, matched with an intermittent renewable option, may suggest to readers its combination with secondary energy sources such as hydrogen, which affects the transportation and residential sectors.

As a unique aspect of this book, there is peer review of descriptions in subsequent chapters by researchers not only in the same field but also in other disciplines, such as life-cycle engineering, process engineering, and innovation management. Experts and peers in the same discipline have sometimes shared norms and opinions regarding their technology and its societal potential, which lack rigid logic and evidence. To best improve the neutrality and reliability of the descriptions, reviewers have commented from a wide range of aspects such as competition with other technologies, potential risks, self-consistency, and feasibility. Also, comments are geared toward understandability for readers in other disciplines while keeping expertise in the field.

2 Constructing the Roadmap with Uncertainty and Risks

A technology roadmap is a map illustrating the future direction of technological development. It aims to direct future paths of advancing technologies and foster understanding among stakeholders. One of the functions of the technology roadmap is to support understanding of focal technology by articulating technological development and plans. It is also used to visualize relationships between technology, products, service, systems, markets, plans, and required resources. Another objective is to accelerate research and development by enabling a concentration of effort toward technological options selected in the roadmap. Explicit statements regarding technology bottlenecks facilitate the overcoming of challenges and thereby accelerate development. In addition, a comprehensive view enables engineers and managers in product/service/system design to make decisions and form a consensus. An attractive future is expected to be realized if the development proceeds as illustrated in the roadmap, which would enhance positive expectations of that roadmap. This book illustrates each technology roadmap within the common framework shown in Fig. 1.

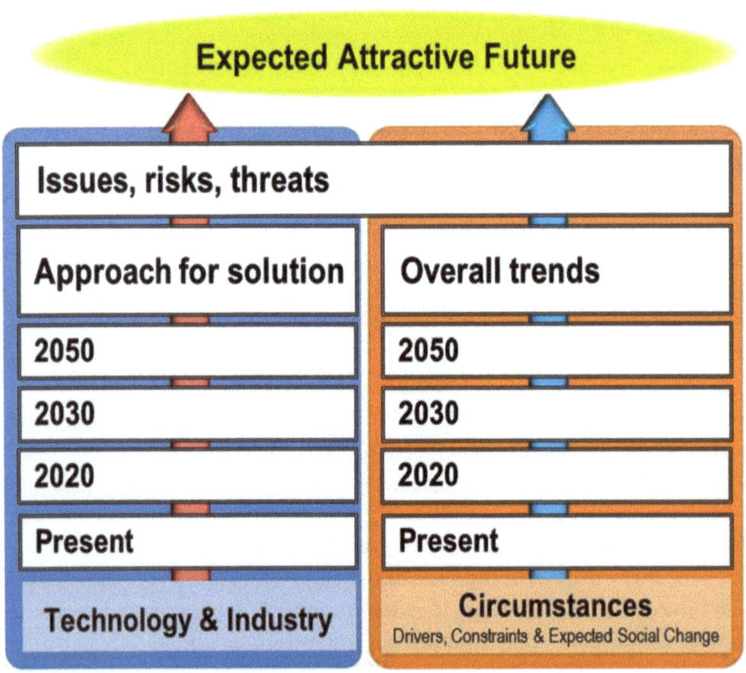

Fig. 1 Framework of the roadmap in this book

Technology roadmaps have been used to promote understanding among stakeholders, accelerate research and development, and support decision making. However, these roadmaps represent an opinion or plan rather than facts. Presently, there are a variety of roadmaps, visions, and scenarios. It is rare for different scenarios to be compared and assessed against each other. There are substantial discrepancies in their reliability and feasibility. It is sometimes difficult to judge whether a roadmap is based on solid mechanistic estimates, feasibility assessment and strategic planning, or simply a manifestation of an author's opinion and dream. Obviously, the future is uncertain; no one can fully predict it. Despite this, we should maximize our capability to envision the future. Although this does not mean that such a future is realized as imagined, that future will seldom be realized if not imagined. We must envision the future in not only attractive and salient ways but also in reliable, feasible, and legitimate ones.

For example, futures with artificial intelligence (AI) have been illustrated in a variety of movies including *2001: A Space Odyssey and Blade Runner*. But the future is still (and in principle) uncertain. It has positive and negative sides. AI can assist and improve our lives by machine translation and diagnose of diseases and as companions. AI might autonomously develop them and be out of our control by threatening our physical and psychological existence, privacy, jobs, lives, and society. A vision and technological breakthrough in AI have attracted imagination for future and investment in research and development, but it has accompanied decline of investment after bubbles. It also has provoked a sense of fear. Expectation and its control are essential means of future innovation and its management for orchestrating challenges and activities that address uncertainty. Expectation has a positive aspect in that it attempts to overcome a sense of crisis with respect to uncertainty, by encouraging the actions of others. Innovation accompanies risk, and its realization is uncertain. Despite the uncertain outlook, we bet on selected options and collect investment and resources for their development. Optimistic expectations enable collective and orchestrated efforts toward the selected option. However, expectation also has a negative aspect in that it can induce gambling investment and fever and cause unbalanced development through unrealistic expectations. The latter, termed hyper-expectations or hype, has been exemplified in various fields, including biotechnology, nanotechnology, information technology, and energy. It can be said that hype is a side effect of any technology roadmap and innovative advocacy.

It is not a rudimentary task to assess the reliability of statements on future perspectives and to avoid hype. As a challenge to our roadmapping process, we explicitly rate the reliability of statements within three ranks, solid mechanistic estimates, extrapolation of trends, and controversial values including personal perspectives (Table 1). The envisioned future has both objective and subjective bases. Some paths are depicted by forecasting past trends via solid mechanistic estimates; others are extrapolations of observed trends from the past; and still others are representations of author opinions and dreams, lacking justification.

Even when based on solid mechanistic estimates, the aforesaid descriptions inevitably include uncertainty, because they result from a projection of plausible

Table 1 Rating of author confidence level

Rating	Technology parameters
★★★	Solid mechanistic estimates
★★	Extrapolation of trends
★	Personal perspective, competition with existing technology (e.g., controversial value)

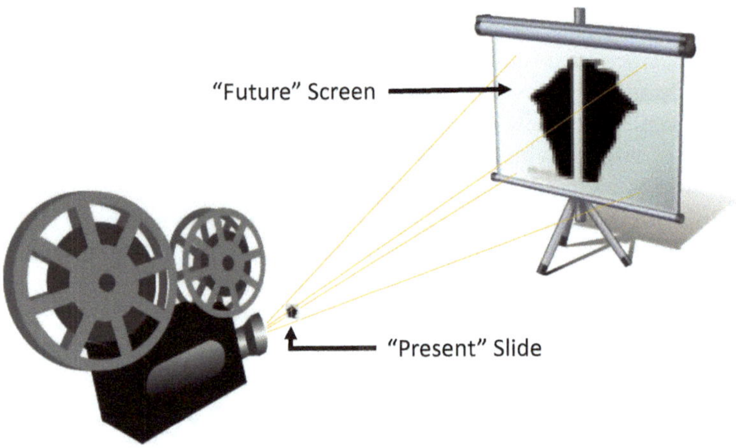

Fig. 2 Projecting to a screen (Published with kind permission of © Michihisa Koyama and Takuya Hasegawa 2015. All rights reserved)

futures based on past trajectories. As schematically illustrated in Fig. 2, the main objective of the projection is not to discern the exact future status but to project the potential future on a screen and provide directions on which we should act now. Although the rank itself and its criteria shown in Table 1 are not strict, we believe that it can help readers understand authors' assessment of the reliability and feasibility of their statements.

A technology roadmap is not a retrospective description of trends but a prospective plan and strategy. It is normative rather than descriptive or analytical. The goal reflects a norm and is shown as what is desirable to attain. Roadmap construction is not just a simple forecast from the present to the future, but also involves a backcasting process from the goal to present status. A roadmap lacking projection is not realistic or feasible. A roadmap depicting a future along the present trend will be realized with high probability, but is not attractive for posing technological challenges to overcome current limitations. Balance and iterative feedback between forecasting and backcasting are essential.

Along the time horizon, the roadmap must be scientifically informed and fact based, with additional efforts to describe what specific years (e.g., 2020, 2030, and 2050) mean. For example, 2030 can be $2020+10$, 2030 ± 3, or 2030 ± 10, with background calculation that socioeconomically supports specific-year accuracy.

However, we admit that the authors in this book do not necessarily intend to give a specific year with accuracy. The authors cite years without such accuracy, just as with indexes of technological "difficulty" for each technology.

Authors show not only attractive futures beyond the paths illustrated in the roadmap but also issues, risks, and threats along every point in those paths. They include technological feasibility, uncertainty in research and development, safety of materials, scarcity and diversity of natural resources, manufacturing capability of products, development of competitive and complementary technologies and infrastructures, the market, and social acceptance.

To fully use the roadmap, we must take care during the first step, because otherwise we make our best efforts using the wrong combination of button and button hole if that step is not carefully examined (Fig. 3). Even when the logic of the second button from the top is coherent, there may be discrepancies between attractive technology of the future and associated risks, supply-side expectation and demand-side/social perceptions, planned development paths and their feasibility, hidden assumptions and a master equation, fact and data as represented fact, and dream and reality. In any case, there is room for learning and self-development by focusing on the above discrepancies or gaps. We hope that this book will be of assistance in the design of future energy systems, identification of the aforementioned gaps that are worthy of investigation and development, and realistic thinking and creative imagination.

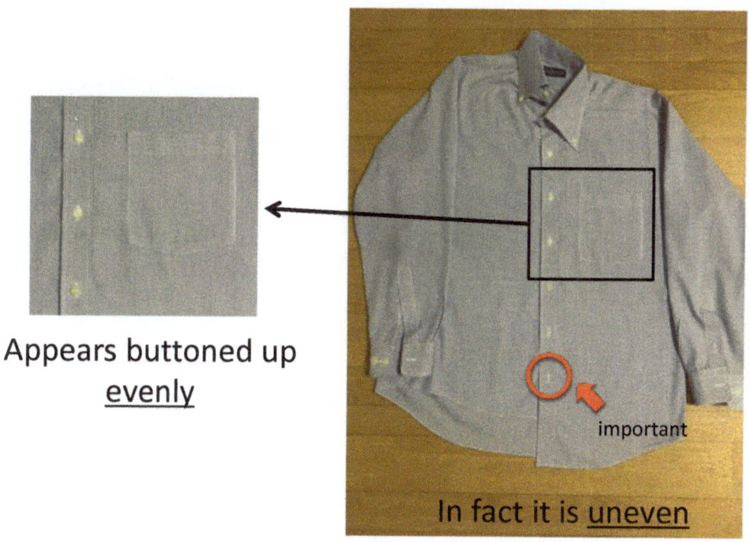

Fig. 3 Illustration of uneven buttoning (Published with kind permission of © Takuya Hasegawa 2015. All rights reserved)

Part II
Multiple Aspects of Energy Systems in Japan: Present and Future Perspectives

Yuya Kajikawa

Energy systems and policy are affected by political, economic, social, and technological factors. Because Japan has a few fossil fuel resources, the country's energy system is heavily dependent on imports. Thus, it is considered important to diversify energy resources, to replace fossil fuel with other energy resources, and to increase efficiency of energy use in electricity generation, industry, transportation, and the commercial and residential sectors. Reflecting this situation, nuclear power is regarded as a quasi-domestic electricity supply. This is because of the long-term use of uranium through reprocessing, and such power is compatible with environmental safeguards including climate change mitigation. However, the accident at Fukushima nuclear power plant caused by the earthquake and tsunami in 2011 changed public attitudes toward nuclear power and provoked public debate on future energy systems. This chapter addresses the background, current status, and future direction of Japan's energy system and related issues. The design of future energy systems and energy policy involves diverse stakeholders, with varying priorities for energy security, the environment, economic efficiency, safety, and industrial and economic development. Alternative guiding principles for energy systems and policy design are needed and discussed herein.

Y. Kajikawa (✉)
Tokyo Institute of Technology, Tokyo, Japan
e-mail: kajikawa@mot.titech.ac.jp

Present Status of Japan's Energy

Yasunori Kikuchi, Seiichiro Kimura, and Michihisa Koyama

Abstract Japan is at a momentous turning point in world history and faces a wide range of social issues that call for proactive solutions, such as the control of environmental loads, countermeasures against rapid population aging and stagnant birthrates, and the construction of sustainable energy systems. These issues have resulted from its mature society, which means that all other countries may confront the same or similar issues in the future. Japan has an opportunity to become the first country to solve them and share its methods.

Realizing sustainable energy use has been an important target. Toward an ideal or favorable future for Japan's energy, understanding the present status and future available energy options is an initial step, followed by a discussion of issues related to each option. The aim of this chapter is to concisely review the current state of Japan and its energy through a statistical investigation. Japan has a fixed estimate of decreasing population in the future. Most of the primary energy in the country has been imported from specific countries, with an associated cost increase for that energy.

With its cutting-edge technologies and cultural creativity, Japan is expected to create new demand and revitalize its socioeconomic affluence, and have a leading role in showcasing solutions to common problems in the world.

Keywords Energy statistics • GDP • Energy per capita • Resource import

---

Y. Kikuchi (✉)
The University of Tokyo, Tokyo, Japan
e-mail: kikuchi@platinum.u-tokyo.ac.jp

S. Kimura
The Matsushita Institute of Government and Management, Kanagawa, Japan
e-mail: s-kimura@mskj.or.jp

M. Koyama
Kyushu University, Fukuoka, Japan
e-mail: koyama@ifrc.kyushu-u.ac.jp

Y. Kato et al. (eds.), *Energy Technology Roadmaps of Japan*,
DOI 10.1007/978-4-431-55951-1_3

1 Status of Japan: Confronting Issues Resulting from a Mature Society

Japan is at a momentous turning point in world history and faces a wide range of social issues that call for proactive solutions, such as the control of environmental loads, countermeasures against rapid population aging and stagnant birthrates, and construction of sustainable energy systems (Fig. 1). These issues have resulted from its mature society, which indicates that other countries may confront the same issues. Each issue shown in Fig. 1 has interconnections. For example, the consumption of fossil resources decreases the self-sufficiency ratio of primary energy and increases greenhouse gas (GHG) emission. The stagnant birthrate is one of the major causes of the increase in the ratio of elderly within the population. Japan has an opportunity to become the first country to solve them and share its methods [1].

Key features of the world in the twenty-first century, which may be regarded as a mature society, can be categorized into four essential trends [1]:

1. Remarkable growth of wealth
2. Longevity resulting from the growth of wealth
3. Saturation of products or amenities they provide
4. Shifts from fossil to renewable primary energy resources

The first feature, the growth of wealth, first occurred in developed countries. In Fig. 2, GDP per capita is shown from the years 1000 to 2008. This demonstrates that GDP per capita increased substantially after the advent of the Industrial Revolution,

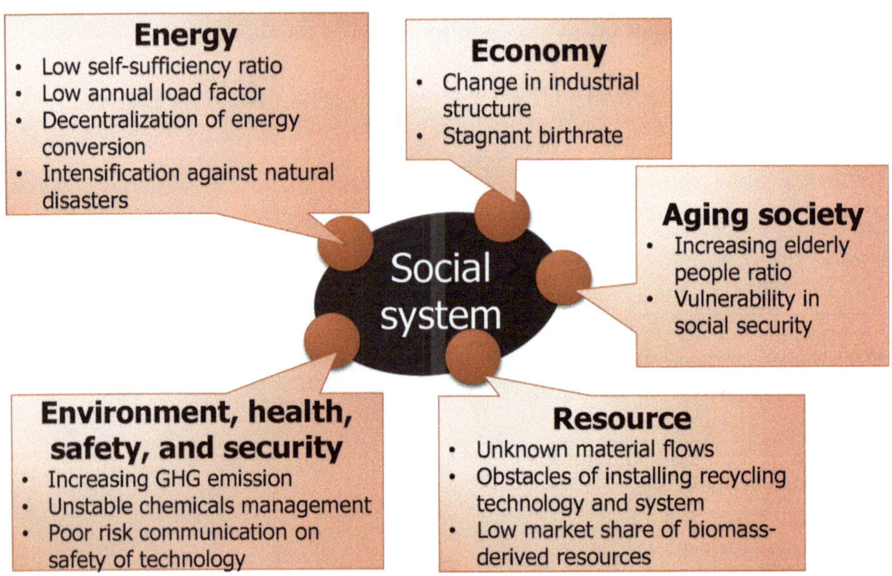

Fig. 1 Example of social issues in Japan

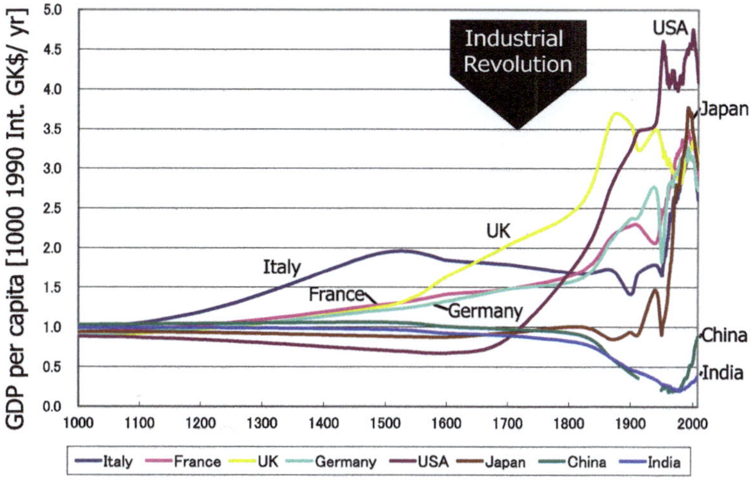

Fig. 2 GDP per capita of selected countries extracted from the literature [1] (Data from Angus Maddison [2] and The Conference Board Total Economy Database [3])

where first the United Kingdom and then the United States, France, Germany, Italy, and Japan had nearly exponential growth in wealth per capita. After fluctuations caused by depressions and wars, the developed countries reached a plateau of GDP per capita in recent years. Meanwhile, developing countries such as China and India have had increasing curves in recent years. They can be expected to reach the status of developed countries soon. In developed countries, people have adequate mobility and access to food, clothing, housing, and information.

The second feature of a mature society is longevity resulting from the growth in wealth. The 1999 life expectancy for OECD countries and the world average exceeded 70 or 65 years old, respectively [1]. Elderly people should be able to find satisfaction and motivation in life within a mature society. Figure 3 shows a population estimation for Japan [4]. That population is now decreasing and will drop to less than 100 million by 2050. The ratio of people over 65 years of age is currently increasing and will reach ~40 % by that year. This is because of a stagnant birthrate in the country and a small number of immigrants. Longevity is definitely desirable, but it should be carefully addressed given the stagnant birthrate.

The third feature is the saturation of products or the saturation of amenities provided by those products. As discussed in the literature, cement production in developed countries has already reached saturation [1]. According to Japanese statistics, the penetration of home electronic appliances such as air conditioners, television sets, refrigerators, personal computers, and mobile phones has reached near 100 % [5]. Electricity consumption per family unit has become saturated [6]. In lieu of products, services have become important in developed countries, which is manifested by decoupling of economic development and energy

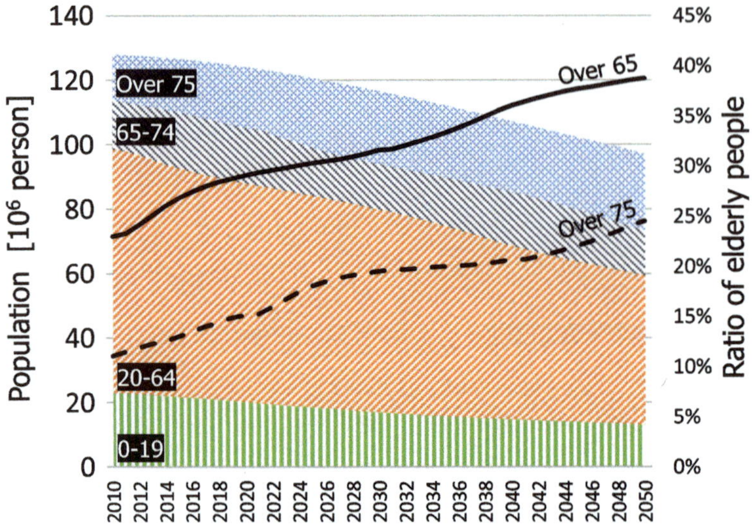

Fig. 3 Estimation of population by age bracket in Japan (*colored hatching*) with the ratio of elderly to total population (*dashed* and *solid black curves*) [4]

consumption. Industrial structures are now shifting from product-based to service-based activities.

The fourth feature is a shift from fossil to renewable resources. Fossil resources were increasingly used from the eighteenth century, when the Industrial Revolution began. Coal, petroleum, and natural gas became indispensable resources all over the world. In recent decades, the impacts on the environment of these resources have constrained human activities, and their use may be limited in the near future. In the twenty-first century, means for shifting from fossil to renewable resources have been explored for constructing sustainable energy systems. Because a rapid shift to such systems is infeasible, the transition should be managed by combining all potential technology options.

As mentioned above, the four features of a mature society are emerging in developing countries or will do so in the near future. In those countries, resource and energy issues may be critical for implementing countermeasures against other issues. A combination of technology options should be deployed in a prompt but appropriate manner to overcome the difficult obstacles toward a sustainable future.

2 Japan's Energy Situation

2.1 GDP vs. Energy Consumption

Figure 4 shows energy ladders for Japan, the USA, and other regions. The plots indicate the primary energy consumption per capita to GDP per capita in 1990, 2000, 2005, 2010, 2011, and 2012, which are arranged from left to right for all countries and regions. The dotted line shows an extrapolation of world average plots. As shown in the plots for the EU (OECD), USA, and Japan, the slopes are negative for 2010, 2011, and 2012. Although the bankruptcy of Lehman Brothers may have caused the decrease of energy consumption from 2005 to 2010, the decrease in 2011 and 2012 may mean that energy efficiency for GDP has increased in the EU (OECD) and USA. The Great East Japan Earthquake on 11 March 2011 had a greater impact on energy consumption in Japan than the aforementioned bankruptcy. It is not easy to distinguish the contributions of various influences on the increase in efficiency of energy consumption per GDP, but the severe power supply deficiency caused by the earthquake may have affected the energy-use attitudes and stimulated the implementation of more efficient systems. One sees that Oceania is in a plateau state, whereas an increase in GDP is associated with increased energy consumption in countries in the other regions. These differences can be recognized as a transition to the decoupling of economic development and energy consumption [8]. In the mature society, the decoupling of the economy and environment has occurred, e.g., in the EU (OECD), USA, and Japan.

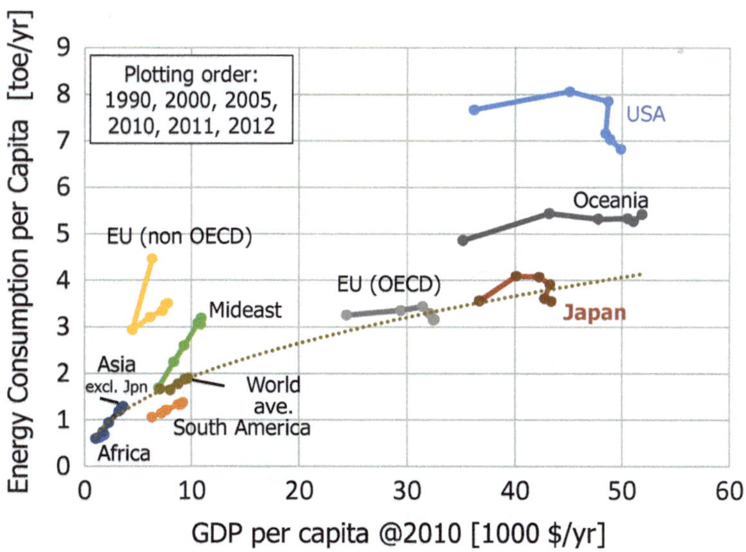

Fig. 4 GDP per capita and energy consumption [7]

2.2 Energy Demand and Flows

Figure 5 shows Japan's energy flow in fiscal year (FY) 2013. It is based on an analysis of data in *Comprehensive Energy Statistics* for FY 2013, referring to the energy flowchart released each year by Lawrence Livermore National Laboratory (LLNL) in the USA [9]. Primary energy resources on the left are converted to electricity in part and then used in residential, commercial, industrial, and transportation sectors. Units in this chart are exajoules (EJ or 10^{18} J), and each number has an error range of approximately ± 1 %. In terms of energy efficiency, actual values were used for thermal power generation. For residential and commercial sectors, 65 % efficiency is used, equivalent to the assumption in the LLNL chart, with 80 % for the industrial sector and 21 % for transportation as assumed values. Energy efficiency of 100 % is used for power generation, except thermal power. The results show that Japanese energy consumption is ~21 EJ, 40 % of which is supplied from petroleum and 30 % each from coal and natural gas. It is also revealed that 45 % of the supplied primary energy is used for power generation. The electrification rate in energy consumption of the residential and commercial sectors is ~50 %, whereas it is ~20 % in the industrial sector and only 2 % in transportation.

The same analysis was conducted for each year after 1990 to distinguish the primary energy supply. Primary energy procurement cost for imports was also analyzed using foreign trade statistics. The results are shown in Fig. 6.

After the Fukushima Daiichi Nuclear Power Plant accident in 2011, Japan made strict regulations on the restart of nuclear power plants. Natural gas, coal, and

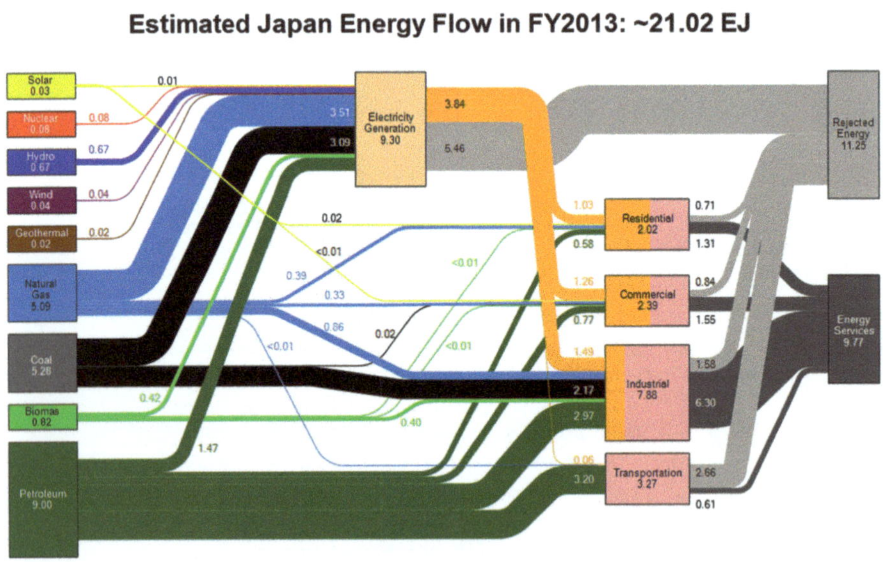

Fig. 5 Estimated Japan energy flow in FY2013

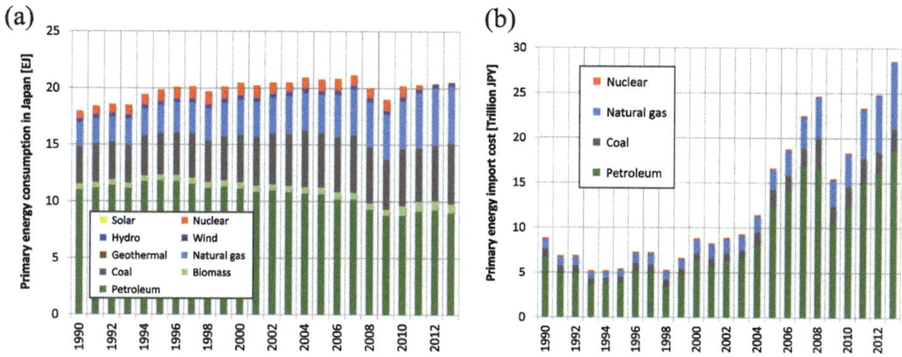

Fig. 6 Primary energy resource consumption in Japan. (**a**) Primary energy consumption in Japan, FY 1990–2013 (HHV: higher heating value base) [10]. (**b**) Energy import cost of Japan, FY 1990–2013 [11]

petroleum have made up for the shortage of electricity supply caused by the shutdown of those plants. Therefore, the nuclear energy supply has decreased since 2011, and, on 16 September 2013, the country was completely without nuclear-produced electrical power. Nuclear energy constituted ~5 % of primary energy sources in Japan, which was ~30 % of the total electric power supply. The thermal power reserve margin was effective because the country has not had a serious power shortage, because the national energy supply structure has been changing since 2000 from petroleum to coal and liquefied natural gas (LNG) (Fig. 6a). That is, Japan established more coal and LNG power plants without operating oil-fired power plants but maintaining their facilities, which helped maintain the power supply when it was confronted with the power shortage. Nevertheless, the energy transition from petroleum to the other energy resources was interrupted after the accident.

The situation regarding energy procurement costs changed in 2004 (Fig. 6b). These costs did not exceed 10 trillion JPY before then, but they increased because of a rise in petroleum prices beginning that year, approaching 25 trillion JPY in FY 2007. Later, procurement cost was reduced by global economic stagnation because of the collapse of Lehman Brothers, but increased again after the Fukushima accident to obtain natural gas for power generation. Therefore, purchasing cost has reached 28 trillion JPY. This is approximately one third of total import cost in a year, and energy procurement is the biggest deficit factor in Japan's trade balance.

Figure 7 shows import partner countries for coal, petroleum, and LNG. As seen from the self-sufficiency ratio for these three fossil resources, Japan is strongly dependent on imports from other countries for fossil fuel. Among the import partners, the dependency of petroleum on the Middle East is as high as 84 %. Strong dependence on certain regions or countries reduces the energy security of the country [12]. Thus, the diversification of import partner countries and of fuel type and an increase of domestic primary energy including renewable resources are important.

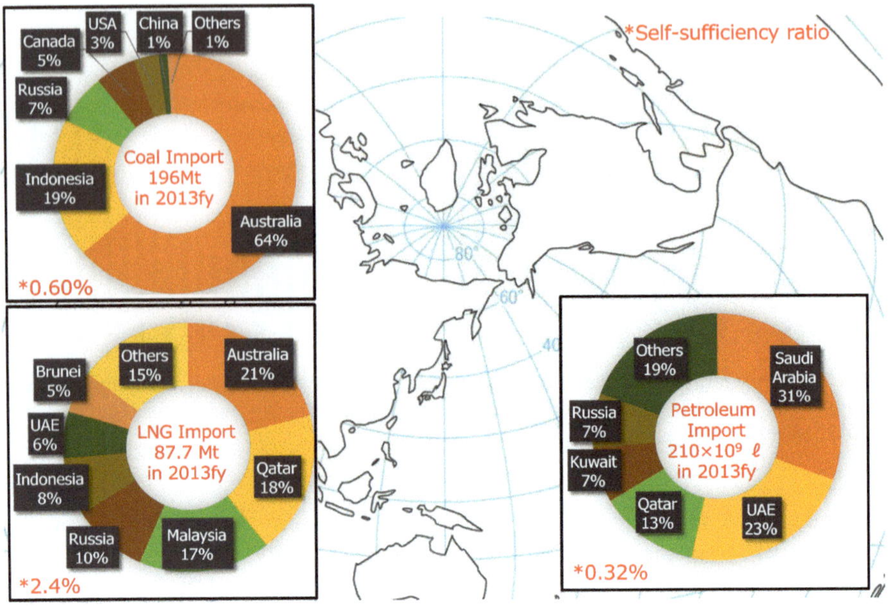

Fig. 7 Import partner countries for coal, petroleum, and liquefied natural gas (LNG) [7]

Figure 8 gives an overview of Japan's energy consumption, in which the contributions in divisions and subdivisions of the energy demand sectors are shown. The largest demand sector is primary and secondary industries, in which the contribution of chemical products and iron and steel production to total energy demand is much larger than the other subdivisions of the primary and secondary industrial sectors. The second largest demand sector is the commercial sector including tertiary industry, in which wholesale and retail trade use the greatest amount of energy. The loss of energy from centralized power generation in the commercial sector is larger than that of the primary and secondary industrial sectors. This is because of greater electricity demand in the commercial sector. In the primary and secondary industrial sectors, many decentralized combined heating and power systems have been implemented, which partly explains the smaller energy loss in those two sectors. The residential sector shows a tendency similar to the commercial sector, because substantial electricity is also used in the former sector. Almost all primary energy in the transportation sector is derived from petroleum oil. Although little loss is shown in the bar graph of that sector, much energy loss occurs during transport. The efficiency of automobiles or other types of fleets should be considered in this sector. As a nonenergy use of fossil resources, petroleum converted to chemicals and coal used as reductants of ferric oxide are in the majority.

We have briefly summarized Japan's energy situation; a detailed energy review may be found in the literature [13].

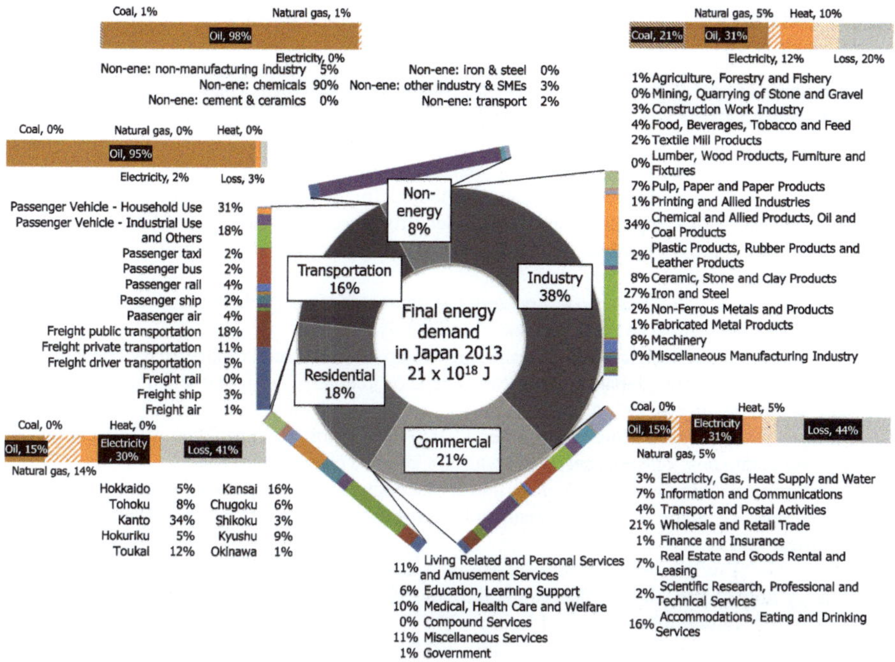

Fig. 8 Overview of energy demand in Japan [10]

3 Challenge for Designing Future Energy Systems

An important task in Japan is the successful redesign of energy systems with feasible technology options. Solutions must be harmonized not only with constraints on energy systems such as safety and energy security but also with social aspects stemming from a mature society. Part of such socioeconomic performance or other aspects can be quantified using existing methods or assessments. Life cycle assessment (LCA) of environmental impacts [14], life cycle costing or social LCA [15] for socioeconomic efficiency [16], risk assessment or process safety analysis for safety [17], and indicators developed for energy security [12] are examples of available quantification approaches. Based on such scientific analyses, technology options can be examined toward a sustainable society. With its cutting-edge technologies and cultural creativity, Japan is expected to create new demand and revitalize its socioeconomic affluence and has a leading role in showcasing solutions to common problems around the world.

References

1. Komiyama H (2014) Beyond the limits to growth. Springer Open, Tokyo
2. Maddison A (2006) The world economy: a millennial perspective/historical statistics. OECD Development Centre, Paris
3. The Conference Board Total Economy Database, "Total economy database". https://www.conference-board.org/data/economydatabase/index.cfm?id=25667. Accessed in 20 July 2015
4. National Institute of Population and Social Security Research, Japan, "Population projection for Japan". http://www.ipss.go.jp/site-ad/index_english/population-e.html. Accessed in 20 July 2015
5. Statistics Japan, "Ownership of consumer durables in Japan". http://www.stat.go.jp/data/zensho/2004/taikyu/gaiyo1.htm. Accessed in 20 July 2015
6. The Federation of Electric Power Companies of Japan, "The trend of electricity consumption per family unit in Japan". http://www.fepc.or.jp/enterprise/jigyou/japan/sw_index_04/. Accessed in 21 July 2015
7. The Energy Data and Modeling Center (2015) Handbook of Japan's & world energy & economic statistics. The Energy Conservation Center, Tokyo
8. e.g., European Commission, Policy review on decoupling: development of indicators to assess decoupling of economic development and environmental pressure in the EU-25 and AC-3 countries, 2005
9. Energy and Environmental Security Mission, Lawrence Livermore National Laboratory, "Energy flow charts". https://flowcharts.llnl.gov/. Accessed in 15 Aug 2015
10. Agency for Natural Resources and Energy, METI, Japan; "Table of the FY2013 comprehensive energy statistics". Accessed in 21 July 2015
11. Ministry of Finance, Japan; "Trade statistics of Japan", http://www.customs.go.jp/toukei/info/index_e.htm. Accessed in 21 July 2015
12. Martchamadol J, Kumar S (2013) An aggregated energy security performance indicator. Appl Energy 103:653–670
13. Koyama M, Kimura S, Kikuchi Y, Nakagaki T, Itaoka K (2014) Present status and points of discussion for future energy systems in Japan from the aspects of technology options. J Chem Eng Jpn 47(7):499–513
14. Hellweg S, Milài i Canals L (2014) Emerging approaches, challenges and opportunities in life cycle assessment. Science 344(6188):1109–1113
15. UNEP Life Cycle Initiative, 2009, "Guidelines for social life cycle assessment of products". http://www.unep.org/pdf/DTIE_PDFS/DTIx1164xPA-guidelines_sLCA.pdf. Accessed in 21 July 2015
16. Bartezcko-Hibbert C, Bonis I, Binns M, Theodoropoulos C, Azapagic A (2014) A multi-period mixed-integer linear optimisation of future electricity supply considering life cycle costs and environmental impacts. Appl Energy 133:317–334
17. Kolluru RV, Bartell SM, Pitblado RM, Stricoff RS (1996) Risk assessment and management handbook for environmental, health, and safety professionals. McGraw-Hill, New York

Sustainable Production and Stable Transportation of Energy Resources: Measures Toward 2050

Naohito Okumura

Abstract Energy must be supplied sustainably to end consumers to support modern industries and sustain modern, convenient, and safe daily life. Most prominent outlooks indicate that global society requires sufficient and sustainable supplies of major fossil resources, such as oil, natural gas, and coal. However, there will be much more difficulty in exploration, production, and transport of those resources, technically, geopolitically, and naturally. State-of-the-art technologies with understanding of international circumstances must be developed to achieve sustainable production and secure transportation of energy resources, as a greatly shifted global paradigm. Sustainable, peaceful global society may also be created through cooperation of the sustainable global energy supply. Furthermore, future sustainable global society might identify the necessary types and volume of energy supply beyond 2050 on the contrary.

Keywords Sustainable production • Introduction of state-of-the-art technologies • Secure production and transportation • Energy conservation • Proactive cooperation to build future peaceful global society

1 Introduction: Overview of Energy History and Outlook

1.1 Historical Overview of Energy

Human beings used horses, cows, whale oil, and renewable energies such as wind, water, wood, and hot springs until the mid-nineteenth century. Modern, convenient, and comfortable daily lifestyles have been supported by large supplies of coal, oil, and natural gas since the Industrial Revolution. Coal, natural gas, and particularly oil are very convenient energy resources that have accumulated for hundreds of millions of years and cannot be reproduced in a short period. Saving and conserving these fossil energy resources are necessary for our descendants as human

N. Okumura (✉)
Security and Strategy Research Institute of Japan, Tokyo, Japan
e-mail: okumura@ssri-j.com

Y. Kato et al. (eds.), *Energy Technology Roadmaps of Japan*,
DOI 10.1007/978-4-431-55951-1_4

properties. There are limited natural domestic reserves of these resources in Japan, which 40 years ago was the only industrialized country in Asia that consumed large amounts of energy. Japan has striven to explore new fossil resources, improve energy efficiency, and use nonfossil alternative energies as much as possible, for example, in the sunshine projects and moonlight projects, since 1974.

1.2 Current Status of Energy: Total Primary Energy Supply, Fuel Share, and Reserves

The world's total primary energy supply (TPES) was 6,105.86 million tons oil equivalent (Mtoe) in 1973 and 13,371.03 Mtoe in 2012, a 2.19-fold increase over the past 40 years. In 1973, 86.7 % of TPES was supplied by fossil fuels and 81.7 % in 2012 [1]. Organization for Economic Cooperation and Development (OECD) member countries consumed 61.3 % of the world's TPES in 1973 and 40.5 % in 2011 [1].

Fossil fuel shares of TPES in 2013 were 90.4 % in China, 86.4 % in the United States, 88.5 % in the Russian Federation (Russia), 91.7 % in India, 93.4 % in Japan, 82.7 % in the Federal Republic of Germany, and 52.8 % in France. Indeed, the fossil fuel share of TPES is very high in most countries [2].

Supplied primary energy has been consumed partially as electricity. World-generated electricity was 6,129 terawatt hours (tWh) in 1973 and 22,668 tWh in 2012. Global electric power generation by fossil fuels was 75.2 % in 1973 and 67.9 % in 2012 in the world [1].

There are various qualities of oil and reservoir conditions. There has been great volatility of oil prices in the daily market. Most oil, natural gas, and coal reserves have been nationalized, and the confidentiality of reservoir data has been strategically strengthened. British Petroleum (BP plc or BP), however, has published that at current production levels, crude oil reserves will be supplied for 53.3 years, natural gas reserves for 55.1 years, and coal reserves for 113 years [2].

1.3 Major Energy Outlooks to 2035 and 2040

There are three world energy outlooks, prepared by the OECD/International Energy Agency (IEA), BP, and US Energy Information Administration (US-EIA), which have been updated periodically following data gathering and assumption changes. The share of fossil fuel will be 74 % in 2040 as estimated by the IEA, 81 % in 2035 by BP, and 74 % in 2040 by the US-EIA. These are described in detail below [3–6]:

1. IEA World Energy Outlook 2014. There are three scenarios, and a new policy scenario is represented as the central position. Total primary energy demand in 2040 will be 15,629–20,039 Mtoe, which will be 36.9 % higher than in 2012

with the new policy scenario and 50.0 % higher than that year with the current policy scenario. The share of fuels produced in 2040 will be 26.0 % (4,761 Mtoe) from oil, 24.3 % (4,448 Mtoe) from coal, 24.2 % (4,418 Mtoe) from natural gas, 18.9 % from renewables, and 6.6 % from nuclear [3–4].

2. BP Energy Outlook 2035 (2014). Total energy demand in 2035 will be 41 % higher than the current level. The share of fuels in 2035 will be 28 % (109 Mb/day) produced from oil, 27 % from coal, 26 % (500 Bcf/day) from natural gas, 7 % from renewables, and 12 % from others [5].
3. US-EIA International Energy Outlook 2013. The world's total gross domestic product (GDP) will increase by 3.6 % annually, and the world's energy use in 2040 will be 56 % higher than in 2010. The share of fuels in 2040 will be 28 % produced from oil (including biofuels), 27 % from coal, 23 % from natural gas, 15 % from renewables (excluding biofuels), and 7 % from nuclear fuel [6].

1.4 Technological Goals and Effective Measures

Stable supplies of energy are accomplished not only through sustainable production but also by secure long-distance transportation. Further advanced technologies must be suitably progressed to develop frontier resources, ensure security for stable production and reliable transportation, treat and upgrade lower-quality natural resources, and maximize utilization of natural fossil resources. Five technological goals and five effective measures are listed below:

1. Technological Goals:
   - Maximum ultimate recovery of oil, gas, and coal resources
   - Proactive utilization of unconventional oil, gas, and coal
   - Optimum oil, gas, and coal supply systems in producing countries
   - Reinforced safety of oil, gas, and coal production operation
   - Optimized transportation of oil, gas, and coal
2. Effective Measures:
   - Observation of international laws
   - Peace building and peacekeeping
   - Trade liberalization
   - Secure marine routes
   - Environmental conservation

1.5 Problems and Risks

Typical problems and risks listed below should be recognized and addressed in both producing and mining countries and for transportation routes.

- Political and geopolitical risk increases in oil-, gas-, and coal-producing countries
- Increased geopolitical risks of marine transportation and pipelines of oil, gas, and coal
- Combined natural disasters such as earthquakes, tsunamis, typhoons, and volcanic eruptions
- Global expansion of pollutants, such as air/water pollutants, GHGs, and nuclear waste

2 Present Status

2.1 Production

Typical present statuses of oil and gas productions are listed below. Almost all coal mining is opencut, with comparatively less difficulty.

- Recovery of conventional resources is generally low.
- Development and production of unconventional resources are high cost.
- Energy resource production and supply systems are not reliable as complete systems.
- Energy resource production and supply systems are inefficient as complete systems.
- There are still many operational accidents at producing fields.

2.1.1 Overview of Coal, Oil, and Natural Gas Technological Development

Developing technology consists of various component technologies. Industries have introduced many timely state-of-the-art technologies through multinational cooperative team projects for many years. Typical new technologies shown below have contributed to many notable development projects.

- Lighter offshore structures have been constructed by replacement of carbon steel plates with high-tensile strength steel (HTSS) plates. This has helped widen and deepen exploration and production activities in offshore fields.
- Introduction of flexible hoses originating from subsea kill and choke lines of blowout preventers (BOPs) has also contributed to deeper water fluids communication as umbilical risers between subsea completion and floating production platforms.
- Acid gas removal, liquefaction, and membrane technologies have made it possible to use a large volume of flared associated gases.

- High-alloy pipes have increased sour hydrocarbon production, particularly in Middle Eastern and Caspian regions.
- Light gas turbines designed for aircraft have been installed on offshore production platforms, resulting in greater availability of electric power than by diesel engines as prime movers.
- Direct current (DC) control systems have made possible the flexible control of drilling machinery.
- Microprocessor chips have been installed inside bottom-hole assemblies (BHAs) as measurement while drilling (MWD) units, which have facilitated drill-extended reach wells and horizontal wells.
- Helicopter-assisted rigs and swamp barges have been introduced for new exploration opportunities deep in tropical rainforests.
- Drilling automation has been attempted with artificial intelligence, but was terminated during formulating the knowledge base for reduction of research and development (R&D) expenditures.
- Computerized material control and maintenance management systems have achieved great cost savings and greater reliability of operations.
- Supervisory control and data acquisition (SCADA) and distributed control systems (DCSs) have achieved reduction of assigned operators through remote monitoring at large offshore production fields.
- Intelligent pigging (IP) systems have realized continuous diagnosis of pipeline inner corrosion and erosion to maximize flow capacity and increase reliability.
- Advancement from minicomputers to capable work stations has facilitated models to simulate and optimize reservoir management with more frequent history matching.
- Cracking technologies such as fluid catalytic cracking (FCC) have been installed close to oil production fields to produce lighter fuels for local domestic use. Environmental guidelines applied to production fields have been stronger than those in many OECD countries.
- Fischer-Tropsch technology for gas to liquid has been introduced commercially, but the substantial energy loss must be reduced.
- Unmanned mining and automated infield transportation have been introduced at large-scale coal fields.

R&D activities, particularly regarding unconventional resource development, have come to be reinvested, owing to increased benefits from higher oil prices. Costly horizontal drilling and fracturing technologies may be used with accumulated experience to produce unconventional resources such as shale gas and shale oil.

2.1.2 Oil and Natural Gas Reserves and Production Costs

Resources and reservoir conditions vary and production costs range widely. Introducing various state-of-the-art technologies can potentially reduce production costs. IEA-published reserve estimation and production costs of crude oil and natural gas are shown below (all items in this subsection per reference [7]):

1. Crude oil. Total global potential oil resources (conventional plus unconventional) are roughly 5.9 trillion barrels (tb). Some 1.2 tb has already been produced, mostly at a cost less than 30 US dollars (USD) per barrel.

   (A) Enhanced oil recovery (EOR): EOR may increase recovery of as much as 500 billion barrels (bb). CO_2 flooding recovers 300 bb in a cost range of 20–70 USD per barrel. Thermal flooding and chemical flooding cost more than CO_2 flooding.
   (B) Ultra-deep sea (deeper than 1,500 m): An estimated 160 bb will be produced. Production cost is estimated to be high, 70–90 USD per barrel.
   (C) Arctic sea: 90 bb of crude oil and 44 bb of natural gas liquid (NGL) may be delivered. Production cost may be from USD 40 to USD 100 per barrel.
   (D) Extra-heavy oil and bitumen: 1.47 tb oil equivalent (Tboe) will be used. Production cost will vary from 50 to 90 USD per barrel.
   (E) Kerogen oil: 1,070 bb may be technically recoverable. Production costs are likely to be in the range of 40–100 USD per barrel.

2. Natural gas. The global potential gas reserve is estimated at ~790 trillion cubic meters (tcm). Some 105 tcm has already been produced, flared, and vented to the atmosphere, at costs reaching 8 USD per million British thermal unit (Mbtu). More than 1.5 tcm has been flared worldwide in the last decade, and only 5 % has been used (8 USD per Mbtu equates to 46.4 USD per boe). The most easily accessible part of the remaining conventional gas resources (220 tcm) is produced at a cost from 0.20 to 9 USD per Mbtu. H_2S- and CO_2-rich natural gas (160 tcm) could be produced at costs from 2 to 11 USD per Mbtu.

   (A) Arctic sea: 30 tcm could be produced at costs from 4 to 12 USD per Mbtu.
   (B) Deepwater sea: 50 tcm could be produced at costs from 5 to 11 USD per Mbtu.
   (C) Unconventional resources: Tight gas (80 tcm), shale gas (200 tcm), and coalbed methane (CBM, 50 tcm) could be produced at costs from 3 to 10 USD per Mbtu.
   (D) Pipeline transport: Capital expenditure plus operational expenditure will be 0.30–1.20 USD per Mbtu and 1,000 m.
   (E) Liquefied natural gas (LNG) transportation: Total costs for liquefaction, transportation, and regasification will be 3.10–4.70 USD per Mbtu.

2.1.3 Coal Reserves and Production Costs

The global share of coal production by OECD countries declined from 55.6 % in 1973 to 26.0 % in 2012 [1]. Data collection of reserves and mining project status in many countries has become more difficult than those of oil and gas generally. The

IEA estimates that the current level of global hard coal production (6.0 gt) will continue for almost 17 years (this and all items in this subsection per reference [7]).

1. **Reserves and Mining**. Mining costs vary widely, such as from low cost in China to high cost in Canada. The IEA estimates average mining costs at ~43 USD per ton.
   - First 100 Gt: Cost is from 42 to 170 USD per ton coal equivalent (tce) at free on rails (FOR), equal to 0.84–3.45 USD per Mbtu.
   - Next 500 Gt: An additional 83 years at current coal mining levels. Mining cost is 71–272 USD per tce at FOR, equal to 1.42–5.44 USD per Mbtu.
   - Remained 128 Gt: Hard coal to cover 120^{+} years at current mining levels. Mining cost 71–800^{+} USD per tce at FOR, equal to 1.41–15^{+} USD per Mbtu.
2. **Land Transportation**. Total costs are typically based on field locations and distances to cities and ports. Transportation costs currently make up >50 % of total cost for some export coal in China and Russia.

2.2 *Transportation*

Current typical statuses for the transportation of oil, gas, and coal are listed below.

- Many carbon steel pipelines have reduced transportation capability because of internal erosion and internal/outer corrosion. International cooperative measures against various terrorist attacks have become more important.
- Piracy attacks have disturbed substantial marine transportation. These have been reduced in the case of Somalia by international cooperative operations.
- Some countries have increased naval activities and constructed artificial islands with offshore structures and airstrips across boundaries, along oil, gas, and coal transportation sea lanes. Waiving of established international laws is serious.
- There are greater natural disasters such as typhoons, earthquakes, tsunamis, and volcanic eruptions on and near oil, gas, and coal sea lanes.
- There are large open deposits of sulfur waste separated from natural gas in producing countries.

2.2.1 Critical Risks for Energy Supply System

Oil, gas, and coal are supplied in a chain with various facilities, plants, pipelines and tankers from oil fields, gas fields, and coal mines to end users. This is accomplished through transforming plants to produce secondary energies such as refineries, power plants, and others (Fig. 1).

Each component has its peculiar and characteristic risks. Political risks, environmental risks, geopolitical risks, and natural disastrous risks are great across all energy supply systems.

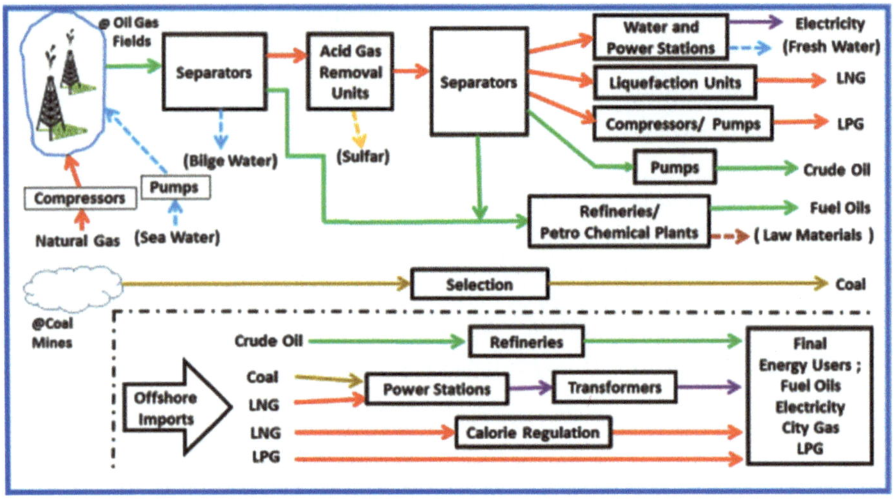

Fig. 1 Typical major energy supply system

Most countries have production in domestic oil fields and coal mines. Strong defense of oil-producing fields, coal mines, pipelines, refineries, and electric power plants by police, coast guard, armies, and navies has been a top-priority issue.

Geopolitical and natural risks are so serious that proper measures should be sought and implemented internationally and cooperatively.

There are also severe natural disaster risks such as earthquakes, tsunamis, typhoons, and heavy rains, both in producing countries and along transportation routes.

Stable supplies of major energy resources such as coal, oil, and natural gas typically require both secure production and safe transportation over more than 30 years of their project lives. Various risks are evaluated at investment decisions, but forecasting of sales prices and exchange rates of foreign currencies has been very difficult.

2.2.2 Various Risks for Secure Transportation

Systematic defense for transportation facilities such as loading terminals, liquefaction plants, vessels, offloading terminals, and international pipelines is indispensable in producing countries.

The US Energy Information Administration studied world oil transit chokepoints and published in 2014, with recent situations of the Strait of Hormuz, the Strait of Malacca and Bashi Channel, the Lombok and Makassar Straits, and the Panama Canal shown below from 1 to 4 (all items in this list per reference [8]).

1. **The Strait of Hormuz**. In 2013, 17.0 million barrels per day (Mbpd) of crude oil and petroleum products was transferred through this strait. There is no

alternative sea lane and alternative bypass pipelines have volumetric limitations. The capacity of a bypass pipeline owned by the United Arab Emirates (UAE) is 1.5 Mbpd, and throughput was 0.6 Mbpd in 2013. The capacity of the Petroline owned by Saudi Arabia is 4.8 Mbpd and current throughput is 2.0 Mbpd.

2. **The Strait of Malacca and Bashi Channel**. The transfer volume through both these passages is made up of 90 % crude oil and 10 % petroleum products. In 2013, 15.2 Mbpd of crude oil was transferred. The Lombok and Makassar Strait route is an alternative sea lane. There have been several proposals for bypass pipeline installation.
3. **The Lombok and Makassar Straits**. These straits are 250 m deep, which is deeper than the Strait of Malacca and Bashi Channel.
4. **The Panama Canal**. In fiscal 2014, 0.877 Mbpd of crude oil and petroleum products was transferred, of which 0.748 Mbpd was refined products. The Trans-Panama Pipeline (TPP) is an alternative route, with a current capacity of 0.6 Mbpd.

Violations of international law have become to be serious for secure international transportations. Several chokepoints on sea lanes toward Japan of major energy resources have been existing currently (Fig. 2). There are several geopolitical difficulties, including changing territorial sovereignties, piracy, development of weapons of mass destruction, and radical naval activities.

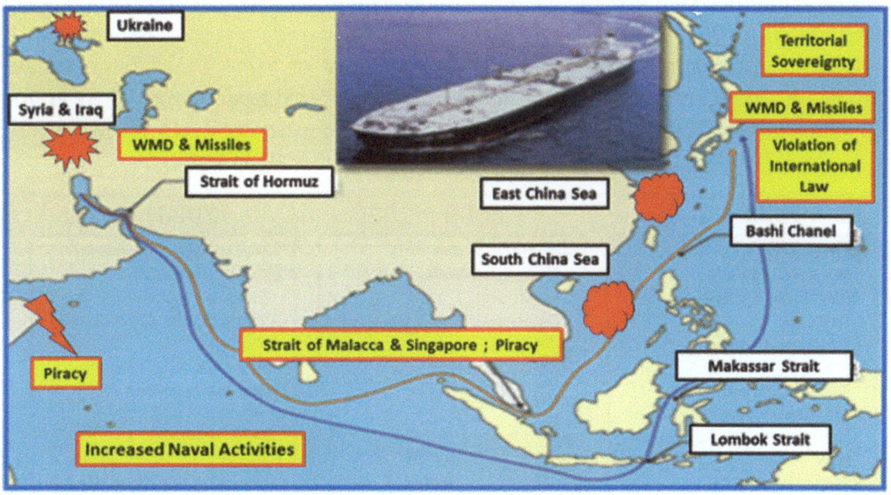

Fig. 2 Major energy resources supply sea lanes and geopolitical chokepoints

2.3 International Conditions

General international conditions currently affecting the production and transportation of oil, gas, and coal are listed below. It may be pointed that most major coal beds are generally reserved in countries with fewer political and geopolitical risks.

- Armed stress and conflict of territorial change have been increasing.
- Piracy has been increasing along energy resource transport sea lanes.
- Terrorist attacks associated with religious fundamentalism have increased.
- Potential ethnic and religious struggles have been recently clarified.
- Lack of transparency in international trade may disturb free international trade of oil, gas, and coal [9–11].

3 Technology Roadmap

3.1 Overview

Fossil resource production and transport technologies have been developed for many years, with proactive introduction of a wide range of state-of-the-art technologies worldwide. Technological R&D has been changed, mainly following market price volatility (Fig. 3).

Oil prices remained less than 40 USD per barrel of West Texas Intermediate (WTI) for 24 years (1980–2004) and less than 20 USD per barrel of WTI for 13 years (1986–1999). Industries have reduced large-scale R&D projects, and

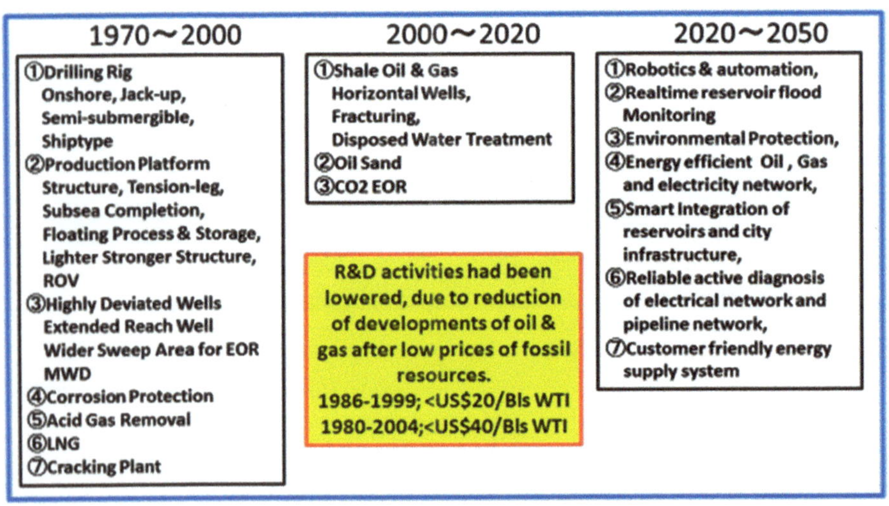

Fig. 3 Technological trends for developing oil, gas, and coal

many experienced experts left those industries because of a reduction in field application activities then. Some experts in international oil industries have been forced to move toward limited numerical modeling and simulation works, far away from real fields.

New state-of-the-art technologies are again required for active field operations with higher oil prices. Key terms for technical direction are needed from 2020 to 2050. International collaborations seem to be effective to achieve higher goals.

3.2 *Production and Mining*

Further technological development for increased recovery, reliable and efficient operation, cost reduction, environmental conservation, and safety improvement for oil, gas, and coal production projects must be planned and executed stepwisely. These are listed below.

- **Conventional resources**. Enhanced recovery and frontier development (2025), medium recovery and frontier production (2035), and maximum recovery and frontier increased recovery (2050)
- **Unconventional resources**. Increased production and technological development (2025); medium production, cost reduction, and new technology application (2035); and greater recovery, further cost reduction, and applied technology optimization (2050)
- **Reliability**. Improvement of existing energy systems (2025); partial upgrades to reliable systems (2035); and replacement by reliable, complete systems of oil/gas production, and urban energy supply (2050)
- **Efficiency**. Improvement of existing energy systems (2025), partial upgrades to efficient systems (2035), and replacement by efficient complete systems consisting of oil/gas production, and urban energy supply (2050)
- **Operational safety**. Accident reduction and Kaizen method (2025); partially automated safer operation (2035); replacement of dangerous work sites by robotics (2050).

3.2.1 Maximizing Recoveries and GTL: Conventional Resources

1. **EOR**. Compared with natural gas and coal, crude oil production is much more difficult in general. Stepwise various measures are used to maximize recovery, with reducing costs. Reservoirs are so diverse that appropriate measures must be selected. Suitable monitoring such as 4-D seismic surveys, modeling, and historical matching reduce water production for minimizing residual oils. Flooding by nitrogen, hydrocarbons, or CO_2 is suited to carbonate reservoirs. Polymer, surfactant and micellar, thermal and combustion, and thermal and

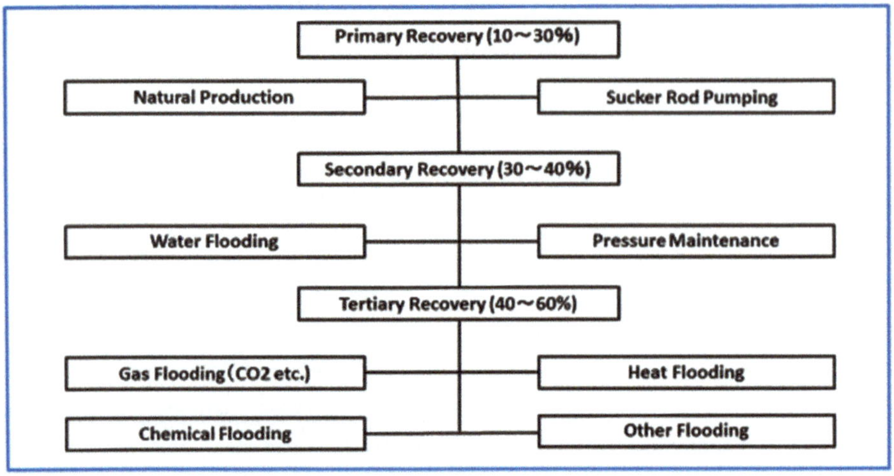

Fig. 4 Stepwise crude oil recovery with improved technologies (examples)

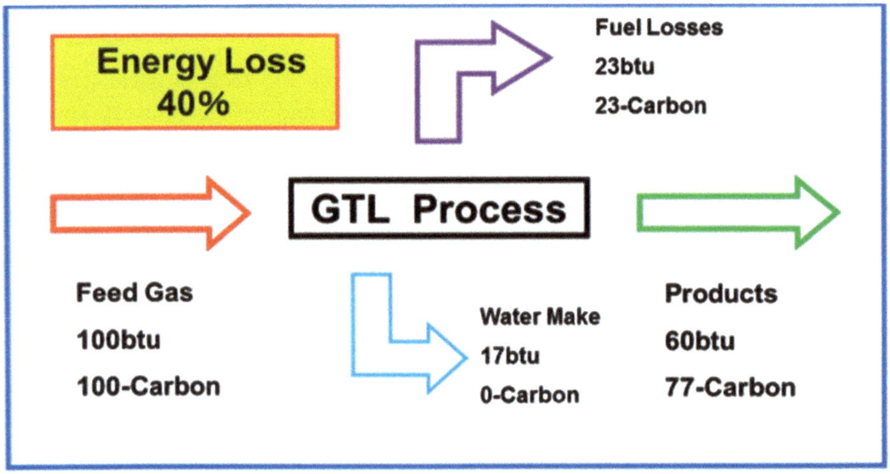

Fig. 5 Current energy loss levels of gas to liquid technology (example) (Original sources from Japan Oil, Gas and Metals National Corporation)

steam EOR are suited to sandstone reservoirs. Further technological development is required for application to various fields (Fig. 4)

2. **GTL**. Nearly 40 % of heat is lost in the process of direct conversion of gas to liquids (GTLs), as presented at a technical symposium by the technology research center of the Japan Oil, Gas and Metals National Corporation. Improvement of energy efficiency is expected (Fig. 5)

3.2.2 Accelerating Frontier Fields Development: Conventional Resources

Further state-of-the-art technologies are necessary for stable, safe, and economical production in frontier fields. A succession of technologies is particularly critical for severe field production operation, considering the oil spill accident in the Gulf of Mexico.

1. **Deep ocean**: The IEA reported that the remaining recoverable conventional oil is 2,700 bb, and 45 % of it is under the ocean bottom. Three hundred bb is under the deep sea at >400 m depth. Most proven reserves are in the Gulf of Mexico and offshore of Brazil, Angola, and Nigeria. Ultra-deep water is defined as depths >1,500 m [7]. Flexible risers, subsea completion, turret mooring and dynamic positioning systems, semisubmersible floating production units, tension-leg platforms, spar and floating production, and storage and offloading (FPSO) have been developed. Further technological developments are required to develop below deeper sea bottom and secure safe and environmentally friendly production operation with automation and introduction of robotics.
2. **Arctic**: The IEA reported that 30 % of global undiscovered natural gas may be in the Arctic. This region has 90 bb of crude oil, 44 billion boe of NGL, and 47 tcm of natural gas [7]. There are five large basins in this region. Gravity-based structure platforms will be most suitable to produce oil and gas in frozen seas (Fig. 6). Effective heat insulation prevents from freezing of many water pipelines and storage tanks on production platforms and supply boats. Icebreaking tankers, supply boats, and various work boats are necessary for production operations in the Arctic. Further technological developments are, also, required

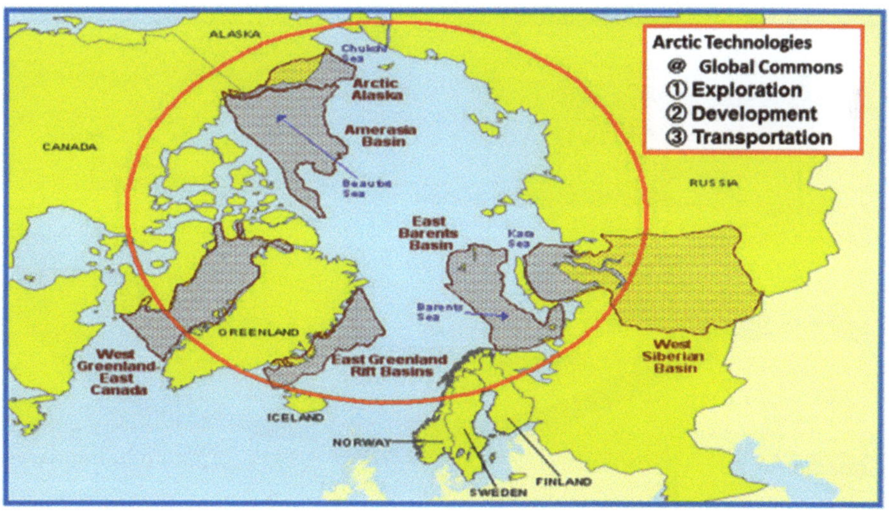

Fig. 6 Arctic prospects for oil and gas reserves (Original sources from the US Energy Information Administration)

to develop Arctic sea area and secure safe and environmentally friendly production operation with automation and introduction of robotics [12].

3.2.3 Cost Reduction of Unconventional Resource Production

Unconventionally producible resources cannot be produced through permeable reservoir rocks as conventional production procedures. Kerogen oil and light tight (shale) oil are recognized as unconventional oils. Unconventional coal-bed methane, tight sand gas, shale gas, and methane hydrates have large gas potentials (Fig. 7). Location and features of each reservoirs are various and development technologies, also, various. Fully skilled and experienced engineers are indispensable for environmentally friendly safe operations. Systematic human resources developments are very important.

1. Shale oil and gas: Developments of shale oil and gas require a large number of producing wells, a large quantity of freshwater, and an extensive land. Development is so costly that major cost reductions are necessary for commercial competitiveness. Environmental conservation and restoration technologies must be upgraded smartly because destruction of environments will be very costly for a long period of time (Fig. 8)
2. Heavy oil: Proven reserves are 175 bb in Canada and 220 bb in Venezuela. Near-solid bitumen in oil sands can be mined. Bitumen below 75 m needs significant stimulation to flow [7, 13, 14, 16, 17]. Opencut and steam-assisted gravity drainage (SAGD) are used generally. Large volumes of water, steam, light hydrocarbons, and hydrogen are required for upgrading and diluting.

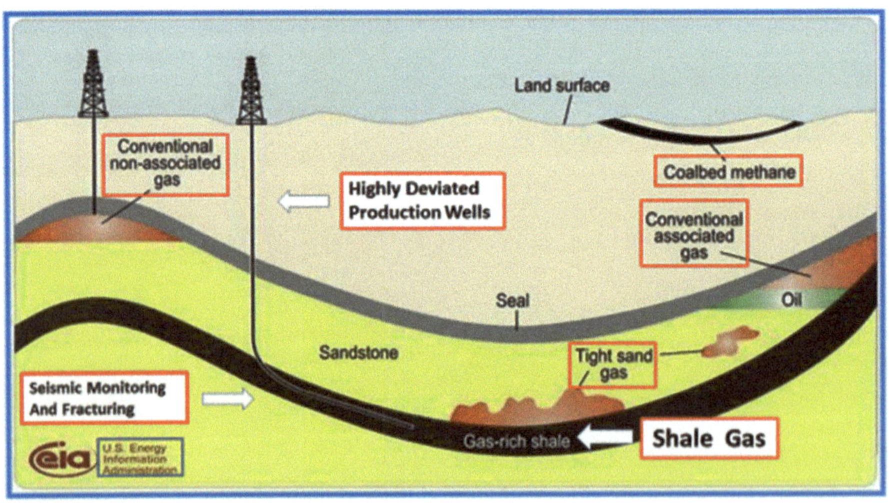

Fig. 7 Various nonconventional natural gases (Original sources from US Energy Information Administration)

Fig. 8 Drilling site and production wells for shale oil and gas production (Original sources from the US Energy Information Administration)

Improvement of efficient processes, organization of international synthetic crude oil supply networks, and state-of-the-art environmental protection technology are necessary.

3. Methane hydrate: This is located in both onshore and offshore under a certain balance of pressure and temperature [7, 15, 17–19]. Locations of offshore reservoirs are in many cases so deep that major cost reductions are necessary for attaining commercial competitiveness, following the design of proper production wells.

3.2.4 Maximizing Reliability and Efficiency as Local Energy Networks

1. Oil and gas supply systems in producing countries: Producing fields consist of production wells, separation plants, storage tanks, and offloading facilities. Production system is connected to refineries, petrochemical plants, water and power stations, and gas stations. Each component has been gradually constructed and integrated through over many years. Sufficient reliability and efficiency need to be improved and optimized totally, with timely and cost-effective replacement.
2. Optimization of reliability as local city infrastructure: Oil and gas production systems are linked with city infrastructures near producing fields usually. Real-time diagnosis, maintenance, and replacements are necessary for reliable production. Total reliability (R) of a tandem system (R_{tan}) is calculated by $R_{\mathrm{tan}} = R_{\mathrm{a}} \times R_{\mathrm{b}} \ldots R_{\mathrm{n}}$, and that (total reliability) of a parallel system (R_{par}) is calculated by $R_{\mathrm{par}} = 1 - (1 - R_{\mathrm{a}})(1 - R_{\mathrm{b}}) \ldots (1 - R_{\mathrm{n}})$. R_{n} represents an individual reliability of component n. A real system, however, is so complicated and

frequently modified that network models are effectively used. Those models, however, should be simplified to allow frequent changes and accesses by many operators. Continual total system management capability must be developed with intuitive power. Sufficient field trainings must be indispensable at human resources developments activities.

3.2.5 Introduction of Robotics at Hazardous Production Fields

1. Toxic content: Crude oil, natural gas, and coal in reservoirs contains CO_2 and toxic hydrogen sulfide (H_2S) usually. Fossil resources with low CO_2 and H_2S contents have been developed and used firstly for many years. However, remaining resources with higher such contents must be developed and used gradually in future. Process automation is effective, and field operators should be replaced by explosion-proof robots to operate in hazardous areas.
2. Frontier development: Future oil and gas production fields will be extended into areas of harsh conditions, such as the arctic and deep offshore. Explosion-proof robots should be developed for oil and gas fields and processing plants. Observation and maintenance robots should be developed at the same time.

3.3 Transportation

Maximum energy efficiency and minimum GHG emission across the entire oil, gas, and coal supply chains are important. Five stepwise reference targets are estimated as below. The majority of coal seems to be transported by rails and bulk marine tankers at sea. Major stepwise targets for transportation are listed below.

- Pipeline transportation: Optimized corrosion-protective efforts for current pipelines (2025), partial installation of new material pipelines (2035), and optimized new reliable pipeline networks (2050)
- Maritime transportation: Lower-risk transportation (2025), international cooperation reliable transportation (2035), and minimum risk transportation (2050)
- Efficiency and reliability: Evaluation of current transportation system (2025), improved efficiency and reliability of current transportation chains (2035), and efficient, reliable, and reasonable transportation at lower cost (2050)
- Natural disasters: Prediction (2025), preventive measures against natural disaster (2035), and enforced protective measures against natural disaster (2050)
- Environment: Reduced separated acid waste disposal (2025), increased export of plaster board for housing (2035), and maximum plaster board utilization (2050)

3.3.1 Pipeline Transportation

There are various pipelines for oil and gas transportation, such as infield flow, oil trunk and gas trunk lines, city infrastructure, and international export lines. Long-time secure operation and maintenance for pipelines will be much more difficult than designing and installation of pipelines because pipelines are very long and inner condition of pipelines are invisible during operation. Every pipeline requires adequate preventive maintenance against outer corrosion usually by paint, anodes, and electrodes. Pipelines installed on salty ground and shore landing portions of underwater pipelines are most corrosive, and special maintenance is required. Inner corrosion and preventive measures vary with flow material and flow-line shape.

1. Infield flow lines: Continuous pressure controls of producing wells following pressure changes are indispensable to reduce water cut and/or gas cut for maximized ultimate recovery. High-alloy pipes are used from wellhead towers to gathering platforms, because any injection of corrosion inhibitors and pigging operation is impossible in this portion. Inner corrosions from gathering platforms to separation platforms are so serious that corrosion inhibitors and biocides should be injected at certain dosage rates, because oil, gas, and saltwater flow together inside this flow line. Moreover, erosion by liquid gas turbulent flow, sediment deposits, and possible scratching by blush pigging are so serious that the flow rate must be properly controlled, and blush pigs for pipeline material must be optimized.
2. Oil trunk lines: Internal corrosion is so serious that proper dosages of corrosion inhibitors and biocides are required at cost-effective dosage rates. Inner diagnosis of pipelines is done by acoustic and electromagnetic intelligent pigs. More accurate identification of corroded points is urgently required for proper installation of emergency leakage prevention clumps outside the pipeline.
3. Gas trunk and city gas lines: Serious corrosion can be avoided with proper dehydration inside gas trunk lines. Acid removal and calorie adjustment for city gas lines are done to suit end-user requirements in both producing and importing countries.
4. International export lines: These are so long that there are various risks for interruption of full flow. Maximizing flow rate in each segment is important. Secure supplies of energy to pump stations and compressor stations are necessary. International cooperative teamwork for operation and maintenance are also indispensable. Geopolitical risks and severe natural disasters risks are much higher than floating vessel transportations risks because pipelines are long, difficult to be removed shortly, and difficult to be identified damaged portions and repaired, particularly subsea pipelines, during short period of time. International cooperation under international laws should be important to mitigate and eliminate geopolitical risks and tensions.

3.3.2 Marine Transportation

Oil tankers, petroleum product tankers, and LNG and LPG carriers are typically used. Both international crude oil and coastal liquid tankers should reduce voyage resistance. LNG transportation needs acid removal plants and liquefaction plants. Costs have been reduced by minimizing safety factors and spare lines, but optimization with reliability is indispensable.

3.3.3 Local Energy Supply System Optimization

Energy supply systems have become similar in many countries. The primary energy supply in Japan was 20,810 × 10^{15} J, and the final energy use was 14,347 × 10^{15} J in 2012, indicating a total energy loss of 31.1 %. Energy conversion and transmission losses at electric generation and transmission points amounted to 59.1 %. Corresponding losses were 39.4 % for steam generation and coal chemical production and 18.1 % for oil refining and petroleum chemical production and relatively negligible for city gas systems (Fig. 9) [21]. Introduction of more efficient power generators and transmission power loss reductions will be largely effective in ameliorating total energy loss. Furthermore, we should reduce energy consumption and GHG emission at the construction stage of manufactures, of transportations, and of energy plants. Substantial total energy loss is not confined to Japan, being found in most oil-, gas-, and coal-producing countries. Proactive cooperation between producing and consuming countries (such as Japan) is effective to improve energy efficiency and to increase reliability and safety through specified energy supply chain, at the time of designing measures by exchanging best practices.

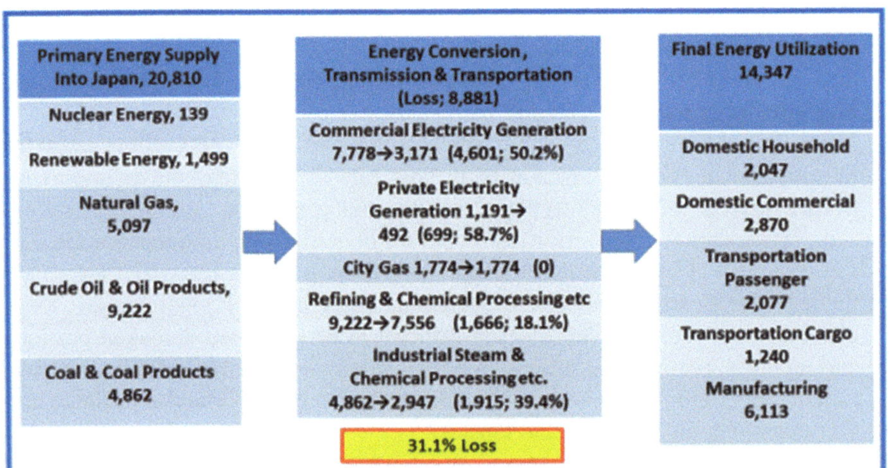

Fig. 9 Domestic energy flow balance in Japan in 2012 (10^{15} J) (Original sources from the White Paper 2014 by Ministry of Economy, Trade and Industry of Japan)

3.3.4 Optimized International Energy Transportation Chains

Lifecycle energy efficiency improvement and lifecycle GHG emission reduction should be considered to take optimized measures for the purpose of achieving global energy security and global environmental protection in their own natures.

1. Energy efficiency and GHG emission: Energy consumption and CO_2 (GHG) emissions are substantial at both natural gas liquefaction (LNG) plants in producing countries and at coal-fuel electric power stations in consuming countries (Fig. 10). Minimizing energy efficiency and minimizing CH_4 and CO_2 (GHG) emission throughout overseas oil, gas, and coal supply chains are required. Suitable fuel selections are important for reducing total life-cycle energy loss.
2. Environmental conservation: Minimizing environmental destruction throughout the energy supply system is necessary. Marine environmental protection, natural conservation, and global warming prevention are crucial, followed by universal energy access to support clean, safe, and comfortable lifestyles, toward 2050.

3.3.5 Stable Supply of Materials to Produce Energy-Efficient Products

Globalization of supply chains has caused lower final product prices. Modern, convenient, and energy-efficient products such as hybrid automobiles contain various special materials such as rare earths. Stable supplies of energy resources and rare materials are essential to complete energy-efficient final products [20].

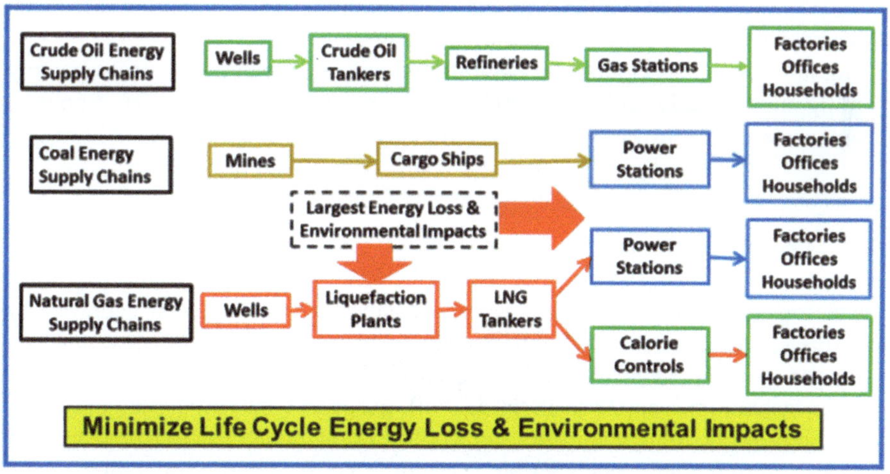

Fig. 10 Life cycle energy chain optimization (Minimum energy loss and GHG emissions)

3.3.6 Measures Against Natural Disaster and Environmental Conservation

Earthquakes severely impact pipeline transportation across faults. Tsunamis strongly affect marine transportation, as do typhoons. Typhoons can be forecasted usually, with associated damage reduction. More reliable prognoses of earthquakes and subsequent tsunamis are required.

3.4 *International Considerations*

There are stepwise reference targets in the roadmap of international cooperative measures for sustainable production and transportation of oil, gas, and coal. These are listed below.

- Territorial expansion: Prohibition of forced territorial expansions on both onshore and offshore (2025) and observance of international law (2035)
- Commercial cooperation: Dialogues between natural gas-exporting and natural gas-importing countries (2025)
- Multiracial and multireligion countries: Proactive cooperation for achieving harmony in these countries (2035) and proactive cooperation for creating harmonious societies in these countries (2050)
- Natural disaster and environment: Proactive international cooperative preparation, rescue, and reconstruction for natural disasters (2025); proactive international cooperation for environmental preservation regarding air, water, and various wastes (2035); and proactive promotion for elimination of social waste, i.e., "Mottainai," toward global sustainability (2050)
- Human security and global cooperation: Proactive cooperation to assist poverty mitigation globally (2035) and proactive cooperation for eradication of poverty and social unrest toward global sustainability (2050)

3.4.1 Project Leadership and Management

Technology development in oil, gas, and coal industries has been achieved by widespread introduction of various technologies from other industries, continuously and globally. Objective and broad viewpoints are indispensable to discover specific technologies suited to individual oil, gas, and coal production and transportation projects. Many engineers and workers of many nationalities have worked together throughout engineering design, construction, operation, and decommissioning, typically for more than 30 years' project lives. Project leadership and management skill must be modified to fulfill these requirements, in a timely and continuous manner.

3.4.2 Proactive International Cooperation

Cooperative joint operation will be indispensable to develop reservoirs across national border. Cargo ships traverse several territorial seas and international waters. Observance of international laws by involved countries and cooperative measures such as against pirate attacks are important for unimpeded voyages. International cooperation at various levels is indispensable [21–23].

4 Benefits and Prospective Future Vision

Technology and international cooperation will bring benefits and an attractive future for society toward 2050, as listed below.

- Guarantees of safe, clean, comfortable, and convenient daily lifestyles worldwide
- Strengthening of reconstruction capabilities after any combined major natural disasters
- Human security establishment through poverty eradication by electricity supply
- Cooperative establishment of a peaceful world through international energy supply and conservation activities

4.1 Safe and Comfortable Daily Lifestyles

The development of safe and comfortable daily lifestyles is realized by maintaining stable production and transportation of major energy resources such as oil, natural gas, and coal. A reliable supply of fuels and electricity is a top-priority activity in emergency response programs for rescuing injured victims, supporting urban lifestyles, and urban reconstruction following natural disasters such as major typhoons, tsunamis, and volcanic eruptions [24–27].

4.2 Contribution to Sustainable Global Society

Eradication of poverty and improvement of sanitation in remote areas by sustainable supplies of fuel and electricity help develop and maintain a global sustainable society. Minimizing emissions of CH_4, CO_2, and various pollutants throughout the international major energy supply system achieves global conservation of nature.

4.3 Building a Peaceful and Sustainable Global Society

Humans have made various efforts to overcome differences of race, custom, language, and religion. Various noble works have been achieved since the end of the reconstruction period for 70 years after the Second World War. These include *The Limits to Growth* by the Club of Rome in 1972, *The End of History and the Last Man* by Francis Fukuyama in 1992, the Rio Declaration on Environment and Development and Agenda 21 in 1992, *The Clash of Civilizations and the Remaking of World Order* by Samuel Huntington in 1996, and the UN Millennium Declaration and Development Goals for 2015 in 2000. The Sustainable Development Goals of 2015, following the Millennium Development Goals, is in preparation. However, there have recently been several instances of parties insisting on the differences and superiority of their own beliefs, hindering mutual cooperation. Cooperative projects to supply fuel and electricity sustainably can represent fruitful opportunities to overcome differences of civilization and strengthen mutual familiarization.

Japan, surrounded by deep and rough oceans, has experienced many and various natural disasters, such as earthquakes, tsunamis, typhoons, and volcanic eruptions. Japanese culture, in which people have historically assisted others and shared with them in the face of such disasters, is unique and peaceful. Accordingly, cultures such as "Mottainai," harmony with nature, patience, and peace can contribute to a sustainable and peaceful world in the future [28–32].

References

1. The International Energy Agency (2014) Key world energy statistics. International Energy Agency, France
2. BP p.l.c. (2014) BP statistical review of world energy. BP p.l.c., The United Kingdom
3. The International Energy Agency (2014) World energy outlook. International Energy Agency, France
4. Birol F (2014) World energy outlook. Foreign Affairs, the Council of Foreign Relations, Washington, DC
5. (2014) BP energy outlook 2035. BP p.l.c., The United Kingdom
6. Sieminski A (2013) International energy outlook. U.S. Energy Information Administration, Washington, DC
7. (2013) Resources to reserves. International Energy Agency, France
8. (2014) World oil transit chokepoints. U.S. Energy Information Administration, Washington, DC
9. Nakatani K (2012) Horumuz kaikyo to kokusai-hou, vol 7. Law Review by Law School of the University of Tokyo, Tokyo
10. Fujioka T (2013) Indoyouno hakeno meguru chu-in no arasoi. Security and Strategy Research Center of Japan, Tokyo
11. Shoji T (2014) Minami Shinakai mondaio meguru anzen-hosho to keizaino kosaku. Journal, the Japan Institute of International Affairs, Tokyo
12. Ikeshima D (2014) Global commons to shiteno hokyoku-kai to anzen-hosho. The Japan Institute of International Affairs, Tokyo

13. Ichikawa M (2014) Extra-heavy oil and bitumen in the world. Journal, the Japan Institute of Energy, Tokyo
14. Matsuoka T, Honda H (2014) Potential of unconventional fossil fuels. Journal, Japan Society of Energy and Resources, Tokyo
15. Urano T (2014) Recent field technology for developing shale gas in Canada. Journal, Japan Society of Energy and Resources, Tokyo
16. Okui T, Yao Y (2014) Current situation of coal seam gas development in Australia. Journal, Japan Society of Energy and Resources, Tokyo
17. Namekawa T (2014) Development of flocculation and magnetic-separation to produced water treatment. Journal, Japan Society of Energy and Resources, Tokyo
18. Mohammed Ali Sergie (2014) U.S. Energy Exports. Foreign Affairs, the Council of Foreign Relations, the United States
19. McNally R, Levi M (2014) Vindicating volatility. Foreign Affairs, the Council of Foreign Relations, Washington, DC
20. Kurasawa A (2012) Shigen no senso. Iwanami Shoten, Tokyo
21. Energy white paper (2013). The Ministry of Economy, Trade and Industry, Japan
22. Martinage R (2015) Under the sea – the vulnerability of the commons. Foreign Affairs, the Council of Foreign Relations, Washington, DC
23. Polak P (2015) How to beat Goliath. Foreign Affairs, the Council of Foreign Relations, Washington, DC
24. Sawada Y (2014) Kyodai saigai risk to keizai. Nikkeibook, Tokyo
25. Miyakoshi K (2013) Damage and restoration of the refinery attacked by the Great East Japan Earthquake. Journal, the Japan Society of Energy and Resources, Tokyo
26. Takahiro S (2013) City gas and the Great East Japan Earthquake. Journal, the Japan Society of Energy and Resources, Tokyo
27. Shumuta Y (2013) Emergency restoration support technologies of electric power transmission and distribution equipment. Journal, the Japan Society of Energy and Resources, Tokyo
28. Huntington SP (1993) The clash of civilizations? Foreign Affairs, the Council of Foreign Relations, Washington, DC
29. Fukuyama F (2012) The clash of ideas- the future of history. Foreign Affairs, the Council of Foreign Relations, Washington, DC
30. Chevalier J-M, Derdevet M, Geoffron P (2012), Masuda T (2013) L'avenir energetique: cartes sur table. Sakuhinsha, Japan
31. Tamura Y (2014) Sekiyu-bunmeiwa naze owaruka. Toyo Shuppan, Tokyo
32. The Economist Newspaper Ltd (2012) Megachange: The world in 2050 The economist. The Economist Newspaper Ltd, Bungei Shunju Ltd, Japan

Fukushima Nuclear Power Plant Accident and Thereafter

Tadashi Narabayashi

Abstract Many lessons can be learned from the Fukushima Daiichi Nuclear Power Plant (NPP) accident. First, if an isolation condenser (IC) had continued to operate, the accident would have been terminated soon. Reactor core isolation cooling (RCIC) steam turbines also stopped because loss of battery power in Units No. 2 and No. 3 and temperature and pressure in each primary containment vessel (PCV) were so high that the accident management water injection took too long. After the loss of emergency core cooling system (ECCS) and IC core cooling, fuels in the core melted down. Leak of fission product and hydrogen began because of high-temperature damage to the PCV packing. A hydrogen explosion occurred in the upper floor in the reactor building in Units 1, 3, and 4. The Nuclear Regulation Authority (NRA) enforcement of the New Regulatory Requirements was based on the concept of "defense in depth," for commercial nuclear power reactors from July 8, 2013. It is hoped that the lessons learned from this accident will improve the safety of nuclear power plants worldwide.

Keywords Fukushima Daiichi Nuclear Power Plant • Isolation condenser • Emergency core cooling system • Hydrogen explosion • New regulatory requirements • Defense in depth

1 Introduction

On March 11, 2011, Tokyo Electric Power Company's Fukushima Daiichi Nuclear Power Station (NPS) was struck by a tsunami caused by the Tōhoku-Pacific Ocean Earthquake, resulting in nuclear accidents in Units 1 through 4 [1, 2]. With the aim of improving the safety of nuclear power plants (NPPs) worldwide, we summarize the lessons learned following a thorough analysis of the event and make specific proposals for improving the safety of such facilities. The author has been involved in investigating accident causes and developing countermeasures for other NPPs in Japan as a member of the Committee for the Investigation of Nuclear Safety of the

T. Narabayashi (✉)
Hokkaido University, Sapporo, Japan
e-mail: tnaraba@eng.hokudai.ac.jp

Y. Kato et al. (eds.), *Energy Technology Roadmaps of Japan*,
DOI 10.1007/978-4-431-55951-1_5

Atomic Energy Society of Japan. He is an advisory meeting member of the Nuclear and Industrial Safety Agency (NISA) and Nuclear Regulation Authority (NRA) with regard to technical lessons learned from the Fukushima Daiichi NPP accidents and a safety evaluation member of NISA for the other NPPs in Japan [3–5].

2 Investigation of Accidents

Figure 1 compares flooded areas at each NPP. Although other NPPs such as Fukushima Daini, Onagawa, and Tokai Daini were also struck by the tsunami, they all were able to safely terminate operation, until the cooldown condition. The Fukushima Daini NPP succeeded in safe shutdown, even though Unit 1 was affected by water flooding through hatches and an emergency diesel generator (EDG) air intake. AC power was restored by changing the power cable, and the seawater pump motors were replaced by bringing in new motors from the Toshiba Mie Works and Kashiwazaki-Kariwa NPP by helicopter. At the Fukushima Daiichi NPP, Unit 5 was brought under control by using EDG power from Unit 6 [3].

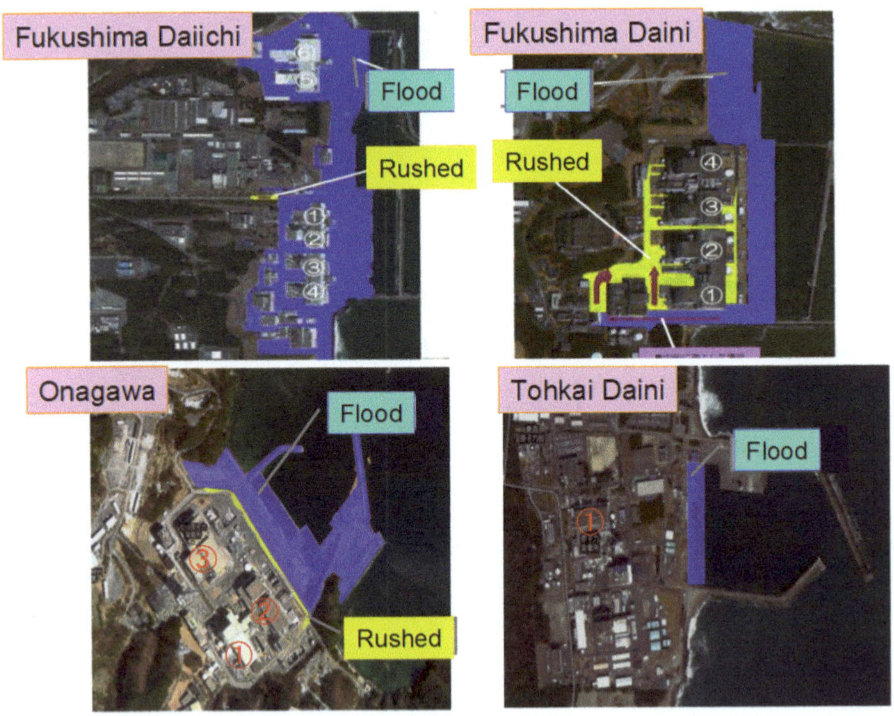

Fig. 1 Comparison of flooded areas at each NPP [3]

Figure 2 shows a comparison of the flood damage to EDGs. At Units 1 through 4, there was a complete loss of both AC power from the EDGs and DC power, and this was the main cause of the ensuring severe accidents [3] (Table 1).

At Unit 1, DC battery power was lost in the main control room. This caused motor-operated (MO) isolation valves to undergo fail-close action, thereby cutting off the isolation condenser (IC), as shown in Fig. 3. This was a fail-dangerous system under such situation. If the IC had continued to operate, the situation would soon have been brought under control [1].

After the loss of both the emergency core cooling system and IC core cooling, primary containment vessel (PCV) pressure increased. Water level measurement drifted because of water evaporation in the reference leg (Fig. 4). Radiation level increased at a turbine building (T/B). There was a hydrogen explosion the after suppression chamber (S/C) wet venting.

As shown in Fig. 5a, both Modular Accident Analysis Program (MAAP) code analysis results and actual data suggest that depressurization of the reactor pressure vessel (RPV) began before its bottom failed. This might have been caused by the melting of traversing in-core probe (TIP) tubes in the core. It was confirmed that the TIP room radiation level is very high even now. Figure 5b shows that the measured water level measurement drifted by more than 4 m owing to water loss in the reference leg. This is likely to have been caused by the high-temperature superheated core. Water should have been supplied to the water level reference leg through instrumentation piping.

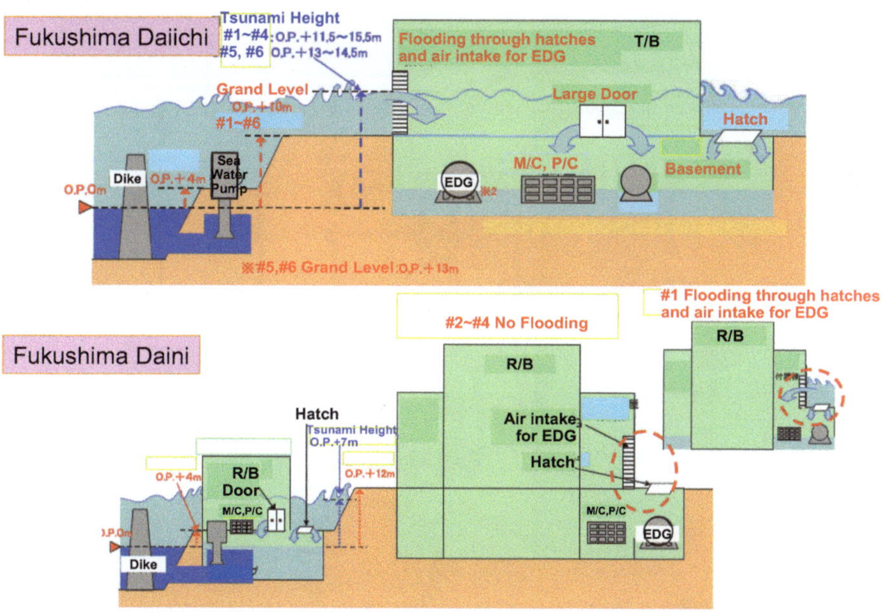

Fig. 2 Comparison of flood damage to emergency diesel generators for Fukushima Daiichi and Daini NPPs [3]

Table 1 Components damaged by tsunami at each unit of Fukushima Daiichi NPP [3]

	#1	#2	#3	#4	#5	#6
DG	A:NG B:NG (T/B B1)	A:NG (B1) B:OK (FP/B 1F)	A:NG B:NG (T/B B1)	A:NG (T/B B1) B:OK (FP/B 1F)	A:OK->NG B:OK->NG (T/B B1) Water Cooling	A:OK->NG (R/B B1) Water Cooling B:OK (DG/B 1F)
Metal-Crad Swich	NG (T/B B1)	NG (T/B B1)	NG (T/B B1)	NG (T/B B1)	NG (T/B B1)	Barely (R/B B2F)
Power Center	NG (T/B B1)	Barely (T/B B1)	NG (T/B B1)	Barely (T/B 1F)	Barely (T/B 2F)	Barely (R/B B2F)
DC Buttery	NG (C/B B1)	NG (C/B B1)	OK (T/B BM1)	NG (C/B B1)	OK (T/B BM1)	OK (T/B BM1)
ECCS RCIC	HPCI:NG IC:OK(FC)	NG RCIC:OK	HPCI:OK RCIC:OK	(No Fuels in RPV)	–	HPCS:OK (R/B B1)

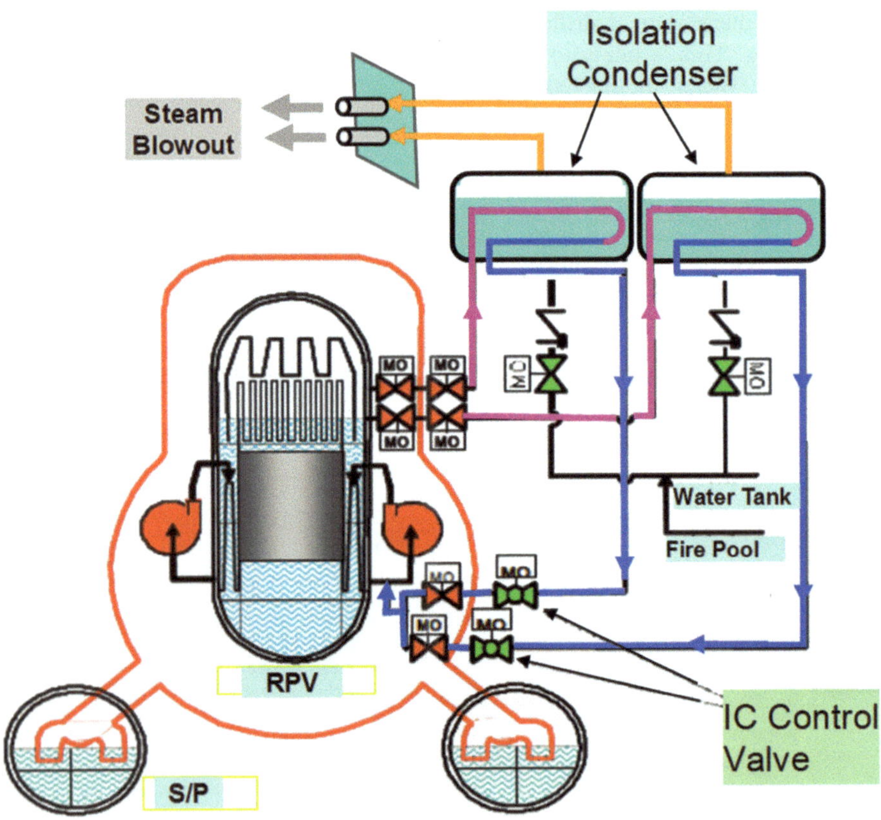

Fig. 3 Isolation condensers in Fukushima Daiichi Unit 1 [1]

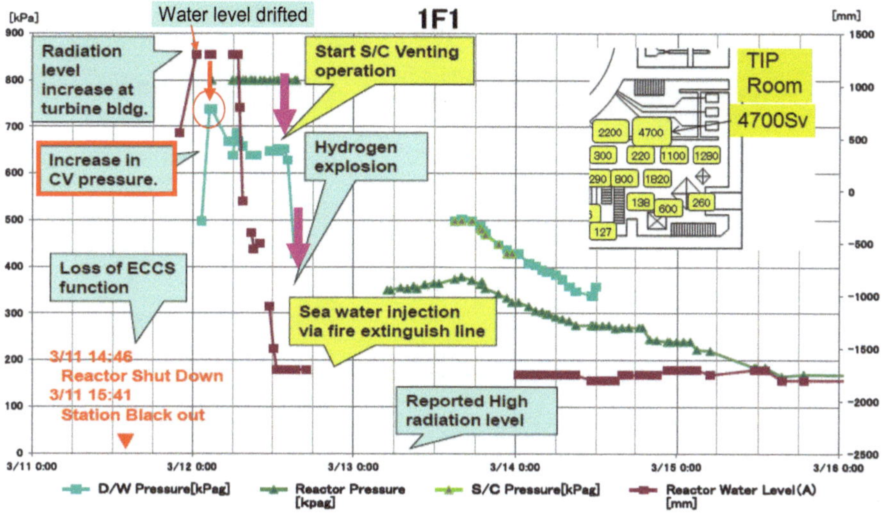

Fig. 4 Measured pressure and water level in RPV and CV of Unit 1 [4, 7]

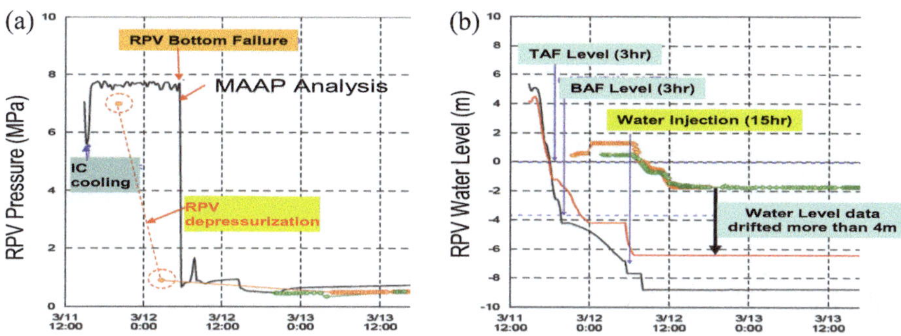

Fig. 5 MAAP analysis results compared with actual plant data for Unit 1 [3]. (**a**) Pressure in RPV. (**b**) Water level in RPV

At Unit 2, reactor core isolation cooling (RCIC) continued to function for about 3 days. Figure 6 shows that soon after the loss of RCIC water injection, the water level in the RPV declined. The safety relief valve (SRV) was opened and seawater injection started. But, RPV pressure shows fluctuation due to water evaporation and metal-water reaction in core. Dry well (DW) pressure increased from 400 to 750 kPa (abs), and PCV top flange leak began through silicon rubber O-ring. It was an initiation of severe contamination around the NPS. In the afternoon on March 15, wind blew toward Iidate village. Melted core relocation into the lower plenum caused the RPV bottom CRD pipe failure and PCV pressure and radiation level increased (Fig. 7). The radiation level was measured by containment atmospheric monitoring system (CAMS) [15].

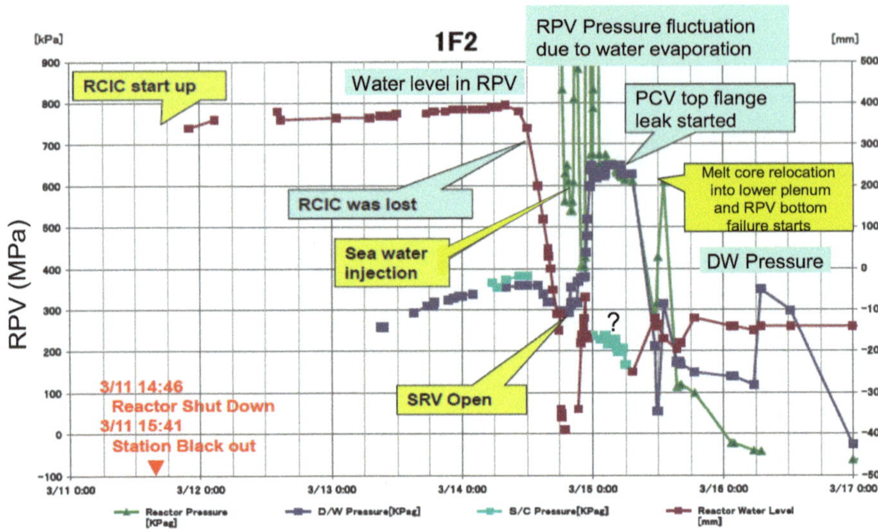

Fig. 6 Measured pressure and water level in RPV and PCV of Unit 2 [4]

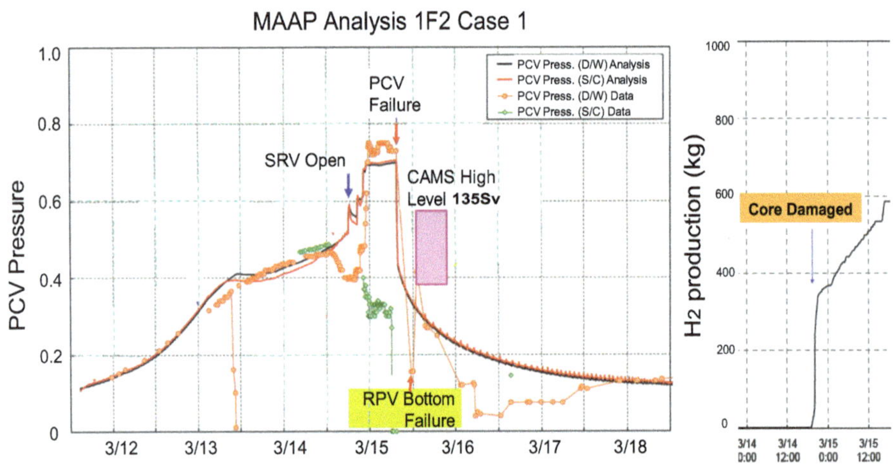

Fig. 7 MAAP analysis results compared with actual plant data for Unit 2 [15]

Figure 8 shows H_2 explosion in Unit Nos. 1, 3, and 4. Upon the occurrence of Unit 3's detonation, the blowout panel of Unit 2 was opened. Hydrogen in Unit 2 was released through the opened blowout panel and there was no explosion/detonation in that unit. The sound of an explosion was reported near the S/C of Unit 2. However, examination of the data showed that this was due to a hydrogen detonation in the reactor building (R/B) of Unit 4. Soon after this detonation, DW pressure in Unit 2 decreased (Fig. 6). Figure 9 shows trends in monitored radiation

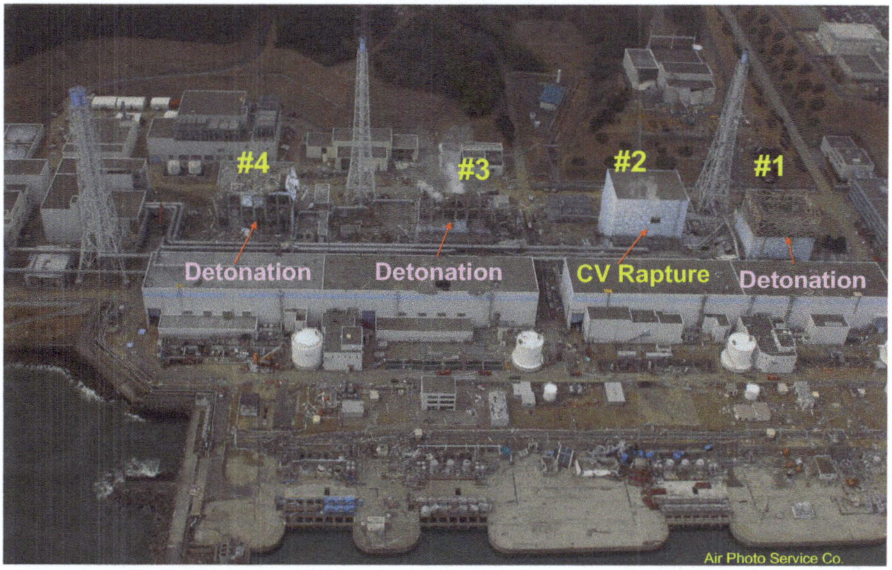

Fig. 8 H_2 detonations occurred after vent operations (Units 1, 3, and 4) [11]

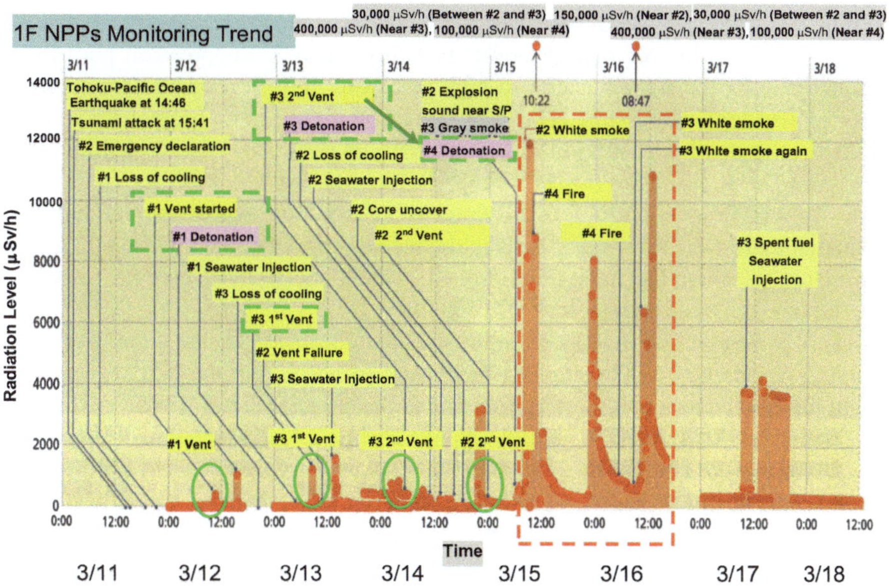

Fig. 9 Monitored radiation levels for Fukushima Daiichi Units 1, 2, 3, and 4 [2]

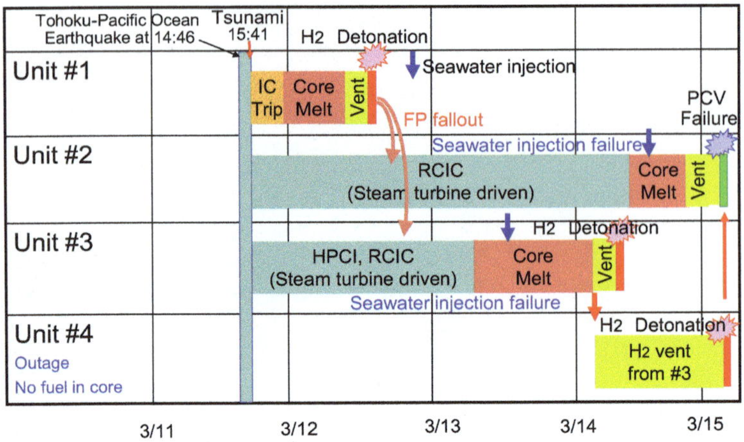

Fig. 10 Chain of major events at Units 1 through 4 causing severe accidents at Fukushima Daiichi NPS [11]

dose levels for Units 1, 2, 3, and 4, which can be compared with events illustrated in Fig. 10.

It appears that the explosion occurred after venting operations. The radiation level increased soon after the Unit 2 PCV rupture on March 15. A loss of core cooling occurred because of the IC trip in Unit 1, and the RCIC steam turbine also tripped owing to loss of battery power in Units 2 and 3. Suppression pool (S/P) temperature and pressure became so high that water injection actions for accident management took a long time. This was the reason for the chain of severe accidents in the four units of the Fukushima Daiichi NPS, as shown in Fig. 10 [11].

Figure 11 shows that the PCV top flange and hatches can act as leakage pathways. Hydrogen and FP flow upward by way of stairways and hatches. Although there were no nuclear fuels in the reactor core of Unit 4, hydrogen flowed from Unit 3 through the stack line into Unit 4 and underwent reverse flow through the standby gas treatment system (SGTS) filters (Fig. 12) [3, 6].

There was a strong hydrogen explosion that occurred in the Unit 4 reactor building on March 14. The author pointed out to NISA that the harden vent line might have acted as a means of hydrogen and fission product (FP) leakage through SGTS and HVAC lines (Fig. 13). As shown in Fig. 14, it was confirmed that the SGTS filters were contaminated and all MO valves were open because of the fail-open design in Units 3 and 4. Seats of the butterfly valves were made of neo-plane rubber and damaged by iodine. This might have caused hydrogen detonation in Unit 4, where there were no nuclear fuels in the reactor core, because hydrogen and FP could have flowed back into each room through the exhaust gas ducts. The vent lines of each NPP should have been independent of the SGTS/HVAC line.

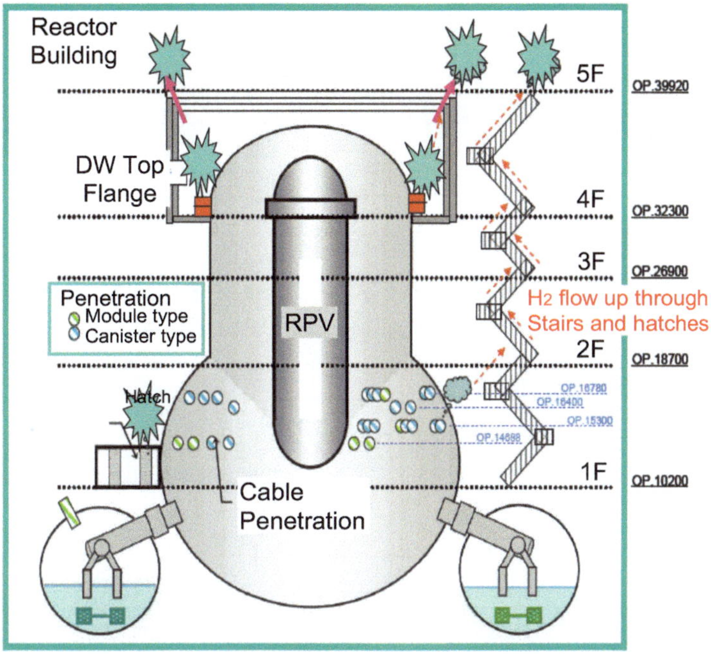

Fig. 11 Estimated leak path from dry well by overpressure and high temperature [3, 6, 11]

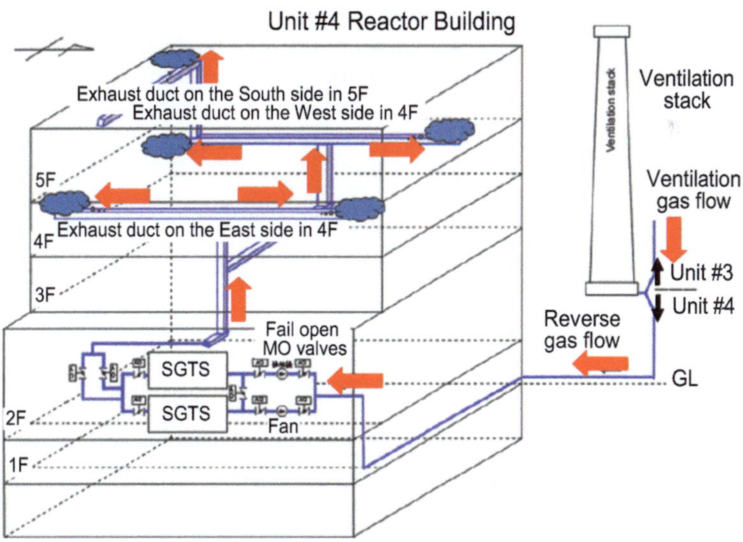

Fig. 12 H_2 gas flow into Unit 4 reactor building from Unit 3 [3, 6, 11]

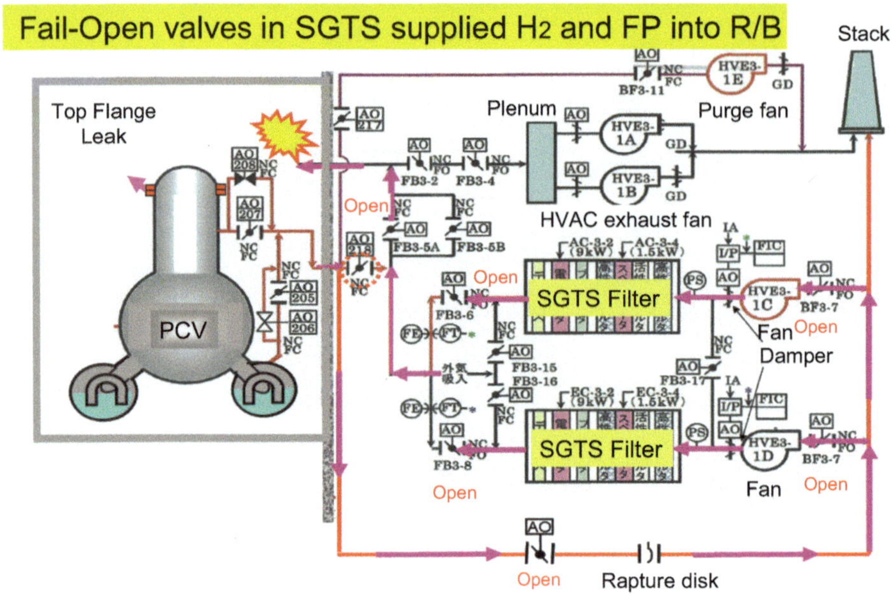

Fig. 13 Flow diagram of SGTS/HVAC and added hard vent system [11]

NISA ordered that the licensees make a new independent vent line for filtered vent system. This is one of the lessons learned. There were no accident reports about Fukushima Daiichi pointing out the vent system fault and potential risks. The causes of severe accidents and countermeasures are shown in Fig. 15, in which (P) means protection and (R) resilience action. Corium cooling was very effective in achieving the cold shutdown cooling, even after the containment failure at the Fukushima Daiichi NPPs.

3 Measures for Severe Accidents Installed in Western NPPs

There are many good practices of countermeasures to prevent FP release in the world. Based on the "defense-in-depth" (DiD) concept (Fig. 16), essential safety features were incorporated in the third layer for design basis accident (DBA) and prevention of simultaneous loss of all safety functions owing to common causes, such as tsunamis. Mobile safety features for the fourth layer such as mobile fire pumps should be deployed for core and containment cooling or corium cooling (Fig. 17) [9].

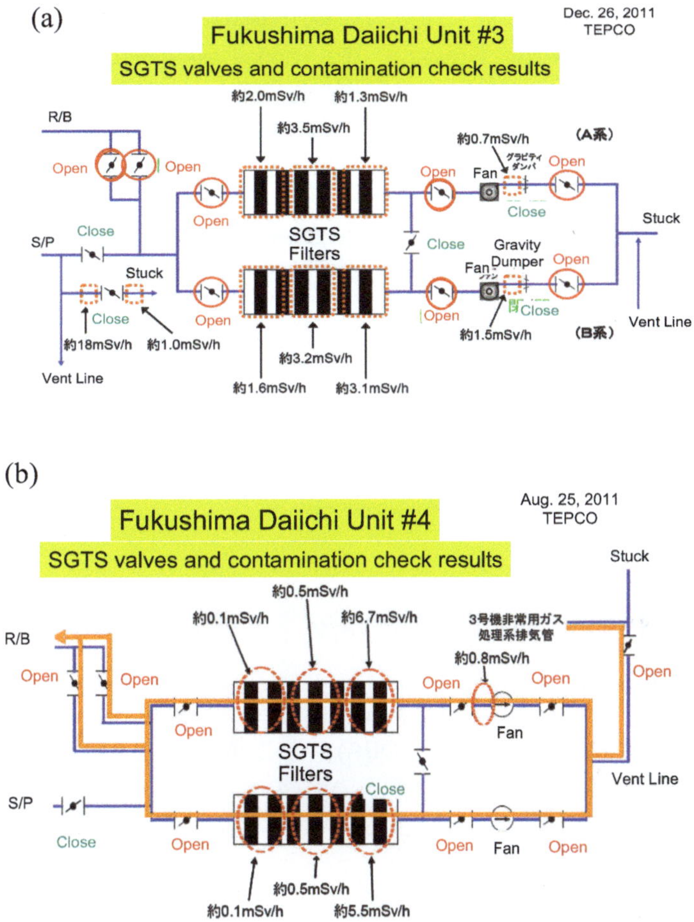

Fig. 14 Results of SGTS valve open/close status and filter contamination [2]. (**a**) Fukushima Daiichi Unit 3. (**b**) Fukushima Daiichi Unit 4

3.1 Filtered Containment Venting System

As shown in Fig. 18, after the Three Mile Island (TMI) Unit 2 and Chernobyl Unit 4 NPP accidents, countries such as France, Germany, Switzerland, Finland, and Sweden decided to install filtered containment venting systems (FCVS) to protect against radioactive material exhaust (Figs. 19 and 20) [8].

Figure 21 shows a schematic diagram of the FCVS installed in the Leibstadt NPP. Venting is automatically initiated when the CV pressure reaches the pressure set for the rupture disk. An operator who wishes to vent early can easily open the vent valve using a hand wheel drive shaft [8].

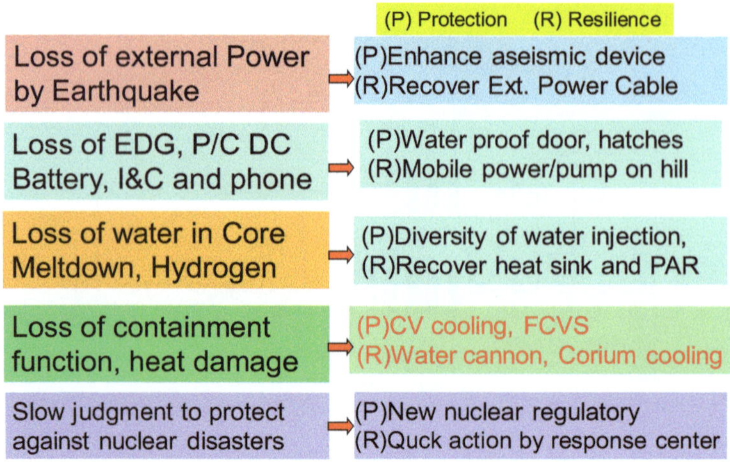

Fig. 15 Causes of severe accidents and countermeasures

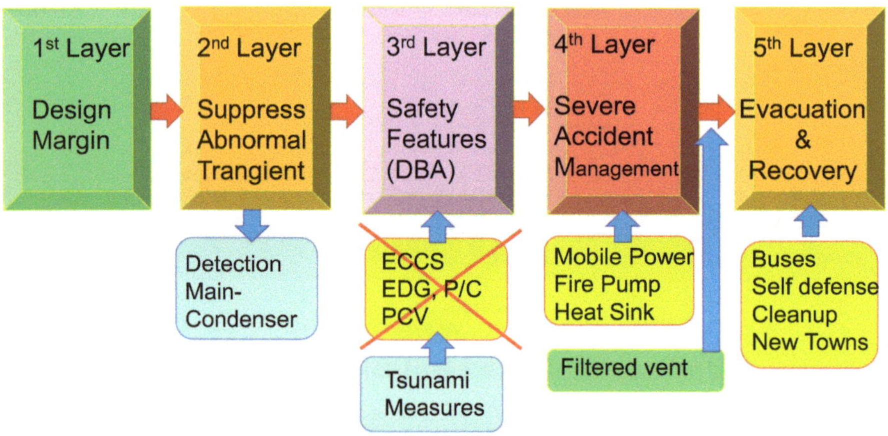

Fig. 16 Concept of "defense in depth" to terminate accidents

In the Fukushima Daiichi NPP accidents, operators should have closed numerous valves in the SGTS system and then opened the vent valve with an air compressor and connecting tubes, because of the station blackout condition. If a FCVS had been installed in the Fukushima Daiichi NPPs, environmental contamination by FP could have been avoided. The decontamination factor is about 1000 for aerosols and 100 for I_2.

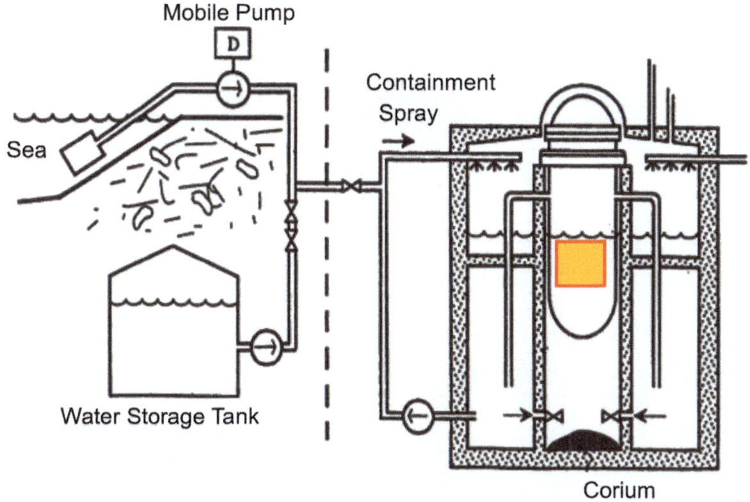

Fig. 17 Mobile safety features for severe accidents in DiD fourth layer [9]

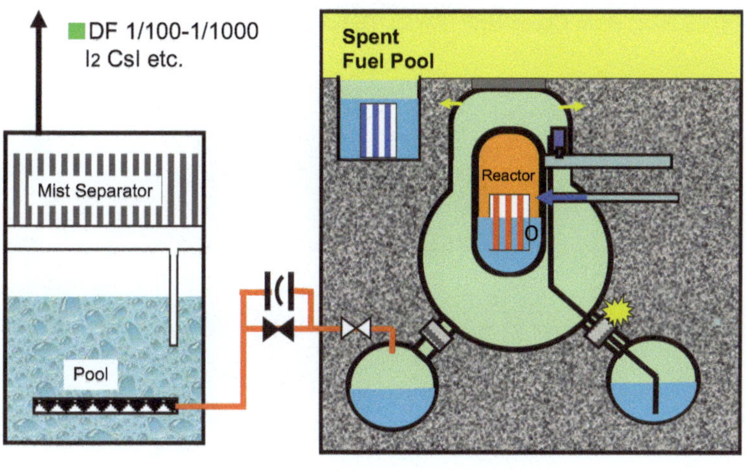

Fig. 18 Filtered containment venting system [8]

After the TMI-2 accident in 1979, Kernkraftwerk Leibstadt (KKL) backfitted the Leibstadt NPP with additional CV cooling (DiD 3) and a mitigation system for severe accidents (DiD 4). The backfitted system was called the Special Emergency Heat Removal (SEHR) system. This system was required by the Swiss regulatory body of Federal Nuclear Safety Inspectorate (ENSI) and Swiss Federal Office of Energy (HSK) in the late 1970s, shortly after the start of project planning, so it was the first backfitting in the present design of KKL.

Fig. 19 FCVS installed in Chooz NPP (PWR), France [8]

Fig. 20 FCVS in Leibstadt NPP (BWR), Switzerland [8]

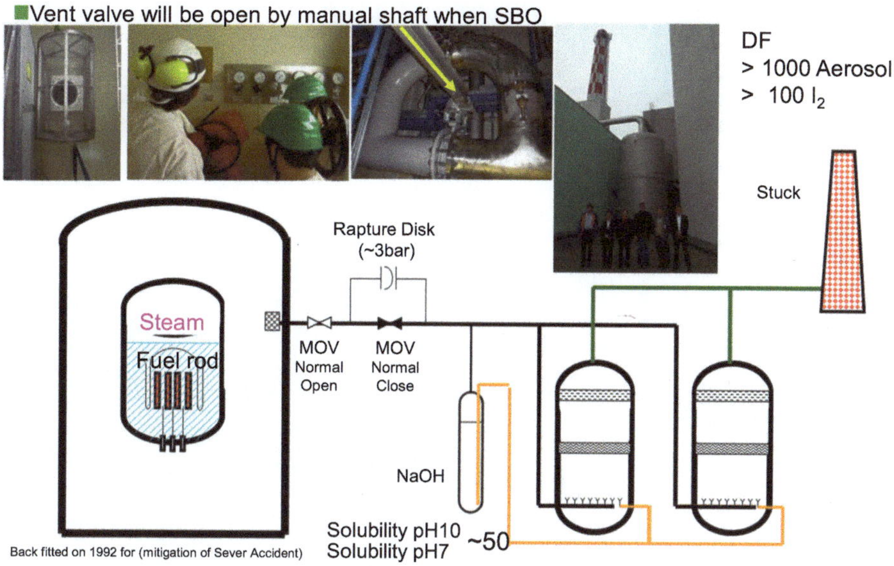

Fig. 21 Schematic diagram of FCVS in Leibstadt NPP [8]

3.2 Special Emergency Heat Removal System

Figure 22 shows SEHR system has two trains of heat removal system. The system was installed to remove a minimum of 36.3 MW (estimated decay heat: 1 % of nominal power). The system has two special EDGs and a huge underground well for water heat sink. The system is able to cool both the core and the CV, using the heat exchanger [8].

3.3 Tsunami Protection

When the Fukushima Daiichi NPP was attacked by the tsunami, all AC and DC power was lost because of damage to the EDGs, power center, metal clad switchgear, and seawater pump motors. At the Fukushima Daini NPP, AC power was able to restore seawater pumps by changing power cable and installing new pump motors. Therefore, it is very important to prevent the seawater flow into important areas. As shown in Fig. 23, at Diablo Canyon NPP in California, USA, the seawater pump motors are equipped with waterproof hatch-type doors and snorkel air ventilation piping for pump motor cooling.

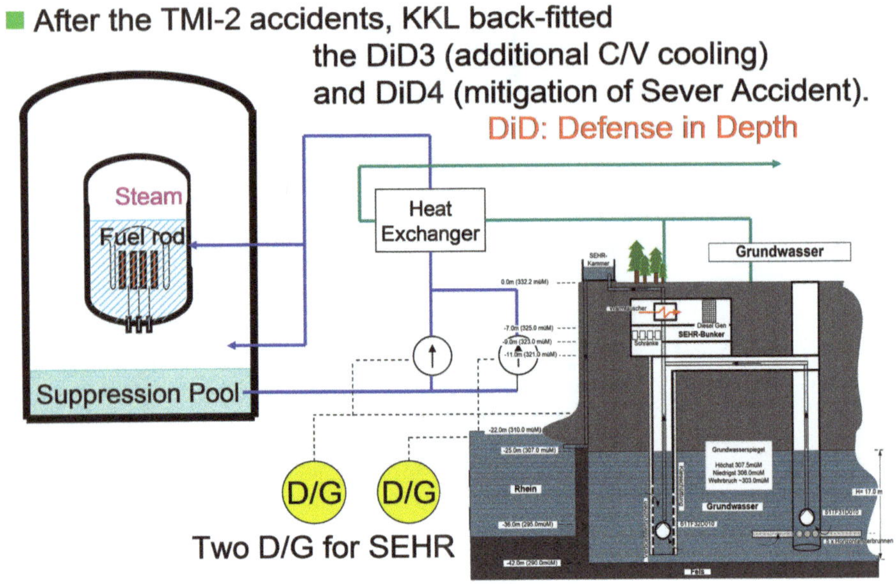

Fig. 22 Special Emergency Heat Removal (SEHR) system [8]

Fig. 23 Tsunami protection at Diablo Canyon NPP, USA [8]

3.4 Force for Action of Nuclear Rapid Response

According to French Nuclear Safety Authority (ASN) requirements, a Nuclear Rapid Response Force (FARN) was established by Électricité de France (EDF) (Figs. 24 and 25) [12]. The FARN should be able to deploy specialized teams and mobile equipment to respond in less than 24 h on station. As shown in Table 2, deployment began in late 2012 and will be completed in late 2015.

"Hardened safety core" equipment and organization are intended to prevent a severe accident or limit its development in case of extreme natural hazards beyond design basis conditions. The ASN requested that EDF define for each plant "a hardened safety core" of equipment and organizational measures needed to control the basic safety functions in emergency situations, before June 30, 2012. Main ASN decision letters for post Fukushima actions already consisted of 36 requirements issued on June 26, 2012.

"Hardened safety core" (HSC) equipment and organization and 16 associated and new requirements were issued in January 2014. The ASN requested in complementary safety assessment (CSA) installation of additional electrical supplies for double-wall containment venting and control room venting systems. Figure 26 shows sodium tetraborate baskets in reactor building sumps to reduce iodine release (to be studied), a passive autocatalytic recombiner (PAR) to limit releases in the event of core meltdown, robustness, and efficiency of filtering existing FCVS seismic reinforcement. EDF showed proposition; SMHV (millennial earthquake) filtration efficiency (iodine filtration) sodium tetraborate baskets will be added to

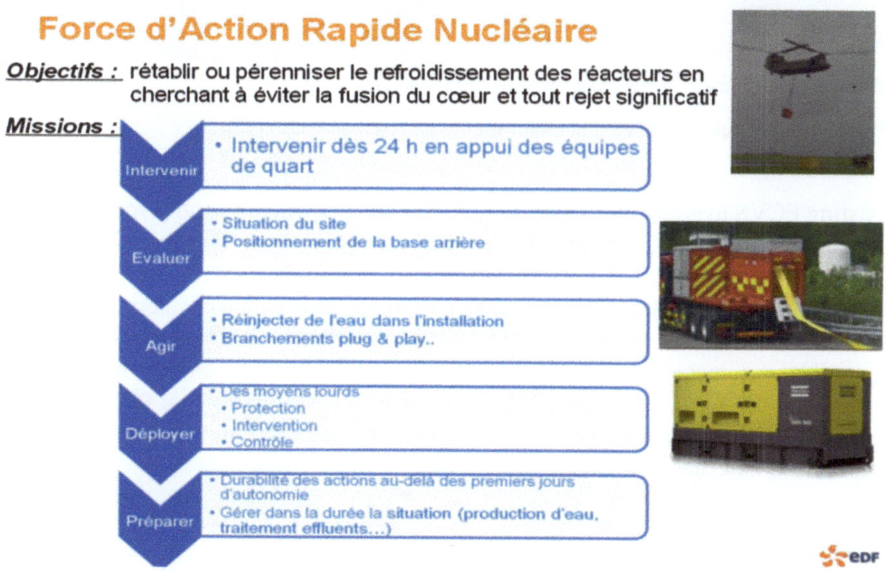

Fig. 24 Nuclear rapid response force (FARN) established by EDF [12]

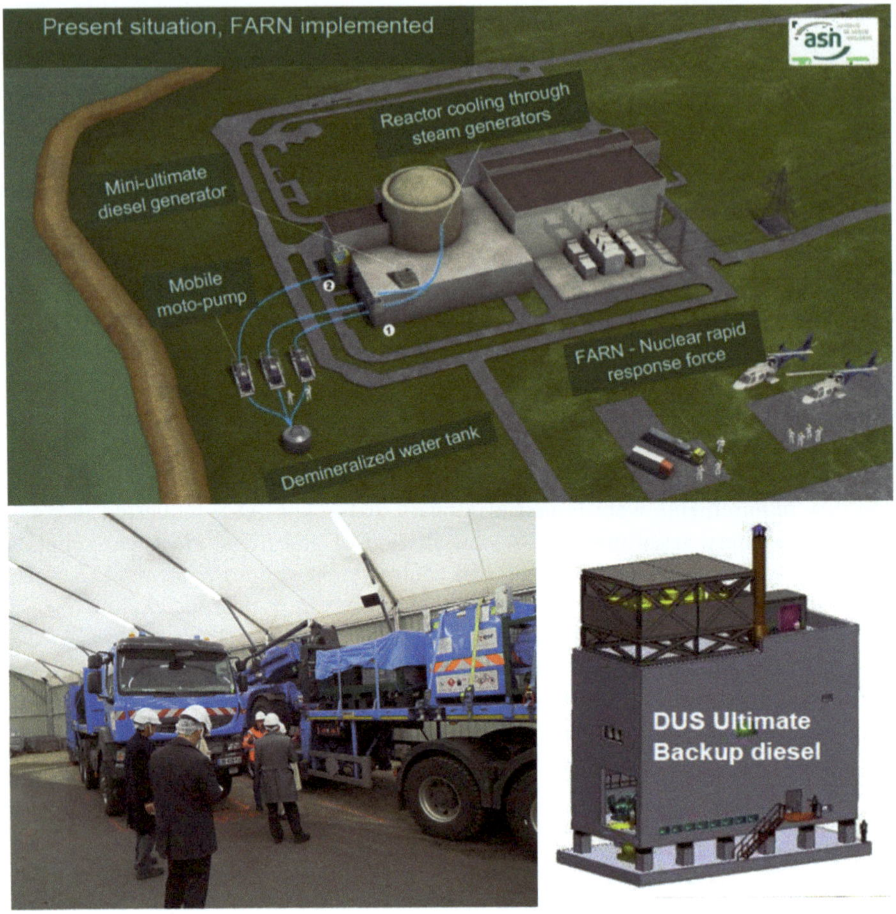

Fig. 25 FARN equipment, mobile trailer trucks, and backup diesel [12]

existing FCVS available for basic design situations. The main control room and site habitability are under study, and new on-site emergency control center will be built. In order to protect/prevent of groundwater pollution in the event of melt through of the reactor vessel by the corium during core melt, core catcher system for current PWR severe accident management (SAM) is under study. This involves core melt spread in the reactor pit to facilitate cooling (dry cavity), basement thickness increase with special concrete, HSC water injection system to cool the core melt, and an HSC containment cooling system (residual heat extraction with recirculation system and external heat exchanger).

Table 2 FARN hardened safety core deployment and completion schedule [12]

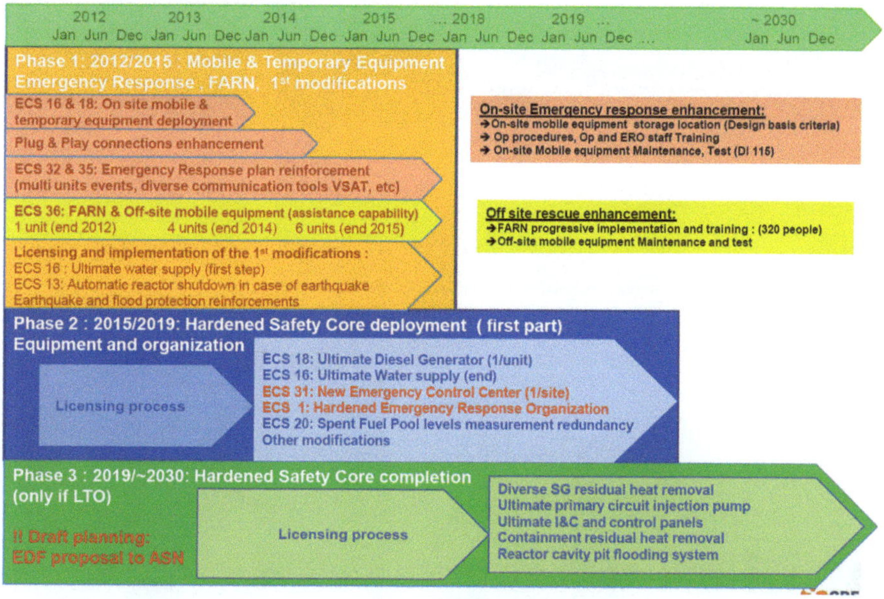

4 Countermeasures Based on New Regulatory Enforcement

4.1 New Nuclear Regulatory Requirements in Japan

A new nuclear regulatory body, the Nuclear Regulation Authority (NRA), was established on September 19, 2012. The NRA performed a complete review of safety guidelines and regulatory requirements [10].

On July 8, 2013, new regulatory requirements for commercial power reactors came into force. These requirements stipulate that all Japanese utilities conform to the regulatory requirement before restarting NPP. Design requirements must treat natural phenomena, such as volcanoes, tornados, and forest wildfires (Fig. 27). The new regulatory guidelines (Fig. 28) require direct deployment of mobile power, mobile pumps, fire engine, and installation of tsunami protection. Design requirements should be prepared to protect against cable fire between the reactor building and main control room and against internal inundation by waterproof areas of important safety components and systems.

New design requirements for severe accidents require measures such as mobile powers, mobile pumps, fire engines, and water tanks or reservoirs to protect core

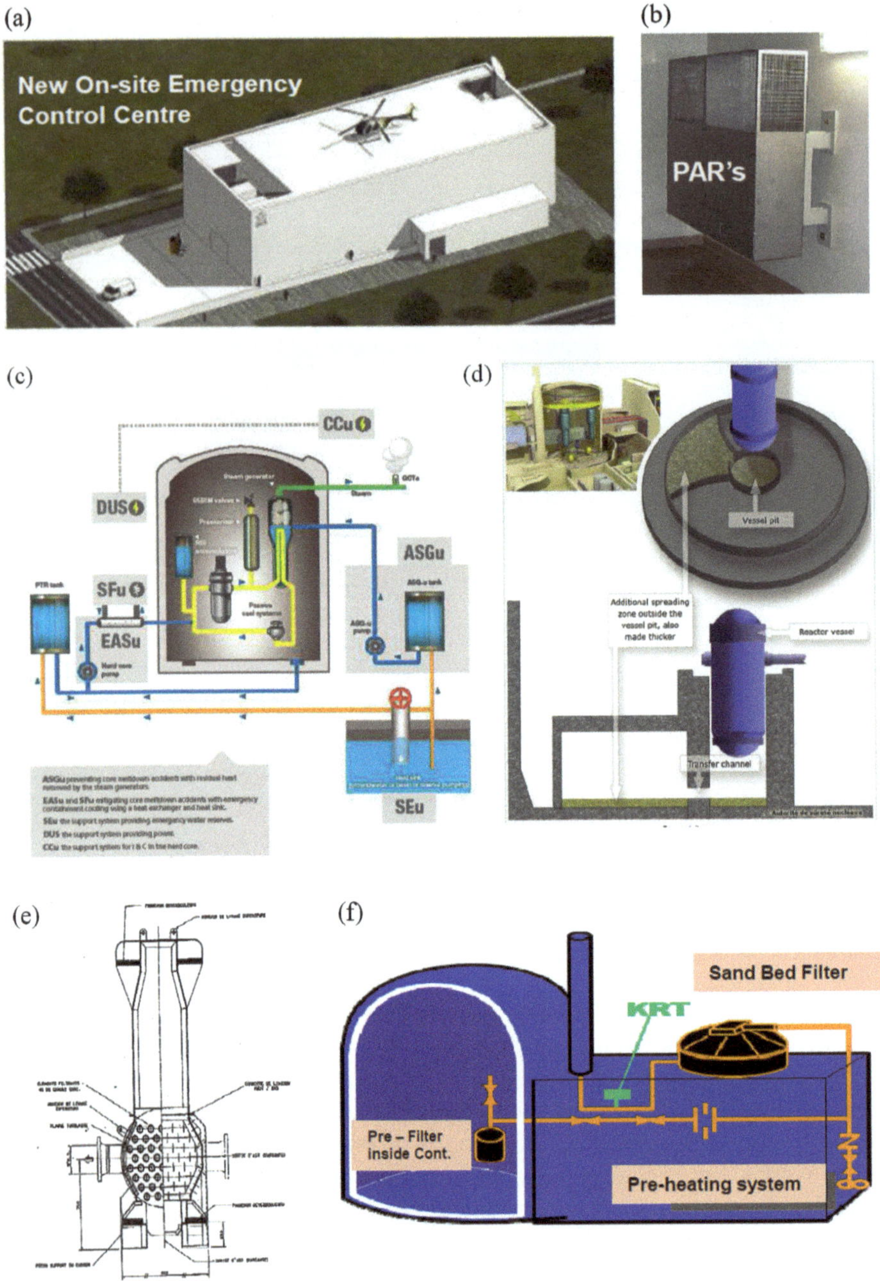

Fig. 26 Severe accident countermeasures by EDF [12, 13]. (**a**) New on-site emergency control center. (**b**) PAR (Auto passive catalytic recombiner). (**c**) Core and SG water injection cooling system. (**d**) Core catcher for current PWRs. (**e**) Pre-filter with sodium tetra-borate baskets. (**f**) Seismic reinforced FCVS with iodine pre-filter

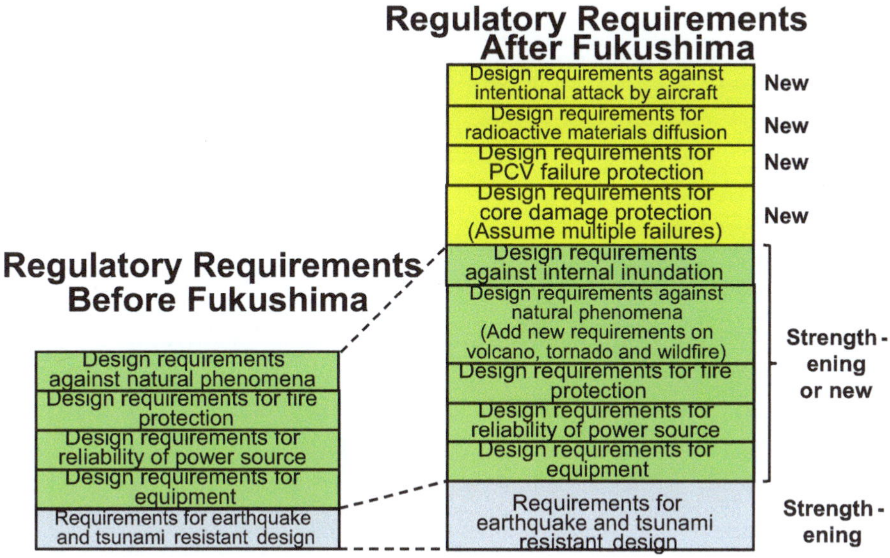

Fig. 27 Regulatory requirement comparison for before and after Fukushima [10]

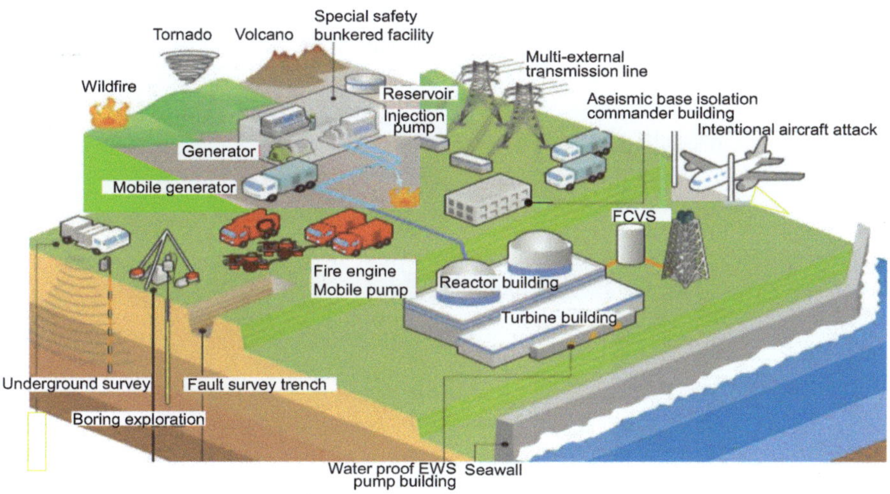

Fig. 28 Enforcement of new regulatory requirements, from July 2013 [19]

damage and to protect containment vessel failure. A bunker-type underground building (Fig. 29) should be constructed to control reactor cold shutdown against intentional attack by aircraft, within a 5-year grace period after approving construction permission.

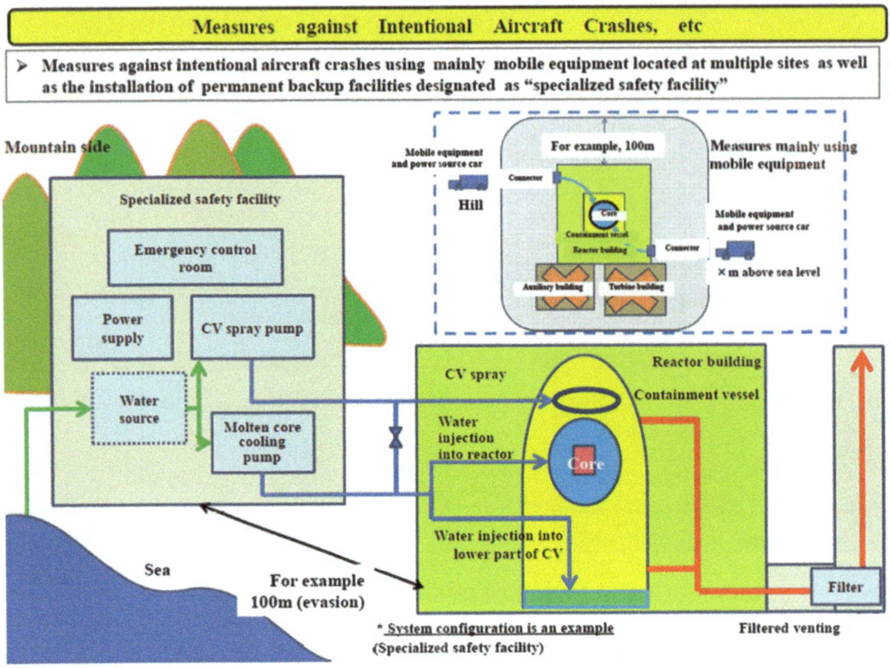

Fig. 29 Measures against intentional aircraft crashes [10]

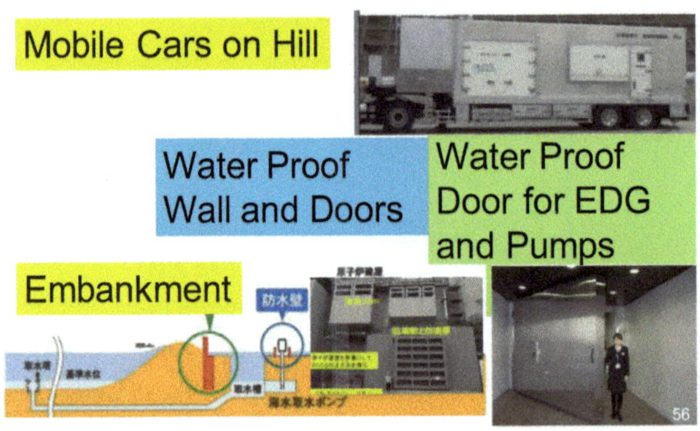

Fig. 30 Tsunami protection concept based on the defense in depth [11]

4.2 *Tsunami Protection Examples*

Tsunami protection is very important in preventing and protecting against severe accident such as what occurred at Fukushima Daiichi. Figure 30 shows tsunami protection concept based on the DiD. Embankments/seawalls are the primary tsunami protection, waterproof walls and doors at entrances to the reactor building

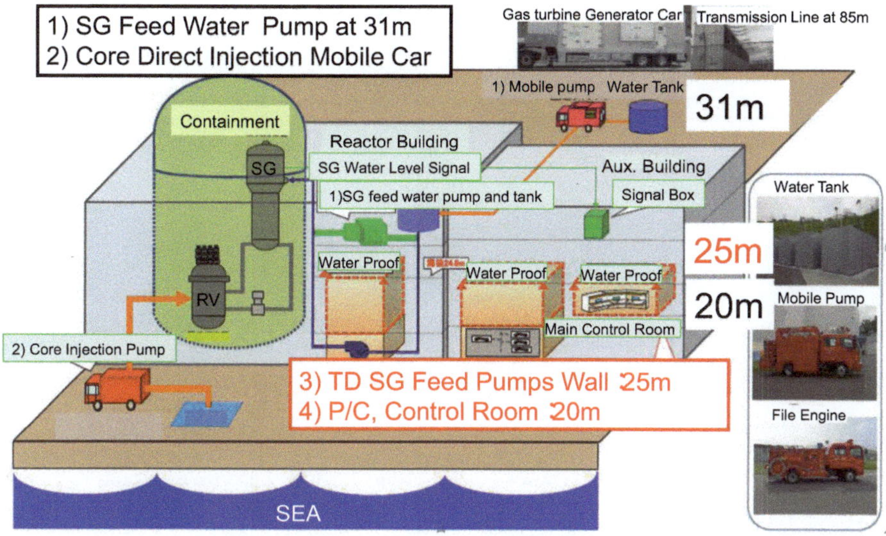

Fig. 31 Tsunami protection measures at Tomari NPS (PWR) [14]

are second, waterproof doors for EDG and pump rooms are third, and mobile gas turbine cars on hill are fourth.

Figure 31 shows tsunami protection measures at Tomari NPS of Hokkaido Electric Power Co. Inc., including a large mobile gas turbine generator and steam generator (SG) feed water pump in a parking area of 31 m height from sea level. Tsunami-proof doors and balconies have been installed in the reactor and auxiliary buildings. Turbine-driven feed pumps for SG are protected against tsunami of 25 m height. The power center (P/C) in the main control room is protected against those of 20 m height.

Figure 32 shows tsunami protection measures at Hamaoka NPS of Chubu Electric Power Co. Inc. A large tsunami wall of 22 m height and 1.6 km length has already been constructed. There is a snorkel building for EWS pumps and 400 kVA (3.2 MW) gas turbine generator at 25 m height. A cross-sectional view of tsunami measures and water reservoir is shown in Fig. 33.

Figure 34 shows examples of tsunami protection measures at NPSs, such as gas turbine generators, an oil tank, and snorkel building at Shimane NPS. At Kashiwazaki-Kariwa NPS, there are a large wall, door and balconies, mobile cooling car, and fire engines. Table 3 shows typical examples of measures that meet requirements at Shimane and Kashiwazaki-Kariwa NPS.

4.3 Tornado Protection Examples

Figure 35 shows a tornado evaluation method and measures for seawater pumps. Tornado wind streamlines were calculated using a model designed by Dr. Fujita.

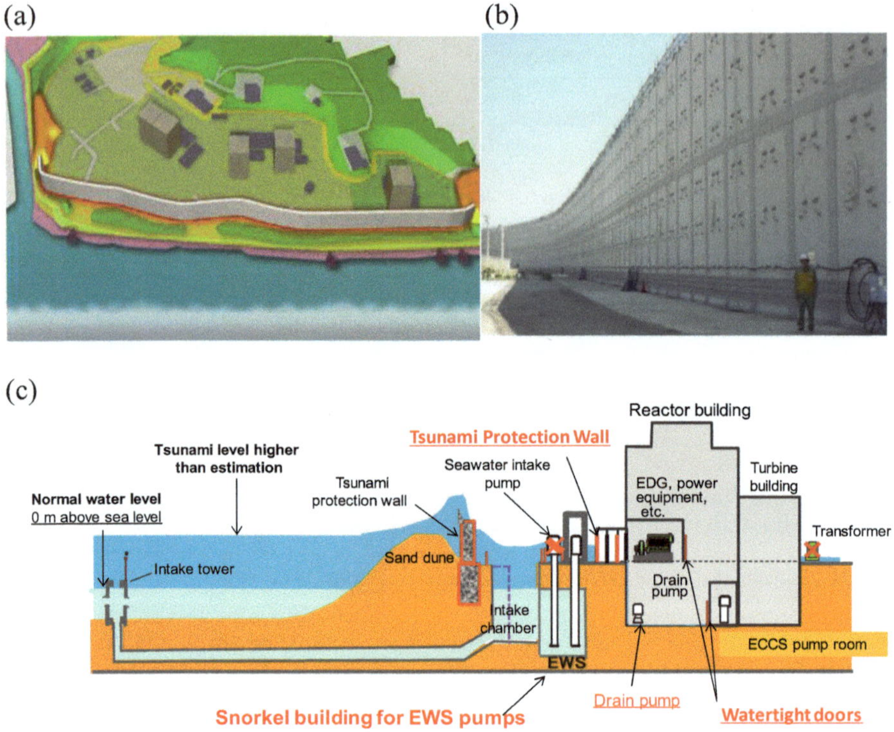

Fig. 32 Tsunami protection measures at Hamaoka NPS (BWR) [14]. (**a**) Tsunami wall (22 m × 1.6 km). (**b**) Complete tsunami wall construction. (c) Cross sectional view of tsunami measures

Based on the wind field, a missile analysis was conducted for both ground level and 40 m height in particular. Final missile speed was used to estimate its kinetic energy for shield strength to protect a seawater pump, reactor building, or other safety-related components.

4.4 PWR NPS to Meet the New Requirements for Restart

In September 2014, Sendai NPS Units 1 and 2 (Kyushu Electric Power, Co.) received NRA permission to upgrade their safety systems as required by the post-Fukushima regulations. This was followed by a construction permit and detailed design review by the NRA. Figure 35 shows the complete cooling strategy for station blackouts (SBOs) to protect against severe accident (SA) for PWR. Fuels in the core are cooled by SGs. A water supply for the secondary side of SG is used to cool the primary side of U tube. The core on the primary loop side is cooled by natural recirculation driven by gravity. In the inverted U tube, the outlet cold-side

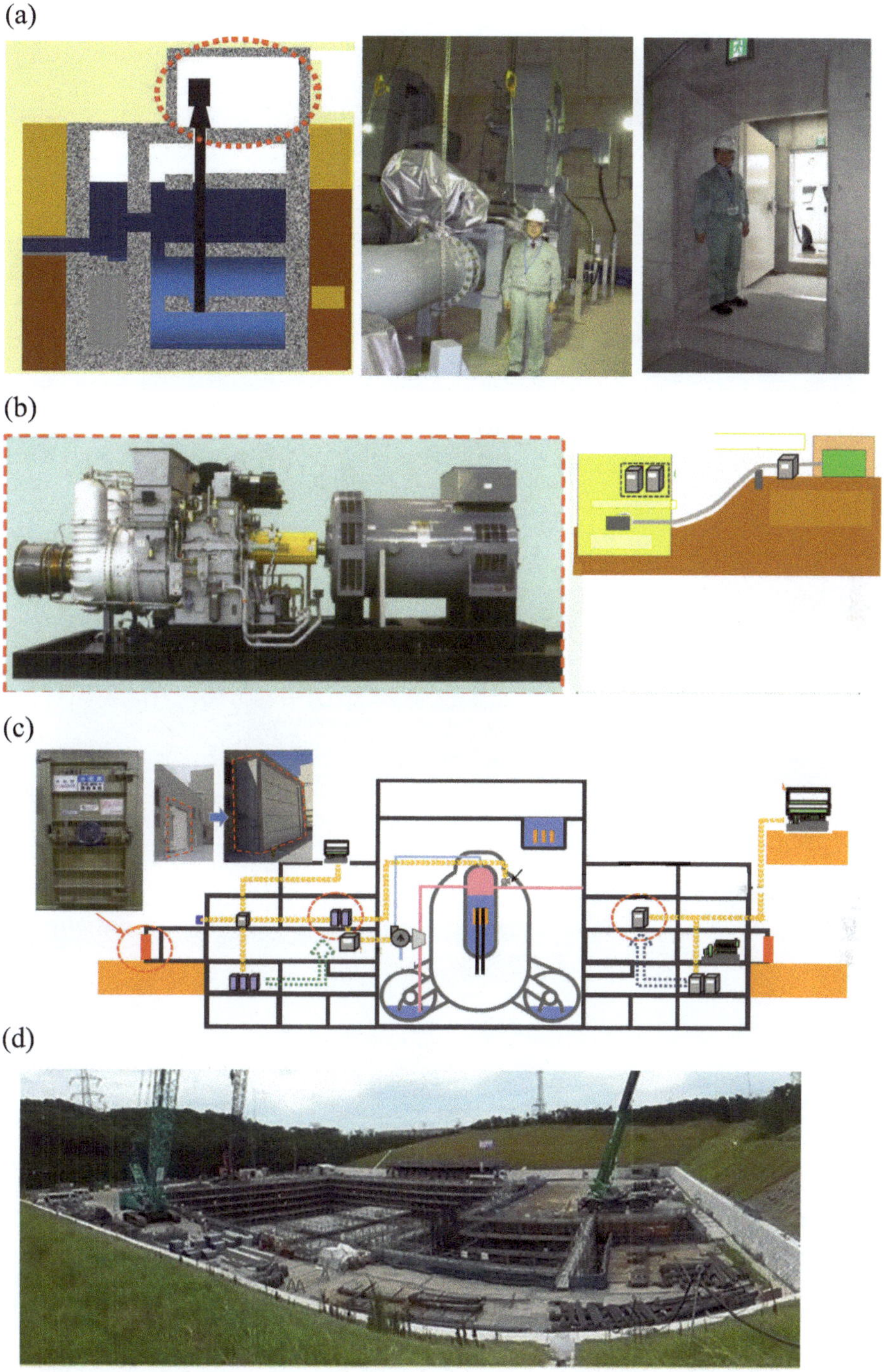

Fig. 33 Tsunami protection measures under construction at Hamaoka NPS [14]. (**a**) Snorkel building for EWS pumps. (**b**) 3.2 MW gas turbine generator installed at 25 m height. (**c**) Cross-sectional view of tsunami measures. (**d**) Water reservoir

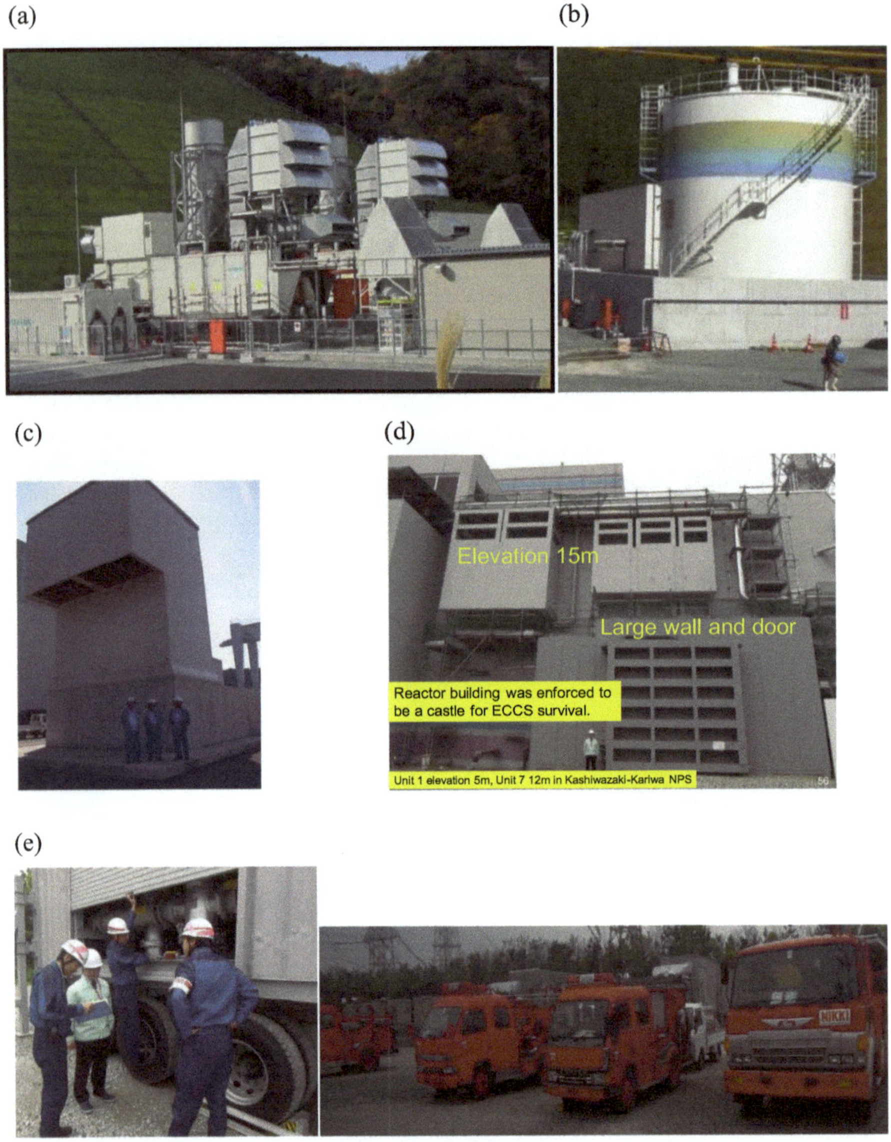

Fig. 34 Examples of tsunami protection measures at NPSs [14]. (**a**) Gas turbine generator at Shimane NPS. (**b**) Oil tank at Shimane NPS. (**c**) Snorkel building at Shimane NPS (**d**) Large wall, door and balconies at Kashiwazaki-Kariwa NPS. (**e**) Mobile cooling car and fire engines at Kashiwazaki-Kariwa NPS

Table 3 Typical examples meeting requirements

Purpose	New requirement	Typical examples which meet the requirements
Protection against flooding	A plant shall be able to withstand against design tsunami.	Installation of protection walls and doors against design tsunami
Ensuring redundancy of power supply	Supply of electricity	Deployment of permanently installed or portable backup AC power supply, enhancement of permanently installed DC power supply, deployment of portable DC power supply, etc.
Ensuring redundancy of core cooling	Cooling function under high reactor pressure	Deployment of buttery for RCIC control, preparation of operation procedure, training etc.
	Depressurization function of reactor	Deployment of buttery for ADS actuation, preparation of operation procedure, training etc.
	Cooling function under low reactor pressure	Permanently installed coolant injection equipment, portable coolant injection equipment, preparation of operation procedure, training etc.
	Ultimate heat sink for prevention of severe accidents	Deployment of vehicle for backup ultimate heat sink, preparation of operation procedure, training etc.
Reinforcement of Safety measures applicable during severe accidents	Prevention function for containment break due to excessive pressure	Installation of filtered containment venting system, preparation of operation procedure, training etc.
	Emergency response center	Ensuring earthquake and tsunami resistant emergency response center, radiation protection, logistics etc.

head is larger than that of the outlet hot side. Core injection using ECCS pumps driven by diesel power supply car and containment vessel (CV) spray is possible, even if under SBO. Spent fuel-water injection is possible because of easy access routes for fire engines (Fig. 36).

Figure 37 shows an additional emergency water injection pump installed on seismic base isolation rubber on the top floor of a turbine building. Discharge piping is via a flexible bellows pipe. The pump can supply water to the core or containment spray as a severe accident resilience action. Personnel in the Genkai NPS are trained to be able to connect the flexible pipe within a very short time.

There are numerous safety reinforcement measures at Genkai Units 3 and 4 (Fig. 38). It is very important to prevent CV failure from overpressure, over temperature, and hydrogen detonation. For this purpose, CV recirculation cooler and CV spray, PAR, and igniter for hydrogen combustion with oxygen in CV are available. A water cannon to suppress radioactive material diffusion has already been deployed to meet requirements of intentional aircraft impact or tornado-driven missiles on the reactor building and CV. These measures are based on the DiD concept and use diverse strategies.

As shown in Fig. 39, many mobile pumps driven by diesel/gasoline engines are deployed at Ikata NPS. Such devices have already been deployed at all NPSs in Japan.

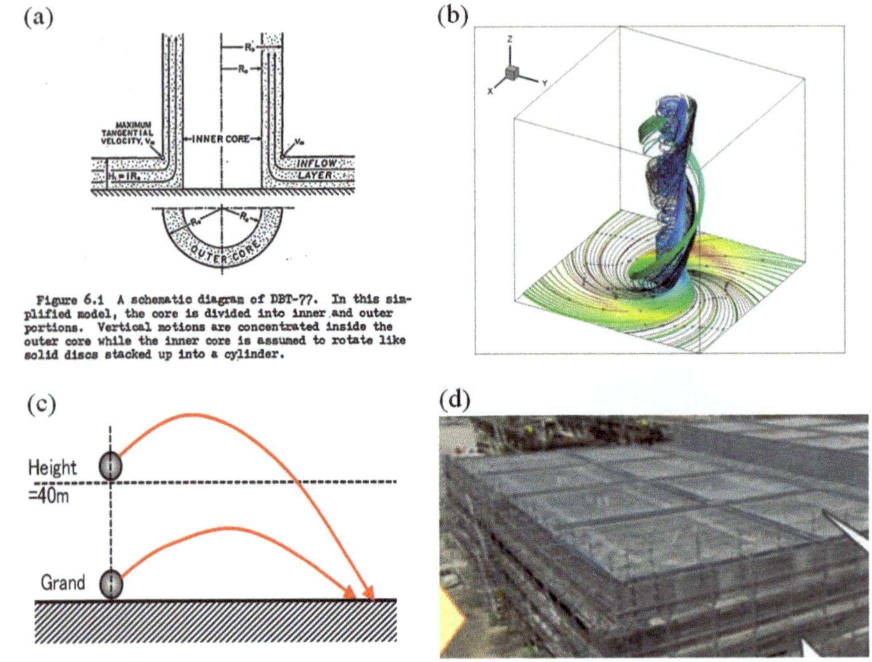

Fig. 35 Tornado evaluation method and measures for seawater pumps. (**a**) Tornado wind field (Dr. Fujita). (**b**) Tornado streamlines from Dr. Fujita model. (**c**) Tornado trajectories for missile analysis. (**d**) Missile shield for sea water pump

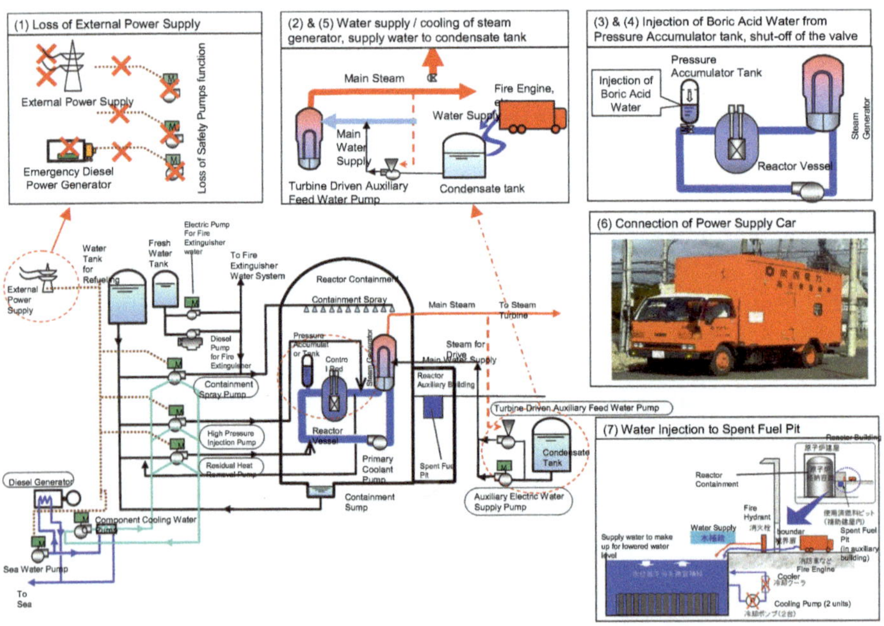

Fig. 36 Full cooling strategy for SBO to protect against SAs for PWR [14]

Fig. 37 Water injection pump

4.5 BWR NPS to Be Reviewed for the New Requirements or Restarting

Following the design reviews of 12 PWRs, nine BWRs with enhanced safety measures were in ongoing design review for restarting.

As shown in Fig. 40, by using seawater cooling system, suppression pool water in suppression chamber will be cooled, and core injection system can remove decay heat from the core to the pool via a main steam safety relief valve (SRV). The system can be operated using mobile generators and heat exchanger car, even for a natural disaster such as a major earthquake or tsunami or sudden flooding. It is very important to cool the suppression pool water by a connected mobile cooling car using plate-fin heat exchanger cooled by seawater pump to maintain an ultimate heat sink.

Upon occurrence of a SA, vent gas with radioactive fission products is blown out to a scrubbing pool through numerous venturi nozzles (Fig. 41). Mist in steam moves upward to a metal fiber filter through a multi-hole baffle plate. After the mist is removed by that filter, radioactive methyl iodide (CH_3I) is captured on the surface of a molecular sieve or AgX, made from zeolite particles with silver coating [16, 17].

Figure 42 shows the FCVS pit at Hamaoka NPS of Chubu Electric and the installing of FCVS at Kashiwazaki-Kariwa NPS of TEPCO, respectively.

Figure 43 shows the FCVS visualization test facility at Hokkaido University. An AgX filter is used downstream of the scrubbing pool and metal fiber filter. This study was conducted by Kakenhi (B) funded No. 2436038802. The thickness of AgX filter is a very important parameter to obtain enough decontamination factor (DF). As shown in Table 4 of TUV test result in Germany, the DF for the radioactive iodine exceeds 10,000 at bed depth (AgX filter thickness) greater than 75 mm [16].

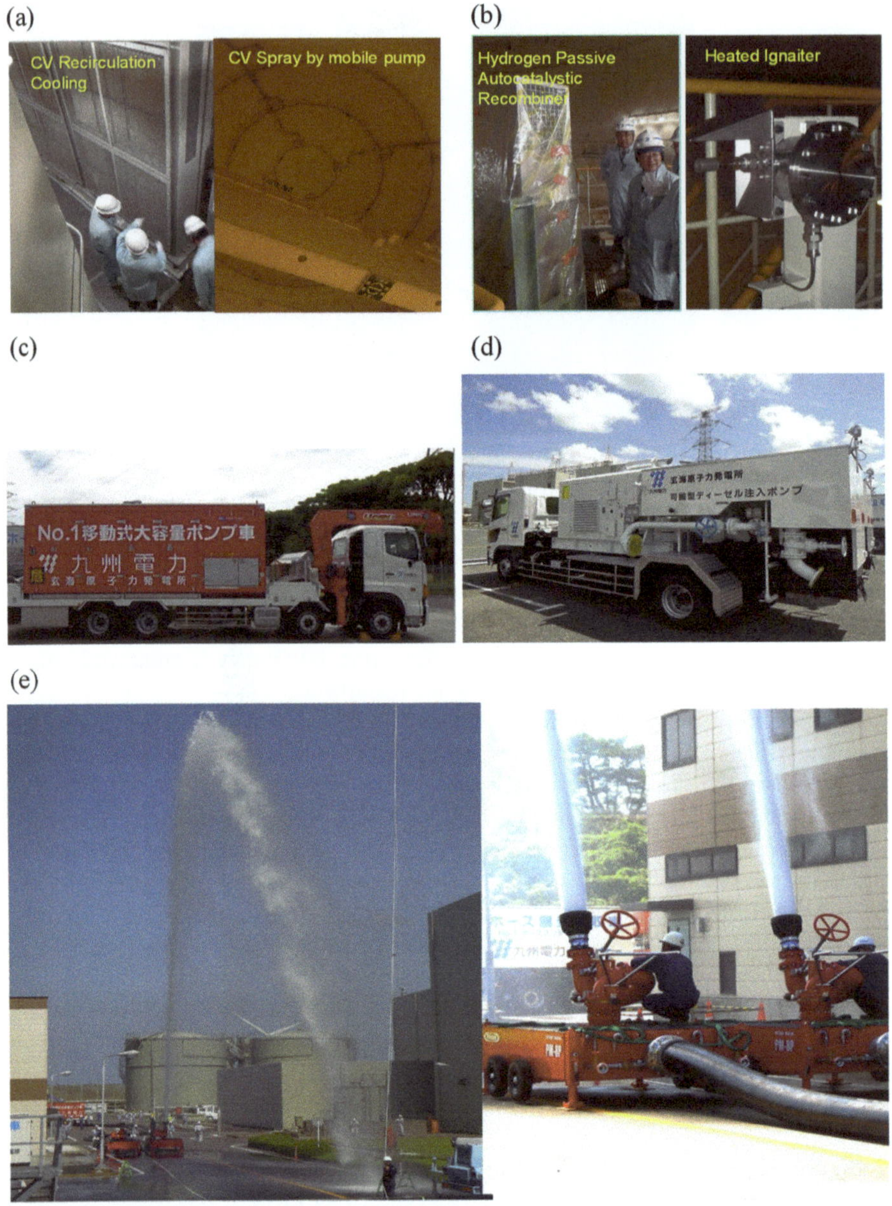

Fig. 38 Safety reinforcement measures at Genkai Units 3 and 4 for SBO. (**a**) CV recirculation cooler and CV spray. (**b**) PAR and ignaiter for hydrogen. (**c**) Mobile motor-driven pump. (**d**) Mobile diesel engine-driven pump. (**e**) Water cannon to suppress radioactive material diffusion

Fig. 39 Mobile pumps deployed in Ikata NPS [14]

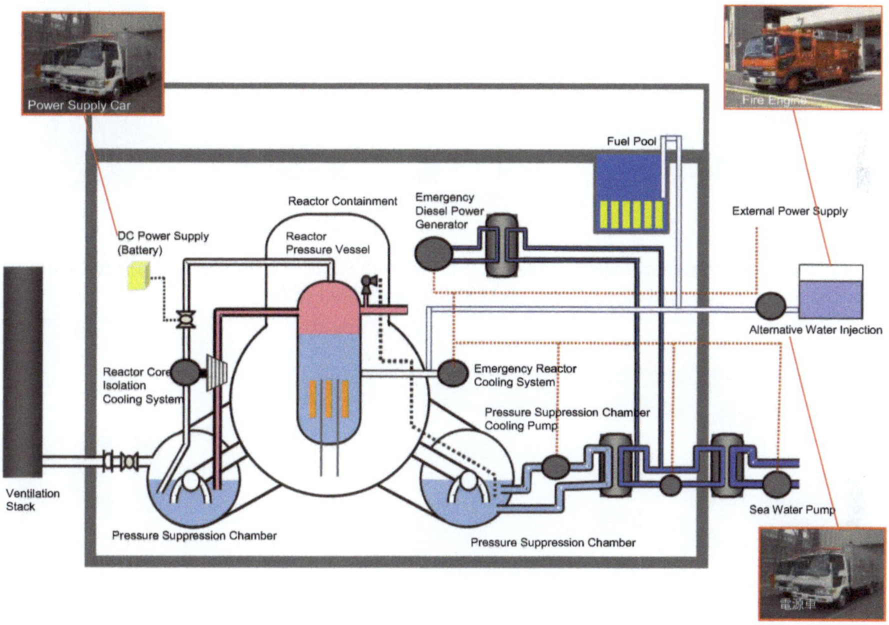

Fig. 40 Core injection heat removal by suppression pool cooling system [14]

Figure 44 shows the full cooling strategy for SBO to protect against SA for BWR [16]. Fuels in the core are cooled by direct water injection via ECCS line of feed water line. Core injection, PCV spray, PCV head flange cooling, suppression pool cooling, and pedestal water injection using MUCW pumps are driven by gas turbine generator, mobile power supply car, and mobile ultimate heat sink car.

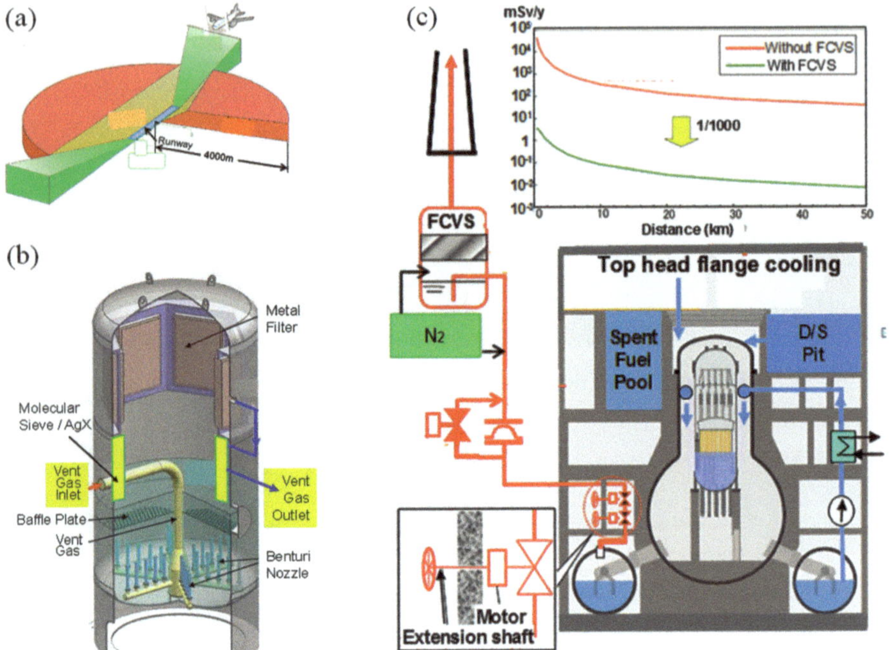

Fig. 41 Filtered containment venting system (FCVS) with silver zeolite [16]. (**a**) Aircraft landing zone. (**b**) Internal structure of vent filter. (**c**) Outline and effect of FCVS

(a) (b)

Fig. 42 Installation of filtered containment venting systems (FCVS) [16]. (**a**) FCVS pit at Hamaoka NPS. (**b**) Installation of FCVS at Kashiwazaki-Kariwa NPS

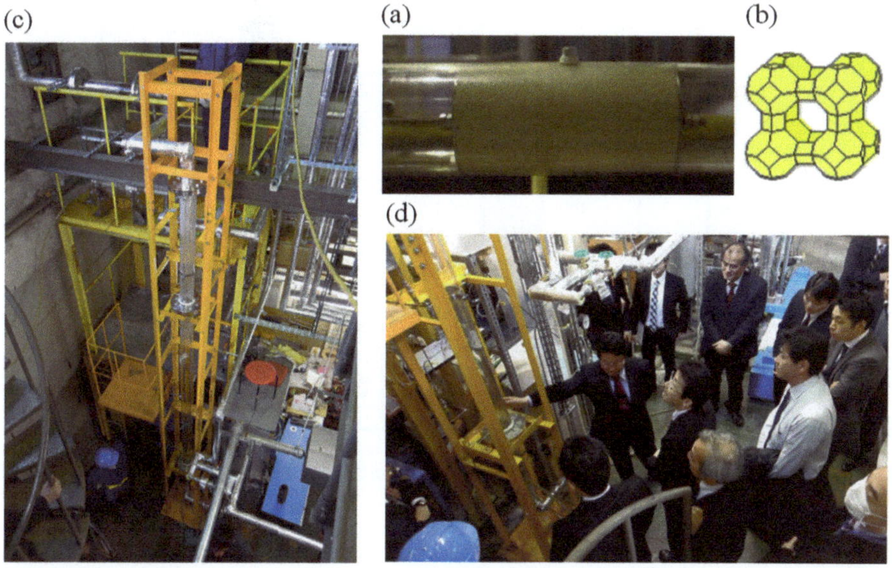

Fig. 43 Development of high-DF FCVS using silver zeolite at Hokkaido University [16]. (**a**) AgX filter test section. (**b**) Zeolite model. (**c**) FCVS test facility at Hokkaido Univ. (**d**) Review meeting of FCVS-WG/JSME

Table 4 Test result of adsorption efficiency of CH_3I under various bed depths [17]

Bed depth (mm)	Residence time (sec.)	Adsorption efficiency of CH_3I (%)
50	0.246	99.967
75	0.369	> 99.999
100	0.492	> 99.999

Testing conditions
Temperature: 130 °C, Pressure: 399 kPa, LV: 20 cm/sec.;
Relative humidity (RH): 95 %, CH_3I concentration: 1.75 mg/m^3 (I-131).

Figure 45 shows a new TWL-RCIC by using a steam turbine-driven, water-lubricated pump. The pump is compact and has a head of 900 m under 183 m^3/h and does not need an oil lubricant pump. The required DC power supply is about half that of current RCIC pumps. These new types of high-pressure water injection devices will provide substantial diversity under SBO conditions. Some BWR utilities will install the TWL-RCIC pump. R&D items in progress to improve safety and resolve problems are shown in Table 5.

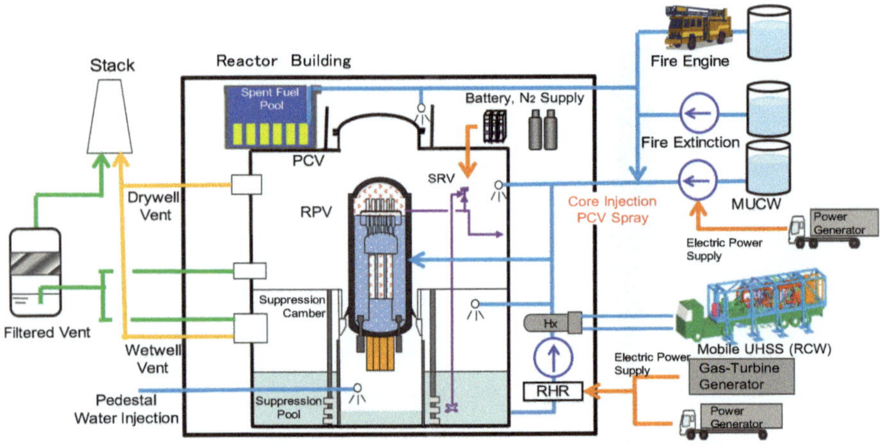

Fig. 44 Full cooling strategy for SBO to protect against SA for BWR [14]

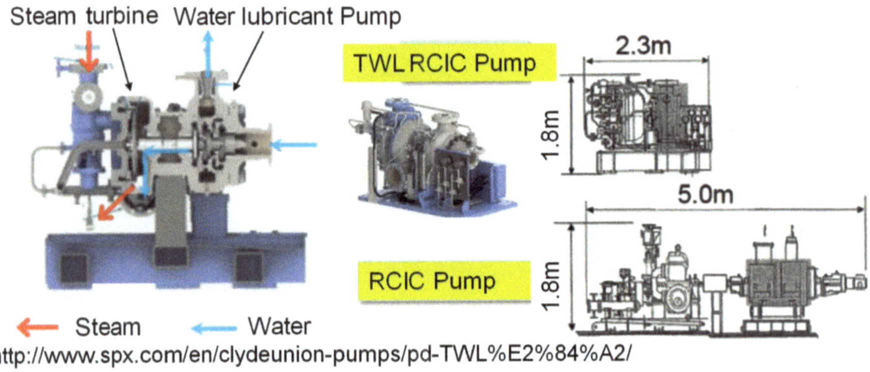

Fig. 45 TWL-RCIC using a steam turbine-driven water-lubricated pump [19]

5 Activities Toward Decommissioning Fukushima Daiichi

5.1 Current Status of the Reactors at Units 1 Through 4

Figure 46 shows the current status of the reactors in Units 1 through 4. It is assumed that the reactor cores of Units 1 through 3 are melted and that some portion dropped onto the pedestal floor of the CV. It is very important to survey the melted core debris distribution from the core to pedestal floor. Some of debris may be attached on the surface of the control rod drive outer casing. Table 6 shows the temperature and water injection flow rate. Units 1 through 4 are in a stable and safe condition because of controlled water cooling. About 4 ton/h of water are injected into the

Table 5 R&D items in progress to improve safety and resolve problems [19]

Measurement value	Problems	Review items (BWR)	Review items (PWR)
Reactor water level	• The water level could not be obtained due to vaporization of water in reference legs. • Multiplexing was ineffective because the same problem happened in all the places.	• Procedures to prevent vaporization of water and refill water in reference legs • Multiplexing with different types of measuring methods	• Reference legs are sealed in existing differential pressure type water level gauges. • Water level can be estimated from the temperature on the outlet of the reactor core.
Dry well water level	• The water level could not be obtained because the measurement points were limited.	• A measuring method to obtain the water level at every height down to the bottom of a dry well	• The water level of the recirculation sump can be monitored. The environmental effect should be investigated.
Dry well hydrogen concentration	• The measuring method was the sampling method that did not work due to loss of power and coolant.	• A measuring method without sampling	• Installation of hydrogen concentration meters in containment vessels. • A measuring method without sampling
Reactor building hydrogen concentration	• There was no hydrogen concentration meter in reactor buildings.	• A measuring method for hydrogen concentration in reactor buildings	• Installation of hydrogen concentration meters in annulus.

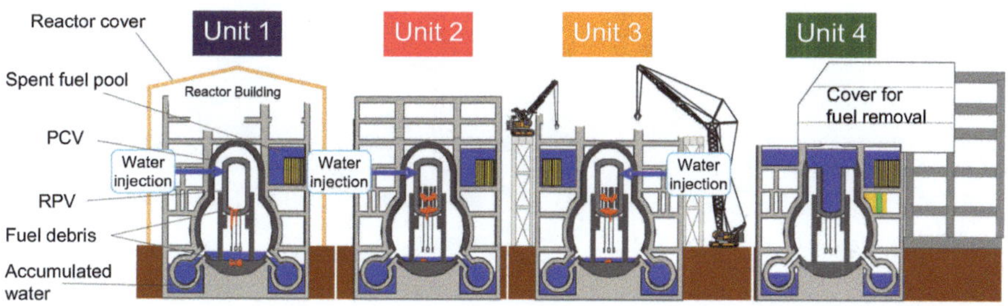

Fig. 46 Current status of reactors in Units 1 through 4 [18, 19]

Table 6 Temperature and water injection flow rate of reactors in Unit 1 through 4

	Unit 1	Unit 2	Unit 3	Unit 4
RPV bottom temp.	About 29℃	About 36℃	About 34℃	-
PCV internal temp.	About 29℃	About 37℃	About 34℃	-
Fuel pool temp.	About 26℃	About 22℃	About 21℃	About 22℃
Reactor cooling water injection volume	About 4.6m^3/h	About 4.5 m^3/h	About 4.3 m^3/h	-

cores of units 1 through 3, and temperature in the pressure vessel is maintained around 30–40 °C. Temperature in the spent fuel pool (SFP) is maintained around 30 °C [18, 19].

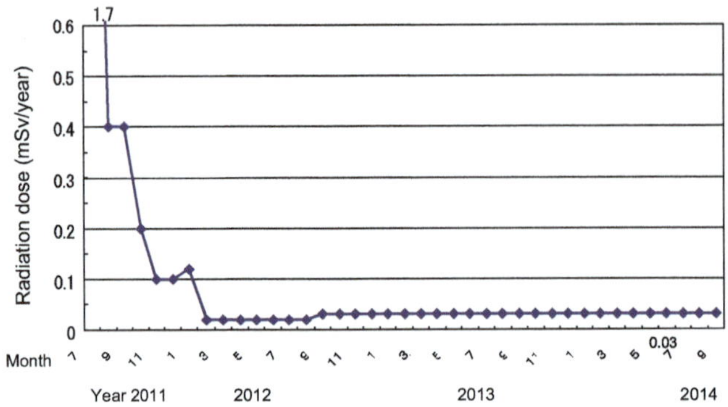

(Reference)

* Concentration limit in the air of environment surveillance area :
[Cs-134] : 2× 10^{-5}Bq/cm^3
[Cs-137] : 3× 10^{-5}Bq/cm^3

* Dust concentration in the area surrounding 1F site boundary :
[Cs-134] : ND (Detection limit: approx. 1× 10^{-7}Bq/cm^3) ,
[Cs-137] : ND (Detection limit: approx. 2× 10^{-7}Bq/cm^3)

Fig. 47 Annual radiation dose at site boundary of Fukushima Daiichi NPS [18, 19]

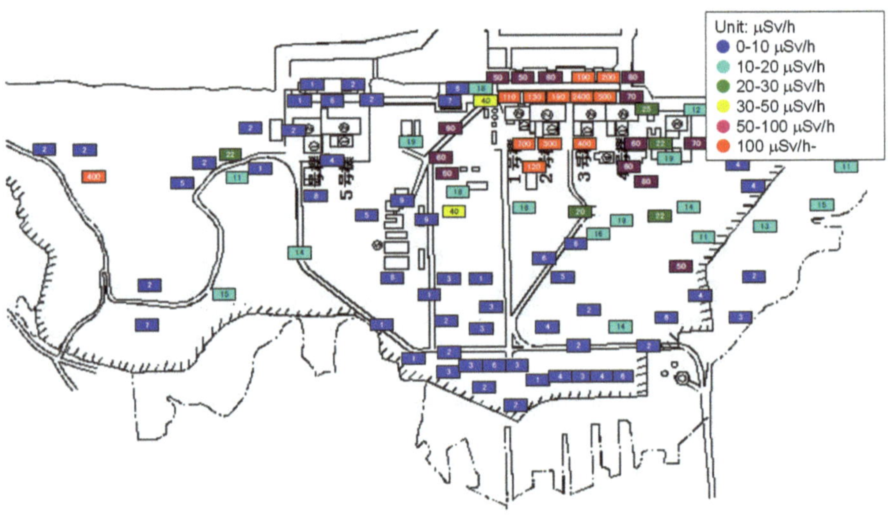

Fig. 48 Exposed dose distribution at Fukushima Daiichi NPS [18, 19]

Figure 47 shows the annual radiation dose from radioactive materials at the site boundary of Fukushima Daiichi NPS, released from Units 1 through 4. At the time of the accident, a vast amount of these materials were released, but the maximum annual dose is now 0.03 mSv/y equivalent to 1/70 that from background radiation.

Figure 48 shows that an approximate half area dose is <10 μSv/h (87.7 mSv/y), but the dose in the reactor building is >100 μSv/h (877 mSv/y).

Just after the accident, contaminated water accumulated in the reactor and turbine buildings leaked into the plant harbor through an underground tunnel. The leak was stanched at the end of October 2014. The current concentration of

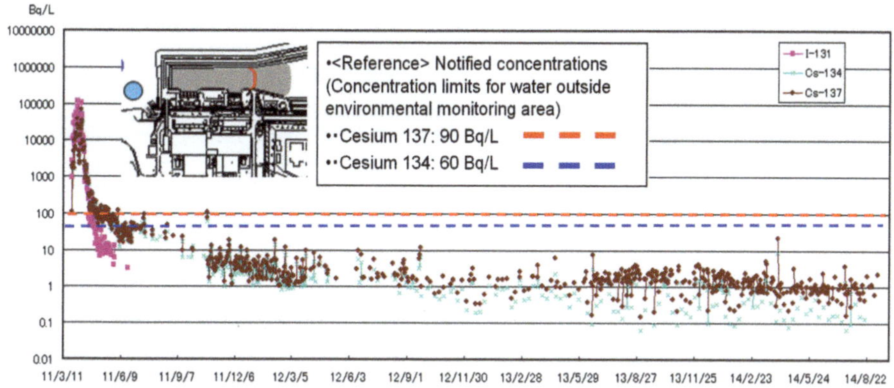

Fig. 49 Concentration of radioactive materials in seawater at the north side of water outlets in Units 5 and 6 [18, 19]

Table 7 Mid- and long-term road map [18]

Dec 2011 (development of roadmap)	Nov 2013	First half of 2025 at the earliest	After 30 - 40 years
Approach toward stabilization	Phase 1	Phase 2	Phase 3
<Accomplishment of cold shutdown> •Cold shutdown condition •Suppression of radiation emission	Period to the start of fuel removal from the spent fuel pool (within 2 years)	Period to the start of fuel debris removal (within 10 years)	Period to the end of decommissioning (after 30-40 years)

radioactive materials in the seawater is below WHO's (World Health Organization) standard for drinking water (Fig. 49).

Table 7 shows the mid- and long-term road map, developed by the government and TEPCO. The decommissioning work will be undertaken in three phases. Phase 1 is fuel removal from spent fuel pool (SFP), Phase 2 is fuel debris removal, and Phase 3 is plant dismantling.

Figure 50a shows the reactor building of Unit 1 after the hydrogen explosion on March 12, 2011. The cover structure was constructed with a controlled ventilation system, to minimize the dispersion of radioactive material. To remove the spent fuel from the SFP, rubble should be removed, and cover should be removed after ensuring countermeasures for local communities during the removal operation.

Figure 50b shows the current status of Unit 2. Its reactor building is undamaged, because the blowout panel was opened owing to the hydrogen explosion in Unit 3. Therefore, the dose rate in the Unit 2 reactor building is very high. Considering this fact, the process for removing SFP fuels is under study.

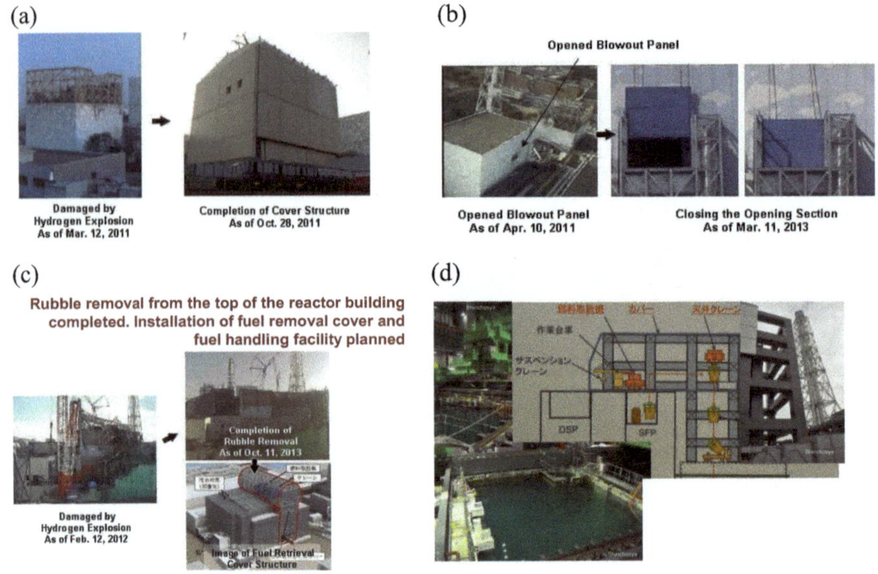

Fig. 50 Current status of reactors; Units 1 through 4 [18, 19]. (**a**) Unit 1, (**b**) Unit 2, (**c**) Unit 3, (**d**) Unit 4

Figure 50c shows the current status of Unit 3. Rubble removal at Unit 3 was completed in October 2013. Preparations are continuing for installation of a temporary cover, to be used while fuels are removed from the SFP by remote-controlled handling equipment. The current level of radiation on the operating floor remains high for workers to safely install the necessary equipment. Therefore, the next step is to reduce the radiation to an acceptable dose.

Figure 50d shows the current status of Unit 4. A cover structure was constructed and fuel-handling machine and heavy-duty ceiling crane were installed, and spent fuel removal from Unit 4 was completely ended by the end of December 2014. The cover structure is supported by a huge cantilever structure, which enables to support the handling of the fuel into transport cask with the ceiling crane.

5.2 *Finding Contaminated Water Leak Path for Leak Shutdown of PCV*

Figure 51 shows the estimated leak path of fission products from the PCV. After recirculating water injection to the core, melted core debris was cooled and reached cold shutdown. However, contaminated water still leaked from the damaged PCV and flowed out of the reactor building.

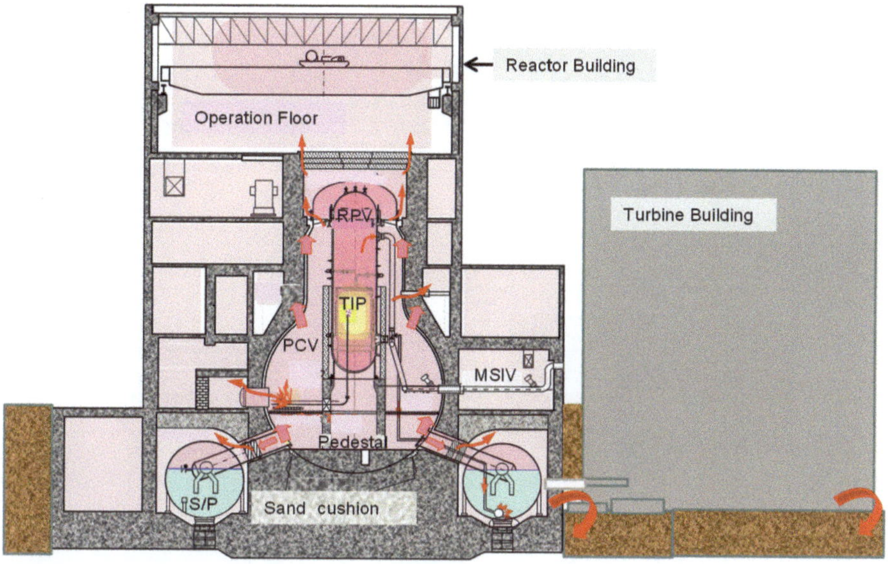

Fig. 51 Estimated leak path of fission products from primary containment vessel

Therefore, the leak location was found by using robots. Figure 52a shows boat-type robots with a camera checking water flow out of the bottom of the PCV area in a torus room on the basement floor of Unit 1. Video from the boat robot discovered water flow from a sand cushion drain pipe [18, 19]. This indicates some failure at the bottom of the PCV wall and that water leaked through the sand cushion drain. Sand cushions are used to mitigate the stress on the bottom flange of the PCV.

Figure 52b shows another type of robot, which can walk on the external catwalk on the torus of the suppression chamber and survey water leakage on the upper side of that chamber and piping. Such robot found leakage from the protection cover of a vacuum breaker expansion joint in Unit 1. This is consonant with the discovery of the water flow on the surface of the suppression chamber torus by the boat-type robot.

Figure 53 shows the effort in Unit 2 to find the water level in PCV. A water level probe with thermocouple was inserted through X-53 penetration. This found a water depth of 300 mm, which is near the overflow height into vent pipe. This indicates a leakage path at the vent pipe or suppression chamber torus. This leak can be stopped using balloon plugs, which are used for main steam line plugging during the annual maintenance of main steam isolation valve (MSIV) [15].

As shown in Fig. 54, a video camera was used to find PCV water leak in the reactor building at Unit 3. The camera was inserted from the second floor and discovered leakage from the expansion joint of main steam line pipe D from the PCV in the MSIV room on the rector building first floor.

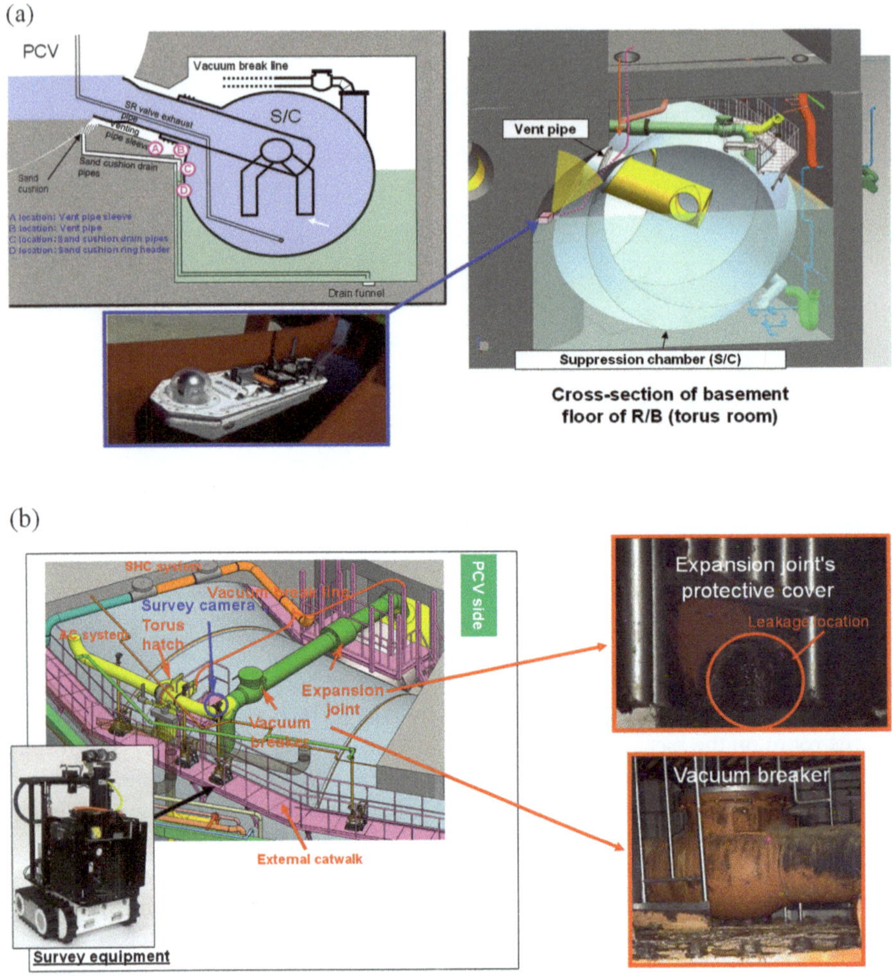

Fig. 52 Survey robot to discover contaminated water leakage in Unit 1 [18, 19]. (**a**) A boat-type robots with a camera checking water flow. (**b**) Survey robot walking on catwalk outside of SC

5.3 Isolation of Groundwater Flow from Contaminated Water

There are several trenches for pipelines and power lines between buildings and pumps in the seaside area, in which extremely contaminated water was discovered. To isolate groundwater mixed with contaminated water flow from the buildings to the trenches, a groundwater bypass system was installed and began operation in May 2014 (Fig. 55). With this system, the volume of water flow into the buildings

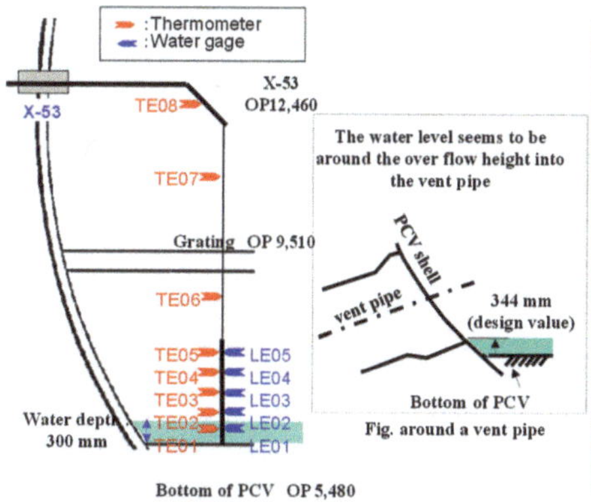

[As of 8:00 6/11]

Monitoring results				Height
Temp.[°C]		Water level		
TE08	33.7	–	–	OP.11,920
TE07	33.7	–	–	OP.10,690
TE06	33.5	–	–	OP.8,100
TE05	33.5	LE05	OFF	OP.6,430
TE04	33.5	LE04	OFF	OP.6,230
TE03	33.6	LE03	OFF	OP.6,030
TE02	35.0	LE02	OFF	OP.5,830 *350mm from the bottom of PCV
TE01	35.8	LE01	ON	OP.5,630 * 150mm from the bottom of PCV

Fig. 53 Water-level checking at bottom of PCV of Unit 2 [15]

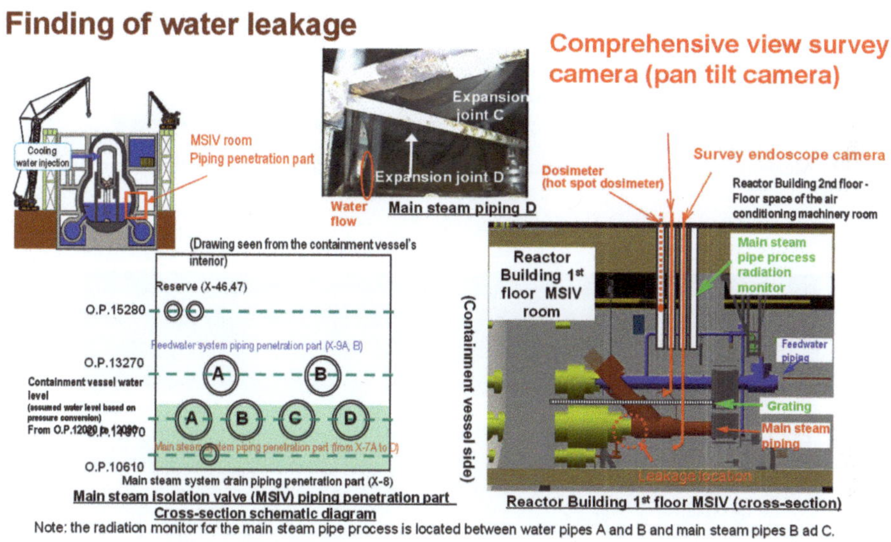

Fig. 54 Investigation to find leakage location in Unit 3 [15]

was reduced from 350 to 80 t/day. As a result, the volume of contaminated water was also reduced. The next fundamental countermeasure is to build an impermeable wall to prevent contaminated water flowing into the sea (Fig. 56). Further, TEPCO has been preparing equipment of landside water shielding wall and subdrain pump-up system to greatly diminish groundwater flow into the contaminated buildings.

The landside impermeable wall composed of frozen soil (ice wall) surrounds the buildings of Units 1 through 4 to block the groundwater from getting contaminated (Fig. 57). After a small-scale demonstration test succeeded, the large ice wall

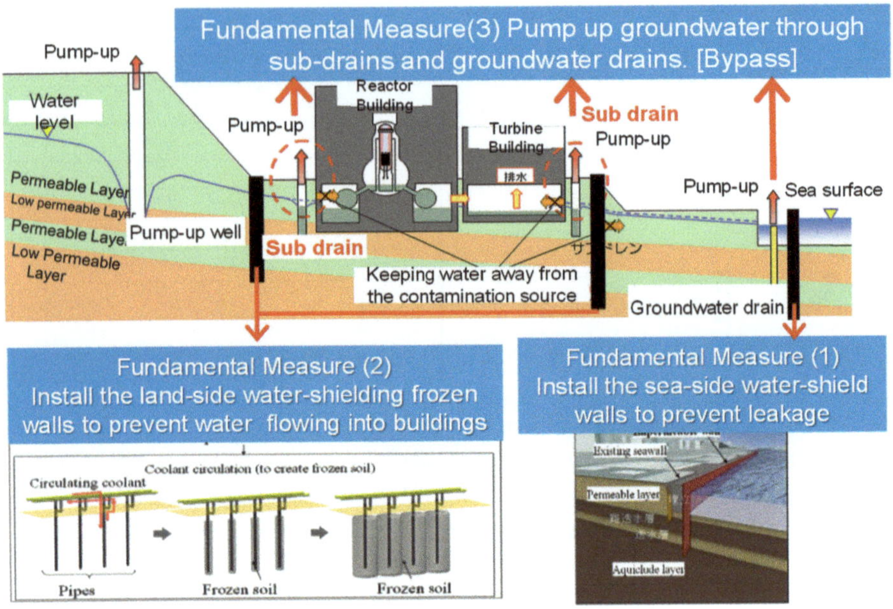

Fig. 55 Measures to prevent contaminated water flowing into the sea by isolation of groundwater [18, 19]

Fig. 56 Impermeable wall to prevent contaminated water flowing into the sea [18, 19]

construction began. There are 1500 bores of 30 m depth at 1 m intervals, for installation of coolant-circulating pipes. Construction should be complete by June before the hot summer in 2015. After ice wall establishment, daily groundwater flows into the buildings of the four units will be reduced to near zero, as will contaminated water flowing out from the ice wall.

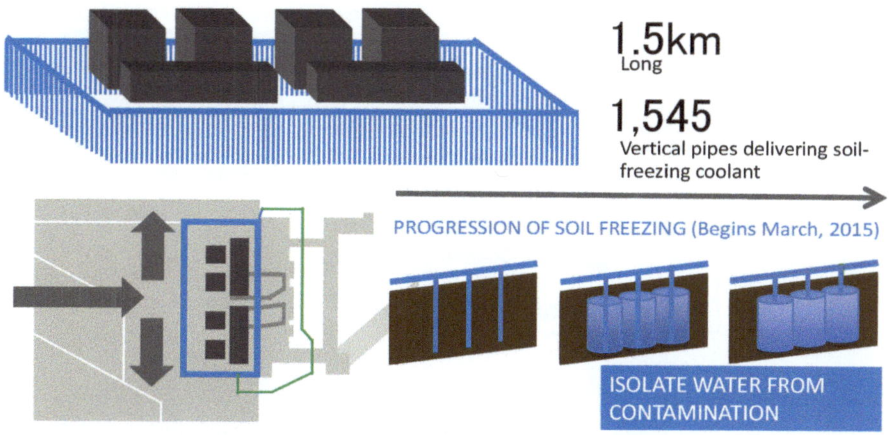

Fig. 57 Landside impermeable "ice wall" [18, 19]

5.4 Contaminated Water Management

Contaminated water management is presently the most important and urgent issue at Fukushima Daiichi NPS. Contaminated water is transferred to cesium removal systems and then to desalination systems of reverse osmosis or evaporation type. Then the desalinated water is injected into the reactors as coolant. The concentrated water after desalination is still radioactive. Therefore, it was transferred to storage tanks at a rate of 400 t/day before the establishment of the groundwater isolation system. There are a huge number of tanks at Fukushima Daiichi. Therefore, an advanced liquid processing system (ALPS) was developed to remove various radioactive materials including strontium, which were then safely stored in the storage tanks. Currently, a total of more than 500,000 t of contaminated water is stored in the tanks. Current tank capacity is ~600,000 t, and this will be increased to 800,000 t by the end of March 2015.

Figure 58 shows the contaminated water processing system with the cesium removal system (SARRY), desalination system (KURION), and ALPS. The ALPS was designed to remove all radionuclides except tritium. The first system consists of three independent lines. Total treatment capacity is 750 t/day. To increase total processing capacity, a second system was constructed. A high-performance ALPS was designed to reduce the volume of secondary waste and will increase the rate of radionuclide removal to 2000 t/day with the first installed system plus additional and high-performance systems. About 120,000 t of contaminated water was processed by the end of October 2014. The ALPS has the important role of water decontamination.

Figure 59 shows the capabilities of the three water processing systems for removing radioactive materials. The cesium and strontium concentration level at the ALPS outlet is reduced by 10^{-8} relative to the original concentration. However, tritium concentration is unchanged. Tritium processing remains to be solved.

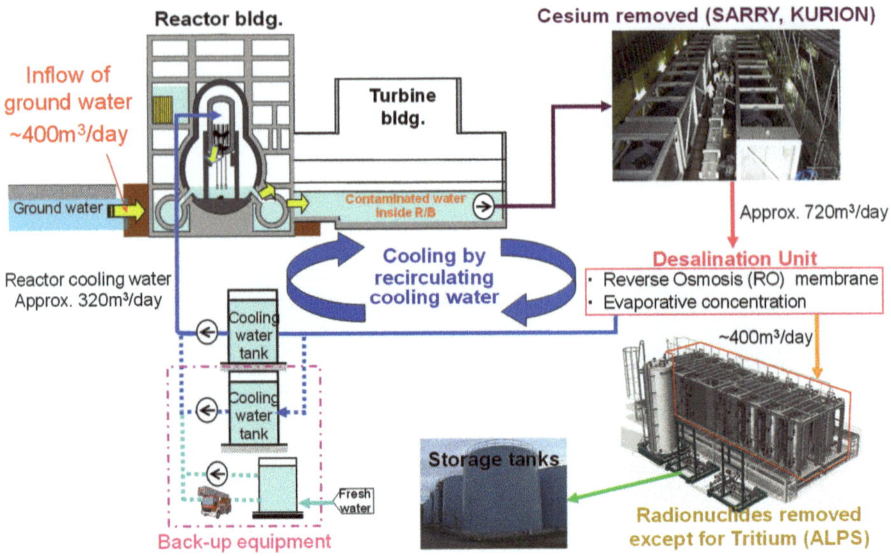

Fig. 58 Contaminated water processing system with SARRY, KURION, and ALPS

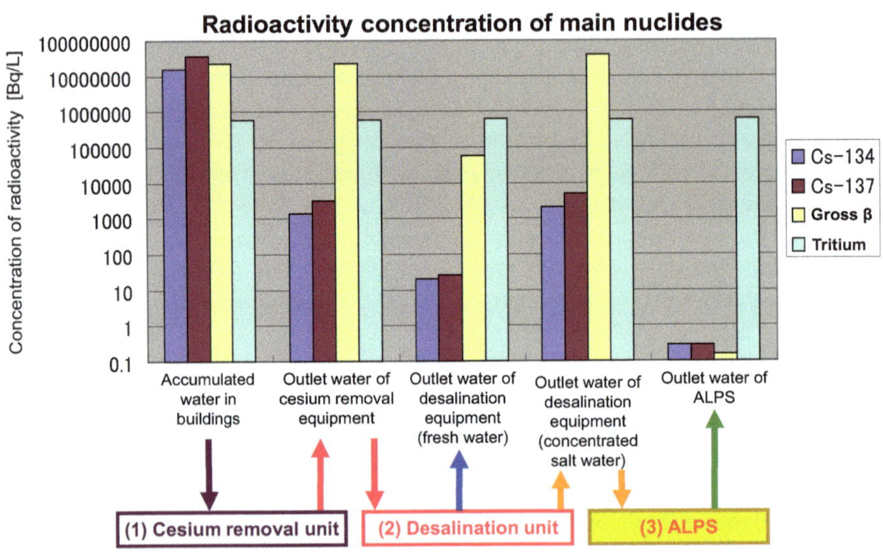

Fig. 59 Capabilities of various water processing systems to remove radioactive materials [18, 19]

5.5 *Preparation for Fuel Debris Removal*

To find melted debris under the RPV, a transformer-type robot was developed by Hitachi GE Nuclear [17, 18]. As shown in Fig. 60, the robot will investigate debris inside the pedestal and estimate the location of fuel debris dropped from the core in

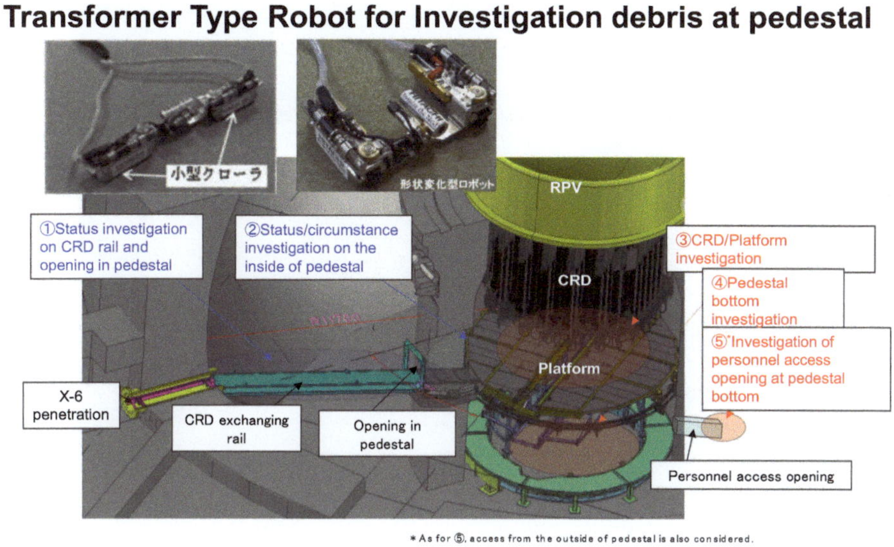

Fig. 60 Transformer-type robot for investigating debris at the pedestal in Unit 2

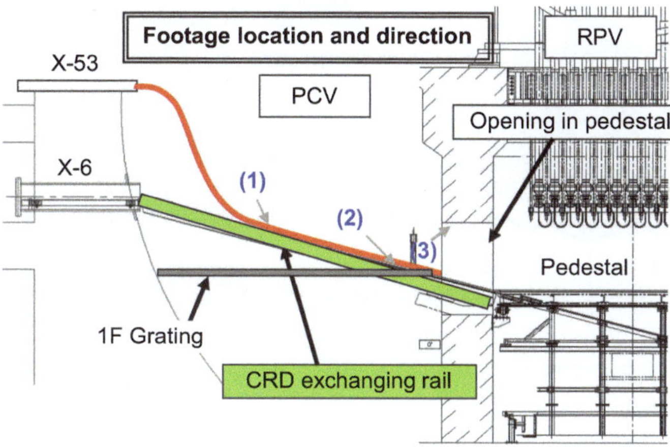

Fig. 61 Robot investigation route into the pedestal using X-6 large penetration

Unit 2. Figure 61 shows that the robot will move into the pedestal using X-6 large penetration, which was used like a snake for CRD exchanges through the penetration pipe. After passing the penetration pipe, the robot transforms its shape from a snake to an "E"-shape vehicle, which has a radiation-proof CCD camera at the center of front flame "E." TEPCO has a plan to investigate in several steps. The preliminary investigation will identify an access route into the pedestal via CRD rail and survey the status of PCV inside the pedestal. A full-scale investigation for

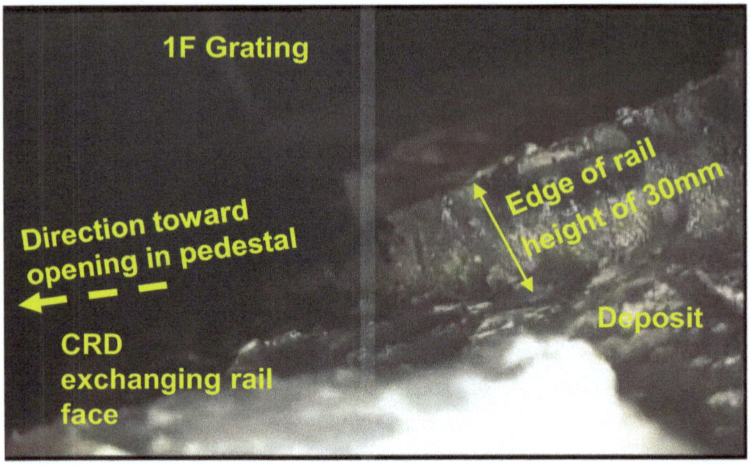

Fig. 62 Video capture at position (2) on CRD exchange rail [18, 19]

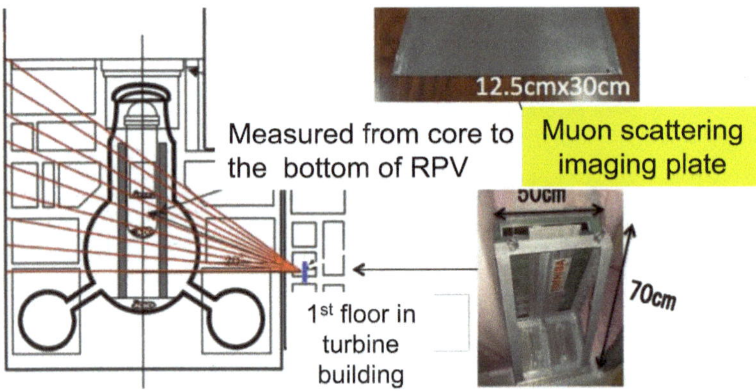

Fig. 63 Muon radiography to find fuel debris [20]

identifying the location of fuel debris will then follow. This is the result of the first preliminary investigation, using X-53 penetration. Figure 55 is video capture at position (2) on the CRD exchange rail, in Fig. 62. It is possible to identify an edge of rail and some deposit on the rail. However, the robot could not enter the pedestal because of some blockage by an unknown deposit on the CRD rail.

In March 20, Nagoya University and Toshiba released to the press results of muon radiography to find fuel debris. Figure 63 shows a layout of the muon detector imaging plate and measurement area from the core in the RPV to the RPV bottom.

Figure 64 compares muon radiography results between Units 2 and 5. No fuels were found in the core or lower plenum at the RPV bottom. All fuels melted downward through the melted piping of the control rod drive (CRD) mechanism and fell to the pedestal concrete floor.

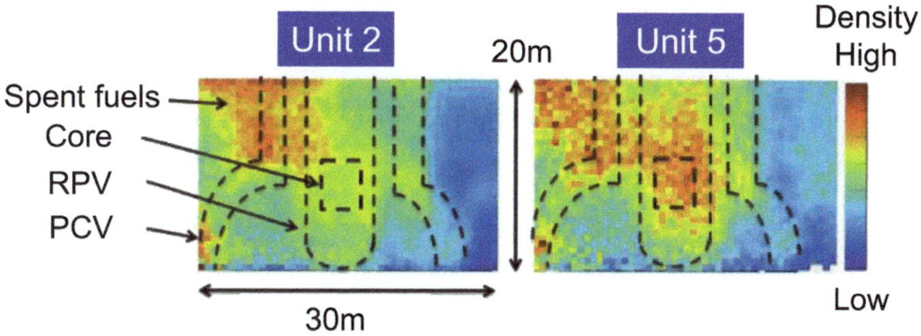

Fig. 64 Comparison of muon radiography results between Unit 2 and Unit 5 [20]

6 Concluding Remarks

There are numerous effective countermeasures to prevent the FP release and tsunami in the world. TEPCO and NISA should consider such systems.

The Fukushima Daiichi NPP accident could have been quickly brought under control if adequate countermeasures had been installed, such as waterproof doors and mobile power sources.

In Europe, as a result of lessons learned from the TMI and Chernobyl accidents, heat removal systems and FCVS have already been installed.

Using the lessons learned from this analysis of the Fukushima Daiichi accident, the author wishes to contribute to achieving first-class nuclear safety worldwide.

In response to the Fukushima Daiichi nuclear disaster, measures for the safety of Japanese NPPs have been strengthened or are being greatly bolstered. It has been pointed out that DiD in Japan before Fukushima was "Level 3" and now it is enhanced to "Level 4."

Therefore, the author submits proposals as shown below to provide scientific and technological support. The lessons can be incorporated in measures taken by institutions and government agencies, thereby enhancing the safety of the many nuclear power plants in operation worldwide.

1. Enhance seismic electric device to prevent loss of external power by earthquake; SF_6 gas-insulated switchgear (GIS) and flexible insulators should be installed for transmission line.
2. For station blackout (SBO) caused by wetting of EDG, P/C, DC battery, I&C and cell phones, waterproof door or hatches, and mobile power should be installed on hills.
3. To prevent core meltdown by loss of water injection, diversification of water injection and heat sinks is very important.
4. To prevent loss of containment function by overheating damage, CV cooling and FCVS should be installed independently from hard vent systems.

5. Pedestal water injection during severe accident is very important to avoid PCV bottom failure that causes water pollution around the reactor building.

Design reviews of 12 PWRs and 9 BWRs with enhanced safety measures are ongoing for restarting. Sendai Nuclear Power Station Units 1 and 2 of Kyushu EPCO received the NRA permission in September 2014 and Unit 1 has restarted since August 2015.

The plant licensing period of 40 years should be determined based on technical aging assessment.

Acknowledgment The author sincerely thanks Mr. Norimichi Yamashita of TEPCO for substantial information on the recovery action plans of Fukushima Daiichi.

List of Nomenclature

AC	Alternating current
ALPS	Advanced liquid processing system
AM	Accident management
AO	Air-operated valve
BWR	Boiling water reactor
CRD	Control rod drive
CV	Containment vessel
DBA	Design base accident
DC	Direct current
DF	Decontamination factor
DiD	Defense in depth
DW	Dry well
EDF	Électricité de France
EDG	Emergency diesel generator
ENSI	National regulatory body with responsibility for the nuclear safety and security of Swiss nuclear facilities
FARN	Nuclear rapid response force established by EDF
FCVS	Filtered containment venting system
FP	Fission product
GIS	Gas-insulated switchgear
HSK	Swiss Federal Office of Energy
HVAC	Heating, ventilation, and air conditioning
IC	Isolation condenser
JNES	Japan Nuclear Energy Safety organization
M/C	Metal clad switchgear
METI	Ministry of Economy, Trade and Industry
MEXT	Ministry of Education, Culture, Sports, Science and Technology

MO	Motor-operated valve
NISA	Nuclear and Industrial Safety Agency
NRA	Nuclear Regulation Authority
P/C	Power center
PAR	Passive autocatalytic recombiner
PWR	Pressurized water reactor
R/B	Reactor building
RCIC	Reactor core isolation cooling system
RPV	Reactor pressure vessel
RV	Reactor vessel
S/C	Suppression chamber
S/P	Suppression pool
SA	Severe accident
SAM	Severe accident management
SBO	Station blackout
SEHR	Special Emergency Heat Removal
SGTS	Standby gas treatment system
SRV	Safety relief valve
T/B	Turbine building
TEPCO	Tokyo Electric Power Company Inc.
TIP	Traversing in-core neutron probe
WW	Wet well

References

1. Narabayashi T, Sugiyama K (2011) Fukushima Daiichi NPPs accidents caused by the Tohoku – Pacific Ocean earthquake and tsunami. AESJ Atoms 53(6):387–400
2. Nakajima H, Narabayashi T (2011) Start from level 7. Nikkei Science, July. (in Japanese). http://www.nikkei-science.net/modules/journal/index.php?vol=201107
3. NISA (2012) Technical knowledge of Fukushima-Daiichi NPP's accidents and countermeasure (Interim report). (in Japanese). http://www.meti.go.jp/press/2011/03/20120328009/20120328009.html
4. NISA, JNES (2011) The 2011 Pacific coast of Tohoku Pacific Earthquake and the seismic damage to the NPPs
5. JNES (2011) IC performance and transient analysis for Fukushima Daiichi NPP unit 1 accidents
6. TEPCO (2011) Fukushima-Daiich NPPs accidents investigation meeting's report (Interim report of Fukushima Nuclear Accidents Investigation Committee). http://www.tepco.co.jp/en/press/corp-com/release/11120205-e.html
7. Report of Japanese Government (2011) The accident at TEPCO's Fukushima Nuclear Power Stations, IAEA Ministerial conference on nuclear safety
8. Narabayashi T (2012) Lessons of Fukushima-Daiichi NPP's accidents for achievement of the 1st class safety in the world, Fukushima severe accident dose management & global lessons

learned in occupational dose reduction, 2012 International ISOE ALARA symposium, Fort Lauderdale
9. Gavrilas M et al (2000) Safety features of operating light water reactors of western design, CNES
10. Nuclear Regulation Authority (2013) Nuclear regulation authority enforcement of the new regulatory requirements for commercial nuclear power reactors. http://www.nsr.go.jp/data/000067212.pdf
11. Narabayashi T (2012) Lessons learned from the Fukushima Daiichi Nuclear Power Plant accident, turbulence, heat and mass transfer 7, Begell House, Inc. www.gepr.org/en/contents/20121231-01/KeynoteDrNarabayashi-THMT-12r2.pdf
12. EDF'S Commitment to Enhance Safety (2014) The post-Fukushima modification programme is part of the continuous improvement initiative for EDF nuclear fleet
13. EDF PWR fleet overview and developments experience and challenges, Japan Society of Mechanical Engineers (JSME) & EDF meeting Paris (November 12, 2014)
14. FVCS working group report, Power and Energy System Division, JSME. (Dec 2013). http://www.jsme.or.jp/pes/Research/A-TS08-08/index03.html
15. TEPCO (2014) Second progress report for Fukushima Daiichi. http://www.tepco.co.jp/cc/press/2014/1240099_5851.html
16. Narabayashi T et al (2014) Development of high efficiency filtered containment venting system by using AgX, Short paper of ICMST-Kobe
17. Wang J, Kobayashi T, Endo K (2014) Properties and applications of silver zeolite (AgX), short paper of ICMST-Kobe
18. Yamashita N (2014) Activities towards the decommissioning of Fukushima-Daiichi (TEPCO), ICMST Kobe
19. Narabayashi T (2015) Lessons learned from the Fukushima Daiichi accident to establish resilience technology for nuclear plants based on the defense in depth philosophy (invited plenary), North American ISOE ALARA symposium, Ft. Lauderdale
20. Nagoya Univ. (2015) Press release of muon radiography results at Fukushima Daiichi. https://www.youtube.com/watch?v=qw_445aOefs, http://www.nagoya-u.ac.jp/about-nu/public-relations/researchinfo/upload_images/20150320_esi.pdf

Energy Policy and Perspectives

Yuya Kajikawa

Abstract This chapter illustrates energy policy in Japan before and after the disaster associated with the Fukushima nuclear power plant accident. In the Strategic Energy Plan approved by the Cabinet in June 2010, it was decided to increase dependence on renewable energies and nuclear power for electricity to about 70 % by 2030. After the accident, the plan was completely abandoned, reflecting the concern of the public toward the safety of nuclear power generation. Nonetheless, there has been a growing concern regarding the current status of energy system in which very little nuclear power generation is operated. Complete dependence on fossil fuels has threatened energy security and reduced national welfare. The government has been unable to provide effective and consistent policies that are acceptable to the majority of people. The introduction of the feed-in tariff for renewables accelerated the deployment of photovoltaic cells, but this has leveled off to maintain grid stability. The nuclear safety authority and regulation system was refurbished, but it still struggles to gain public acceptance. Feasibility and effectiveness of energy-related R&D programs and electricity system reformation are uncertain. To make energy policy and the policymaking process credible, salient, legitimate, and feasible, we need (1) innovation and integrated design of energy technologies and systems; (2) war of ideas among technologies, institutions, and social systems; and (3) roadmaps illustrating long-term technological and policy trends to realize a sustainable society, not only to assure energy security but also to achieve environmental safeguards, economic efficiency, safety, and industrial and economic development.

Keywords Energy policy • Policymaking • Roadmaps

Y. Kajikawa (✉)
Tokyo Institute of Technology, Tokyo, Japan
e-mail: kajikawa@mot.titech.ac.jp

Y. Kato et al. (eds.), *Energy Technology Roadmaps of Japan*,
DOI 10.1007/978-4-431-55951-1_6

1 Introduction

1.1 Strategic Energy Plan and Fukushima Nuclear Power Plant Accident

The Japanese government has adopted policies aimed at improving energy efficiency and reducing petroleum demand, especially after the oil shock in the 1970s. An energy-efficient economy has contributed to the vitalization and competitiveness of industry. Industry has taken up voluntary business activities for reducing energy use and flexibility of planning [1]. In 1997, Japan hosted and played the key role at the Third Conference of the Parties (COP3) to the United Nations Framework Convention on Climate Change (UNFCCC), at which the Kyoto Protocol was declared and adopted. Japan targeted a 6 % reduction of emission compared with 1990.

The low-carbon policy is very compatible with energy security policy. Japan is vulnerable in terms of energy supply security compared with other OECD nations, because its energy self-sufficiency ratio was 4 % (without nuclear power) and 18 % (with nuclear power) in 2007 [2]. The chronic and extreme scarcity of fossil fuel resources in the country has meant that methods to improve national energy security have been long-standing challenges.

Nuclear power is regarded as a quasi-domestic electricity supply because of the long-term use of uranium through reprocessing and is compatible with environmental safeguards including climate change mitigation. To improve energy security, Japan plans to increase the fraction of electricity supply from nuclear and solar power generation systems. After the 1997 Kyoto Protocol and a recognition of the serious risks of global warming, there was a period of nuclear renaissance, characterized by renewed interest in nuclear energy as a non-carbon-based energy source. However, such a policy faces challenges after the accident and ensuing disaster from the Fukushima Daiichi Nuclear Power Plant, caused by the Great East Japan Earthquake and Tsunami of 11 March 2011. Such challenges are not limited to Japan but are also faced in most developed countries [3].

Energy policy cannot be represented by a single document or law. The policy consists of laws, regulations, and programs, but their directions are under the influence of the Strategic Energy Plan (SEP) approved by the Japanese Cabinet. The SEP was established based on the Basic Act on Energy Policy enacted by the Diet, which is "the general guiding direction for Japan's future energy policy" [4]. The Ministry of Economy, Trade and Industry (METI) was tasked with formulating a draft of the basic energy plan and seeking cabinet approval before reporting it to the Diet. A comprehensive review paper on energy policy in the country has already been published elsewhere [5]. Therefore, this chapter focuses on recent policy and its context.

The first SEP was developed and adopted in October 2003 and revised in March 2007. Because Japan has scarce energy resources, energy policy and planning have been crucial issues not only since the oil shock but even before World War II. For energy policy, the government published the Long-term Prospect of Supply and Demand of Energy, but this does not determine policy. In 2002, however, the Japanese government created a more systematic and comprehensive energy policy planning structure [6].

In May 2007, Prime Minister Shinzo Abe of the Liberal Democratic Party announced the initiative "Cool Earth 50," aimed at drastic reduction of greenhouse gas emissions in Japan and the world. In September 2009, the new government led by the Democratic Party of Japan announced an ambitious goal of reducing those emissions by 25 % over the 1990 level by 2020. Researchers even indicated possible 70 %–80 % reductions by 2050 [7, 8].

Given the above context, the third SEP was established and approved by the Cabinet in June 2010. That SEP is aimed at increasing the share of renewables from 9 % in 2007 to 21 % by 2030. Petroleum, natural gas, and coal will be decreased to 2 %, 13 %, and 11 % shares, from 13 %, 28 %, and 25 %, respectively. It was also decided to increase dependence on nuclear power for electricity to ~53 % (electricity generation base) by 2030, which was 26 % in 2007 [9]. The plan called for at least 14 new nuclear reactors to be constructed; however, because of the accident, the government abandoned these plans.

The accident at Fukushima Daiichi caused by the earthquake and tsunami led to a catastrophic disaster and was a serious event for Japan's energy security [10, 11]. In February 2011, the month before the accident, nuclear power supplied ~31 % of Japan's electricity [12]. However, nuclear reactors across the country have been shut down sequentially for periodic inspection, and they have faced difficulty in restarting. Thus, the share of nuclear power fell to 12.4 % in August 2011 and zero in May 2012 [12].

Given the substantial loss of nuclear electricity, the government, electricity suppliers, industries, citizens, and society implemented various countermeasures and actions to bridge the supply-and-demand gap for preventing unexpected power outages [13–15]. The decline in supply capacity of electricity was especially significant in the Kanto region (including metropolitan Tokyo), because not only the Fukushima nuclear power plants but also many thermal power plants in the disaster area have become incapable of supplying electricity to that region. The challenge in supplying peak demand during summer 2011 was overcome by collective measures, including urgent enhancement of power supplies, reduction of peak demand through lifestyle and behavioral changes, and others.

A large portion of electricity supply loss was compensated by recovering disaster-affected thermal power plants and hibernated old ones. In addition, the Thai government offered two gas turbines as aid. Imports of oil and natural gas as alternatives to nuclear energy have increased, increasing Japan's dependence on fossil fuels as a power source, from 60 % before the earthquake to 90 %. Owing to these increased imports of fossil fuels, in 2011, Japan's trade balance turned into a deficit for the first time in 31 years. In 2012, the trade deficit expanded and, in 2013, reached a record high of 11.5 trillion JPY [16]. Therefore, a reduction of fossil fuel consumption is urgently needed, and it is not a simple task to find relevant alternatives. The reason for the accident is often partially attributed to "Japan's nuclear village" building advocacy coalition [17], which heavily damaged public trust in nuclear power. It appears difficult to regain public acceptance of nuclear power because of the long-lasting influence of the Fukushima accident on large populations in Japan.

1.2 Energy Policy After Fukushima Accident

Radical reforms and reviews of Japan's energy and environment strategy have been required because of the serious accident and disaster. The Japanese government established several committees for strategic reform, including the Energy and Environment Council (EEC), Committee for Verifying Power Plant Costs (CVPPC) in the EEC, a committee under the Agency for Natural Resources and Energy (ANRE) of METI, a committee under the Ministry of Environment (MOE), and a committee under the Atomic Energy Commission. Within the CVPPC, during cost evaluation, not only data but methods and assumptions became subjects of debate. Therefore, the CVPPC released a spreadsheet on their website that allows calculation of the cost of electricity generation options under explicit assumptions.

Several "Options for Energy and the Environment" were developed by the Fundamental Issues Subcommittee of the Advisory Committee for Natural Resources and Energy for the EEC [18]. The major focus of these options is the share of electricity generation by nuclear power plants. These include 0 %, 15 %, and 20–25 % nuclear power scenarios in 2030. A nationwide debate was held during July and August 2012, which included a call for public comments, public hearings in 11 locations throughout Japan, and a deliberative poll. Despite these processes, the *Innovative Strategy for Energy and Environment* (ISEE) [19] approved by the Cabinet in September 2012 did not refer to the aforementioned options and declared the maintenance of the nuclear fuel cycle, without any changes from conventional strategy. By the election of December 2012, the government regime changed from the Democratic Party of Japan to the Liberal Democratic Party, which essentially reversed the legitimated position of the ISEE, leaving a presumed energy mix scenario to continuing political debates.

2 Present Status

The SEP was revised into the fourth SEP, which was approved in April 2014. This SEP referred to the Fukushima accident and stated that the government and nuclear operators were trapped by a "safety myth" of nuclear plants, resulting in a failure to adequately deal with serious accidents and prevent disasters.

The SEP confirmed the following "3E+S" principles for energy policy: energy security, economic efficiency, environment, and safety. In addition, the SEP indicated the importance of a global perspective including a global supply chain and co-procurement, global climate change, and global development. As the final principle of energy policy, it also put forth industrial and economic development through strengthening competitiveness of Japan's energy industry and boosting its presence in the global market.

Regarding the above final principle, nuclear power is still regarded as an important option for a situation in which the Japanese economy is stagnating and promotion of social infrastructure export is a priority for industrial policy. For example, an assessment of economic impacts of an alternative "Options for Energy and the Environment," synthesized by independent groups, was done using an energy-economy model [20]. According to their assessment, GDP in 2030 will decrease by 7.4 % within 0 % nuclear energy scenarios, and this decrease would be 4.4–4.9 % in scenarios assuming that Japan relies for 15 % and 25 % of its electrical energy on nuclear power. Development of the economy by a reduction of electricity prices and expected improvement of industrial competitiveness is a persuasive theme, because of decades of economic sluggishness. Further, expansion of the nationwide nuclear industry and the potential export of nuclear reactors and associated parts were core growth strategies in the country [21]. Plant makers like Toshiba, Hitachi, and Mitsubishi Heavy Industries have built global partnerships with Westinghouse, GE, and AREVA to strengthen their capabilities.

In the following, some recent policy initiatives reflecting SEP are illustrated by focusing on:

1. Feed-in tariff for renewables
2. Nuclear safety and regulation
3. Innovative technologies for low-carbon society
4. Electric system reformation

1. Feed-in tariff for renewables

Renewables had a relatively minor role before the Fukushima accident [22]. In Japan, subsidies for the deployment of renewables began in the mid-1990s, followed by the Renewable Portfolio Standard (RPS) in 2003. However, photovoltaic (PV) cells and wind power accounted for only 0.21 % and 0.24 % of electricity production in 2008, respectively [23]. Geothermal projects were installed in the 1970s and 1990s (533 MW in total), but growth has slowed since 1999 [24]. To accelerate the diffusion of renewable energy, the feed-in tariff (FIT) was introduced in 2009 for PV cells, using a net-metering system with procurement price 48 JPY/kWh. In 2012, the FIT scope was extended to cover other options such as nonresidential PV, wind, geothermal, small-size hydropower, and biomass. Because of social expectations for rapid installation of renewables after the accident, procurement prices were set high, e.g., 42 JPY/kWh for nonresidential PV systems. Such high prices attracted much more capital investment in PV systems than expected. This resulted in an abrupt announcement by the Hokkaido Electric Power Company in April 2013 on halting the contract for grid connection of PV systems. A similar hold of the contract for grid connection occurred in Kyushu. Other electric power companies followed these announcements, so it became difficult for further PV installation under the FIT scheme. This is regarded as unfair to new entrants to the market for installing PV, because their supply cannot be purchased at the expected volume. The entrants should have considered the possibility of abrupt public and corporate policy changes. This also imposes a financial burden on electricity users of 0.75 JPY/kWh [16]. These situations and unintended consequences have given impetus to policymakers for reconsidering the current FIT operation [25].

2. Nuclear safety and regulation

The principal nuclear regulatory authority in Japan before the Fukushima accident was the Nuclear and Industrial Safety Agency (NISA). However, it lacked independence as a regulatory body. This was because NISA was operating under the METI, which promoted nuclear energy as an export industry and an energy security option. After the Fukushima accident, the Nuclear Regulation Authority (NRA) was established under the Ministry of Environment. The NRA adopted a "no return rule" of officers to other ministries and agencies promoting nuclear power. If an officer comes from METI (the ministry promoting nuclear power) as a tentative transfer and is expected to return there after a period working for the NRA, he/she may behave as a representative of the interests of METI. The no return rule is necessary to avoid such conflict of interest (COI). The rule also regulates employment for organizations with COI, including nuclear operator companies.

The government has also promoted research on nuclear safety and training of talent for reactor safety and risk management. The Nuclear Risk Research Center was launched within the Central Research Institute of Electric Power Industry. This center has the role to assist nuclear operators and the nuclear industry to continually improve the safety of nuclear facilities by developing and implementing modern methods of probabilistic risk assessment, risk-informed decision-making, and risk communication. The Japan Nuclear Safety Institute (JANSI) was also established as a new entity for training of nuclear operators.

In the late May 2012, no nuclear power reactors were operating in Japan, the first time since 1966. However, this was not because of the new safety rules adopted by the NRA. In Japan, safety regulations require shutting down reactors every 13 months to check their status and safety. This was responsible for the nonoperational status in May 2012, 13 months since the earthquake of March 2011. After routine shutdown for inspection, the plants must pass new safety rules to restart. But even if these rules were met, they could not restart because of political and social issues and legal judgments. For example, the Fukui district court made a judgment that blocked the restart of the Oi nuclear power plant in Fukui Prefecture. This plant had the greatest electric power capacity of Kansai Electric Power Company and was the first nuclear plant restarted after the Fukushima accident, in July 2012. Plants have been routinely inspected since September 2013. The Fukui district court accepted a petition raised by residents and made the decision to prevent the restart. However, the Oi plant is under examination by the NRA regarding safety under the new safety rules. The court decision was valid because of the independence of the three branches of government (legislative, administrative, and judicial), but it raises a fundamental issue in public decision-making. The dramatic change to the current inactive nuclear power state (relative to proactive nuclear power policy prior to the disaster, i.e., the nuclear renaissance) was not the result of a dramatic change in Japanese national policy, but a consequence unintended by the government. Therefore, the future path of nuclear power in Japan remains uncertain.

In addition to the current difficulty in restarting light-water reactors (LWRs), nuclear waste and plutonium recycling with eventual commercialization of fast-breeder reactors (FBRs) has been a fatal issue in nuclear policy [26]. This is because the status of a fuel recycling plant in Rokkasho village of Aomori Prefecture was settled with the promise that it is for fuel recycling and not for final disposal of nuclear waste. In other words, that village is a production center of nuclear fuel but not a dumping ground for nuclear waste. Decommissioning of LWRs and abandonment of domestic nuclear power entails the loss of a role in fuel recycling. Nuclear safety, public trust, and LWR acceptance are therefore regarded as key issues for all nuclear power policy.

3. Innovative technologies for low-carbon society

A *New Low Carbon Technology Plan* (NLCTP) identified 37 "innovative technologies" [27]. The NLCTP categorized these as technologies for short-/medium-term development (to be developed by ~2030) and medium-/long-term development (to be put into practical use after ~2030).

The former category includes the following in the production and supply sector: high-efficiency coal-fired power generation, high-efficiency natural gas-fired power generation, wind power generation, solar energy, geothermal power generation, ocean energy, and nuclear power. Short- and medium-term technologies in the consumption and demand sector include next-generation automobiles, innovative structural materials, innovative devices, energy management, and energy-efficient houses/buildings. Those in the distribution and supply/demand integration sector are fuel cells, high-performance electricity storage, heat storage, and insulation technologies. The medium-/long-term technologies include CO_2 capture and storage, artificial photosynthesis, biomass utilization, and hydrogen production, transport, and storage.

The NLCTP also indicated the importance of measures required for global expansion and diffusion of innovative technologies from Japan, including promotion of the joint crediting mechanism and international standardization, and the strategic use of public funds.

The government identified the above technological categories as targets of innovation, and most have been conventionally targeted in national R&D programs. Examples include the Sunshine (1974–1992, 532.2 billion JPY), Moonlight (1978–1992, 129.7 billion JPY), and New Sunshine (1993–2002, 354.7 billion JPY) projects. The NLCTP is a rough sketch of the national R&D program and project planning, based on which individual funding agencies invest in each technology. However, the plan lacks detailed analysis and design of each technology.

4. Electricity system reformation

The national electricity market has been practically a monopoly in each region, vertically integrated with respect to electricity supply, transmission, and distribution. Electricity prices are set to fully cover expenses with a certain margin and are expensive (or at least not cheap) relative to other countries [28]. Such a system enables electricity delivery throughout the country at the same price in the area where each electric power company covers so-called universal service. However, such a closed and integrated system is regarded as one of the causes of elevated electricity prices. A fixed price during daytime might hamper the introduction of PV, whose peak electricity production fits the demand in summer. If the price is not fixed, strong electricity demand during the summer peak might drive PV introduction. Liberalization of the electricity market in Japan began in 1995, facilitating an Independent Power Producer (IPP) on the supply side. On the demand side, since 2000, customers with contracted power capacity (CPC) >2,000 kW have been able to choose their electricity supplier. This capability was extended to customers with 500 kW in 2004 and 50 kW in 2005. Such stepwise reformation has contributed to the reduction of electricity prices as a result of competition between IPPs. A planned electricity system reformation in April 2016 will open the market to wider customers whose CPC is <50 kW and moves toward dynamic pricing and exclusion of universal service. There remain several important issues under discussion, such as standard offer service, universal service to small islands, wheeling charges, unbundling of supply and transmission, and design of transition to the new institution.

It is uncertain whether such electricity system reformation will succeed. It is expected to lower average electricity prices, decrease overall electricity demand, save on expenses, and accelerate the introduction of renewables. Full market liberalization including the residential sector was discussed in 2007 and 2008, but the decision was made not to fully open the market because of insufficient competition with market segments that had already been opened [5]. Design and operation regarding the remaining issues above will affect the consequences. Considering recent "failure" in renewables and nuclear policy, the system reformation is a great challenge for energy policy and policymakers in Japan.

3 Perspectives

Reflecting on the aforementioned situation and contexts, what future energy system and society can we envision? How can we find the future path? We need a roadmap, perspective, and path, not only for energy technologies but also energy policy and policymaking.

Discourse on energy policy raises complex issues, because of (1) multiple objectives, (2) involvement of diverse stakeholders, and (3) needs for long-term strategic planning. Therefore, underlying policy design and logic should be credible and appropriate for the purposes, policymaking should be legitimate and democratic, and strategy and plans should be salient and coordinated among components, thereby constituting a grand design of policy. Where credibility involves the scientific adequacy of technical evidence and arguments, salience deals with relevance of assessments to the needs of decision-makers. Legitimacy reflects the perception that the production of information and technology has been respectful of stakeholders' divergent values and beliefs, unbiased in its conduct, and fair in its treatment of opposing views and interests [29]. Although it is rare that the above criteria are fully met, the recognized failure of Japanese energy policy before the disaster provoked public interest in that policy, the policymaking process, and the decision-making in the presence of trade-off and uncertainty. The failure also raised concern for the credibility, saliency, legitimacy, transparency, and fairness in the policy and its formulation. These aspects are also important for the government to gain public trust in their policy and policymaking.

It is clear that both energy policy and its formulation should be reformed to meet the aforementioned criteria. This reformation should consider and resolve the following issues.

1. Innovation and integrated design

Energy policy must have multiple objectives, including energy security, economic efficiency, economic development, the environment, and safety. There are often trade-offs among these, which makes policymaking difficult. For example, PV is a desirable option for environmental sustainability relative to fossil fuel-based power generation systems, but it may be relatively expensive in Japan when we consider the excess cost to stabilize its intermittent electricity output. Nuclear power is regarded as a powerful means to ensure energy security, although it has caused grave safety concerns. Therefore, decision-makers must first note the existence of trade-offs, then make continuous efforts toward innovation and integrated design.

One way to overcome trade-offs is integrated design. For example, let us assume a situation in which utilization of a local resource like biomass from forestry residue is environmentally sustainable, but not economically feasible within a local energy system. Symbiotic design integrating various systems like energy and forestry can allow for both environmentally sustainable and economically feasible utilization of the resource. The benefit of cultivating forest ecosystems is not only for forestry; it is for energy resources, agroforestry, and other integration of various systems.

Another important method for surmounting trade-offs is innovation. Innovation means value creation with something new. Innovation is not competition for a limited pie; it represents value added to the current system. Means for innovation are not limited to new technology and include new institutions, business models, lifestyles, and social systems. Discourse in the policy arena tends to focus on "Options for Energy and the Environment," such as zero or 15 % nuclear and targets for renewables. This situation is conducive to the representation of specific interest groups, and discussions become controversial and political rather than constructive. Salient innovation and integrated design are means to avoid such situations with their trade-offs between different objectives.

2. Unique voice and war of ideas

The energy system is complex and consists of numerous components and options. There is no expert with expertise in all components and options, but expertise is necessary to maintain saliency. A tendency for experts to focus on topics in which they have proficiency hampers comprehensive design and fosters specialization and segmentation [30]. The Diet lacks expertise, and ministries with responsibilities associated with a topic are expected to prepare reports providing policy direction regarding issues related to that topic. Each ministry organizes committees with several subcommittees. A subcommittee has working groups consisting of members with expertise on the topic and stakeholder representatives. A working group summarizes discussion on the issues and reports on it, with tentative conclusions and consensus on the topic. Conclusions are legitimated as a unique voice of the working group. However, such a bottom-up decision-making process cannot determine shares of electricity generation options as expected by the EEC. Further, experts generally believe in the advantages and importance of the technology within their focus and often consciously or unconsciously emphasize that importance. This gives undue weight for that technology against competing ones. At best, the experts can furnish credible knowledge within their expertise but not outside it, which sometimes delegitimizes the unique voice and makes it controversial.

More fundamentally, future energy systems will be equipped with innovations that resolve trade-offs among objectives and concerns. Each innovation inevitably involves uncertainty. We need to reduce uncertainty of scientific statements to facilitate more appropriate decisions. If such statements can be described as physical and chemical mechanisms, it is likely to reduce (or at least assess) uncertainty so that there will be negligible concern for troublesome risks. However, this involves social mechanisms and future events, so substantial uncertainty usually persists. Therefore, it is very difficult to realize a "unique voice" regarding future energy systems. Such a voice regarding an entire energy system, if it can be reached by consensus in current policymaking in Japan, often lacks expertise on such systems and therefore credibility and saliency. Moreover, such a voice on specific components is associated with policy coalitions and risk when relied upon and therefore lacks legitimacy. Instead of or at least prior to achieving such a unique voice, "war of ideas" is desirable or indispensable for making blueprints more credible and salient. This, I believe, will be an appropriate way to take steady steps toward future energy systems.

3. Roadmap for long-term technological and policy trends

Stakeholder involvement, roadmap development, and stakeholder commitment to that roadmap are effective toward reducing its uncertainty. In the science and technology (S&T) roadmap, the following definition is frequently referred to and shared: "a consensus articulation of a scientifically informed vision of attractive technology futures" [31]. The roadmap can enhance coordination and strong investment in specific options specified within it. For example, a roadmap for electricity reform enables firms to invest in potentially required technologies associated with that reform. An FIT roadmap would attract investment in renewables and related technologies. A roadmap for nuclear safety and commitment to it by the NRA would break down barriers to investment in safety technology and nuclear power plants, meeting target criteria described in the roadmap. Investment in energy technologies and infrastructure typically has decadal payback periods. Such long periods can hamper investment, to avoid shortages of funds at hand. Investment in R&D also has similar characteristics. An S&T roadmap is expected to guide stakeholders toward methods described in the roadmap, to suggest orchestration of R&D activities and strategies, and to attract investment.

The roadmap is expected to at least have a role in accelerating investment and innovation but includes risks that are often neglected. In existing roadmaps, plausible risks are not usually stated in a comprehensive and explicit manner, but attractive technology futures are thoroughly visualized. This means that these roadmaps recommend promising future paths while ignoring risks inherent in those paths [32]. There should be both pro-innovation and cautionary outlooks in policy processes [33]. The roadmap should have legitimacy, which, as defined above, reflects stakeholders' divergent values and beliefs, opposing views and interests [29].

4 Sustainability: An Alternative Guiding Principle of Energy Policy

Energy policy involves diverse stakeholders with varying priorities for energy security, the environment, economic efficiency, safety, and industrial and economic development. The policy represents diverse ideas, perspectives, concerns, motives, strategies, and conflicts. It also has wide coverage, including science, technology, markets, industries, regulations, institutions, and international relationships. Given this complexity of social and technological facets of energy systems, it is not easy to reach agreement on the structure of future energy systems and methods to realize them. It is especially difficult to reach consensus on the ratios of individual energy supply options. Each stakeholder insists on the importance of options from which they realize benefits or of which they believe should have priority. There is no credible technological assessment or salient integrated design, or at least they are seldom referred to during political debate. Domestic issues aside, global climate change may become a focal point of international politics and increase tensions and conflict between countries, especially regarding caps on greenhouse gas emission and economic development.

There are diverse guiding principles for energy policy. Each people and community stands on each principle. Here, innovation and integrated design are necessary to resolve trade-offs between principles. Such innovation and design will be triggered by war of ideas regarding innovative technologies and institutions. In initial phases, it is difficult to reach a unique voice on future energy system and energy technologies toward their prioritization and fulfillment. However, this voice gradually emerges through ideas and deliberations. Then, roadmaps advance technologies and institutions with the commitment of stakeholders, with regard for risks and uncertainty.

I believe that innovation is especially important, and, aside from the above principles of energy policy, sustainability and innovation must be emphasized. Herman Daly advocated the following guiding principles for sustainability: For renewable resources, the rate of harvest should not exceed that of regeneration; rates of waste generation from projects should not exceed the assimilative capacity of the environment; for nonrenewable resources, their depletion should require comparable development of renewable substitutes for them [34]. It is not easy to preserve the principle of sustainability in compatible ways among other principles, especially when innovative technologies are not ready for use in a feasible manner. An alternative approach for current society in such a situation is to invest in and develop promising options, which are described elsewhere in this book. This is ethical behavior for supporting subsequent generations and realizing sustainable energy systems and societies in the near future.

References

1. Wakabayashi M (2013) Voluntary business activities to mitigate climate change: case studies in Japan. Energy Policy 63:1086–1090
2. ANRE (2010) Energy in Japan. http://www.enecho.meti.go.jp/en/brochures/
3. Hayashi M, Hughes L (2013) The Fukushima nuclear accident and its effect on global energy security. Energy Policy 59:102–111
4. International Energy Agency (IEA) (2008) Energy policies of IEA countries: Japan 2008 review. IEA, Paris
5. Takase K, Suzuki T (2011) The Japanese energy sector: current situation, and future paths. Energy Policy 39:6731–6744
6. Duffield JS, Woodall B (2011) Japan's new basic energy plan. Energy Policy 39:3741–3749
7. Fujino J, Hibino G, Ehara T, Matsuoka Y, Masui T, Kainuma M (2008) Backcasting analysis for 70% emission reduction in Japan by 2050. Clim Pol 8:S108–S124
8. Ashina S, Fujino J, masui T, Ehara T, Hibino G (2012) A roadmap towards a low-carbon society in Japan using backcasting methodology: feasible pathways for achieving an 80% reduction in CO_2 emissions by 2050. Energy Policy 41:584–598
9. METI (2010) Establishment of the strategic energy plan of Japan. http://www.meti.go.jp/english/press/data/20100618_08.html
10. Vivoda V (2012) Japan's energy security predicament post-Fukushima. Energy Policy 46:135–143
11. Hayashi M, Hughes L (2013) The policy responses to the Fukushima nuclear accident and their effect on Japanese energy security. Energy Policy 59:86–101
12. ANRE (2011) Electricity statistics. http://www.enecho.meti.go.jp/statistics/electric_power/ep002/ (in Japanese)

13. Fukushima Y, Kikuchi Y, Kajikawa Y, Kubota M, Nakagaki T, Matsukata M, Kato Y, Koyama M (2011) Tackling power outages in Japan: the earthquake compels a swift transformation of the power supply. J Chem Eng Jpn 44:365–369
14. Fujimi T, Chang SE (2014) Adaptation to electricity crisis: businesses in the 2011 Great East Japan triple disaster. Energy Policy 68:447–457
15. Wakiyama T, Zusman E, Monogan JE III (2014) Can a low-carbon-energy transition be sustained in post-Fukushima Japan? Assessing the varying impacts of exogenous shocks. Energy Policy 73:654–666
16. METI (2014) Strategic energy plan. http://www.enecho.meti.go.jp/en/category/others/basic_plan/pdf/4th_strategic_energy_plan.pdf
17. Fam SD, Xiong J, Xiong G, Yong DL, Ng D (2014) Post-Fukushima Japan: the continuing nuclear controversy. Energy Policy 68:199–205
18. Energy and Environment Council of the Japanese Government (2012) Options for energy and environment. http://www.cas.go.jp/jp/seisaku/npu/policy09/sentakushi/pdf/Report_English.pdf
19. Energy and Environment Council of the Japanese Government (2013) Innovative strategy for energy and environment. http://www8.cao.go.jp/cstp/sonota/kankyoene/keikaku.pdf (in Japanese)
20. Honma T, Akimoto K (2013) Analysis of Japan's energy and environment strategy after the Fukushima nuclear plant accident. Energy Policy 62:1216–1225
21. Cabinet Office, Government of Japan (2010) On the new growth strategy. http://www.meti.go.jp/english/policy/economy/growth/report20100618.pdf
22. Huenteler J, Schmidt TS, Kanie N (2012) Japan's post-Fukushima challenge – implications from the German experience on renewable energy policy. Energy Policy 45:6–11
23. IEA (2011) OECD Electricity and heat generation from renewables, IEA renewables information statistics. International Energy Agency (IEA), Paris
24. Sugino H, Akeno T (2010) Country update for Japan. In: Proceedings world geothermal congress. Bali, pp 25–29
25. METI (2014) Reconsideration of operation of feed-in-tariff toward the maximum introduction of renewable energy (in Japanese). http://www.meti.go.jp/press/2014/12/20141218001/20141218001.html
26. Katsuta T, Suzuki T (2011) Japan's spent fuel and plutonium management challenge. Energy Policy 39:6827–6841
27. Cabinet Office (2013) New low carbon technology plan. http://www8.cao.go.jp/cstp/english/doc/new_low_carbon_tec_plan/nlctp_text.pdf
28. IEA (2015) IEA energy prices and taxes statistics. International Energy Agency (IEA), Paris
29. Cash DW, Clark WC, Alcock F, Dickson NM, Eckley N, Guston DH, Jäger J, Mitchell RB (2003) Knowledge systems for sustainable development. Proc Natl Acad Sci 100:8086–8091
30. Kajikawa Y, Abe K, Noda S (2006) Filling the gap between researchers studying different materials and different methods: a proposal for structured keywords. J Inf Sci 32:511–524
31. Branscomb LW, Keller JH (1998) Towards a research and innovation policy. In: Branscomb LW (ed) Investing in innovation; creating a research and innovation policy that works. MIT Press, Cambridge
32. Kajikawa Y, Kikuchi Y, Fukushima Y, Koyama M (2013) Utilizing risk analysis and scenario planning for technology roadmapping. In: Cetindamar D, Daim T, Beyhan B, Basoglu N (eds) Strategic planning decisions in the high tech industry. Springer, London/New York, pp 231–244
33. Stirling A (2007) Deliberate futures: precaution and progress in social choice of sustainable technology. Sustain Dev 15:286–295
34. Daly HE (1990) Toward some operational principles of sustainable development. Ecol Econ 2:1–6

Part III
Advanced Use of Secondary Energy Media

Yukitaka Kato

Large-scale electrical energy storage systems with electrochemical batteries offer the promise of improved utilization of electricity, with load-leveling and massive introduction of renewable energies from solar and wind power. Sodium-sulfur and redox flow batteries are candidate technologies. Achieving high energy efficiency and cost reduction represents a technical challenge.

Hydrogen production involves considerations of production energy efficiency, carbon dioxide emission (CO_2), and cost. Even when hydrogen is produced conventionally from fossil fuel, a fuel-cell vehicle's well-to-wheel CO_2 emission can be less than that of gasoline engine vehicles.

The importance of the energy carrier for hydrogen produced by renewable energy use has been recognized. Long-term and stable storage of hydrogen is necessary to complete the hydrogen supply chain in the form of liquid hydrogen or chemical hydride, such as ammonia or organic hydride.

Waste heat recovery has great potential in Japan. Thermochemical energy storage has potential for high-temperature storage (>200 °C). Heat transport at <200 °C by latent heat storage has practical possibilities for waste heat utilization. Technology for heat exchange from the gas phase will increase in importance. Consistent design of material, reactor, heat exchanger, and heat utilization systems at low cost is an ideal technology goal for heat recovery.

Y. Kato (✉)
Tokyo Institute of Technology, Tokyo, Japan
e-mail: yukitaka@nr.titech.ac.jp

Large-Scale Electrical Energy Storage Systems

Shohji Tsushima

Abstract Large-scale electrical energy storage systems with electrochemical batteries offer the promise for better utilization of electricity with load leveling and the massive introduction of renewable energy from solar and wind power. In this chapter, an overview of large-scale energy storage systems is presented, together with the current and future states of electricity demand in Japan. The present status and perspectives of NaS batteries and redox flow batteries are discussed as massive electrical energy storage systems. The technical challenges that remain to further achieving high energy efficiency and cost reduction are also described.

Keywords Energy storage • Redox flow battery • NaS battery • Renewable energy • Electricity supply

1 Introduction

Large-scale electrical energy storage systems [1] have garnered much attention for increasing energy savings. These systems can be used for electricity load leveling and massive introduction of renewable energy sources with intermittent output, which contribute to reduced nuclear power generation and less fossil fuel consumption.

According to perspectives on energy published by the government of Japan [2, 3], targeted values of overall installed capacity for photovoltaic power generation and wind-generated electricity are 33 GW in 2020 and 63 GW in 2030. This corresponds to electricity generation at 39.6 TWh in 2020 and 74.8 TWh in 2030, which are 3.7 and 7.3 % of estimated total electricity demand, respectively. To maintain a stable supply of electricity to the electrical grid, large-scale electricity storage systems that operate together with these renewable sources are needed, because both solar and wind power generation are inherently intermittent.

Electricity demand varies considerably throughout a day or year. Thermal power plants that can be operated at large turndown ratio with swift response are generally

S. Tsushima (✉)
Osaka University, Osaka, Japan
e-mail: tsushima@mech.eng.osaka-u.ac.jp

Y. Kato et al. (eds.), *Energy Technology Roadmaps of Japan*,
DOI 10.1007/978-4-431-55951-1_7

used to follow demand. Pumped-storage hydropower plants are also used to satisfy demand by storing off-peak electricity for later use. Response times of these plants are very different.

Massive introduction of solar and wind power for generation of electricity might lead to an unstable electricity supply. In 2014, several electric power companies in Japan announced that they would temporarily halt acceptance of applications for the feed-in-tariff system with solar and wind power generation. Large-scale electricity storage systems can play a central role in this purpose in the coming decade and have been developed worldwide using batteries, compressed air, flywheels, super capacitors, superconducting materials, and others.

Among these, electrochemical battery systems have several advantages, i.e., fast response to charge or discharge electricity in the grid, less limitation on construction location, short installation time, and versatility from large- to small-scale energy storage. However, further improvement in energy efficiency and cost reduction are of primary importance for practical use of these systems.

In this chapter, we focus on sodium–sulfur (NaS) batteries, including sodium ion and redox flow batteries for large-scale (megawatt) electricity storage. We also discuss future perspectives and emerging issues for further R&D.

2 Present Status

2.1 Potential Economic and Environmental Benefits

There are economic and environmental incentives for the introduction of large-scale electricity storage systems. Figure 1 gives a typical electricity demand (generation) profile for a sunny summer day in Japan. Base, intermediate, and peak loads are identified. Base load generation has been primarily by nuclear power plants in Japan but, in 2012, nuclear power plants contributed 1.7 % of

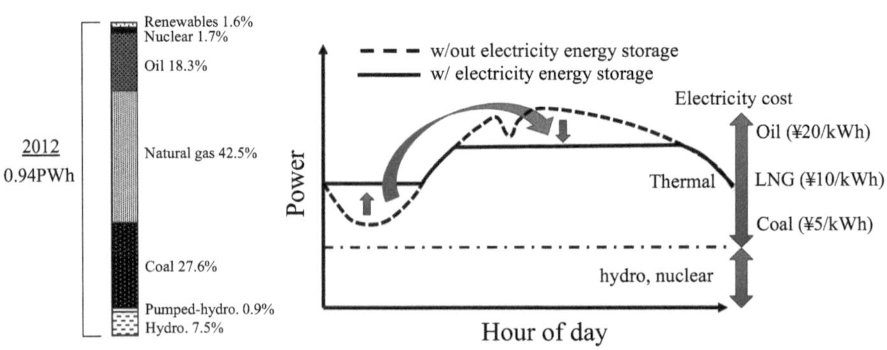

Fig. 1 Electricity supply in 2012 and schematic of typical electricity demand profile for a sunny summer day in Japan

overall electricity supply. Therefore, thermal power plants play a major role in covering the entire demand load [4].

Thermal power plants can be operated using a variety of fossil fuels, oil, liquid natural gas (LNG), and coal. Prices for these fuels vary, so optimal control for operation of these power plants is needed for improved fuel economy. Assuming that electricity prices for oil-fueled, LNG-fueled, and coal-fueled power plants are ¥20/kWh, ¥10/kWh, and ¥5/kWh, respectively, oil-fired power plants should be operated only for peak loads, because of their poor fuel economy.

By introducing large-scale electricity storage systems, we can increase electricity generation by operating low-cost power plants at night instead of high-cost power plants in daytime for peak demand. Differences in electricity cost depending on the type of power plant can generate economic benefits attributable to large-scale electricity storage, although roundtrip efficiency during the storage and supply of electricity is $<100\,\%$.

Under present conditions, pumped-storage hydropower plants are widely used as large-scale electrical energy storage. In Japan, the total capacity of these plants was estimated at ~20 GW [5], and almost 1 % of total electricity supply was provided by the plants in 2012 (Fig. 1).

Regarding environmental impacts, lowering fossil fuel consumption and reductions in CO_2 emissions are feasible by introducing large-scale electrical energy storage. Obviously, systems for such storage aid massive utilization of renewable energy, which suppresses fossil fuel consumption.

All the above factors motivate installation of large-scale battery systems in the grid. However, state-of-the-art battery technologies do not fully satisfy the demands, so further improvement of energy efficiency and cost reduction is needed.

We have seen a wide variety of affordable battery systems for massive energy storage. In a report released by the Japanese Ministry of Economy, Trade and Industry in 2012 [6], the installation cost of battery technologies was compared with pumped-storage hydropower plants as follows: pumped-storage hydropower = ¥23,000/kWh, NaS battery = ¥40,000/kWh, lead–acid battery = ¥50,000/kWh, nickel metal hydride battery = ¥100,000/kWh, and lithium ion battery = ¥200,000/kWh. As addressed later, the redox flow battery (RFB) is also a candidate for large-scale electricity storage. The installation cost of RFBs of all-vanadium type was estimated as the same or slightly higher than that of an NaS battery. Another cost estimation presented in the United States is shown in Table 1 [7, 8].

Adding to initial installation cost, battery life, and economy during operation and maintenance as well as environmental impact upon disposal become important factors that determine the overall life cycle cost of the system. The International Energy Agency estimates 310 GW of additional grid-connected electricity storage capacity needed in 2050 [1]. As a large-scale electricity storage system, the NaS battery has been already commercialized. Two demonstration projects in which renewable energy (solar or wind power) is integrated with battery systems (lithium ion or vanadium RFB) are underway in Japan, to show their feasibility for stable supply of electricity with those renewables.

Table 1 Electrical energy storage characteristics (Adapted from Ref. [7])

Technology option	Maturity	Capacity (MWh)	Power (MW)	Duration (h)	% Efficiency (total cycles)	Total cost ($/kW)	Cost ($/kWh)
Pumped hydro	Mature	5400–14,000	900–1400	6–10	80–82 (>13,000)	1500–2700	250–270
Sodium sulfur	Commercial	300	50	6	75 (4500)	3100–3300	520–550
Advanced lead–acid	Commercial	200	50	4	85–90 (2200)	1700–1900	425–475
Vanadium redox flow	Demonstration	250	50	5	65–75 (>10,000)	3100–3700	620–740
Li-ion	Demonstration	0.25–25	1–100	0.25–1	87–92 (>100,000)	1085–1550	4340–6200

In the following sections, we address two promising candidates, the NaS battery and RFB, as large-scale battery systems.

2.2 *Sodium–Sulfur (NaS) Batteries*

An NaS battery uses beta alumina ceramic for the electrolyte, with sodium and sulfur as active materials for the negative and positive electrodes. NaS batteries typically operate at temperatures ~300 °C to maintain the electrode materials in a molten state. Figure 2 shows a schematic of an NaS battery cell [9]. The working voltage of the battery is 1.78–2.08 V at 350 °C [9, 10]. NGK Insulators, Inc., has commercialized an NaS battery since 2002 and has installed over 450 MW of the system worldwide. Cell performance is dominated by ohmic resistance of the electrolyte plus activation resistance of the positive electrode, both of which are responsible for 90 % of overall resistance in the cell [11]. Energy efficiency of an NaS battery is 75 % [12]. Comparison of energy efficiency among secondary batteries is by no means straightforward, although values are available as shown in Table 1 [7]. A battery system generally requires balance of plants (BOPs), which need supplemental energy input for operation. The operation of BOPs reduces overall energy efficiency of the battery system. Energy conversion efficiency of AC/DC converters should also be considered. Moreover, energy loss from cell operation is greatly affected by the current density of the cell. Energy efficiency of the system varies greatly with them.

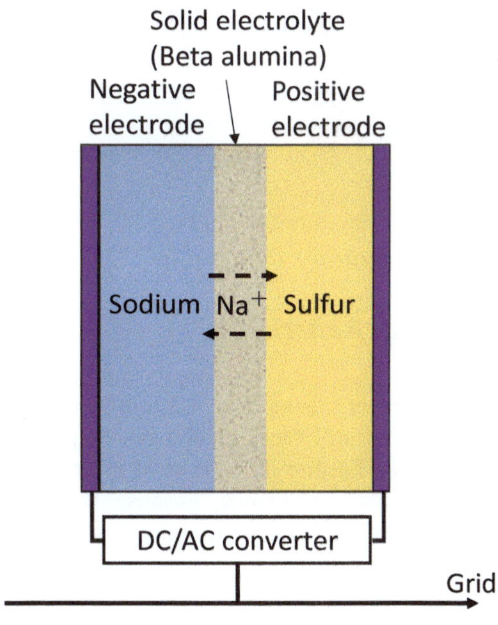

Fig. 2 Schematic of NaS battery cell

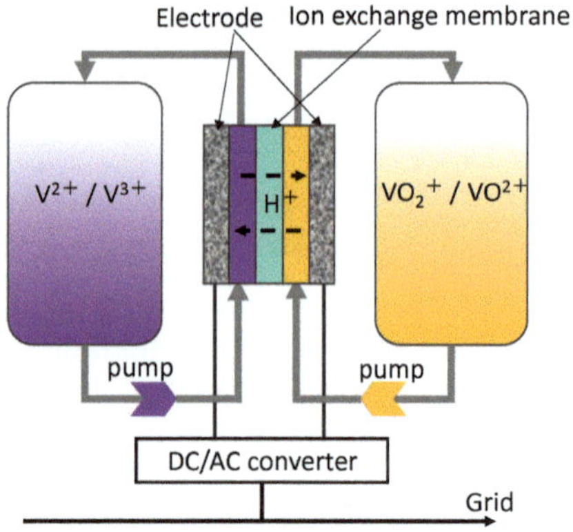

Fig. 3 Schematic of redox flow battery (all-vanadium type)

2.3 *Redox Flow Batteries (RFB)*

RFBs use redox processes of active species in solution that are stored in external tanks (Fig. 3) [13]. They thus provide a unique feature of system scalability, i.e., independent sizing of energy and power, suitable for large-scale applications. A variety of redox pairs have been tested and there have been industrial demonstrations with iron–chromium (Fe^{2+}/Fe^{3+} and Cr^{3+}/Cr^{2+}) or all-vanadium (V^{2+}/V^{3+} and VO^{2+}/VO_2^{+}) systems. An all-vanadium RFB ($E^{eq} = 1.26$ V under normal conditions) shows an advantage, in that the reduction of its capacity caused by crossover of active species across the ion exchange membrane is suppressed. Sumitomo Electric Industries Co. Ltd. has built demonstration plants, such as wind power (30.6 MW), RFB (4 MW × 1.25 h), photovoltaic power (100 kW), and RFB (1 MW × 5 h) [14, 15].

More recently, there has been rapid growth of RFB installation by other companies. A 2009 report by the National Renewable Energy Laboratory [16] gave AC conversion energy efficiency at 72 % and DC conversion efficiency at 80 %.

3 Technology Roadmap

3.1 *NaS Batteries*

An NaS battery using beta alumina ceramic electrolyte has already been commercialized, but still faces challenges of performance and cost for massive market penetration. Further improvement of cell performance would be very effective in achieving these demands, because better cell performance enhances roundtrip efficiency of the battery and reduces stack size, with less cost for materials. Energy loss in an NaS battery is mainly in the ceramic electrolyte and positive electrode. Ionic conductivity of the beta alumina ceramic electrolyte is 0.4 S/cm at 350 °C [17] and can be improved by designing and synthesizing novel ionic conductive materials. Thinning the solid electrolyte is another potential approach to reduce ohmic resistance of the electrolyte, but its mechanical durability should be maintained. Energy loss in the positive electrode can also be reduced. Ohmic and activation polarizations ascribed to the electrode have not been fully explored regarding its structure. The reaction distribution and ionic/electronic transport in the porous electrode under cell operation remain unknown and could be further optimized by modifying electrode structure.

In another type of sodium–beta alumina batteries, the solid-state halide electrode (cathode) is used with metallic sodium (anode) and the ceramic electrolyte. Higher cell potential can be achieved using nickel chloride ($NiCl_2$: E = 2.58 V vs. Na at 300 °C) in the cathode [18]. This type of battery has been developed as Zero-Emission Battery Research Activity cells and is operated at high temperature (~300 °C) to maintain favorable ionic conductivity of the sodium–beta alumina membrane.

More recently, low- and/or room-temperature operation of sodium ion batteries has attracted much attention because of advantages such as low-cost and high-capacity energy storage systems. In NaS batteries, molten sulfur, sodium, and the polysulfide compounds are highly corrosive at elevated temperatures ~350 °C, and containers and seals must be resistant under these conditions [19]. Low-temperature operation of the battery could mitigate this concern and simplify the system with fewer BOPs needed, reducing cost. However, electrochemical reaction and ionic conductivity in the cell are inactive and are reduced at low temperatures.

All-solid-state batteries are another choice, providing an opportunity to be free from flammable electrolytes that cause safety concerns. A sodium battery operating at 200 °C using a sodium (Na) superionic conductor [20] as the electrolyte was recently described [21]. The use of superionic glass–ceramic electrolytes was reported for room-temperature sodium batteries [22]. Cell performance and ionic conductivity of the solid electrolyte remain to be improved for practical applications, thereby offering promising opportunities for further research.

In summary, NaS batteries are promising candidates for large-scale electrical energy storage, with further cost reduction and higher energy efficiency. Meanwhile, low- or room-temperature sodium ion batteries face many challenges, from fundamentals to applications. It is necessary to advance R&D for the enforcement of early proof examination regarding the low-temperature sodium ion battery to explore the R&D target and develop strategic roadmaps in the future.

3.2 RFB

For the RFB, all-vanadium and iron–chromium systems lead in industrialization. For greater introduction, improvement of energy efficiency and cost reduction is required. Improvement of battery energy efficiency leads to downsizing the entire system and can reduce cost. To decrease energy loss in RFBs, porous structure in the electrodes and flow patterns in the cell are influential in determining cell performance, and electrochemical activity of the electrode surface is also vital. In recent years, RFB cell performance has been improved by introducing fundamental knowledge and technologies established in proton exchange membrane fuel cells [23, 24]. Systematic research toward the most suitable electrode for RFBs for reducing kinetic and transport loss in the electrode, with less pressure drop in the battery system, should be promoted. Also encouraged is the exploration of electrode materials for enhanced electrochemical activity, with a wide electric potential window free from hydrogen and oxygen evolution under cell operating conditions.

Further cost reduction for the membrane can be expected by introducing a cheaper hydrocarbon system than the conventional fluorine-based electrolyte membrane. It is necessary for the electrolyte membrane to suppress crossover of active redox species between the electrodes. Durability is also a concern for a hydrocarbon membrane implemented in RFBs for long-term operation.

Implementation of the aforementioned approaches can improve energy efficiency, reduce cost, and advance RFBs, with market penetration by 2020 and greater penetration later.

Easy recycling of active species in RFBs is attributed to the unique geometry of the battery system, in which redox species dissolved in liquid electrolyte are stored in external tanks and supplied to the cell. Although material recycling is not a major concern, vanadium electrolyte still represents a third of total capital cost according to the literature [25]. Therefore, a search is necessary regarding a new reaction system. In recent years, a metal-free organic–inorganic aqueous flow battery system has been proposed [26]. New redox systems not only widen the choice of available materials but also improve electrochemical properties beneficial to RFBs with high power and energy densities.

Regarding the above, it is effective to promote concentrated R&D while gathering knowledge. Further, for the promising reaction system, it is necessary to examine systematization at an early stage. Early-stage demonstration tests give profound insights into system performance and durability, which are readily fed back to fundamental research. R&D by laboratories in the fields of systematic screening materials and cell design offer promise to large-scale electrical energy storage systems.

4 Benefits and Future Vision

A large-scale electricity storage system does not produce energy in itself, but is significant in energy conversion and storage for efficient utilization of electricity generated by fossil fuel consumption and/or nuclear energy. Further implementation of renewable energy in society can be ably supported by such storage systems. Moreover, CO_2 emission could be further reduced. If we assume a life cycle CO_2 emission intensity of 738 g-CO_2/kWh for oil-fired power plants and 474 g-CO_2/kWh for LNG-fired, combined-cycle power plants [27], it is much better to use LNG for electricity generation. Although electrical energy storage systems generate some fraction of energy loss during charge and discharge of electricity, e.g., 30 % loss by pumped-storage hydropower plants, shifting oil-fired to LNG-fired power plants with the electrical energy storage will still reduce overall CO_2 emission.

Energy loss associated with energy conversion and storage by large-scale electricity storage is inevitable. However, various energy technologies including the storage systems provide a versatile choice for a sustainable and low-carbon society. Toward this goal of robust mixes of energy technologies, the large-scale batteries such as NaS and RFBs surveyed in this chapter are expected to be key devices. This is because they can mitigate temporal intermittency and spatial inhomogeneity of renewable energy sources, yielding profound benefits for stable supplies of electricity to the grid. We must also discuss which sectors should be responsible for massive introduction of the large-scale electrical energy storage systems, with due consideration of economic benefits and risks.

We recognize that available material resources, electricity demand affected by the future population and national economy activity of Japan, international competitiveness of Japanese manufacturing, system updates, and the establishment of material recycling and disposal with sustainable growth are potential risks to the future diffusion of large-scale batteries in the country. Given this recognition, it is mandatory to promote more R&D.

Government support and initiatives are required for formulation of an electricity rate system that creates economic advantages during the stage when system cost does not decline sufficiently, as well as a stable supply of electric power with introduction of renewable energy in the early period. Such a system should become widespread as a social infrastructure supporting the smart grid by about 2030, and contributions are anticipated regarding risk reduction for future fossil fuels and rare resource acquisition. Furthermore, the energy carrier should diversify by 2050, and large-scale electricity storage will become a social infrastructure indispensable to electric energy supply. Further expectations are of renewable energy introduction and contributions to the realization of a low-carbon society through sustainable energy supply, plus CO_2 emission reduction in the economy from primary energy diversification via massive battery systems (Fig. 4).

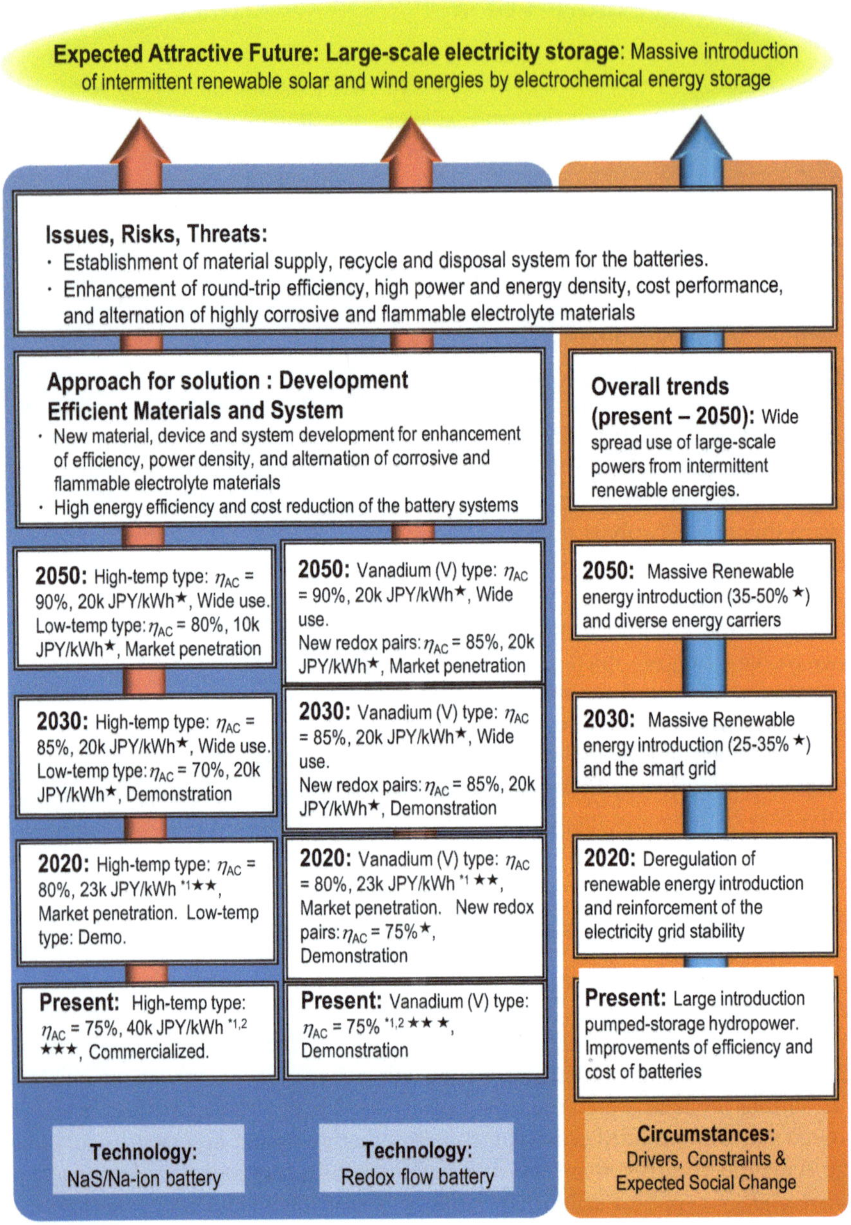

*1 Ref. [6]
*2 Ref. [7]

Abbreviations:
Demo.: demonstration

Fig. 4 Roadmap for NaS/Na-ion and RF batteries

References

1. International Energy Agency (2014) Technology roadmap: energy storage
2. Ministry of Economy, Trade and Industry in Japan (2009) Long-term energy demand perspectives
3. Agency for Natural Resources and Energy in Japan (2012) Energy supply in 2030 on strategic energy plan
4. Agency for Natural Resources and Energy in Japan (2015) Energy white paper 2014
5. Ministry of Economy, Trade and Industry in Japan (2012) Report on Energy demand and supply
6. Ministry of Economy, Trade and Industry in Japan (2012) Storage battery strategy
7. Electric Power Research Institute (2010) Electric energy storage technology options: a white paper primer on applications, costs, and benefits, 1020676
8. Akhil AA, Huff G, Currier AB, Kaun BC, Rastler DM, Chen SB, Cotter AL, Bradshaw DT, Gauntlett WD (2013) DOE/EPRI 2013 Electricity storage handbook in collaboration with NRECA, Sandia Report SAND2013-5131
9. Oshima T, Kajita M, Okuno A (2005) Development of sodium-sulfur batteries. Int J Appl Ceram Tech 1(3):269–276. doi:10.1111/j.1744-7402.2004.tb00179.x
10. Lu X, Xia G, Lemmon JP, Yang Z (2010) Advanced materials for sodium-beta alumina batteries: status, challenges and perspectives. J Power Sources 195:2431–2442. doi:10.1016/j.jpowsour.2009.11.120
11. NGK Insulators, Ltd (2004) Development of NAS battery cells and modules. NGK Rev 60:10
12. Mizutani T (2009) Innovative technological developments in electricity storage, Hannover messes – World energy dialogue
13. Alotto P, Guarnieri M, Moro F (2014) Redox flow batteries for the storage of renewable energy: a review. Renew Sustain Energy Rev 29:325–335. doi:10.1016/j.rser.2013.08.001
14. Shigematsu T (2011) Redox flow battery for energy storage. SEI Tech Rev 73:4–13
15. Shibata T, Kumamoto T, Nagaoka Y, Kawase K, Yano K (2013) Redox flow batteries for the stable supply of renewable energy. SEI Tech Rev 76:14–22
16. Steward D, Saur G, Penev M, Ramsden T (2009) Lifecycle cost analysis of hydrogen versus other technologies for electrical energy storage, NREL/TP-560-46719
17. NGK Insulators, Ltd (2004) Development of solid electrolyte. NGK Rev 60:4–9
18. Sudworth JL (2001) The sodium/nickel chloride (ZEBRA) battery. J Power Sources 100:149. doi:10.1016/S0378-7753(01)00891-6
19. Ellis BL, Nazar LF (2012) Sodium and sodium-ion energy storage batteries. Curr Opin Solid St M 16:168–177. doi:10.1016/j.cossms.2012.04.002
20. Fergus JW (2012) Ion transport in sodium ion conducting solid electrolytes. Solid State Ion 227:102–112. doi:10.1016/j.ssi.2012.09.019
21. Lalère F, Leriche JB, Courty M, Boulineau S, Viallet V, Masquelier C (2014) An all-solid state NASICON sodium battery operating at 200°C. J Power Sources 247:975–980. doi:10.1016/j.jpowsour.2013.09.051
22. Hayashi A, Noi K, Sakuda A, Tatsumisago M (2012) Superionic glass-ceramic electrolytes for room-temperature rechargeable sodium batteries. Nat Commun 3:856. doi:10.1038/ncomms1843
23. Aaron DS, Liu Q, Tang Z, Grim GM, Papandrew AB, Turhan A, Zawodzinski TA, Mench MM (2012) Dramatic performance gains in vanadium redox flow batteries through modified cell architecture. J Power Sources 206:450–453. doi:10.1016/j.jpowsour.2011.12.026
24. Tsushima S, Kondo F, Sasaki S, Hirai S (2014) Efficient utilization of the electrodes in a redox flow battery by modifying flow field and electrode morphology. Proc 15th Int Heat Transfer Conf. doi:10.1615/IHTC15.ecs.009326, IHTC15-9326
25. Carnegie R, Gotham D, Nderitu D, Preckel PV (2013) Utility scale energy storage systems. State Utility Forecasting Group, Purdue University, West Lafayette

26. Brian H, Marshak MP, Suh C, Er S, Gerhardt MR, Galvin CJ, Chen X, Aspuru-Guzik A, Gordon RG, Aziz MJ (2014) A metal-free organic–inorganic aqueous flow battery. Nature 505:195–198. doi:10.1038/nature12909
27. Imamura E, Nagano K (2010) Evaluation of life cycle CO_2 emissions of power generation technologies: update for state-of-the-art plant, CRIEPI Research Report: Y09027

Heat Storage, Transportation, and Transfer

Yukitaka Kato, Hiroshi Suzuki, and Naoki Shikazono

Abstract The potential and contribution of heat storage, transportation, and transfer are overviewed for efficient heat recovery and usage in future society. Waste heat recovery has great potential in Japan. Efficient heat usage needs optimized technologies combining heat storage, transportation, and transfer. Thermochemical energy storage has substantial potential for greater density storage at temperatures over 200 °C. Heat transfer enhancement of materials and reactors is required. Heat transportation at less than 200 °C by latent heat storage has practical possibilities for waste heat utilization. Development of hard-shell encapsulating technology and cost reduction of the shell are important. Heat exchange from the gas phase will be more important. Compatible heat transfer enhancement and cost reduction is key. A consistent, low-cost design for the material, reactor, heat exchanger, and heat utilization system is the ideal technology goal. Such technology developments promote diffusion of the heat utilization systems within society. This diffusion will contribute to energy consumption savings and CO_2 emission mitigation in that society.

Keywords Heat storage • Heat transportation • Heat transfer • Thermochemical energy storage • Latent heat storage • Heat transfer enhancement

Y. Kato (✉)
Tokyo Institute of Technology, Tokyo, Japan
e-mail: yukitaka@nr.titech.ac.jp

H. Suzuki
Kobe University, Kobe, Japan
e-mail: hero@kobe-u.ac.jp

N. Shikazono
The University of Tokyo, Tokyo, Japan
e-mail: shika@iis.u-tokyo.ac.jp

Y. Kato et al. (eds.), *Energy Technology Roadmaps of Japan*,
DOI 10.1007/978-4-431-55951-1_8

1 Introduction

Any energy system is accompanied by waste heat emission. The potential of waste heat for secondary utilization is substantial and would be effective for energy saving and carbon dioxide (CO_2) emission mitigation. Heat utilization technologies have been developed with the expansion of energy utilization in society. However, there is still room for improvement of the technologies. A heat utilization system from a heat source to a heat demand is connected to thermal technologies for heat transfer, storage, and transportation, as shown in Fig. 1. It is important to consider the temporal and location gap between the heat source and demand for a discussion of thermal technology development. Heat storage technologies that store heat as latent, sensible, and chemical thermal energies are needed to adjust temporal and location gaps between heat output from the heat source and heat usage at the heat demand point. Heat transfer technologies that use heat conduction, convection, and radiation are required for heat storage between the heat source and heat storage media and for heat demand. Heat transportation technologies are desirable to adjust location differences between heat source and heat demand. Heat transportation is realized using tube or pipe lines, trucks, cars, trains, and boats. It is possible to depict the heat utilization system in a figure. However, the system has not been sufficiently established, because every thermal technology process is accompanied by some loss of energy and exergy (temperature, Gibbs-free energy) in comparison with electricity transfer, which has only ohmic loss in its transmission.

In Japan, heat sources have diversified recently from industrial waste heat to heat from engines and renewable energy systems. Thus, to establish conventional heat utilization systems, comprehensive development of efficiency for both thermal technologies will be required from standpoints of enthalpy and exergy efficiencies,

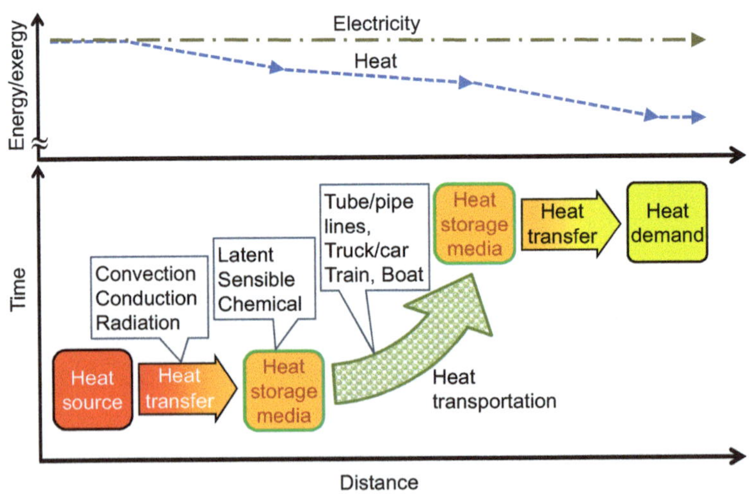

Fig. 1 Thermal technologies in heat utilization system

with greater energy density, higher rates of heat transfer, and chemical reactions. Such development reduces system volume, mass, and cost and promotes diffusion of the heat utilization systems within society. This diffusion will contribute to energy consumption savings and CO_2 emission mitigation in that society.

2 Heat Storage

2.1 Present Status

Waste heat utilization has major potential for energy saving and CO_2 emission mitigation in Japan. Over 70 % of primary energy in the country is used for process heating. Onsite waste heat recovery technologies in energy processes has been well developed in Japan. In particular, after the first oil crisis in 1973, this made a great contribution to energy saving in the country. However, significant waste heat is still emitted into the atmosphere because of a lack of thermal management technologies, especially of the heat storage variety, which adjusts the temporal and location gap between the heat source and demand.

Figure 2 shows industrial waste heats in each temperature range for Japan [1]. Industrial waste heat >200 °C is 1250 PJ/year ($=1.25 \times 10^{18}$ J/year) or 40 % of total industrial waste heat nationwide. High-temperature heat at temperatures >200 °C represents an important target for heat recovery and usage, because it is substantial at present and has relatively high energy quality.

The demand for heat storage technology in vehicle cabin air conditioning is increasing, following improvement of vehicle fuel mileage. The conventional internal combustion engine of a vehicle has sufficient exhaust heat for cabin heating, but hybrid and electric vehicles lack such heat sources for this purpose. It is reported that an electric vehicle needs 30–40 % of the electric capacity of a

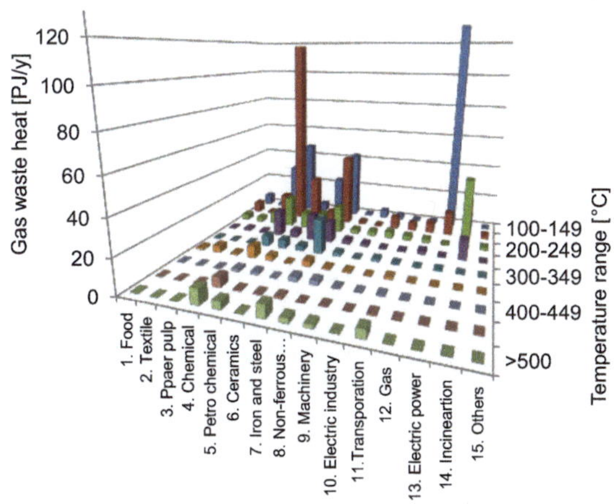

Fig. 2 Industrial gas waste heat in Japan [1]

Li-ion battery for air conditioning [2]. Heat storage technology is capable of assisting air conditioning and reducing the capacity and ultimate cost of the battery. Stabilization of output from renewable energy systems is also required in the future.

Output from a concentrating solar power (CSP) system fluctuates largely from second to second and day to night. CSP is a required heat storage system for stabilized thermal output to energy conversion/consumption systems downstream. Heat storage systems have substantial potential for efficient future heat systems.

2.2 *Technology Roadmap*

Conventional heat storage technologies at less 100 °C have already had a practical market using latent and sensible heat storage. A high-temperature heat storage system for CSP is already in operation, using sensible heat storage of ceramics at the German Aerospace Center (DLR) in Germany. The storage system stores heat at up to 680 °C of 2.25 MW (=8.1 GJ) [3]. High-performance heat storage technologies at higher temperatures are required for future energy systems. Figure 3 shows energy storage densities of chemical and physical changes. Chemical thermal energy storage has greater energy storage density than latent and sensible heat storages. However, a high enthalpy change reaction such as carbon combustion also

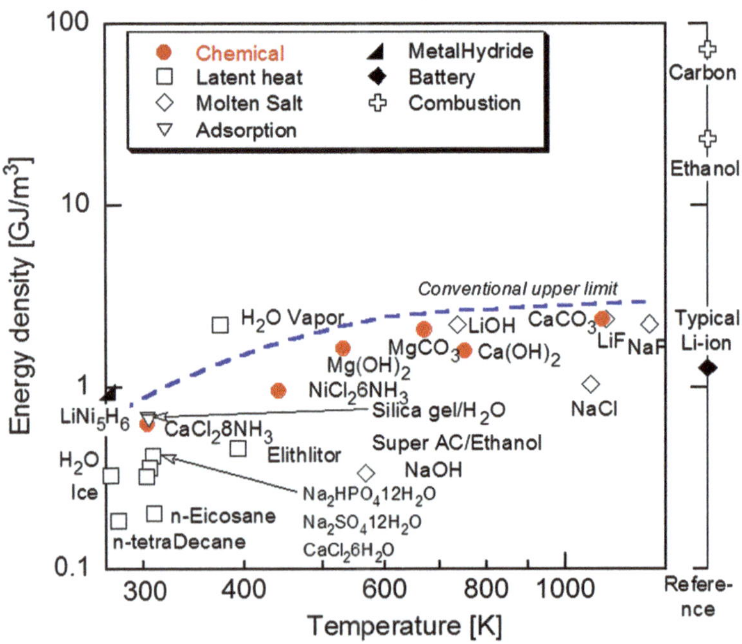

Fig. 3 Map of volumetric density of thermal energy storage of energy materials vs. operational temperature (chemical change is based on the product, considering practical particle vacancy)

has strong Gibbs-free energy change, so it is difficult to use it for chemical storage because of small reversibility. Thus, reversible chemical reactions such as magnesium oxide/water (MgO/H_2O) and calcium oxide/water (CaO/H_2O) are candidates for thermochemical energy storage (TCES). TCES is expected to have heat storage capability for temperatures >200 °C over longer periods, with greater density than conventional heat storage. The operational temperature range of TCES depends on the reaction equilibrium and can be extended by the development of new reversible reaction systems. A gas/solid reaction system is a candidate for TCES. Solid reaction material generally has low thermal conductivity, and heat conduction resistance between the material and heat exchange plate is considerable. Heat transfer enhancement in a reactor bed and the heat exchange surface are important for high-performance TCES and heat storage systems [4]. Chemical reaction in TCES induces material deformation and reduction of kinetic performance of the material during repetitive operation. Development of TCES material with high thermal conductivity and durability is a major consideration and risk of the technology.

2.3 Benefit and Attractive Future Vision

Industrial waste heat at >200 °C of 1250 PJ/year (=40 GW) is emitted in Japan. TCES for heat storage at these temperatures is expected to be developed for solar thermal energy and industrial waste heat, instead of sensible and latent heat storage. If heat output from CSPs with an installed capacity of 3 GW is capable of storage by TCES, 9.5 PJ/year of heat is stored if one assumes that a yearly operation ratio is 15 % for CSP and that two thirds of the CSP heat is stored by TCES. If one assumes that 1 % of waste heat at >200 °C is recovered by TCES in 2050, 13 PJ/year of heat (=400 MW) would be saved. Then, potential heat storage and saving capacities >10 PJ/year would be expected for TCES. However, a TCES market has not been established. Development of a practical TCES system and creation of a new market for heat storage will be required. For practical waste heat recovery, not only heat storage but also heat transfer and transportation technologies are needed. Efficient combination and optimization of these technologies are required for waste heat recovery in future energy systems.

3 Latent Heat Transportation

3.1 Present Status

Latent heat transportation systems containing phase-change particles in water have been developed for thermal energy transportation systems. Latent heat transportation systems can be classified into two categories, namely, tank truck and pipeline.

Tank truck transportation systems are currently used for higher-temperature transportation from 100 to 200 °C, as in industry [5]. In this capacity, polysaccharides such as pentaerythritol are primarily used as the phase-change material. The tank truck system does not require any special infrastructure, and so is low-cost. However, trucks use petroleum for transportation. Therefore, they cannot earn the distance to transport thermal energy from the perspective of balance in consuming petroleum energy. Meanwhile, a pipeline system is used for low-temperature transportation from −5 to 15 °C, as in cooling systems. Latent heat transportation systems with pipelines have high heat density compared with water-sensitive thermal energy transportation systems. Thus, flow rate can be reduced and pumping power savings can be realized, although high-cost pipeline infrastructure is required. For this purpose, ice/water and clathrate hydrate slurries are used in actual systems. The ice/water system uses the cheapest material. In Europe, the system is used for supermarket display systems. However, the ice particles readily agglomerate, and this can cause severe accidents from pipe blockage. To prevent this agglomeration, a brine system is widely used. The addition of brines such as ethanol or ethylene glycol prevents the solidification of water between particles and resulting agglomeration. However, the addition of brines depresses the freezing point. At <0 °C, the efficiency of refrigerators dramatically decreases. The ice/water system has a problem when its phase-change temperature is <0 °C, which is too low for household cooling systems. Conversely, clathrate hydrates have a phase-change temperature suitable for residence cooling [6]. Among these, tetrabutylammonium chloride hydrate is used in actual building cooling systems. Its phase-change temperature ~11 °C is suitable for cooling systems, and particles are well dispersed in water. However, this material has a weak toxicity and is expensive compared with other phase-change materials.

3.2 Technology Roadmap

The tank truck transportation system cannot be expected to be feasible for long-distance transportation. Thus, the pipeline transportation system is anticipated as a future thermal energy transportation system. For ice slurries, antifreeze protein is suggested as a new dispersant of ice particles [7]. Dispersion and controlled particle size can be achieved by adding small amounts of this material. It is very expensive at present because it can only be obtained from the blood of fish living in polar areas, but its synthesis has been investigated and is expected to have lower cost in the near future. Some emulsion systems transporting phase-change materials are also under research for cooling systems [8]. Such systems readily separate the water and oil phases, and emulsion stability has been improved. In Japan, energy consumption for house heating systems is much greater than that for cooling. Thus, it is important to develop latent heat transportation systems for high temperatures (~50 °C). A type of inorganic hydrate system is the only suggestion for this purpose

[9] because particle growth occurs at room temperature, although the particle fraction can be controlled in operation. This causes severe accidents from pipe blockage upon system restart. However, some techniques for preventing particle growth at room temperature have been developed. Encapsulation techniques of phase-change materials have also been elaborated. Although it has a cost disadvantage, the particle growth can be completely prevented, and some toxic and corrosive materials can be contained as phase-change materials. Additionally, the extent of supercooling decreases in a microscale space. Polymer-shell microencapsulated phase-change materials containing alkanes have been recently provided for cooling systems [10]. The polymer shell has mechanical weakness, but material for the capsule shell has been improved. Recently, silica hard-shell microcapsules have been developed [11]. A polymer shell is not useful in a high-temperature environment. However, a silica hard shell can be incorporated in a much higher-temperature latent heat transportation system. At present, this has some problems of cost, production, and erosion in transportation systems. However, it has mechanical and thermal toughness, and so is promising for latent heat transportation in the future.

3.3 *Benefit and Attractive Future Vision*

In Japan, 2.5 EJ at <200 °C of waste thermal energy is exhausted from power plants and industries. The temperature range of such energy is 50–200 °C (Fig. 2). This range is unsuitable for conversion from thermal to electric energy but is suitable for house heating systems. Just as lower-temperature thermal energy for cooling systems can be obtained via heat pumps using medium-temperature thermal energy such as chemical heat pumps and absorption chillers, such waste thermal energy can also be used for residential cooling systems. Total energy consumption for heating and cooling systems in residential use is <2.0 EJ. If all waste thermal energy were used for residential heating/cooling systems, CO_2 could be reduced by 17 % [12]. However, there is also a time and space thermal gap, which is that between supply/demand of thermal energy in time and space. The latent heat transportation system can be used to solve such a thermal gap problem, owing to its high heat density and temperature sustainability (less temperature variation).

There are some district heating/cooling and building air-conditioning systems in Japan. In most of these systems, water is used as the heat transportation medium, and this consumes 400 PJ per year. The pumping energy consumption is about 15 % of total consumption in these systems. If latent heat transportation was used instead of water-sensitive energy transportation in these systems, there could be a minimum 10 % total energy consumption (=40 PJ) reduction. In the near future, the waste thermal energy could be shared with people in Japan (Fig. 4), via a scheme called thermal grids. The latent heat transportation system is key technology for realizing this scheme.

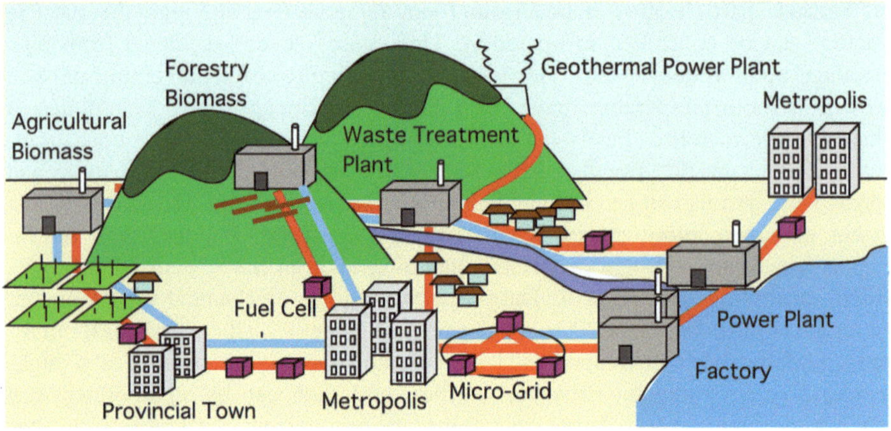

Fig. 4 Thermal grids with latent heat transportation system

4 Heat Transfer

4.1 Present Status

Primary energy in Japan is mostly supplied by fossil fuels, for which the exergy ratio is nearly 100 %. However, final energy utilization in terms of exergy is estimated at only ~1/3 of the primary energy supply. Thus, nearly two thirds of total energy is wasted during energy conversion and transport processes. Fossil fuels are initially combusted to produce heat in most cases, and then the heat is transferred to other heat carriers. It is known that the exergy ratio drastically declines at $T<1500$ °C, which is the typical utilization temperature range of human civilization [13]. Therefore, temperature difference during heat transfer is considered as the intrinsic cause of the exergy loss. Of course, heat will not transfer without a temperature difference, but it is very important to strive toward its reduction. To reduce temperature difference during heat exchange, it is necessary to increase the product of overall heat transfer coefficient K and heat transfer area A [14]. Without K improvement, A and cost will increase. This is why heat transfer enhancement is of great importance [15]. For example, a heat exchanger accounts for around 10–20 % of the total system cost of air conditioners and is one of the most expensive system components along with the compressor and electronic substrate. In general, heat transfer enhancement of gas phase is the major issue, because thermal conductivity of gas is much smaller than those of liquid and solid phases. Figure 5 shows heat transfer enhancement technologies for single-phase flow. Interrupted fins for air conditioners and automobile radiators at low Reynolds numbers, along with turbulent promoters for gas turbine blade cooling at high Reynolds numbers, are widely used. However, for new applications such as waste heat recovery and industrial heat exchangers, heat transfer enhancement techniques

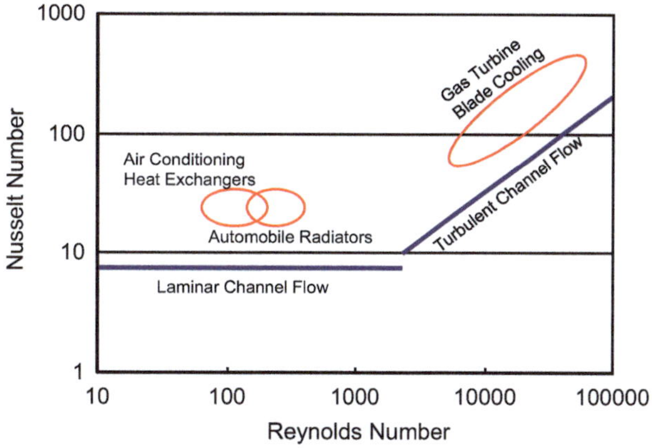

Fig. 5 Heat transfer enhancement technologies

at moderate Reynolds numbers are necessary. Development of novel heat transfer enhancement techniques for such Reynolds numbers is essential.

4.2 Technology Roadmap

Research and development of heat exchanger technologies have been limited to several typical applications such as air conditioning, automobiles, and gas turbines. However, the importance of using heat such as waste and renewable heat, which have not been widely used in the past, is increasing faster than ever. These heats will have conditions different than conventional applications, e.g., temperature, flow rate, and fouling. In addition, energy savings and efficiency will be of even greater importance in the future. For example, the coefficient of performance (COP) of air conditioners is expected to exceed 8 as a stock basis by 2050 [16]. These social needs will require new technologies for heat exchanger design and heat transfer enhancement approaches. Expensive materials such as copper will be replaced by less expensive ones like aluminum. Use of different materials entails different manufacturing processes, which will increase large equipment investment and large-scale mass production. In general, sophisticated heat transfer enhancement methods tend to be less tolerant of fouling, such as scaling, frost, dust, and drainage. Fouling will be one of the remaining challenges for next-generation heat exchangers, and surface treatment and coating technologies will be extremely important. Furthermore, designs using materials with small thermal conductivities such as stainless steel and plastics will become important.

A roadmap for a heat exchanger of an air conditioner is introduced as a typical example of the technology development. Present-specific air-side heat transfer areas are about 1500 and 1200 m^2/m^3 for indoor and outdoor unit heat exchangers, respectively. These values are estimated from specifications of commercial heat exchangers on the market. Then, specific air-side heat transfer areas in the future were estimated based on extrapolation of past technological progress. In 2020, replacement of copper with aluminum will proceed, giving 1500 and 1200 m^2/m^3. By 2030, antifouling technologies against drainage and frost will progress, and 2000 and 1500 m^2/m^3 will be achieved. Finally, by 2050, with novel surface modulation and coating technologies, specific heat transfer areas are expected to increase by 2400 and 1700 m^2/m^3. In addition to heat exchanger compactness, heat transfer enhancement technologies for single-phase laminar flow will progress substantially with the aid of computational fluid dynamics and optimization technologies. Direct numerical simulation with novel optimization algorithms will make it possible to optimize complex three-dimensional surfaces of the heat exchanger. The overall heat transfer coefficient of the indoor heat exchanger will therefore be expected to continuously increase from 80 to 140 $W/(m^2 K)$ between 2015 and 2050. The corresponding increase for the outdoor heat exchanger will be from 50 to 100 $W/(m^2K)$, a very challenging target in frost-forming conditions. These values were also forecast by extrapolating past technological progress.

4.3 *Benefit and Attractive Future Vision*

As described above, there is a trade-off between reducing the temperature difference and cost. The solution for this problem is heat transfer enhancement. This enhancement reduces not only the cost of heat exchanger size but also that of the chassis and housing, which is often more beneficial to the complete system. However, heat transfer enhancement techniques for new applications are not fully mature. In addition, the importance of using less expensive materials is emphasized. New technologies using new materials that have difficulties of application to heat exchangers, such as aluminum, steel, or plastics, will be very important.

5 Summary

Figure 6 shows an energy technology roadmap of heat storage, transportation, and transfer based on the discussion in Sec. 2-4. Waste heat recovery has great potential in Japan. Efficient heat usage requires optimized technologies combining heat storage, transportation, and transfer. New and related technologies are required to supersede conventional heat recovery technologies. In particular, development of heat storage materials and heat transfer enhancement technologies achieving cost-effectiveness is required for sensible and latent heat and thermochemical energy storage. Heat transfer enhancement reduces not only the cost of heat exchanger size but also that of the chassis and housing, which is often more beneficial to the complete system. A consistent design of low-cost innovative materials, reactors, heat exchangers, and heat utilization systems originating from research and development is an ideal technology goal.

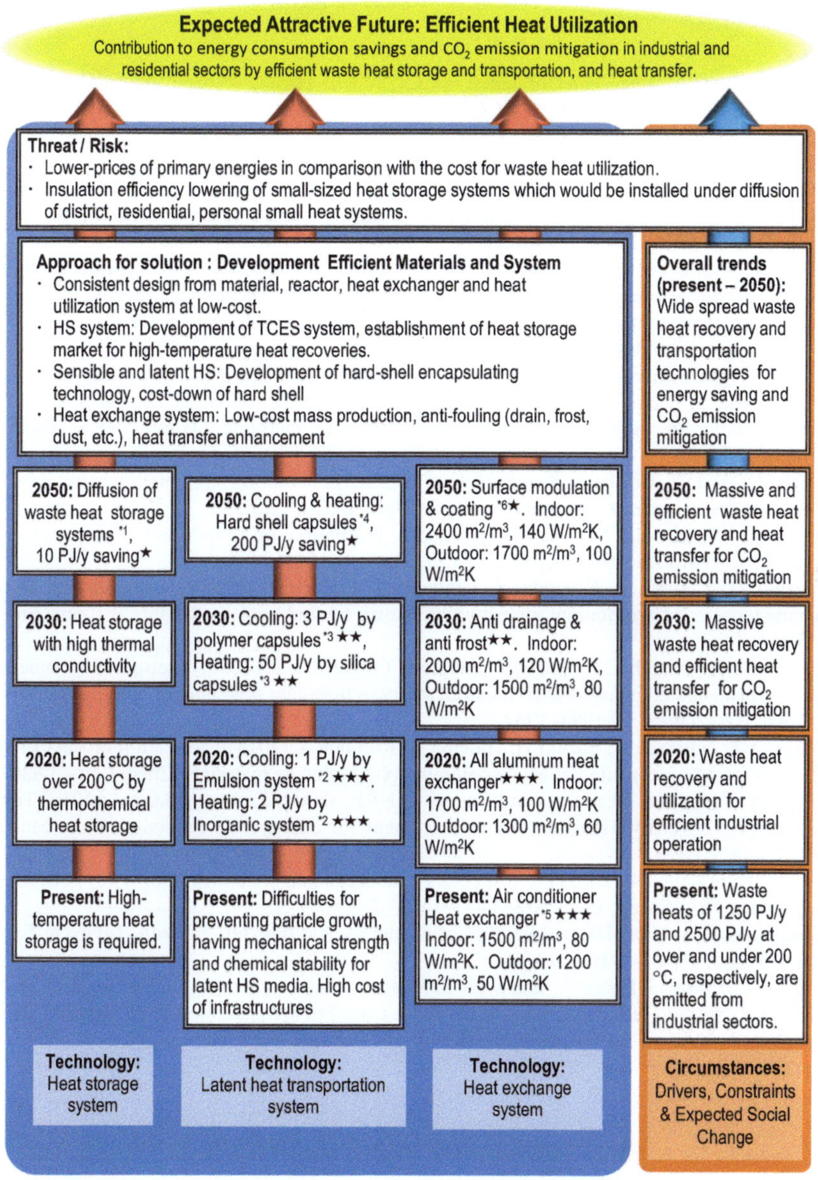

*1 Heat storage for concentrated solar power system of 3 GW. *2 Large water sensitive heat transportation systems will be replaced by latent heat transportation systems. *3 Encapsulated phase change materials will be replaced to for smaller systems as building air-condition systems. *4 Hard shell microcapsules having advantage in mechanical toughness will catch up polymer microcapsules and replace them. *5 Present heat exchanger values are estimated from the specifications of the commercial heat exchangers in the market. *6 Estimated from the specifications of the present heat exchangers in the market.

Abbreviations: HS: Heat Storage, TCES: Thermochemical Energy Storage

Fig. 6 Energy technology roadmap of heat storage, transportation, and transfer

References

1. Energy Conservation Center of Japan (ECCJ) (2004) Project research report for promoting energy conservation technology, ECCJ, Tokyo
2. Ono E (2010) 3rd meeting for market development of environmental friendly car. Ministry of Environment of Japan, Tokyo
3. Koll G, Schwarzbözl P, Hennecke K, Hartz T, Schmitz M, Hoffschmidt B (2009) The Solar Tower Jülich – a research and demonstration plant for central receiver systems. Proceedings of SolarPACES 2009, Berlin, Germany
4. Kato Y, Zamengo M, Fujioka K (2014) Effect of thermal conductivity enhancement of thermochemical energy storage material on unused heat utilization system. Proceedings of The 15th International Heat Transfer Conference (IHTC-15), Kyoto, Japan
5. Hauer A, Gschwander S, Kato Y, Martin V, Schossig P, Setterwall F (2010) Transportation of energy by utilization thermal energy storage technology. ECES Annex 18 Final Rep
6. Fukushima S, Takao S, Ogoshi H, Ida H, Matsumoto S, Akiyama T, Otsubo T (1999) Development of high-density cold latent heat with clathrate hydrate. NKK Tech Rep 166:65–70
7. Inaba H, Inada T, Horibe A, Suzuki H, Usui H (2005) Preventing agglomeration and growth of ice particles in water with suitable additives. Int J Refrig 28:20–26
8. Schalbart P, Kawaji M, Fumoto K (2010) Formation of tetradecane nanoemulsion by low-energy emulsification method. Int J Refrig 33:1612–1624
9. Suzuki H, Konaka T, Komoda Y, Ishigami T, Fudaba T (2012) Flow and heat transfer characteristics of ammonium alum hydrate slurry treated with surfactants. J Chem Eng Jpn 45:136–141
10. Alvarado JL, Marsh C, Sohn C, Phetteplace G, Newell T (2007) Thermal performance of microencapsulated phase change material slurry in turbulent flow under constant heat flux. Int J Heat Mass Transf 50:1938–1952
11. Toyoda T, Narisada R, Suzuki H, Hidema R, Komoda Y (2014) Fabrication process of silica Hear-Shell Microcapsule (HSMC) containing phase-change materials. Chem Lett 43:820–821
12. Agency of National Resources and Energy (ANRE) (2008) Energy in Japan, ANRE, Tokyo
13. Shikazono N (2013) Heat transfer enhancement for exergy loss reduction. Func Mater 33 (7):13–19
14. Shah RK, Sekulic DP (2003) Fundamentals of heat exchanger design. John Wiley & Sons, Inc., Hoboken, New Jersey, ISBN 978-0-471-32171-2
15. Webb RL, Kim N-H (2005) Principles of enhanced heat transfer, 2nd edn. Taylor & Francis, New York, ISBN 1-59169-014-5
16. Fujino J, Hibino G, Ehara T, Matsuoka Y, Masui T, Kainuma M (2008) Back-casting analysis for 70% emission reduction in Japan by 2050. Clim Pol 8(Supplement):S108–S124

Hydrogen Production

Hiroshige Matsumoto, Seiichiro Kimura, Kenshi Itaoka, and Gen Inoue

Abstract Hydrogen production methods to meet hydrogen demand as a future fuel are considered. Current hydrogen production methods are described, and energy efficiency, CO_2 emissions, and cost are discussed. After estimating possible future hydrogen use and demand, various hydrogen production methods meeting future hydrogen demand are addressed and their prospects considered.

A brief conclusion is that future demand for hydrogen fuel cell electric vehicles can be met by conventional fossil fuel-based hydrogen production methods, but novel low-carbon techniques for this production using biomass, renewable energy-based electrolysis, thermochemical methods, and photoelectrochemical water splitting are important to reduce CO_2 emissions. The introduction of hydrogen energy provides benefits of energy saving, renewable energy use, and stabilization of energy security.

Keywords Fossil fuel-based hydrogen • Low-carbon hydrogen • GHG emission • FCEV • Water splitting

1 Introduction

Hydrogen is a multi-aspect chemical species. It is an important raw material in various industries and can work as a fuel. With regard to the latter aspect, hydrogen is a secondary energy, and, accordingly, hydrogen production is a process that converts a primary energy to the chemical energy of hydrogen. Fossil fuel is

H. Matsumoto (✉) • K. Itaoka
Kyushu University, Fukuoka, Japan
e-mail: matsumoto@i2cner.kyushu-u.ac.jp; k.itaoka@i2cner.kyushu-u.ac.jp

S. Kimura
The Matsushita Institute of Government and Management, Kanagawa, Japan
e-mail: s-kimura@mskj.or.jp

G. Inoue
Kyoto University, Kyoto, Japan
e-mail: ginoue@cheme.kyoto-u.ac.jp

Y. Kato et al. (eds.), *Energy Technology Roadmaps of Japan*,
DOI 10.1007/978-4-431-55951-1_9

currently a main primary energy for the production of hydrogen. Hydrogen is produced by reforming hydrocarbons and is generated as a by-product in steel, soda, petroleum refining, and other industries. Renewable energy, such as sunlight, wind, and biomass, can be used as primary energy.

Hydrogen production should be conducted using suitable primary energy with respect to the social background, such as the severity of CO_2 emission, primary energy cost, and others. Hydrogen fuel cell electric vehicles (FCEVs) have recently been put on public sale in Japan. Toyota's FCEV "Mirai" started sales in December 2014 [1], and Honda announced the release of a new FCEV in 2015 [2]. Hydrogen has increased in importance as energy. In this chapter, hydrogen production methods for demand as a fuel and/or energy medium are addressed.

2 Present Status

2.1 Current Hydrogen Usage and Production Methods

Hydrogen is mainly used as a raw material in industries, including petroleum refining, synthesis of ammonia or other chemicals, metal and ceramics, food, semiconductor production, and others. Hydrogen is mostly produced by reforming fossil fuels or as by-products. Water electrolysis is also a production method. The annual amount of hydrogen produced in Japan is $15–20 \times 10^9$ Nm^3 [3, 4]. Most of the hydrogen is consumed captively, i.e., the consumers produce it.

2.1.1 Reforming

Hydrogen is produced by reforming natural gas, naphtha, and other hydrocarbon-based fuels. Steam reforming and partial oxidation are the two major chemical reactions. These reactions can be written as follows:

$$C_mH_{2n} + mH_2O \rightarrow mCO + (m+n)H_2(\text{steam reforming}) \quad (1)$$

$$C_mH_{2n} + 0.5mO_2 \rightarrow mCO + nH_2(\text{partial oxidation}) \quad (2)$$

Steam reforming is endothermic and partial oxidation is exothermic. In both cases, generated CO can be used to produce more hydrogen as

$$CO + H_2O \rightarrow CO_2 + H_2(\text{water gas shift reaction}) \quad (3)$$

Steam reforming of methane proceeds spontaneously at temperatures >700 °C and is done industrially at 850–950 °C and pressures 1.5–3 MPa. Because the reaction is endothermic, there is a need for heating reaction tubes, determining the upper limit of operation temperature. Partial oxidation has large negative free energy change and can thus proceed non-catalytically. Higher reaction temperatures

(1300–1500 °C) and pressures (4–8 MPa) can be used industrially, enabling use of heavy hydrocarbons such as heavy oil and coal. Lower temperatures are favored by the water-gas shift reaction, and 400–600 °C are typical reaction temperatures. Steam reforming and partial oxidation can be mixed in an appropriate ratio for thermal balance, called the autothermal method.

In these reforming processes, hydrogen is produced as a mixture with CO_2, H_2O, and other impurities, and purification is necessary, usually done with pressure swing adsorption (PSA).

Petroleum refining consumes hydrogen in transforming heavier hydrocarbons to lighter species and in desulfurization. Hydrogen is produced by steam reforming of naphtha and LPG. Ammonium synthesis consumes hydrogen, which is typically produced by steam reforming of natural gas.

2.1.2 By-Product Hydrogen

Hydrogen is generated as a by-product in industrial processes, including sodium hydroxide production, reduction of iron ores to produce steel, petroleum chemistry, and others. In many cases, the by-product hydrogen is consumed captively as fuel and as a chemical raw material, and so the use as fuel other than captive consumption requires an alternative energy resource to replace current usage of such hydrogen.

In petroleum refineries, hydrogen is generated during dehydrogenation of alkanes to produce aromatics. However, more hydrogen is needed and is produced by reforming petroleum-based hydrocarbons. In steelmaking, carbon made from coal, i.e., cokes, is used as a reducing agent of iron ore, and gas from the coke oven contains substantial hydrogen (~55 %). About 16×10^9 Nm^3 of coke oven gas (COG) is produced annually. Large amounts of COG are used for electricity generation or as heat sources and are not used in the form of hydrogen [4]. In the soda industry, NaOH is produced by electrolysis of NaCl aqueous solution and hydrogen, and chlorine is produced as a by-product. Annually, ~1.1×10^9 Nm^3 of hydrogen is generated, accompanied by 4×10^6 tons of NaOH [4]. This hydrogen has high purity and is commercially supplied to other companies.

2.1.3 Electrolysis

Alkaline water electrolysis and polymer electrolyte water electrolysis are the commercial methods. In Japan, the former was used in the past in hydrogen production for ammonia synthesis and is thus technically well established. However, the hydrogen production method for ammonia synthesis has been replaced by cheaper steam reforming of natural gas or other hydrocarbons.

2.2 Hydrogen Production Capability of Current Facilities

In steel production, the total amount of hydrogen contained in generated COG is 9×10^9 Nm^3/year in Japan. Only about a tenth is used as hydrogen gas [4], and the remaining 8×10^9 Nm^3/year can be used as fuel. In addition, COG contains methane and CO that can be reformed to produce another 17×10^9 Nm^3/year of hydrogen. If we assume 60 % recovery of hydrogen through purification by PSA, in principle, 5×10^9 Nm^3/year and 10×10^9 Nm^3/year of hydrogen can be produced, respectively. At present, COG is not a surplus but is used as fuel for thermal power and other heat sources. Therefore, COG can be considered economically as a hydrogen source only when cheaper fuel is available to substitute COG utilization.

COG is a product of coal through coke production, which is a CO_2-emiting process and must be accommodated by CO_2 separation to avoid such emission. As mentioned in Sect. 2.1.2, about 1.1×10^9 Nm^3 of hydrogen is generated as a by-product in the soda industry with high purity and can be used as a hydrogen source. Petroleum refineries are equipped with hydrogen production facilities that are capable of excess hydrogen production (i.e., planned hydrogen production capability minus actual hydrogen consumption) using steam reforming of naphtha and LPG, owing to the closing of oil refineries and suspension of ethylene production facilities [5]. The estimated capability is as much as 4–6 $\times 10^9$ Nm^3/year [3, 4]. Steam reforming of natural gas can be used to produce hydrogen, as explained in Sect. 2.1.1. The Mizuho Information and Research Institute estimates that hydrogen that can be newly produced is 7–8 $\times 10^9$ Nm^3/year [4].

The amounts of hydrogen that can be generated by the methods discussed above are summarized in Table 1. Given the hydrogen demand assumed later in Table 6, the hydrogen need of FCEVs would be 13.5×10^9 Nm^3/year even in 2050 and is affordable because of the by-product steam-reformed hydrogen shown in Table 1. These conventional hydrogen production methods can address the assumption of FCEV hydrogen demand. However, this production is accompanied by CO_2 emission, because fossil fuel is the energy source. For reduction of this emission, hydrogen production should shift to non- or low-carbon hydrogen in the future, and related production techniques are explained in the following sections.

Table 1 Hydrogen production/supply capability of by-product hydrogen and steam reforming

Hydrogen source	Hydrogen production capability [10^9 Nm^3/year]
Steel industry	5 (15[a])
Soda industry	1.1
Petroleum refineries	4–6
Steam reforming of natural gas	7–8
Total	17–20 (30[a])

[a]CO and CH_4 contained in COG are considered hydrogen sources

2.3 *Comparison of Hydrogen Production Methods*

To compare energy efficiency, CO_2 emissions, and costs, the following energy pathways from natural gas, crude oil, coal, biomass, and electricity as primary energy for vehicle fuel tanks are considered for hydrogen production. Energy efficiency is defined as process efficiency of hydrogen production. CO_2 emissions and costs should include those associated with all processes from the well to hydrogen production. Hydrogen purification is not considered, because the required hydrogen purity varies with use. Hydrogen compression and filling are also not included in Fig. 1.

2.3.1 Energy Efficiency

Energy efficiency of steam reforming is defined as (hydrogen energy)/[(feedstock energy)+(input energy)] and that of water electrolysis as (hydrogen energy)/[(electricity)+(input energy)]. Values were collected from reference data of the Mizuho Information and Research Institute [6], US Department of Energy Hydrogen Analysis Project (H2A) [7], and Japan Hydrogen and Fuel Cell Demonstration Project (JHFC) [8]. Largest and smallest mean values are summarized in Fig. 2. Data in reference [8] include those under research, and these are included in the figure. In steam reforming, energy efficiencies are higher for reforming of methanol, dimethyl ether (DME), and natural gas than those for reforming of others. Efficiency is lowest for kerosene and Fischer-Tropsch (FT) synthetic fuel. The energy efficiency of alkaline and proton exchange membrane (PEM) water electrolysis is comparable to that of steam reforming.

2.3.2 CO_2 Emissions

CO_2 emissions are defined as the total amount of CO_2 emitted in feedstock production, transportation, and hydrogen production. Greenhouse gases (GHG) other than CO_2 are also taken into account as CO_2 equivalent values. The Institute of Energy Economics Japan [9], Mizuho Information and Research Institute [6], H2A [7], and JHFC [8] have estimated CO_2 emissions with different feedstock, and these are summarized in Figs. 3 and 4. Data in reference [8] include those under research and these are included in the figure. Reference [7] deals with the situation in the USA, but its data are used in the figures because the percentage of CO_2 emissions in feedstock production and transportation of the total are not significantly different from those in the other references dealing with Japan (e.g., in natural gas reforming, CO_2 emissions from mining, and transportation are 12 % in the USA [4] and 14 % in Japan [5]). For water electrolysis, CO_2 emissions for electricity production are included, which is assumed to be conducted in Japan even using the data of H2A [7]. CO_2 emissions of construction and maintenance equipment are not considered.

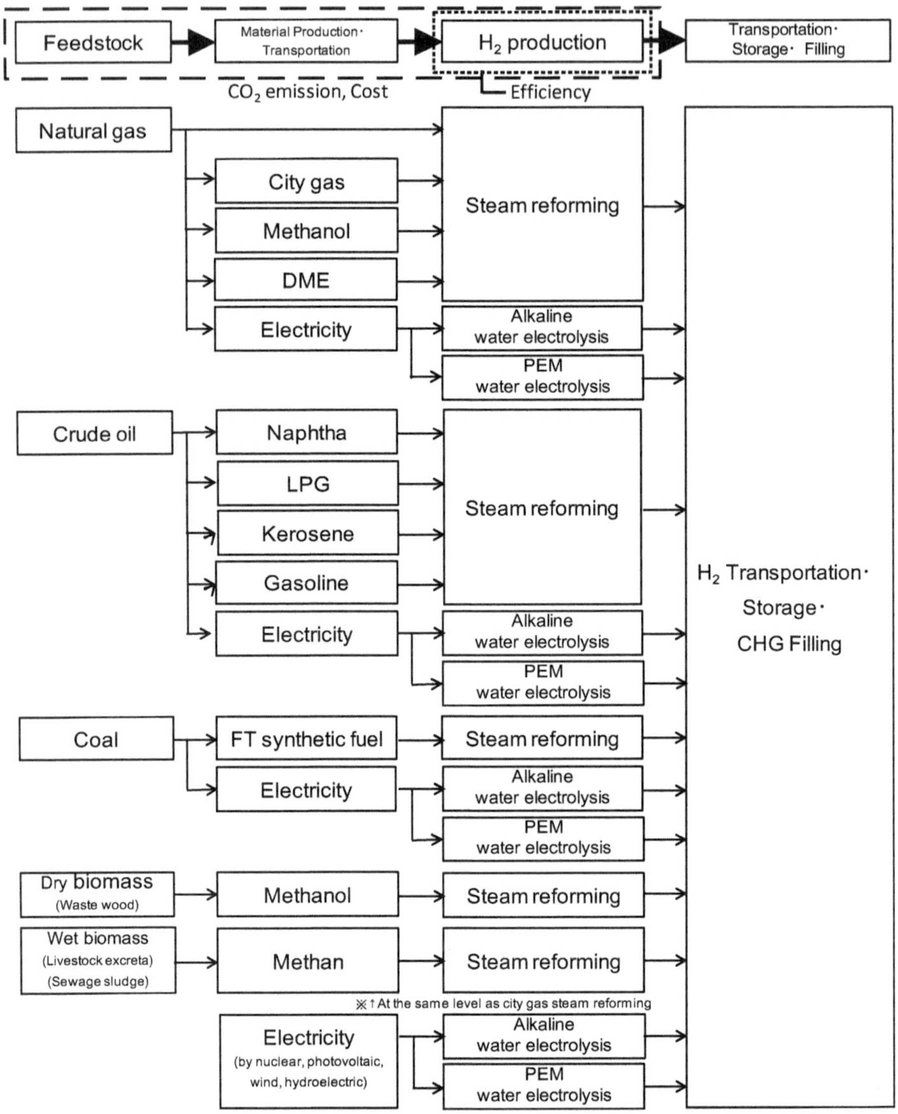

Fig. 1 Energy paths of assumed hydrogen production for comparing efficiency, CO_2 emissions, and costs

2.3.2.1 Steam Reforming

CO_2 emissions for hydrogen production from various feedstocks are shown in Fig. 3. Autothermal reforming is not included. The breakdown of total CO_2 emission, such as natural gas reforming, is ~10 % from material production and transportation, 60 % from derivation from fuel, and 30 % from operation of hydrogen production equipment. In steam reforming, except in the case of biomass, CO_2

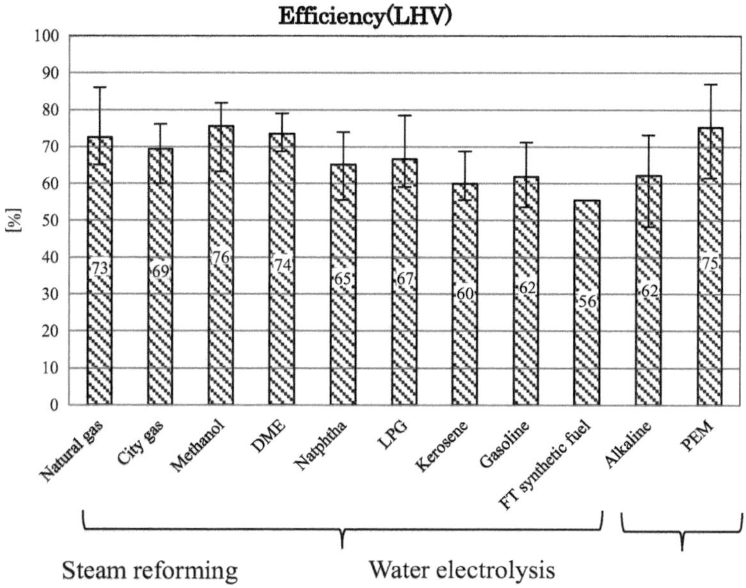

Fig. 2 Efficiency of hydrogen production from various fuel-reforming processes and electrolysis

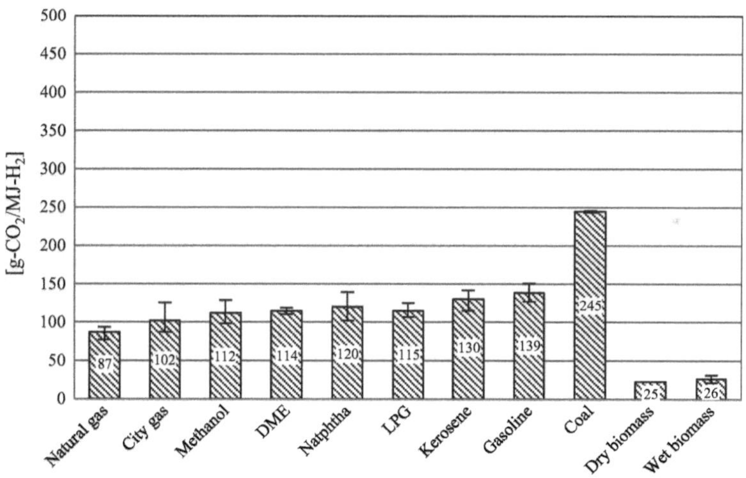

Fig. 3 CO_2 emissions for hydrogen production from various feedstocks by steam reforming

emitted from natural gas reforming is the least, and that from reforming of coal is the greatest. Emissions from biomass are less than from other fuels and are only 30 % of those from natural gas. Challenges include efficiency improvement in hydrogen production, increased biomass use, and technology development in CO_2 capture and storage for further emissions reduction.

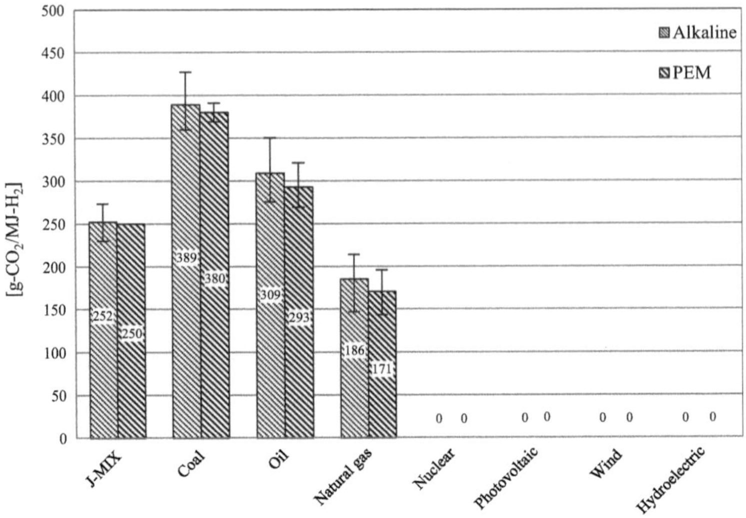

Fig. 4 CO_2 emission from electricity for unit water electrolysis, derived from various fuels/ primary energy sources

2.3.2.2 Water Electrolysis

CO_2 emissions from water electrolysis using electricity derived from various fuels/ primary energy sources are shown in Fig. 4. For J-Mix, the Japanese average electricity mix in 2013 reported in Electricity Generation by Source (2014) [10] is assumed to consider the effect of cessation of nuclear power plants. The breakdown of CO_2 emissions, such as alkaline water electrolysis derived from natural gas, is ~60 % from electricity production from natural gas, 10 % from derivation from fuel, and 30 % from operation of hydrogen production equipment. In water electrolysis, CO_2 emission is slight compared with steam reforming if electricity from renewable energy is used.

2.3.3 Costs

In the present study, cost is based on hydrogen production at 70 MPa onsite and a 300 Nm^3/h direct-filling commercial hydrogen station operated by JHFC [11]. Tables 2, 3, and 4 show presumed hydrogen production conditions. Costs related to hydrogen compression and filling are not included.

Table 2 Presumed conditions

Scale of hydrogen station[a]	Onsite 300 Nm^3/h
Operating condition[a]	365 day/y, 13 h/day
Annual hydrogen production volume[a]	1,538,396 Nm^3
Construction cost of hydrogen production equipment (including PSA)[a]	186 million yen
Hydrogen purity[a]	99.99 % (4 N)

[a]JHFC [11]

Table 3 Variable costs

		Unit price		Annual usage	
City gas[a,b]		114	Yen/kg	441,890	kg
Clean water[a]		300	Yen/ton	3428	ton
Waste water[a]		200	Yen/ton	1714	ton
Electricity[a]				1,141,251	kWh
	J-Mix[c]	16	Yen/kWh		
	Wind[d]	14	Yen/kWh		
	Hydro[d]	21	Yen/kWh		
	Solar[d]	36	Yen/kWh		

[a]JHFC [11]
[b]Calculated from major contracts of city gas (Osaka gas, Tokyo gas, Saibu gas)
[c]Calculated from the Federation of Electric Power Companies of Japan [10]
[d]Calculated from the Energy and Environment Council [12]

Table 4 Fixed cost

Labor[a]	7 million yen annually per person, 1.5 persons employed
Book value[a]	55 % of construction cost
Depreciation[a]	10 years
Repair[a]	3 % of construction cost
Insurance[a]	0.77 % of book value
Fixed assets tax[a]	1.4 % of book value
Overhead[a]	0

[a]JHFC [11]

Table 5 compares estimated costs for hydrogen production via steam reforming and alkaline water electrolysis with various operation rates. Hydrogen produced via water electrolysis is more expensive than via steam reforming. For this electrolysis, percentages of electricity versus total costs are 72 % at 100 % operation rate and 61 % at 60 % operation rate. The price of the electricity is a key factor determining the cost of this electrolysis and is the reason for its higher cost of hydrogen production relative to steam reforming.

Table 5 Estimation of cost for hydrogen production via steam reforming and electrolysis

Technology	Operation rates (100 % = 365 day/year, 13 h/day)	Production cost [yen/Nm^3]
Steam reforming (city gas)	100 %	Approx. 60
	80 %	Approx. 66
	60 %	Approx. 76
Alkaline water electrolysis	100 %	Approx. 109 (J-Mix)
		Approx. 99–208 (wind, hydro, solar)
	80 %	Approx. 116 (J-Mix)
		Approx. 107–215 (wind, hydro, solar)
	60 %	Approx. 130 (J-Mix)
		Approx. 118–226 (wind, hydro, solar)

3 Technology Road Map

3.1 Perspectives on Future Hydrogen Demand in Transportation and Large-Scale Energy Conversion Sectors

Use of hydrogen as energy will increase in the future, with FCEVs as an example. FCEVs have recently been introduced in Japan. Assuming that an FCEV is driven 12,000 km/year (an estimate from ordinary cars in the country) and consumes 1 kg of hydrogen per 100 km distance driven [13], the amount of hydrogen is estimated at 1.35×10^9 Nm^3/year/million-FCEVs. The Fuel Cell Commercialization Conference of Japan (FCCJ) published a scenario for the commercialization of FCEVs [14]. The scenario specifies two million FCEVs and construction of ~1000 hydrogen stations by 2025. Table 6 shows an estimation of hydrogen demand as fuel for FCEVs by 2020, 2030, and 2050. The numbers of FCEVs are estimated from the FCCJ scenario.

Another use of hydrogen is for electricity generation. Economic efficiency may be a problem, however. Hydrogen is considered a potential option for use as a fuel for thermal power plants in the form of either a mixture with other fuels or pure hydrogen [3, 4]. There is an estimate that hydrogen demand will increase by another 22×10^9 Nm^3/year by 2030, assuming that all newly built thermal power plants adopt a fuel mixture containing 50 % hydrogen [4]. If this is realized, such hydrogen demand for thermal power would exceed the demand of FCEVs (Table 6). However, hydrogen use by thermal power plants results in the reduction of CO_2 emission, so hydrogen should be produced from low-carbon energy sources or accompanied by carbon capture and sequestration (CCS).

Table 6 Estimated hydrogen demand in 2020, 2030, and 2050

Year	Hydrogen demand [10^9 Nm^3/year]	Breakdown
2020	0.5–1.1	
	0.5–1.1	0.4–0.8 million FCEVs[a]
2030	2.7–27.0	
	2.7–5.4	2–4 million FCEVs[1]
	0–21.6	Hydrogen turbine (mixed at 50 %)[b]
2050	13.5–35.1	
	13.5	10 million FCEVs[a]
	0–21.6	Hydrogen turbine (mixed at 50 %)[b]

[a]FCEV: 120 kg-H_2/car/year, giving 1.35×10^9 Nm^3/year/million-FCEVs
[b]According to a report from the Mizuho Information and Research Institute [4]

3.2 *Hydrogen Production Methods Under Development*

3.2.1 Membrane Reactor Reformer

For technical advancement of reforming, a membrane reformer equipped with a palladium-based alloy membrane is an advanced method for steam reforming of methane (Fig. 5) [15]. The aim is to conduct steam-reforming reactions and hydrogen separation processes simultaneously, without shift converters and purification systems. This results in greater energy efficiency (~80 %) and more compact equipment size than those of conventional technologies and is in the validation phase.

3.2.2 Biomass

Biomass, i.e., wood from forestry, arboricultural activities or from wood processing, agricultural residues, food waste, industrial waste and coproducts from manufacturing and industrial processes, and sewage sludge, can be used as an energy source for hydrogen production. Figure 3 shows that hydrogen derived from biomass has low CO_2 emissions relative to fossil fuel-based hydrogen. In the case of woody biomass, autothermal reforming is used in hydrogen production. Anaerobic fermentation is conducted to produce methane followed by steam reforming.

There is an estimate for the domestic hydrogen production potential of wood, livestock waste, and agricultural residue that indicates ~1300 hydrogen stations can be operated based on these types of biomass across Japan [16]. Assuming 1000 FCEVs per hydrogen station, the estimated number of stations could manage 1.3 million FCEVs. Dispersion of locations and small amounts of generated biomass per location are the major drawbacks of this biomass. Accordingly, mass production of hydrogen from biomass corrected for a wide area will increase cost because of transportation of the biomass feedstock. Onsite hydrogen production from sewage

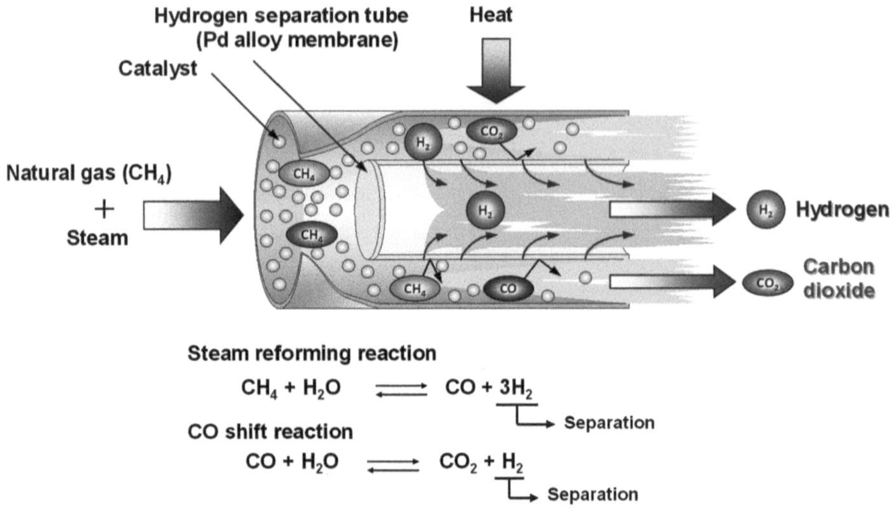

Fig. 5 Principle of a hydrogen separation reformer [15]

sludge for hydrogen stations will be an effective use of biomass. The Ministry of Land, Infrastructure, Transport and Tourism estimates a hydrogen production potential ~0.13×10^9 Nm^3/year [17]. There is an ongoing validation plant of hydrogen production from sewage sludge in the city of Fukuoka [13], which is an effective means of supply to hydrogen stations by locating stations near sewage treatment plants; the estimated hydrogen costs are 80–85 yen/Nm^3.

3.2.3 Thermochemical Water Splitting

One-step thermodynamic water splitting needs an extremely high temperature. The same molar amounts of steam, hydrogen, and oxygen are chemically in equilibrium at ~4100 °C [18]. However, the sequence of multiple chemical reactions combined with hydrogen separation enables complete water splitting at much lower temperatures [19, 20]. The iodine-sulfur (IS) process is one example and works when 900 °C heat is available (Fig. 6) [20, 21]. The Japan Atomic Energy Agency (JAEA) is verifying a combined system of the IS process and a high-temperature gas-cooled reactor (HTGR), which is a nuclear reactor using helium as the coolant [21]. A test plant with hydrogen production capability 200 NL/h has been constructed and tested. In *Nuclear Energy Vision 2100* proposed by the JAEA, hydrogen produced by an HTGR will be supplied to industries from around 2040 [22]. An HTGR with thermal output of 600 MW can produce 0.6×10^9 Nm^3/year of hydrogen.

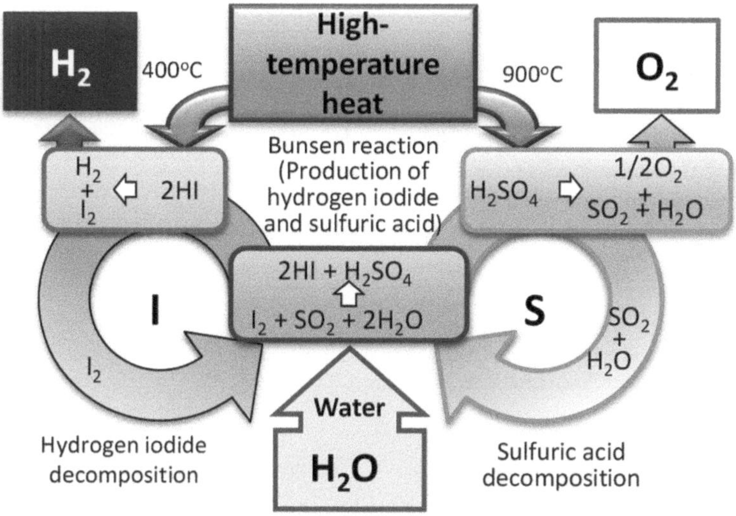

Fig. 6 Iodine-sulfur process. The full process is divided into three sections, Bunsen, H_2SO_4, and HI [21]. $SO_2 + I_2 + 2H_2O \rightarrow H_2SO_4 + 2HI$ (Bunsen reaction). $H_2SO_4 \rightarrow H_2O + SO_2 + 0.5O_2$ (sulfuric acid decomposition). $2HI \rightarrow H_2 + I_2$ (hydrogen iodide decomposition)

3.2.4 Water/Steam Electrolysis

Comparing CO_2 emissions from various hydrogen production methods (Fig. 4), water electrolysis can produce low-carbon hydrogen when low-carbon electricity is used. Therefore, water electrolysis based on low-carbon or renewable electricity is an important direction for future hydrogen production. A power-to-gas project in Germany is based on such a concept, using excess electricity for the production of hydrogen and methane [23]. In this case, the operation rate cannot be sufficiently high to affect the hydrogen production cost, because electrolyzers function only on sunny or windy days. In the case of Japan, as mentioned in Sect. 2.3.3, the price of electricity is the most significant in hydrogen production cost via water electrolysis; 61–72 % of that cost is from electricity for operation rates of 60–100 %. However, facility cost is another important contributor at 21 % (operation rate, 100 %) and 30 % (operation rate, 60 %). Further, the price of hydrogen varies with the operation rate (Table 5).

Electrolysis is an electrochemical water-splitting process and occurs on inputting electric energy greater than Gibbs free energy of water formation ($\Delta_f G$). Either liquid water or steam can be electrolyzed depending on the operation temperature. Figure 7 shows a thermodynamic breakdown of the energy for electrolysis conducted at 25 °C and 600 °C, in the cases of liquid water and steam electrolysis, respectively [24]. Alkali water electrolysis and polymer electrolyte water electrolysis are the two technically established methods. These typically operate at around 2 V, corresponding to 75 % energy efficiency (based on high heat value, HHV). There is a NEDO research project for hydrogen utility that has targeted electrolysis

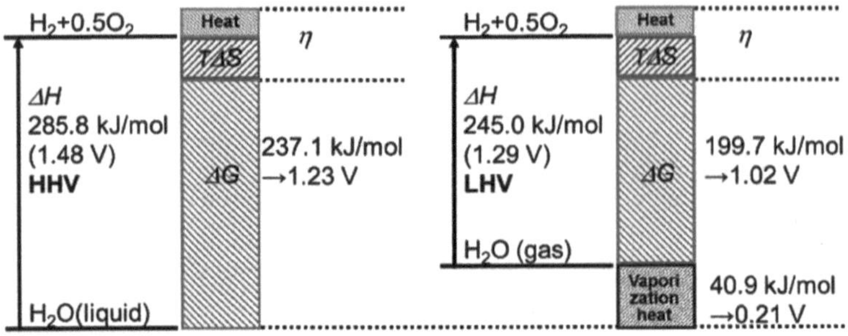

Fig. 7 Comparison of thermodynamics of water/steam electrolysis operated at 25 °C and 600 °C [20]. Minimum required electrolysis voltage is $\Delta_f G/2\,F$, where F is Faraday's constant and is a function of temperature, i.e., 1.23 V at 25 °C and 1.02 V at 600 °C

voltage 1.8 V (at 0.6 A/cm^2) to improve efficiency and thereby reduce hydrogen production cost [25]. As suggested by Fig. 7, steam electrolysis will need less electrolysis voltage. The NEDO project targets 1.6 V for steam electrolysis [25]. This type of effort should make water/steam electrolysis a practical hydrogen production method in combination with renewable energy.

The cost of hydrogen is estimated in Table 5. This cost is primarily a function of the price of electricity and energy efficiency of electrolysis. Therefore, electrolysis will become effective via cost reduction of renewable electricity, and improved energy efficiency of photovoltaic cells is crucial.

3.2.5 Photoelectrochemical Water Splitting

Water splitting can occur on the surface of a semiconductor in contact with water with the incidence of light when the semiconducting material is appropriately selected. This process is similar to photosynthesis that uses light to produce a chemical, i.e., sugar in natural photosynthesis. Since the discovery of photoelectrochemical water-splitting catalysis in anatase TiO_2 [26], extensive research has been conducted. The advantage of photoelectrochemical water splitting is its simplicity. The action of a combination of photovoltaic cells and water electrolyzers can be conducted simultaneously with only water, a catalyst, and sunlight. The drawback is energy efficiency; state-of-the-art efficiency is currently 2 % [27]. Such a low efficiency necessitates large areas of land.

3.2.6 Summary of Hydrogen Production Methods Under Development

The hydrogen production methods addressed in Sect. 3.2 are summarized in Table 7, with an assumed capability of hydrogen production. The most important feature is

Table 7 Low-carbon hydrogen production methods

Methods	Potential hydrogen production [10^9 Nm^3/year]	Remarks
Biomass	1.9[a]	Effective when used for local hydrogen stations
Thermochemical hydrogen	0.6	600-MW thermal HTGR is assumed as thermal source: 0.6×10^9 Nm^3/year/nuclear reactor
Water/steam electrolysis with renewable electricity	NA	Available for renewable electricity hydrogen conversion
Photoelectrochemical water splitting	NA	Direct conversion of solar energy to hydrogen

[a]1.75×10^9 Nm^3/year from wood, livestock waste, and agricultural residue and 0.13×10^9 Nm^3/year from sewage sludge (Sect. 3.2.2)

that they are low-carbon hydrogen production methods. This feature is of primary importance for the sustainability of daily life and is discussed in the next section.

3.3 *Road Map*

A road map for hydrogen production is summarized in Fig. 8 and is discussed here.

An assumed technological goal for hydrogen production is a supply of hydrogen meeting the demand of FCEVs and other hydrogen energy devices. Reduction of CO_2 (GHG) emissions is mandatory for the sustainability of society, so primary energy for the hydrogen production should be shifted from fossil fuel to renewable energy. The right-hand column in the roadmap chart (Fig. 8) explains social status relevant to hydrogen energy. At present, hydrogen energy is viewed from the perspective of energy saving and reduction of CO_2 emissions. Hydrogen-fueled FCEVs have been commercialized since 2014 [1] in Japan, and the number of FCEVs is assumed to increase as mentioned in Sect. 3.1. If we assume ten million FCEVs by 2050, this number will roughly correspond to 15–20 % of total automobiles nationwide. Accordingly, around 5000 hydrogen stations should be operated, assuming an average capacity of 2000 FCEVs per hydrogen station. When the above technological goal is fully achieved, hydrogen will be used in society as an energy medium for transportation and other potential hydrogen devices.

Based on the assumed social situation above, increasing demand of hydrogen as fuel can be defined and is not very different from the assumption of future hydrogen demand outlined in Sect. 3.1 and Table 6. That is, hydrogen is currently produced at $15–20 \times 10^9$ Nm^3/year, and this will need to increase by 100–200 % by 2050 for hydrogen fuel, as mentioned in Sect. 2.1. As addressed in Sect. 2.2 and Table 1, hydrogen needed by FCEVs can be supplied by the conventional types even in 2050, i.e., steam reforming and by-product hydrogen, as long as the supply of fossil

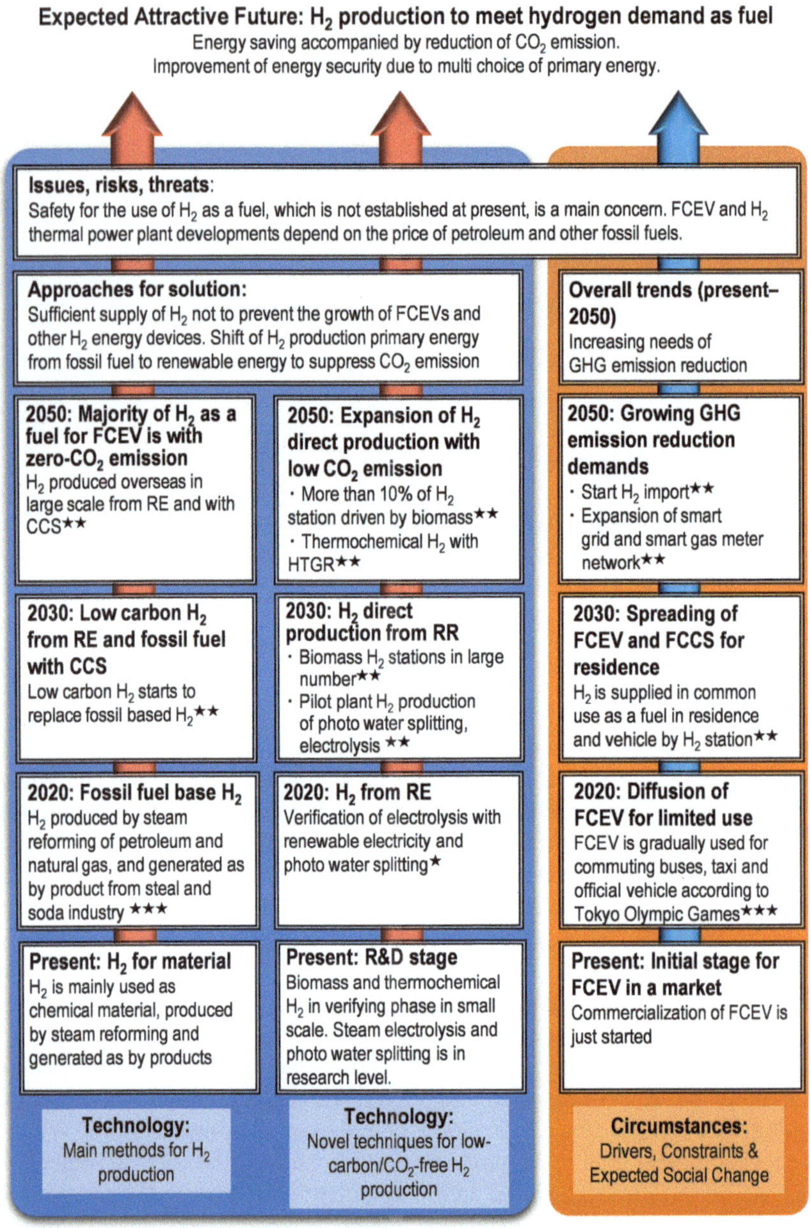

*1 Estimated from FCCJ scenario

Abbreviations:

FCEV: Fuel Cell Electric Vehicle, FCS: Fuel Cell Cogeneration System, HTGR: High Temperature Gas-cooled Reactor, R&D: Research and Development, RE: Renewable Electricity, RS: Renewable Resources

Fig. 8 Roadmap of hydrogen production

fuel is sufficient. However, these conventional hydrogen production methods are fossil fuel-based and accompanied by CO_2 emissions. To reach the technical goal for hydrogen production shown in Fig. 8 and mentioned above, development of production methods of low-carbon hydrogen discussed in Sect. 3.2 is important for reducing CO_2 emissions. In the next 20–30 years, hydrogen may be used in electricity generation in a fuel mixture with natural gas for thermal power plants. In this case, the need for hydrogen is greater than that of FCEVs, and the hydrogen should be low carbon as discussed in Sect. 3.1. Steam reforming of brown coal together with CCS and water electrolysis combined with solar or other renewable electricity can be assumed as overseas hydrogen production schemes [3, 4, 28, 29].

According to the above discussion on the quantity and quality of hydrogen necessary as a future fuel, substantial progress in methods producing low-carbon or CO_2-free hydrogen is mandatory. Hydrogen production methods of this kind with the potential capability of hydrogen production were explained in Sect. 3.2 and Table 7. The statuses of these techniques are different, as described in the following.

Biomass is currently in the verification phase. As mentioned in Sect.. 3.2.2 and shown in Table 7, a hydrogen production capability of 1.9×10^9 Nm^3/year is predictable, corresponding to 1.4 million FCEVs (14 % to the total expected number of FCEVs in 2050). As thermochemical hydrogen production, the IS process with HTGR at a small-scale hydrogen production facility (~200 NL/h) is operating for technology verification, as mentioned in Sect. 3.2.3. An HTGR with thermal output 600 MW can produce 0.6×10^9 Nm^3/year of hydrogen (corresponding to 0.44 million FCEVs), and its utilization from around 2040 is proposed [22].

Water electrolysis (alkaline and PEM) is technically nearly established, and improvement of energy efficiency is being studied [25]. Steam electrolysis and photoelectrochemical water splitting are under research, but could move into the verification stage in the next 10 years. Water/steam electrolysis (Sect. 3.2.4) is the only technique that produces hydrogen from electricity, and will be important in hydrogen production methods using renewable energy, because the main form of energy available from that energy is electricity. Photoelectrochemical water splitting is more beneficial, because it directly produces hydrogen from sunlight (Sect. 3.2.5). For practical use, however, the energy efficiency should be increased from the current state-of-the-art 2 % [27] to ~20 % or higher, comparable to the combination of photovoltaic cells and water electrolysis, which make up ~21 % of hydrogen production. State-of-the-art commercial energy efficiencies of these two technologies are 25 % [30] and 87 % [31], respectively.

Future expansion of these low-carbon hydrogen production methods will reduce CO_2 emissions and is the technological goal of hydrogen production.

4 Benefit and Future Vision

A hydrogen energy society using hydrogen as an energy medium has the following benefits:

1. Energy saving accompanied by reduction of CO_2 emissions
2. Facilitating a CO_2-free energy system and substantial reduction of CO_2 emissions by producing hydrogen from renewable energy
3. Improvement of energy security via multiple choices of primary energy

FCEV is a good example of benefit (1). Even when hydrogen fuel is produced conventionally from fossil fuels, well-to-wheel CO_2 emission can be roughly 0.5–0.6 that of a conventional internal-combustion engine vehicle (based on representative values, comparing a FCEV fueled with hydrogen produced by onsite reforming of town gas and a gasoline or diesel internal-combustion engine vehicle) [3]. Benefit (2) is more important to dramatically decrease CO_2 emission over its present status. CCS is one method to reduce CO_2 emission, but the use of large portions of renewable energy is essential for the sustainability of society. In contrast, hydrogen is an energy medium suitable for the use of renewable energy, which can be enhanced by the spread of hydrogen energy. This characteristic can stabilize the energy situation in Japan, which has long suffered from poor energy resources. We can produce hydrogen within Japan in the future via biomass, water electrolysis with renewable electricity, and photoelectrochemical water splitting, which are advantageous to fossil fuel-based hydrogen from an energy security standpoint.

Such low-carbon techniques of hydrogen production are in various stages toward practical applications and are currently studied and developed both scientifically and technologically. We can even assume the emergence of novel techniques that are presently unknown. As mentioned above, progress of these techniques is required to realize a hydrogen energy society. Resources should be spent on their research and development to realize the benefits of hydrogen energy.

Acknowledgments Prof. Masaki Tajima of Kyushu University and Dr. Yoshiyuki Inagaki of the Japan Atomic Energy Agency are acknowledged for data and figures used in Sects. 3.2.2 and 3.2.3.

References

1. Press release on Nov. 18, 2014 from Toyota Motor Corporation (in Japanese). http://newsroom.toyota.co.jp/en/detail/4197769
2. Press release on Nov. 17, 2014 of Honda Motor Co., Ltd. (in Japanese). http://www.honda.co.jp/news/2014/4141117.html
3. New Energy and Industrial Technology Development Organization (NEDO) Hydrogen energy white paper (*Suiso Energy Hakusho*), July 28, 2014 http://www.nedo.go.jp/news/press/AA5_100292.html

4. Mizuho Information and Research Institute, FY2012 report: a study on present and future status of hydrogen demand and supply http://www.nedo.go.jp/library/seika/shosai_201311/20130000000462.html
5. Agency for Natural Resources and Energy. Hydrogen production, transportation and storage. http://www.meti.go.jp/committee/kenkyukai/energy/suiso_nenryodenchi/suiso_nenryodenchi_wg/pdf/006_s02_00.pdf
6. Mizuho Information and Research Institute. Yusouyou nenryou no well-to-weel hyouka. (2004). http://www.mizuho-ir.co.jp/publication/report/2004/pdf/wtwghg041130.pdf
7. U.S. Department of Energy. Production case studies. http://www.hydrogen.energy.gov/h2a_prod_studies.html. Accessed 7 Nov 2014
8. Japan Hydrogen and Fuel Cell Demonstration Project, Japan Automobile Research Institute (2011) Sougou kouritu to GHG haisyutu no bunnseki houkokusyo. http://www.jari.or.jp/Portals/0/jhfc/data/report/2010/pdf/result.pdf
9. Institute of Energy Economics Japan (2009) Zidousyayou nennryou tositeno suiso enerugi- no gennzyou to konngo no doukou ni tuite. http://eneken.ieej.or.jp/data/2523.pdf
10. Federation of Electric Power Companies Japan (2014) Denjiren kaicyou teirei kaiken yousi. http://www.fepc.or.jp/about_us/pr/kaiken/__icsFiles/afieldfile/2014/05/23/kaiken_20140523.pdf,pp.5
11. Japan Hydrogen and Fuel Cell Demonstration Project (2011) Nenryou denchi sisutemu tou zissyou kennkyuu houkokusyo. http://www.jari.or.jp/Portals/0/jhfc/data/report/pdf/tuuki_phase2_01.pdf
12. Energy and Environment Council (2011) kosuto tou kenshou iinkai houkokusho. http://www.cas.go.jp/jp/seisaku/npu/policy09/pdf/20111221/hokoku.pdf
13. http://www.hydrogennet.dk/fileadmin/user_upload/PDF-filer/Aktiviteter/Afholdte_aktiviteter/Transportworkshop%20d.%201.%20dec%202011/4_Sloth.pdf
14. Fuel Cell Commercialization Conference of Japan (2010) Commercialization scenario for FCVs and H_2 stations. http://fccj.jp/pdf/22_cse.pdf
15. Shirasaki Y, Tsuneki T, Ota Y, Yasuda I, Tachibana S, Nakajima H, Kobayashi K (2009) Int J Hydrog Energy 34(10):4482–4487
16. Tajima M (n.d.) Fukuoka strategy conference for hydrogen energy, presentation slides, p 29. http://www.f-suiso.jp/wp-content/uploads/2012/12/H24bunkakai4-2l.pdf
17. A report from Ministry of Land, Infrastructure, Transport and Tourism. http://www.meti.go.jp/committee/kenkyukai/energy/suiso_nenryodenchi/suiso_nenryodenchi_wg/pdf/006_s02_00.pdf
18. NIST-JANAF Tables. http://kinetics.nist.gov/janaf/janaf4pdf.html
19. Funk JE (2001) Int J Hydrog Energy 26(3):185–190
20. Kameyama H, Sakurai M, Masuda A, Fukui T (2012) Hydrogen Energy System, 37 3-10 (in Japanese)
21. Kasahara S, Tanaka N, Noguchi H, Iwatsuki J, Takegami H, Yan X L, Kubo S (2014) Proceedings of the HTR 2014, Weihai, China, October 27-31, ,Paper HTR2014-21233
22. JAEA. Nuclear energy vision 2100 -towards a low carbon society-, 2008 (in Japanese). http://www.jaea.go.jp/03/senryaku/seminar/08-4-2.pdf
23. Germany Energy Agency. Power to gas strategy platform. http://www.dena.de/en/projects/renewables/power-to-gas-strategy-platform.html
24. Matsumoto H, Kwati Leonard (2014) Catalyst (Shokubai), 56:296–299
25. NEDO, project on hydrogen utiliation. http://www.nedo.go.jp/activities/ZZJP_100068.html
26. Fujishima A, Honda K (1972) Nature 238:37–38
27. NEDO, news release, March 31, 2015. http://www.nedo.go.jp/news/press/AA5_100372.html
28. Yoshino Y, Harada E, Inoue K, Yoshimura K, Yamashita S, Hakamada K (2012) Energy Procedia 29:701–709
29. Okada Y, Sasaki E, Watanabe E, Hyodo S, Nishijima H (2006) Int J Hydrog Energy 31:1348–1356
30. Nayak PK, Cahen D (2014) Adv Mater 26:1622–1628
31. Haryu E et al (2011) Honda R&D Tech Rev 23(2):90–97

Concept of Energy Carrier, Candidate Materials, and Reactions

Koichi Eguchi

Abstract The importance of the energy carrier for the utilization of renewable energy has been recognized in recent years because hydrogen is considered a clean and efficient energy source for the near future. Long-term and stable storage of hydrogen should complete its supply chain in the form of energy carriers. Candidate substances for these carriers are liquid hydrogen, organic chemical hydride, and ammonia. For transportation between continents, storage of hydrogen or hydride in stable liquid form is required. Properties, production, transportation, and utilization of candidate materials for energy carriers are summarized.

Keywords Energy carrier • Hydrogen • Ammonia • Fuel cell • Renewable energy

1 Introduction

The ultimate solution to the energy problem is to establish a long-term vision of the energy resource using renewable energy. In this section, solar energy is taken as an example of a highly attractive renewable source. Generally, candidate production sites of electrical and thermal energies from abundant solar radiation and heat are remote from their consumption sites. The amounts of these energies cannot be controlled and change significantly over time and seasons, irrespective of demands. It is very difficult to solve this major obstacle of spatiotemporal differences between renewable energy supply and electricity demand.

Fuel cell power systems have been recently commercialized in several applications such as household cogeneration units and stationary power sources. In particular, vehicle power sources have been attracting attention as clean and efficient generation systems of the near future. In this situation, effective production, supply chain, and efficient utilization of hydrogen are desired to be developed for the construction of a future energy demand and supply framework. Hydrogen is currently produced by conversion of fossil fuels with well-developed catalytic

K. Eguchi (✉)
Kyoto University, Kyoto, Japan
e-mail: eguchi@scl.kyoto-u.ac.jp

Y. Kato et al. (eds.), *Energy Technology Roadmaps of Japan*,
DOI 10.1007/978-4-431-55951-1_10

reaction technologies, but it is reasonable to transition in the future to energy conversion technologies based on hydrogen produced from renewable sources.

To promote the hydrogen supply based on renewable sources, energy storage media (energy carriers) or hydrogen carriers and their handling systems must be designed and developed. The hydrogen carrier, containing large amounts of hydrogen in molecules, can be transported over long distances and stably stored as liquid. Hydrogen must be easily extracted at consumption sites from the energy carrier. Another possibility is to consume the energy carrier directly as fuel for combustors or fuel cells. In particular, production of energy carriers should be accomplished with utilization of renewable energy as solar heat and power. Battery systems have been developed for energy storage, but their energy density and stability are insufficient compared with the aforementioned energy carrier.

Figure 1 shows a schematic of the flow of materials and energy conversion based on hydrogen. Hydrogen has been produced from fossil fuels by a steam-reforming reaction and coal gasification. The primary energy resource for hydrogen production will shift to renewable energy. Part of the produced hydrogen is converted into energy carrier compounds such as liquid hydrogen, methane, organic chemical hydride, and ammonia for transportation to remote sites with heavy energy demands. The final stage is energy conversion with combustion and fuel cells. There are other possible energy resources, energy carriers, and utilization methods that are not included in the scheme.

Carbon monoxide hydrogenation to hydrocarbon compounds with Fischer–Tropsch (FT) synthesis or the gas-to-liquids (GTL) process has been investigated and partially commercialized as conversion and utilization of fossil fuel during the long history of their development. In the future, however, this technology should be investigated as a possible utilization of renewable energy as energy carrier

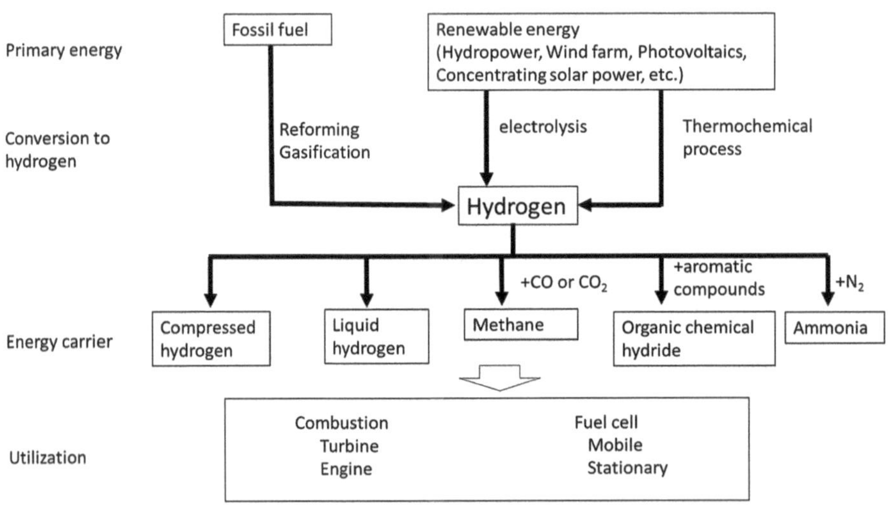

Fig. 1 Production of hydrogen and conversion to energy carriers

production. Various types of energy carrier candidate substances including methanol and methane are expected to find their respective application fields. A direct synthetic route developed recently for dimethyl ether (DME) makes it a candidate for energy carrier. A series of these hydrocarbons and oxygenated compounds are synthesized from $CO+H_2$ synthetic gas or CO_2+H_2 by vapor-phase catalytic processes. Although fabrication of such compounds is currently considered an effective use of fossil fuels, they can be reconsidered as energy storage substances because of their energy density and chemical stability during storage and ease of hydrogen production. However, consumption of this fuel is accompanied by emission of CO_2. Therefore, carbon-free or carbon-neutral technologies without CO_2 emission are the ultimate targets. For such a purpose, liquid hydrogen, organic chemical hydride, and ammonia are beginning to be studied as major candidates for a national project in Japan. Effective synthetic and utilization technologies for such energy carriers were selected as one theme of the Strategic Innovation Promotion Program (SIP) that commenced in 2014 [1].

2 Hydrogen Production

A variety of renewable energies have been considered as possible future resources. Hydrogen production from solar energy is considered the first stage in the design of energy carrier systems. For realizing this system, international development and cooperation with countries in the Sunbelt areas of the world are indispensable. Electrical and thermal energies from sunlight are used for the production of hydrogen and hydrogen carriers. The principal candidate carriers selected in the research program are liquid hydrogen, organic chemical hydride, and ammonia. These energy carriers are transported as liquid by marine transportation to consumption locations such as Japan. Effective conversion of carriers into electrical energy or hydrogen extraction is another important topic. Energy carriers transported to Japan are used in conversion systems such as fuel cells, combustors, and hydrogen-fueled devices.

For the conversion of solar light, photovoltaic cells are commonly used, but the collection of solar heat is attracting attention as concentrated solar power (CSP) [2]. CSP systems have been developed in areas with abundant flat space and sufficient sunlight to produce heat and electricity, through a combination of numerous mirrors and steam turbines. This technology has recently attracted attention for hydrogen production, using abundant electricity and heat. In tower-type power generation by CSP, angles of a number of plane mirrors are computer controlled via heliostats to track sun motion and concentrate sunlight at a collector atop a tower. Thermal energy is transferred to a heating medium and accumulated. The second typical CSP system is a combination of elliptic mirror (parabolic trough) and heating tubes located at the focus. A flowing medium in the pipe is heated by focused sunlight. Other CSP methods have been proposed and developed (Fig. 2). Electricity is generated by a steam turbine from provided heat. Unfortunately, few

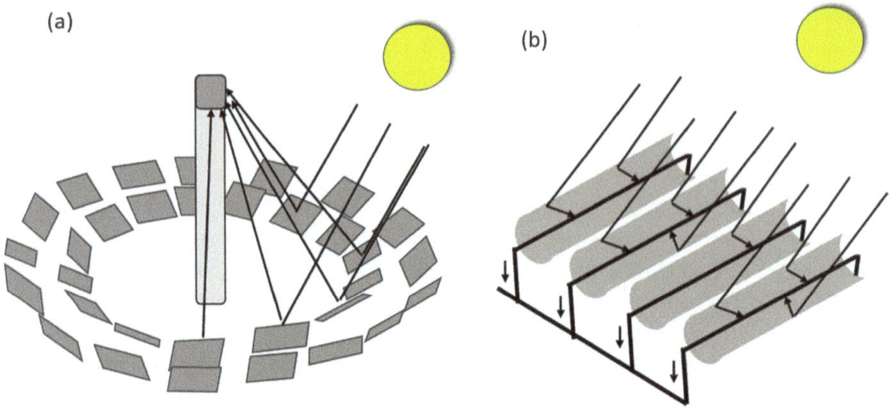

Fig. 2 Types of concentrated solar power systems [2]. (**a**) Power tower, (**b**) parabolic trough

areas have sufficient solar heat for this purpose in Japan. There is no large-scale proof-of-concept example in the country, but investigation continues in European countries such as Spain, in the United States, and in other countries. Overseas resources are also attracting attention regarding the use of wind energy.

Thermochemical water splitting and water electrolysis have been considered hydrogen manufacturing processes using renewable energy. Thermochemical processes constructed with sulfur and iodine compounds have been studied in the energy carrier project (Fig. 3). These processes were developed earlier for utilization of heat from nuclear reactors, including gas-cooled fast reactors. Suitable processes have begun to be pursued for high-temperature thermal energy from renewable sources. The iodine–sulfur (IS) process is composed of multiple processes through conversion of iodine (I) and sulfur (S) compounds. This produces hydrogen (H_2) and oxygen (O_2) by splitting water in multistage reactions, with inputs of water (H_2O) and high-temperature heat during the process [3].

Bunsen reaction, exothermic

$$I_2 + SO_2 + 2H_2O \rightarrow 2HI + H_2SO_4 \tag{1}$$

Decomposition of hydrogen iodide, weakly endothermic

$$2\,HI \rightarrow H_2 + I_2 \tag{2}$$

Decomposition of sulfuric acid, endothermic

$$H_2SO_4 \rightarrow H_2O + SO_2 + 0.5\ O_2 \tag{3}$$

Hydrogen production by the electrolysis of water is another candidate process for hydrogen production from renewable energies. Electrolysis of concentrated aqueous alkaline solution is widely done, using low-carbon steel or nickel as

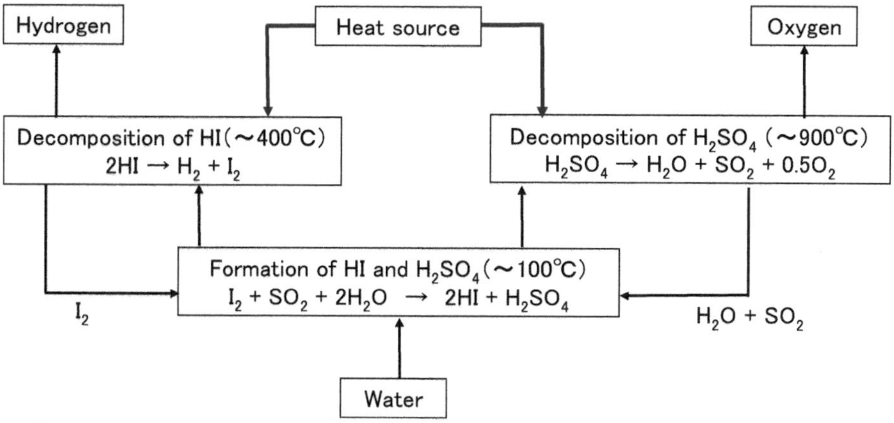

Fig. 3 Outline of thermochemical IS process [3]

electrodes [4]. The metal electrodes are immersed in concentrated KOH solution in the electrolysis batch, in which a porous diaphragm is located to avoid mixing formed hydrogen and oxygen at cathode and anode sides, respectively. As gaseous bubbles are vigorously generated with enhanced electrolysis current, gas diffusion resistance becomes dominant.

Anodic reaction

$$2\,OH^- \rightarrow H_2O + 1/2O_2 + 2\,e^- \quad (4)$$

Cathodic reaction

$$2\,H_2O + 2\,e^- \rightarrow H_2 + 2\,OH^- \quad (5)$$

Water electrolysis using a proton-exchange membrane has been developed in accordance with commercialization of proton-exchange membrane fuel cells (PEMFCs). Polymer electrolyte fuel cells (PEFCs) are popular. In this section, however, the term PEMFCs is used to distinguish them from anion-exchange membrane fuel cells (AEMFCs). Water electrolysis using PEM has been commercialized for practical use. Small hydrogen production devices, e.g., for gas chromatography, are popular, and larger systems are in operation. Iridium-based alloy electrodes are attached on the PEM because of rapid degradation of the Pt-based electrodes. The produced hydrogen and oxygen, being separated by the PEM film, are less mixed with each other than in the conventional electrolysis method. Highly concentrated hydrogen of >99.999 % purity can be produced without further purification.

Anodic reaction

$$H_2O \rightarrow 2\,H^+ + 1/2O_2 + 2\,e^- \quad (6)$$

Cathodic reaction

$$2\,H^+ + 2\,e^- \rightarrow H_2 \quad (7)$$

Pressurized hydrogen is also produced electrochemically by this method. The PEM should be reinforced by a metallic mesh to tolerate the strong differential pressure that develops between the hydrogen and oxygen chambers. The Nernst potential corresponding to the pressure difference is supplied to the electrodes, which is equivalent to the electrochemical potential estimated from that difference.

In recent years, solid oxide fuel cell (SOFC) systems have been commercialized using the yttria-stabilized zirconia (YSZ) solid electrolyte. The operational temperature of the household cogeneration system is set to ~750 °C. Electrolysis cells with reverse operation of SOFCs have also been actively investigated. Solid oxide electrochemical cell (SOEC) systems are superior to other electrolysis processes in their greater reversibility near the open circuit. The high operational temperature of SOEC also reduces the theoretical decomposition voltage $E°$; since $E°$ originates from the free energy, $\Delta G°$, from the water decomposition reaction of $H_2O \rightarrow H_2 + 1/2\ O_2$, is $\Delta G° = nFE°$. Thus, $E°$ decreases with increasing electrolysis temperature. A large margin of the endothermic region is thus created in the I–V characteristics of SOEC. Therefore, a combination of SOEC with heat storage material is considered an effective regenerative cell.

3 Hydrogen Storage and Carrier

Among the variety of energy storage technologies, some with different concepts are competing. For storage of electricity, secondary batteries are considered important for transportation, stationary systems, and portable applications. The application area of hydrogen carriers partially overlaps with that of batteries. A high-pressure hydrogen container is put to practical use in fuel cell electric vehicles (FCEVs) and batteries and in capacitors for electric vehicles. Metal hydrides are also developed for hydrogen storage and hybrid tanks of such containers. A series of secondary batteries has been used as portable devices for electricity storage. The most suitable device or energy storage system is chosen from a variety of methods to achieve convenience and suitability. Selection and differentiation between energy carrier storages are possible for a typical usage, but there are several competing energy storage methods, including energy carriers, in many application fields. Therefore, introduction of an energy carrier system should be carefully considered based on introduction period, location, application, and cost. Some candidate materials for hydrogen carriers are compared in Fig. 4 for their gravimetric and volumetric

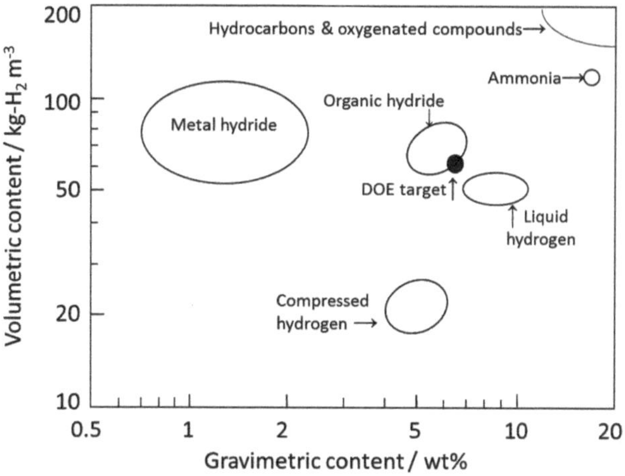

Fig. 4 Volumetric and gravimetric hydrogen contents of various compounds. x and y axes have logarithmic scales

contents of hydrogen. Either volumetric or gravimetric content is high for hydrocarbons and their oxygenated compounds, such as methanol, ethanol, and dimethyl ether. However, these compounds have been developed for effective utilization of fossil fuels. These fuels are accompanied by the formation of carbon dioxide with hydrogen production. Ammonia, organic chemical hydride, and liquid hydrogen demonstrate acceptable hydrogen contents compared with US Department of Energy (DOE) targets. Metal hydrides and compressed hydrogen are inferior because of their gravimetric and volumetric contents, respectively.

Some example compounds for hydrogen are described in Table 1 along with liquid and pressurized hydrogen. Hydrocarbons and oxygenated compounds produce a large amount of hydrogen via the reforming reactions. Steam reforming of higher hydrocarbons generates a greater volumetric production amount per unit volume of liquid. These processes are adopted in petroleum refining industries.

In the production of methanol and DME, solid catalysts are important in their synthesis at high pressures. Hydrogen production from these compounds involves steam reforming of methanol and dimethyl ether over solid catalysts. For reforming of DME, water is mixed with it to produce substantial amounts of hydrogen. DME is attractive as a carrier owing to its lower reforming temperature relative to hydrocarbons [5]. This catalysis series has been developed as clean fuel conversion processes using synthetic gas derived from fossil fuels. Hydrocarbon and methane fuels are produced by FT synthesis and methanation reaction, respectively. Fuels from FT synthesis or the GTL process can be directly used in currently available oil infrastructure.

Ammonia and organic chemical hydride are expected to be compatible with various applications for storage as an energy carrier material. However, efficiencies of the various applications are not clearly understood. These applications include long-distance transportation without degradation. For their acceptance as new energy storage media, life-cycle assessment (LCA) is necessary for choosing

Table 1 Candidate compounds and properties for energy carrier

Energy carrier	Density of hydrogen ($kg\text{-}H_2/m^3$-liq.) In molecule	Density of hydrogen ($kg\text{-}H_2/m^3$-liq.) Formed hydrogen	Boiling point (°C)	Reaction enthalpy, ΔH_r° ($kJ/mol\text{-}H_2$)
Ammonia, NH_3	120	–	−33.3	–
$2NH_3 \rightarrow N_2 + 3H_2$	–	120	–	31
Methylcyclohexane, C_7H_{14}	110.0	–	101	–
$C_7H_{14} \rightarrow C_7H_8 + 3H_2$	–	47	–	68
Methane, CH_4	106		−162	–
$CH_4 + H_2O \rightarrow CO_2 + 3H_2$		212	–	69
Methanol, CH_3OH	99		64.7	–
$CH_3OH + H_2O \rightarrow CO_2 + 3H_2$		148	–	16
Dimethyl ether (DME), CH_3OCH_3	96	–	−24.8	–
$(CH_3)_2O + 3H_2O \rightarrow 2CO_2 + 6H_2$	–	192	–	20
Liquid hydrogen	71		−252	
Compressed hydrogen (70 MPa)	39 (kg/m^3-gas)		–	

from various candidate materials, with consideration of new manufacturing methods, conversion to electricity, energy interconversion, and chemical properties.

Candidate compounds for energy carriers in Table 1 are ammonia, methylcyclohexane, methane, methanol, and dimethyl ether, as well as liquid hydrogen. Among these, ammonia, methylcyclohexane, and liquid hydrogen are not accompanied by the formation of carbon dioxide in hydrogen recovery processes. It is reported, for example, that methylcyclohexane and ammonia are proven as chemical substances of the energy carrier candidates. Characteristic properties of each energy carrier material should be clarified for selection of the most suitable system, from utilization of a renewable source to conversion into consumer energy. Examples of such properties are hydrogen production reaction, energy carrier production, stability during storage and transportation, safety, and conversion into electricity and combustion energy. The boiling point of a carrier compound is important for liquefaction and transportation, because transportation in liquid form over long distances can be effective.

For hydrogen extraction from an energy carrier, catalytic reactions at elevated temperatures are required. For effective energy utilization, fuel cells with various operational temperatures are receiving attention. Typical temperature ranges of catalytic hydrogen production from hydrocarbons and energy carrier materials, hydrogen production by electrolysis, and driving temperature ranges of the fuel

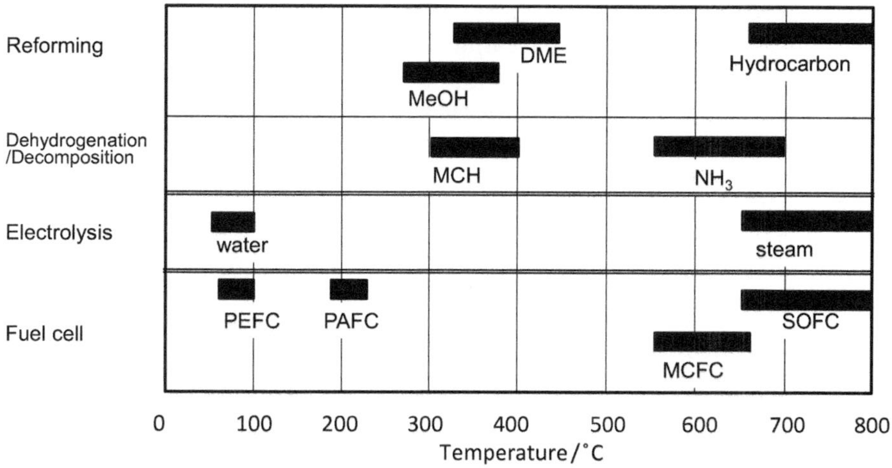

Fig. 5 Operational temperature ranges of hydrogen production and fuel cell systems. *DME* dimethyl ether, *MCH* methylcyclohexane, *PEFC* polymer electrolyte fuel cell, *PAFC* phosphoric acid fuel cell, *MCFC* molten carbonate fuel cell, *SOFC* solid oxide fuel cell

cells are summarized in Fig. 5. For hydrogen production, heat supply in each temperature range is necessary to compensate endothermic heat of the reforming or dehydrogenation reaction and to advance hydrogen production at an energy-consuming site. The source of this heat supply determines the effectiveness of the overall energy carrier system. Further investigation is required for most energy carrier systems to optimize efficiency. For example, endothermic heat in a reformer or cracker is chemically recovered by Joule's heat and thermodynamically emitted heat from a generator with a high-temperature fuel cell, if the temperature range is compatible. A combination of reactors constitutes a system of high efficiency, because combustion and fuel cell generation are exothermic reactions.

Several candidate materials of energy carriers, such as ammonia (NH_3), DME, methanol, and methylcyclohexane, can be considered for their ease in liquefaction, such that large amounts of hydrogen are stored in liquid form. Liquefaction of hydrogen accompanies its high purity because of a low melting point. High-purity hydrogen is attractive as a carrier and for its direct connection to the supply chain. Hydrogen is cooled and liquefied by Joule–Thomson expansion. Hydrogen liquefaction requires substantial energy for cooling. In addition, part of the liquid hydrogen vaporizes to be lost by boil-off, because of the extremely low boiling temperature during transportation. It is crucial to technically minimize this boil-off during long-distance transport. Although the high purity is desirable, the density of liquid hydrogen is generally less than other liquids.

Organic chemical hydrides have been proposed as future energy carriers and have been actively investigated. Because the hydrogenation and dehydrogenation processes use hydrocarbon-based technology, these hydrides are superior for direct use of existing infrastructure in oil storage and transportation. The process initiates catalytic production of cycloparaffin by hydrogenation of an aromatic compound at energy-abundant sites. The cycloparaffin oil is then converted by dehydrogenation for extraction of hydrogen. The energy carrier cycle is withstood either using energy and sources of fossil fuels or solar energy. The key technology for the process is catalytic dehydrogenation of cycloparaffin to obtain hydrogen. Methylcyclohexane and decalin have been recently considered candidates for energy carrier compounds and are called organic chemical hydrides.

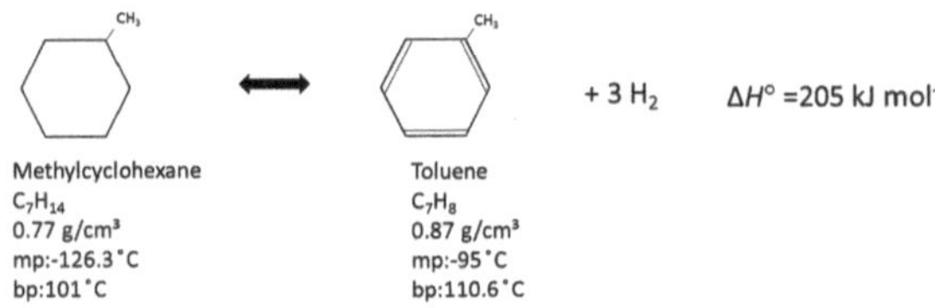

(8)

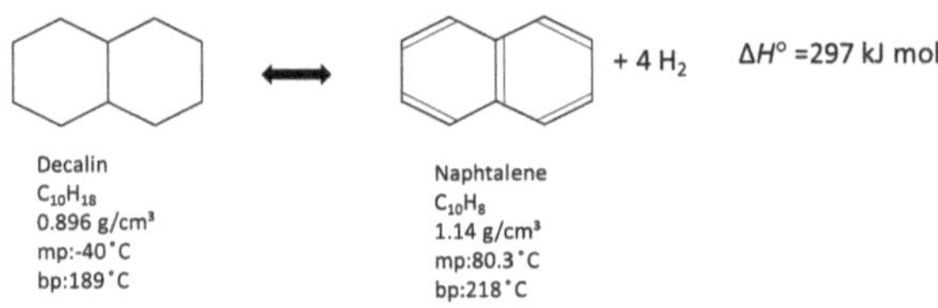

(9)

The Chiyoda Corporation proposed to store methylcyclohexane as organic chemical hydride via toluene hydrogenation [6]. Discovery of a new active catalyst for dehydrogenation advanced organic hydride systems for practical use. A toluene hydrogenation catalyst and reactor have already been developed. As for dehydrogenation, a catalyst with sufficient durability was first developed for implementation. At a hydrogen-consuming site, the heat source to compensate endothermic dehydrogenation is an important factor for realization. Because of the similarity of the physicochemical properties to gasoline or oil, the organic chemical hydride can be adapted to the oil-based infrastructure in operation. This situation is extremely advantageous, since toluene and methylcyclohexane are stored in tanks and stand in for oil. One of the proposed application models is composed of a dehydrogenation reactor for supply to the hydrogen dispenser after on-site purification at the hydrogen stations.

4 Effective Utilization of Energy Carrier

Utilization technology of the energy carrier as fuel must be examined based on efficiency and environmental adaptability. Fuel cells are establishing a position as a stationary generation system of high efficiency. More than 100,000 units of the household fuel cell system with city gas or LPG-reforming reactor were installed by the end of 2014 in Japan, under the brand name "Ene-Farm." Furthermore, it is most

Table 2 Specification of hydrogen for FCV (ISO14687-2)

Hydrogen purity		99.97 %
Impurity	Total hydrocarbons	2 ppm
	Moisture	5 ppm
	Oxygen	5 ppm
	N_2, Ar	100 ppm
	He	300 ppm
	CO_2	2 ppm
	CO	0.2 ppm
	Sulfur compounds	0.004 ppm
	Formaldehyde	0.01 ppm
	Formic acid	0.2 ppm
	Ammonia	0.1 ppm
	Total halogen compounds	0.05 ppm
	Particle	1 mg/kg

desirable for the energy carrier supply chain to be connected to the hydrogen supply network as a hydrogen station to FCEVs. Hydrogen stations are expected to be readily installed, because FCEVs were commercialized in 2014. Therefore, for realization of the energy carrier, the means of connecting this hydrogen supply network is important. However, production sites with abundant renewable energy are generally remote from consumption sites with strong hydrogen demand. High-purity hydrogen should be stored at hydrogen stations at high pressure to supply FCEVs, to be dispensed at up to 70 MPa. Other gas ingredients and impurity density are strictly limited [7]. High-purity hydrogen can be produced by cryogenic separation or pressure swing adsorption (PSA). The impurity levels summarized in Table 2 are considered acceptable for the introductory commercialization period of FCEVs. These levels may include some safety margins.

Hydrogen separation by means of PSA attains high purity, to 99.99 %. However, it is necessary to use purified hydrogen for cleanup of adsorbate, which is contaminated by impure gases. Thus, production efficiency of a practical PSA unit is reduced to ~70 % of high-purity hydrogen which is consumed for purging adsorbed impurities on adsorbent materials for reproduction.

Progress in effective separation technology is expected by the development of membranes with hydrogen permselectivity. Alloy membranes of Pd-Cu or Pd-Ag have been used, which are mechanically reinforced by mounting on porous metallic or ceramic supports. To ensure mechanical strength and a pinhole-free membrane, the thickness of the alloy membrane is determined, but the permeation rate is often insufficient. It is also difficult to fabricate a pinhole-free membrane, but permeation selectivity is weakened by a small number of membrane defects. Hydrogen separation with tubular porous ceramics is another possible membrane separation process because of its high permeation rate. Improvement of permeation selectivity is essential to realization of this separation method. Stability and reliability of the porous filter film should be attained, because weakening of the permeation rate and selectivity of the film are often caused by impurities such as moisture.

The supply chain and effective use of hydrogen should be developed. However, combustion of energy carriers with gas turbines or other types of heat engines, boilers, and furnaces is considered conventionally. A hydrogen engine and the hydrogen combustion gas turbine are also considered.

5 Ammonia as an Energy Carrier

For the development of carbon-free energy systems, ammonia is considered a promising candidate as an energy carrier. The amount of synthetic ammonia exceeds 150 million tons per annum by the Haber–Bosch process. Large quantities of ammonia are consumed as agricultural fertilizer and as a base chemical for various derivative compounds. Haber–Bosch has been a well-established technology for ammonia synthesis from its invention in 1906, for which doubly promoted iron has been used as a catalyst. Since the reactor is compact to function under elevated temperature and pressure, the Haber–Bosch method remains superior to other catalytic systems.

A Ru-based ammonia synthesis catalyst, first discovered by Aika and Ozaki in 1992, is known to be more active than Fe-based catalysts. Recently, active Ru-loaded electrode $[Ca_{24}Al_{28}O_{64}]^{4+}(e^{-})_4$ catalysts were reported to be more active than other ammonia synthesis catalysts [8, 9]. Ammonia has not been synthesized or used as an energy carrier or fuel thus far. An international conference on ammonia fuel is held every year in the United States [10], and ammonia is gradually attracting more attention as an energy carrier or fuel. Clearly, the catalysts are keys in the production and utilization of ammonia fuel.

Direct combustion of ammonia in internal combustion engines has been studied. An early-stage example is an ammonia-fueled engine for a bus, tested in Belgium in 1940. Recently, an Italian automobile company, Marangoni Corporation [10], succeeded in driving a sports car with hybrid LPG/ammonia fuel equipped with an engine made by Toyota [11]. National Institute of Advanced Industrial Science and Technology (AIST) in Japan operated a 20-kW gas turbine with fuel mixture kerosene and ammonia in 2014. As shown by these examples, direct combustion of ammonia by engines is feasible. Further investigation is necessary for other internal combustion engines burning pure ammonia fuel.

One problem of ammonia combustion systems is NO_x formation. Thermal NO_x emitted at high temperature is problematic in hydrocarbon combustion, but fuel NO_x mainly contributes to ammonia combustion in the ignition stage. Fuel NO_x decreases with temperature toward the equilibrium concentration, until the thermal NO_x emission becomes dominant. Furthermore, ammonia is less flammable relative to hydrocarbons. The stable combustion range of air/fuel ratio is narrow, and combustion is slow. Therefore, the combustion mechanism and characteristics are worth studying [12].

When the effectiveness of ammonia as an energy carrier is accepted, it will receive attention for supplying fuel cell generators with higher efficiency. Direct

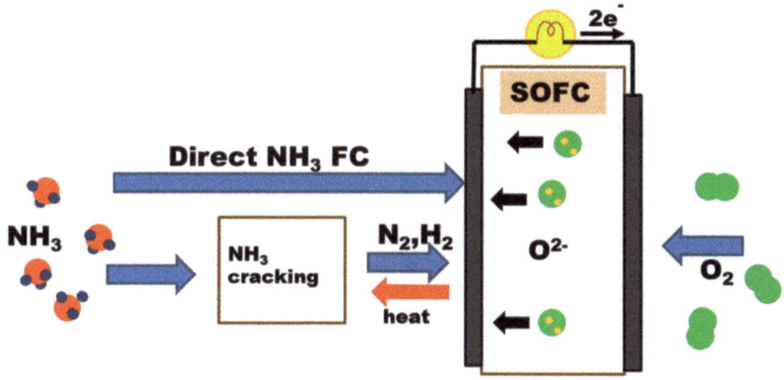

Fig. 6 Solid oxide fuel cell based on internal decomposition of ammonia

supply of ammonia to a fuel cell leads to a simplified system with high-efficiency generation. From the standpoint of compatibility with ammonia fuel, the anion-exchange membrane type of fuel cell, which possesses conductive OH^- ions and a SOFC with conductive O^{2-} ions, is a candidate type [12].

Operational SOFC temperatures are 700–900 °C, and NH_3 decomposition with endothermic heat readily proceeds in a catalytic reactor operated at such high temperatures in the fuel cell chamber with the aid of effective transfer of exothermic heat from the SOFC. A supported Ru catalyst is known to be most active for this decomposition. However, a noble metal catalyst is unnecessary. This is because the supported Ni catalysts demonstrate sufficiently high conversion, owing to the high temperature even with indirect internal decomposition operation (Fig. 6).

As a result of screening of various inexpensive base metal catalysts, supported nickel catalysts showed sufficient activity for decomposition of ammonia in the operational SOFC temperature range. The activity of a Ni/Al_2O_3 system was high because of the large surface area of the support [13, 14].

The high operational temperature and activity of Ni–YSZ cermet enables direct supply of ammonia to the electrode (Fig. 6), which simplified the system. It is also possible to introduce ammonia to the SOFC stack in the direct internal decomposition operation. For direct introduction of ammonia, the following electrochemical process can be expected.

Anodic reaction

$$2\,NH_3 + 3\,O^{2-} \rightarrow N_2 + 3\,H_2O + 6\,e^- \quad (10)$$

Cathodic reaction

$$3/2\,O_2 + 6\,e^- \rightarrow 3\,O^{2-} \quad (11)$$

A possible alternative process is combination of thermal catalytic decomposition and electrochemical oxidation of hydrogen.

Thermal catalytic reaction

$$2\,NH_3 \rightarrow N_2 + 3\,H_2 \quad (12)$$

Anodic reaction

$$H_2 + O^{2-} \rightarrow H_2O + 2\,e^- \quad (13)$$

Cathodic reaction

$$1/2\,O_2 + 2\,e^- \rightarrow O^{2-} \quad (14)$$

The reaction process is estimated from the open-circuit voltage of the aforementioned reactions (Reactions 10 and 11 or Reactions 13 and 14). Actually, ammonia is decomposed catalytically into hydrogen and nitrogen on the electrode surface and subsequently converted by electrochemical oxidation of hydrogen processes. Exothermic heat from the fuel cell unit (electrochemical oxidation) is effectively used, resulting in high efficiency.

A hydrocarbon-based polymer with the quaternary ammonium ion incorporated in the chain is the low-temperature anion-exchange membrane (AEM). A fuel cell with AEM electrolyte has been investigated as an OH^- ion conductive device. Because the operational temperature of AEMFCs is ~50 °C, hydrogen is supplied from an external ammonia decomposition reactor. Ammonia is less harmful to this membrane than a proton-exchange membrane type, and residual ammonia is permitted to some extent. However, because an atomic nitrogen adsorption formed by dissociation of ammonia becomes a catalyst poison for the Pt electrode, concentrated ammonia cannot be introduced directly.

6 Conclusions

Production, transportation, and effective utilization of energy carrier materials are important technologies in the realization of future energy networks using renewable energy. Materials development in conjunction with system design and evaluation is critical in the realization of energy carrier systems using renewable energy.

References

1. Dicorato M, Forte G, Trovato M (2012) Wind farm stability analysis in the presence of variable-speed generators. Elsevier J Energy 39:40–47. doi:10.1016/j.energy.2011.11.050
2. http://www8.cao.go.jp/cstp/gaiyo/sip/
3. http://www.eere.energy.gov/basics/renewable_energy/
4. Terada A, Noguchi H, Takegami H, Kamichi M, Inagaki Y (2011) JAEA Res 2011-041, pp 1–62

5. Mitsushima S, Matsuzawa K (2011) J Hydrogen Energy Syst Soc Jpn 36(1):11
6. Okada O (2008) Fuel Cell 7(4):29
7. Okada Y (2012) Japan Institute of Energy, vol 91 sixth 473
8. Tomioka H JARI Res J 20130806
9. Aika K, Takano T, Murata S (1992) J Catal 136:126–140
10. Kitano M, Inoue Y, Yamazaki Y, Hayashi F, Kanbara S, Matsuishi S, Yokoyama T, Kim S-W, Hara M, Hosono H (2012) Nat Chem 4:934–940
11. http://nh3fuelassociation.org/introduction/
12. http://tyre.marangoni.com/Tuning.aspx#.VBOlYmNTc4M
13. Eguchi K, Kobayashi H (2013) J Hydrogen Energy Syst Soc Jpn 38(4):264–271
14. Muroyama H, Saburi C, Matsui T, Eguchi K (2012) Appl Catal A Gen 443–444:119–124

Part IV
Energy Supply Infrastructure

Seiichiro Kimura

Part IV discusses the infrastructure of energy supply to people and economic activities in Japan, focusing on supplies for city gas, electricity and energy source for automobiles. After imported crude oil, coal and natural gas as primary energy, heating value, component and quality are adjusted to Japanese Industrial Standards at petroleum refineries and part of them are converted into electricity by thermal power plants.

For energy delivery, pipelines and transmission grids are usually used for city gas and electricity, respectively. These infrastructures are forced to change given current trends in electricity and gas market reformation, discussions of infrastructure robustness after the Great East Japan Earthquake and requirements for low-carbon energy use. Regarding automobile fuel supplies, the number of gasoline refueling stations is declining because of improvements in fuel economy and the introduction of alternative fuel and business efficiency of gas stations. The deployment of electric car charging stations and hydrogen refueling stations will progress in near future. Such new fuel stations will be expected to have new features as demand response and energy storage in connection with BEVs, PHEVs and FCVs to further reduce CO_2 emission.

S. Kimura (✉)
The Matsushita Institute of Government and Management, Kanagawa, Japan
e-mail: s-kimura@mskj.or.jp

Electricity Grid Infrastructure

Hiroshi Asano

Abstract After experiencing the Great East Japan Earthquake and nuclear accident at Fukushima Daiichi Nuclear Power Plant in March 2011, Japan has been challenged to drastically change its energy policy toward more renewable energy sources (RES) and less dependence on nuclear energy. Substantial RES are promoted by feed-in tariffs for a low-carbon society. A smart grid aims at comprehensive optimization of power grids and consumers, with respect to power supply reliability, quality, CO_2 emission reduction, and cost. Active participation of distributed energy resources such as controllable loads and battery storage is expected to reduce costs of large penetration of energy generated from RES. Electricity reforms should fully utilize advanced technologies regarding not only conventional power plants and network technologies but also distributed energy resources and smarter energy management systems in the future grid.

Keywords Smart grid • Demand response • Renewable energy • Distributed energy resources

1 Introduction

Japan has few energy resources and is dependent on imports for 96 % of its primary energy supply. The country's energy-supply structure is extremely vulnerable. Following two oil crises in the 1970s, Japan diversified its energy sources through increased use of nuclear energy, natural gas and coal, as well as the promotion of energy efficiency and conservation. There has been an urgent need to implement global warming countermeasures, such as reducing carbon dioxide emissions from fossil fuel use. To ensure a stable electricity supply in Japan under these environmental constraints, it is crucial to establish an optimal combination of power

H. Asano (✉)
Central Research Institute of Electric Power Industry, Tokyo, Japan

The University of Tokyo, Tokyo, Japan

Waseda University, Tokyo, Japan
e-mail: asano@criepi.denken.or.jp

Y. Kato et al. (eds.), *Energy Technology Roadmaps of Japan*,
DOI 10.1007/978-4-431-55951-1_11

sources that can concurrently deliver energy security, economic efficiency, and environmental conservation.

A low-carbon society is a political issue. Substantial renewable energy sources (RES) are promoted by a feed-in tariff (FIT). The Japanese power system is required to integrate large amounts of energy from RES. After the severe Fukushima nuclear plant accident in 2011, operations at most nuclear plants have been halted. The resulting substantial change of the generation mix has caused supply shortages and increases of electricity tariffs. In addition, the public need to choose the type of electricity and its supplier has increased. Thus, market reform will be required to (1) ensure a stable supply of electricity, (2) suppress electricity tariffs to the maximum extent possible, and (3) expand choices for consumers and business opportunities. This reform will be executed in three steps during 2013–2020.

After the Great East Japan Earthquake and nuclear accident at Fukushima Daiichi Nuclear Power Plant in March 2011, Japan has been challenged to drastically change its energy policy toward more renewable resources and less dependence on nuclear energy. The government has been addressing the best mix of energy sources, robust power supply systems, and nuclear power policy suited to the country.

Electricity supplies were tight in Japan until the 1950 Electricity Utility Industry Reorganization Order. This divided the country into nine service areas in May 1951, each served by a vertically integrated electric power company (EPCo) that generated, transmitted, and distributed power to end users. This resulted in nine regional monopolies, or investor-owned General Electricity Utilities (GEUs), Hokkaido, Tohoku, Tokyo, Chubu, Hokuriku, Kansai, Chugoku, Shikoku, and Kyushu EPCos, which remain the dominant players to this day. With the return of Okinawa to Japan in 1972, Okinawa Electric Power Co. became the tenth GEU in the country. Figure 1 shows the service areas of each regional power company.

Until the amendment of the Electricity Business Act in 1995, Japan's electric power industry was composed of the GEUs and the following wholesale electric utilities: Electric Power Development Co., Ltd., fully privatized in October 2004 and referred to below as J-POWER, Japan Atomic Power Co., and various wholesale electric utilities such as joint thermal power-generation companies. All of these have received investment from GEUs and publicly owned hydroelectric power generators.

This arrangement contributed to a reliable power supply in Japan. In the 1990s, prompted by restructuring developments in other parts of the world, the Japanese government decided to introduce competition in the electricity sector [1]. The intent of the 1995 Electricity Business Act, which was the first comprehensive amendment in 30 years, was to reduce electricity prices to internationally comparable levels through competition among stakeholders. The act introduced partial competition in the generation sector by allowing independent power producers (IPPs) to participate in the wholesale market. Since 1995, the electric power market in the country has been liberalized in stages, in 1999, 2003, and 2008 (Table 1). The ongoing regulatory reform since 2013 is described later.

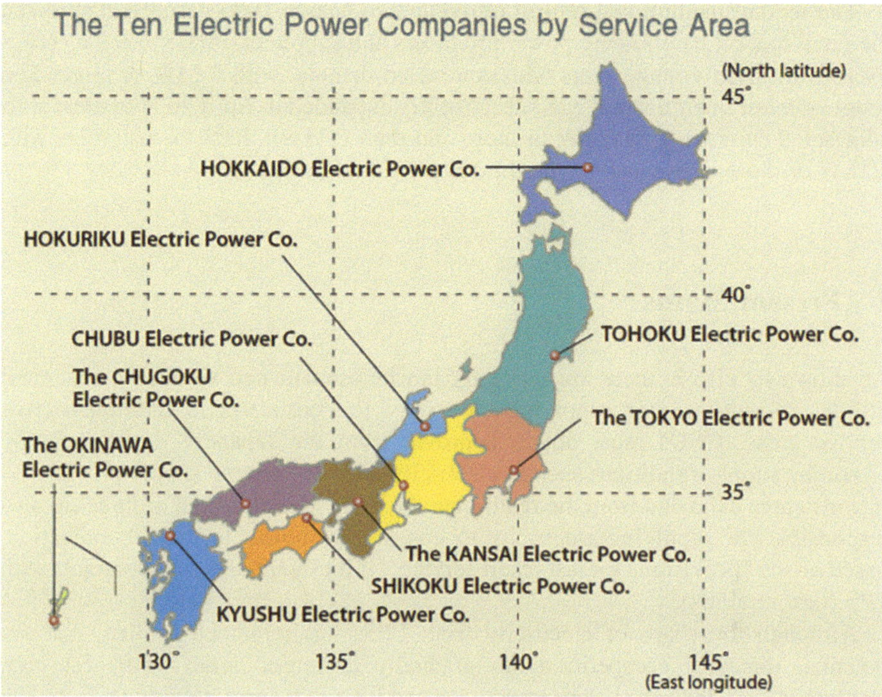

Fig. 1 Ten regional power companies in Japan (Source: Federation of Electric Power Companies of Japan)

Table 1 Historical development of regulatory reform in Japan

1st Institutional Reform: 1995
Introduction of competition into generation sector; auction of new power plant procurement
Foundation of new system that allows new entrants to supply electricity to customers in specified area
2nd Institutional Reform: 1999
Partial liberalization of retail electricity market for extra-high-voltage users with maximum demand of 2,000 kW
Shift from approval system to notification system of rate changes when the rate decreases
3rd Institutional Reform: 2003
Gradual extension of retail liberalization to high-voltage users with maximum demand of 50 kW
Establishment of neutral organization for monitoring of transmission; Electric Power System Council of Japan (ESCJ)
Establishment of wholesale power market; Japan Electric Power Exchange (JEPX)
4th Institutional Reform: 2008
Wholesale power market reform for activating power trading
Improvement of competition conditions on transmission usage for new entrants
No extension of retail market liberalization; reexamination after 5 years

The second institutional reform introduced in March 2000 partially liberalized the retail market by allowing power producers and suppliers (PPSs) to sell electricity to extra-high-voltage users with contracted demand with 2 MW or more. The scope of retail liberalization was subsequently expanded in April 2004 to users with contracted demand of 500 kW or more and then in April 2005 to customers with 50 kW or more, as a result of the third institutional reform.

2 Present Status

As shown in Fig. 2, there are currently ten investor-owned vertically integrated utilities responsible for supplying electricity to consumers in their respective service areas. GEUs must obtain approval from the Japanese government by providing supply conditions such as electricity rates as "general supply provisions" to consumers excluded from the retail liberalization. The regulated utilities are also responsible for supplying electricity to consumers subject to that liberalization, based on the "provisions for last resort service," if they cannot secure contracts with PPS (new retailers).

Although the share of electricity from PPSs has remained limited, just the potential threat of competition has gradually decreased retail tariffs for both residential and industrial consumers. Competitive pressures stemming from the market reforms have forced the incumbent utilities to increase their operational efficiencies while lowering electricity rates and offering a variety of pricing plans. These developments have resulted in a 15 % drop in system average prices over 1990–2011.

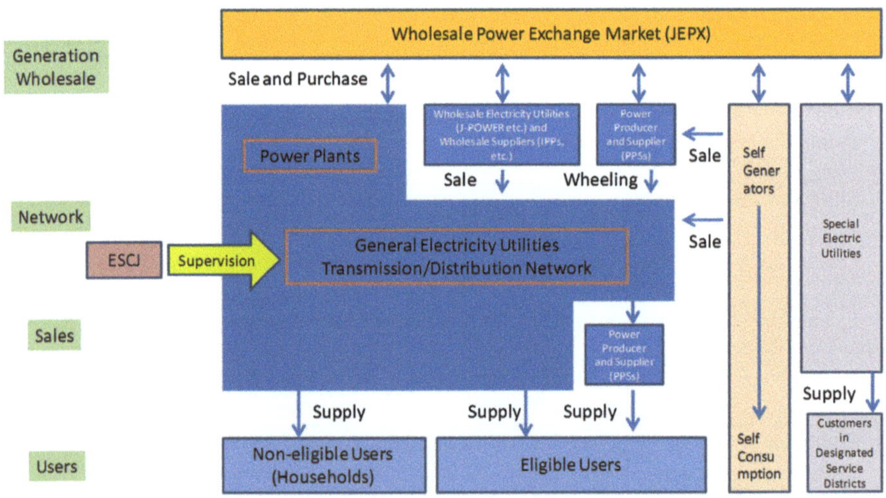

Fig. 2 Current electricity supply system in Japan as of August 2012 (Effective 1 April 2006)

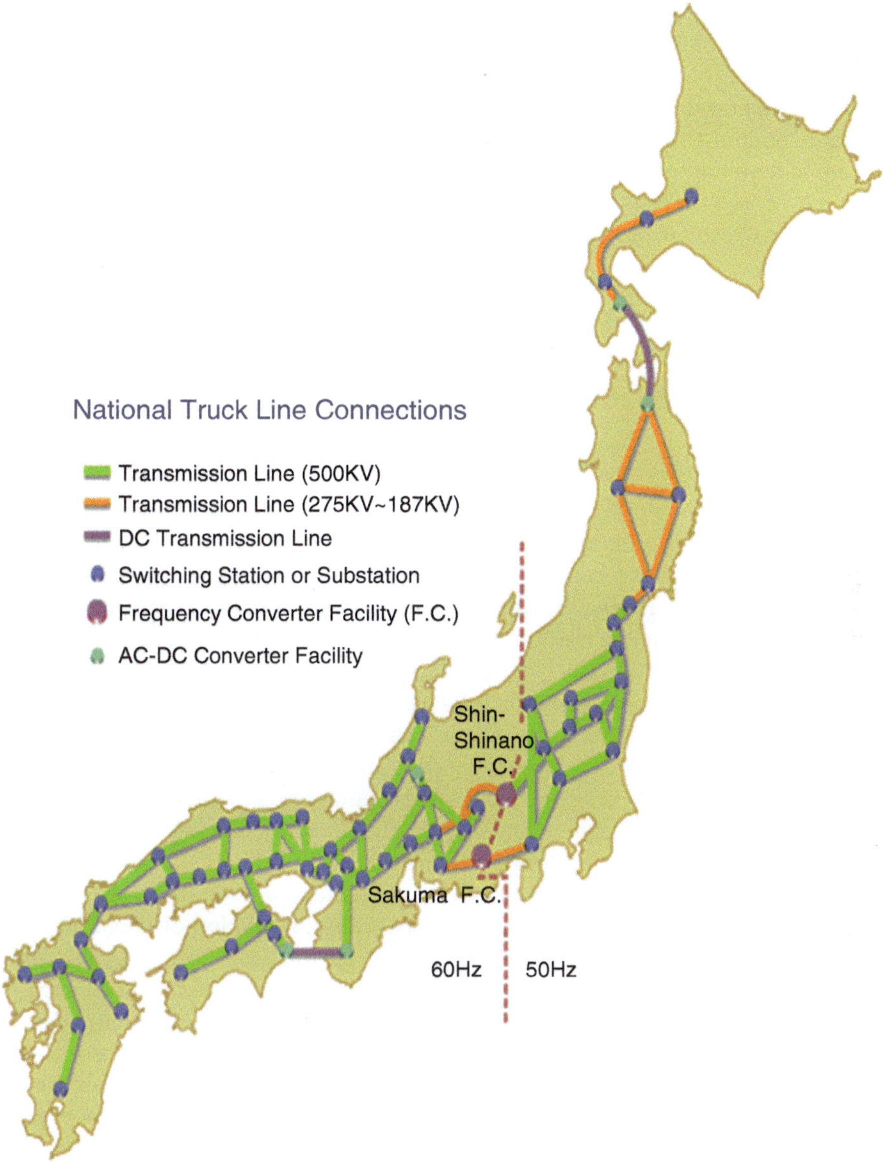

Fig. 3 National transmission grid in Japan (Source: JEPIC 2011)

After the Great East Japan Earthquake in March 2011, eastern Japan faced severe electricity shortages because power plants were damaged by the earthquake and tsunami. Figure 3 shows the national transmission grid, which has a limited capacity of frequency converters (FC) between the 50-Hz eastern grid and the 60-Hz western grid. FC capacity is currently under expansion, from 1.2 to 2.1 GW.

The country faces many challenges on multiple fronts, including the need for institutional reforms to further encourage competition in both wholesale and retail markets, address limited inter-utility transfer capacity, and formulate policies promoting renewable energy integration and demand-side participation.

Large amounts of RES energy are promoted by the FIT. The Japanese power system is required to integrate such amounts. Moreover, the public need to choose the type of electricity and its supplier has increased. Market reform has been discussed to resolve the aforementioned problems in a government committee during 2012.

3 Technology Roadmap

3.1 Smart Grid

To realize the future grid, related technologies such as smart meters have already been demonstrated in the field. A smart grid was discussed before the Fukushima disaster, to integrate renewable energies and activate demand. These meters are being installed by Japanese electric power utilities to improve customer service and enhance operational efficiency.

Reinforcement of interconnection lines has been studied for secure integration of supply and renewable energies. The electricity reforms should fully utilize advanced technologies, not only conventional power plants and network technologies but also distributed energy resources and smarter energy management systems. From the perspective of new technology, R&D in energy and the environment is increasingly important for ensuring a future sustainable society. It may be more effective to increase R&D expenditures on energy and environment for boosting new generation technologies, rather than supporting installation of uncompetitive generation technologies by FIT and emerging technologies for integrated grid control. One example of a roadmap to future technologies is shown in Fig. 4. GHG reduction rates by 2030 were under discussion by the government panel in March 2015. The future grid should utilize demand-side resources such as electric vehicles (EVs) and plug-in hybrid EVs to reduce the additional cost of substantial RES integration through a more integrated approach. R&D for supply-side resources including clean coal technologies and carbon capture and storage are described in other chapters of this book. Day-ahead, intraday, and balancing electricity markets with the participation of producers, resource aggregators, and consumers are significant.

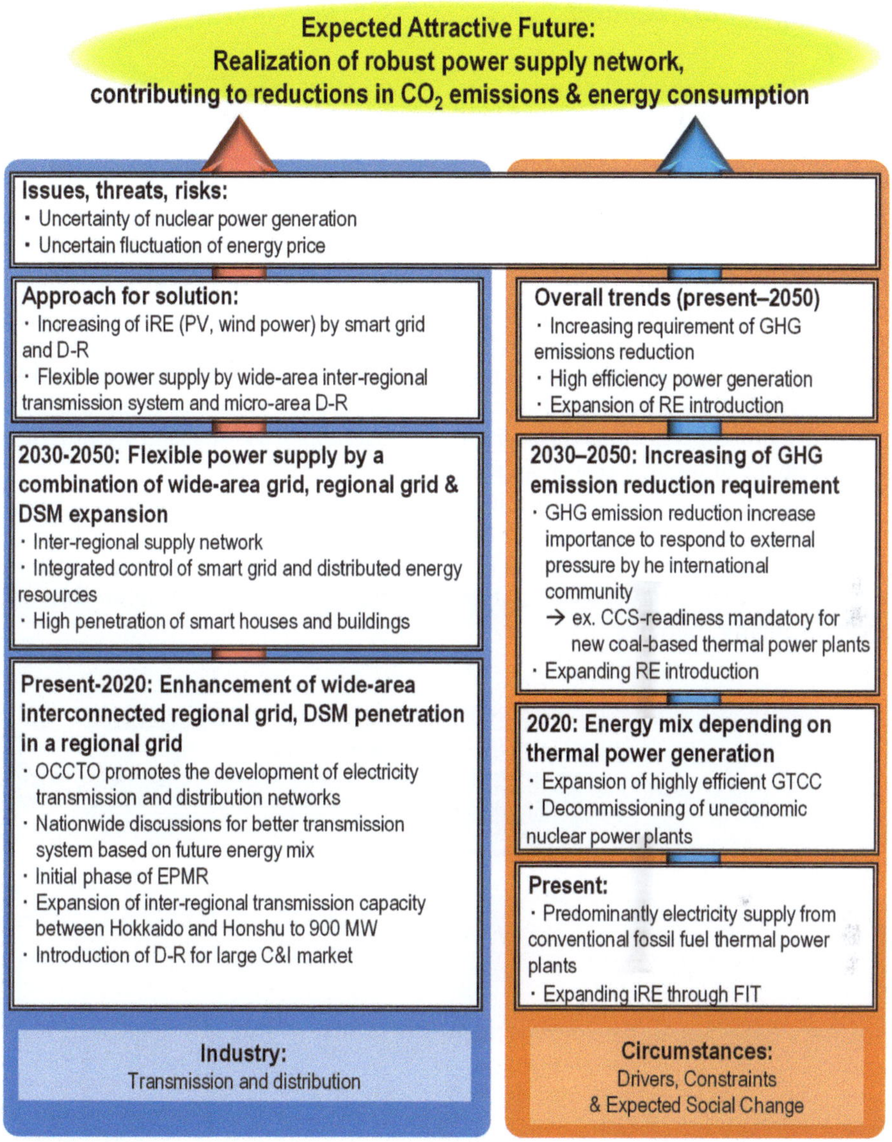

Abbreviations:

CCS: Carbon Capture and Storage, C&I: Commercial and Industry sectors, DSM : Demand Side Management, D-R: Demand-Response, EPMR: Electric Power system Market Reformation, FIT: Feed-In Tariff, GHG: Green House Gases, GTCC: Gas Turbine Combined Cycle, IGCC: Integrated coal Gasification Combined Cycle, iRE: intermittent Renewable Electricity, OCCTO: Organization for Cross-regional Coordination of Transmission Operators, Japan, PV: Photovoltaics, RE: Renewable Electricity

Fig. 4 Technology roadmap for electric power grid

3.2 Regulatory Reform

The regulatory reform will be performed in three steps (Fig. 5) [2]. "Organization for Cross-regional Coordination of TSOs (OCCTO)" will be established in step 1 during 2015. Retail will be fully liberalized in step 2 during 2016. Present utility companies will be legally unbundled in step 3, around 2020. After the reform, OCCTO will optimize nationwide balancing between TSOs. The main roles are (1) formulation of a supply-demand plan and an electrical grid plan, advancing the development of an infrastructure for transmission, such as interties between control areas and nationwide system operations beyond individual areas; (2) under normal situations, addressing coordination from the standpoint of wide-area operation, regarding the supply-demand balancing and frequency adjustment by transmission and distribution sectors in each area; (3) under a tight supply-demand situation caused by problems such as disasters, balancing supply and demand by ordering reinforcement of thermal power sources and power interchange.

Regulatory reform of the gas industry is also under discussion, including unbundling of gas pipelines. The Agency for Natural Resources and Energy (METI) endeavors to treat both this and the electric industries fairly, to transition into an integrated energy industry that can provide more consumer benefits through one-stop energy solutions.

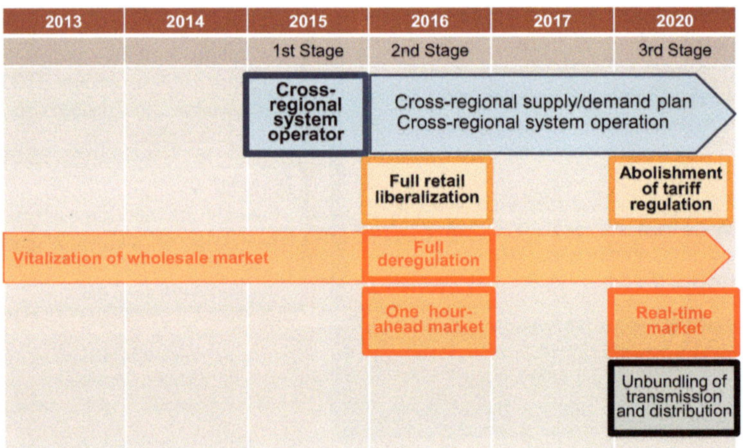

Fig. 5 Roadmap of reform in three steps

3.3 Integration of Renewable Energy Generation

Figure 6 shows challenges and solutions for PV integration. Operational challenges to RES integration are uncertainty, variability of output, and interconnection. Figure 7 shows a diagram of PV over-generation. Some electric utilities have already suspended responses to tremendous numbers of applications for connecting solar PV facilities, owing to a favorable FIT. Because the issue of restriction on grid connection is emerging, it is essential for Japan to introduce a more effective and meticulous output-control scheme to introduce renewable energy to the maximum extent possible. The METI decided to shift from the current system for introducing renewable energy to one with a new output-control scheme and revision of the current operation system for the FIT scheme.

Enhancing interregional operation for wind integration from Hokkaido to Tohoku and Tokyo is one of the solutions.

Ancillary services ensure system reliability. Some ISO/RTOs in the USA, e.g., ERCOT, allow demand-side resources to provide regulation and reserve services by demand-response programs. High penetration of variable generation such as wind power requires more flexible resource capacity.

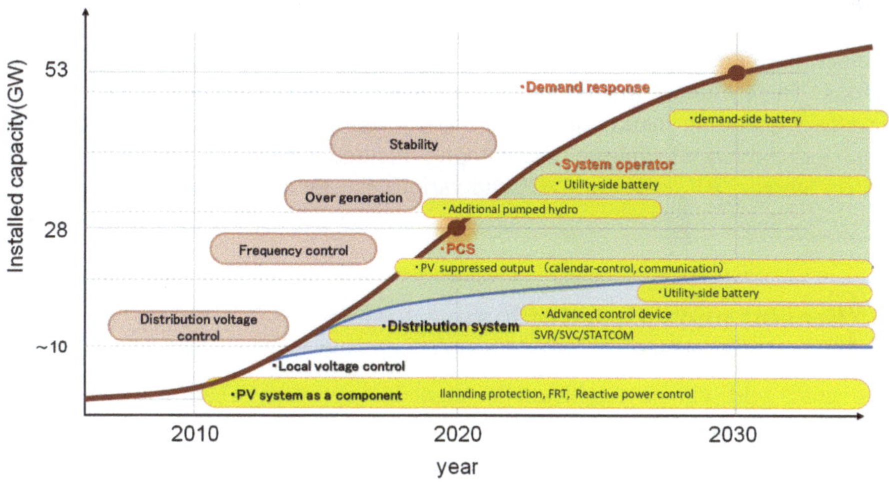

Fig. 6 Challenges and solutions for integration of PV

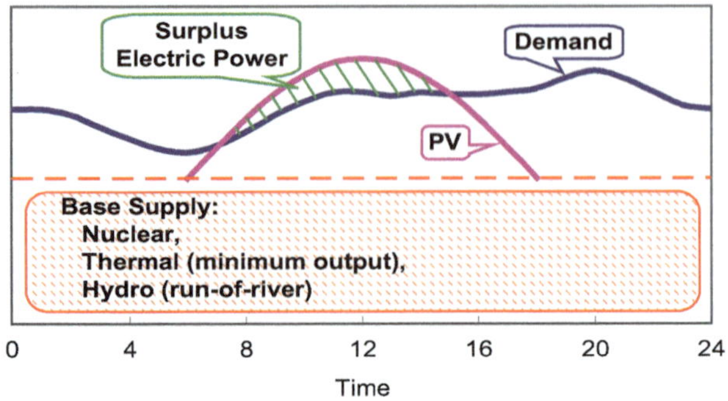

Fig. 7 Over-generation of PV

3.4 Integration of Demand Response

In the present environment, the majority of consumers are fully insulated from the dynamic wholesale electricity markets and actual real-time cost of electricity supply. Another opportunity is the use of distributed energy resources.

In the future grid, customers will be an integral part of the power system. They will help balance supply and demand and ensure system reliability by modifying the way they use and purchase electricity through various demand-response programs. These modifications come as a result of consumers having choices in full retail competitive markets and appear as a capability for load adjustment within the market. These choices involve new technologies, dynamic pricing, new forms of incentives based on market prices, and new information that transform consumer behavior. Advanced end-use technologies include smart appliances, advanced battery storage, EV, efficient CHP, and various energy management systems.

4 Benefits and Attractive Future Vision

The smart grid will be significant in maintaining system reliability, even under substantial integration of variable generation, and realize environmental benefits. Demand response will be critical in the operation of power grids after regulatory reform. Demand-side resources reduce the need for new generation resources and improve system reliability by providing ancillary services in the long run. As more intermittent RES such as PV and wind power are connected to the grid, the need for demand-response resources and balancing capabilities of power grid operators will increase.

It is important to seriously consider new energy policy that is suitable to Japan, paying attention to its unique conditions, i.e., the paucity of fossil fuels. Maintaining a stable power supply is essential for the public and industries that have supported Japanese economic prosperity. It is time to establish this new energy policy by sharing the powerful concept of future energy security across the nation. This is particularly important from a global perspective. Japan must reconsider the role of nuclear power as an option to maintain security of supply and GHG reduction. Furthermore, it is necessary to establish a desirable energy-supply structure for ensuring a stable and robust supply under the restrictive energy conditions in the country. Energy utilities with a reinforced business base contribute to building such a new structure.

References

1. Asano H, Goto M (2013) After Fukushima, the evolution of Japanese electricity market. In: Fereidoon P (ed) Evolution of global electricity markets: new paradigms, new challenges, new approaches. Elsevier, Sioshansi
2. Asano H, Okamoto H, Yokoyama A (2014) Japanese market reform with inter-regional operation, complete retail liberalization and unbundling of the transmission and distribution sector. CIGRE, Paris

Gas Supply Infrastructure

Yasuhiko Urabe, Toshio Kawamura, Takashi Sakanoue, Osamu Uno, and Yoshio Matsuzaki

Abstract In 2012, Japan imported 87 million tons of LNG or 36.2 % of total worldwide LNG imports of 239 million tons. The main sources, accounting for over 70 % of the total, are Asia-Pacific countries such as Australia, Malaysia, Russia, and Brunei. LNG production is dispersed more widely around the world, making it less exposed to geopolitical risk than oil. LNG can be transported in several ways, such as tank lorries and tank containers, to satellite bases for distribution. Gas pipelines are the primary means in Japan, where there are two kinds of gas pipelines. One is for transport of natural gas produced in domestic gas fields; the other is for transport of city gas produced in LNG terminals to areas of demand. For metering gas consumption at customer sites, there is an increasing need for remote meter reading. Furthermore, it is believed that demand will increase for services that emphasize security and safety or for those facilitating monitoring of senior citizens or those living alone via their gas use. Energy saving is another important consideration. To meet all these expectations, a smart gas meter is being developed.

Keywords Natural gas • LNG • LP gas • Pipeline • Smart meter

1 Introduction

The infrastructure of gas, one of the major energies in Japan, is described. As described in the following session, "characteristics comparison of LNG (liquefied natural gas) and LP (liquefied petroleum) gas," the LNG share in the gas market has increased, and LP gas still has an important role. The infrastructure of gas energy, therefore, is described with a focus on natural gas, with a supplemental description of LP gas. According to an energy white paper published by the Ministry of Economy, Trade and Industry (METI) [1], the ratio of natural gas as a primary energy source was ~24.5 % in 2012, the second largest behind oil (44.3 %).

Y. Urabe (✉) • T. Kawamura • T. Sakanoue • O. Uno • Y. Matsuzaki
Tokyo Gas Co., Ltd., Tokyo, Japan
e-mail: urabe@tokyo-gas.co.jp; ts_kawa@tokyo-gas.co.jp; sakanoue@tokyo-gas.co.jp; unoosa@tokyo-gas.co.jp; matuzaki@tokyo-gas.co.jp

Y. Kato et al. (eds.), *Energy Technology Roadmaps of Japan*,
DOI 10.1007/978-4-431-55951-1_12

In Japan, most energy sources are imported. Therefore, gas infrastructure includes production facilities for import, storage, and vaporization of LNG, domestic pipelines for transport from terminals to customers, and a huge number of metering devices at customer sites. This infrastructure is classified into the three categories of production, transmission, and metering, whose present status and future vision are described herein.

Although nationwide construction of pipelines is not adequate in Japan, a shift in primary energy source to natural gas is expected to proceed continuously, with support from the government. In the meantime, LP gas is important as a heat source in areas without sufficient pipelines for city gas and as an emergency fuel. LP gas has four main advantages [2]: (1) relatively small CO_2 emission, (2) established systems of supply and storage, (3) easy transport and storage, and (4) contribution to emergencies. On average in Japan, natural gas has twice the share of LP gas in the gas market on a weight basis.

2 Present Status

2.1 *Need for Robustness Against Disasters*

As shown in Fig. 1, the LNG terminal of the Gas Bureau of the city of Sendai was affected by the tsunami after the 2011 Tohoku Earthquake off the Pacific coast [2]. Because of this disaster, there were initial concerns that the base would suspend its city gas supply to Sendai for about a year, because of the period of restoration of

Fig. 1 Damage to LNG terminal from tsunami caused by 2011 Tohoku earthquake off the Pacific coast (shipping facilities for LNG tank truck) [3]

gas pipelines. However, there was success in providing an alternate supply using a gas pipeline that connects to the base from the Sea of Japan side. LP gas has additional advantages and robustness in emergencies, because there is no need for pipelines for transport to customers. Such robustness is required for disasters.

2.2 Gas Storage

Stockpiles of oil and LP gas are required by Japanese law, among which natural gas is not included. There are two types of stockpiles, private and government.

The required LP gas stockpile for the private sector is 50 days of import volume. The government stockpile also has a capability of several tens of days of such volume. As of September 2014, volumes of LP gas stockpiles were 72.0 and 28.0 days of import volume in the private and government sectors, respectively.

There are many natural gas underground storage facilities in Europe. However, Japan has only five natural gas underground storage facilities in operation, and these are limited to the storage of natural gas produced in domestic gas fields. However, if sufficient transport capacity and a robust network can be developed, wide-area gas pipelines have the potential to act as backup for large-scale demand. Thus, they are extremely important in increasing the robustness of the natural gas supply system.

2.3 LNG Production

2.3.1 Resources to Produce City Gas

2.3.1.1 Outline

City gas began to be supplied in Japan following construction of the country's first city gas plant, at Yokohama in 1872. It was initially produced from coal rather than natural gas. The proportion of naphtha used in production rose in the 1970s. After the introduction of LNG to Japan in 1969, LNG's share of resources used to make city gas continued to grow; it is presently >90 %, making it the primary resource (Fig. 2).

2.3.1.2 Characteristics of LNG

The first LNG carrier arrived in Japan from Alaska in 1969, since which LNG imports have risen drastically.

In 2012, Japan imported 87 million tons of LNG or 36.2 % of total worldwide LNG imports of 239 million tons (Fig. 3). The main sources, representing >70 % of

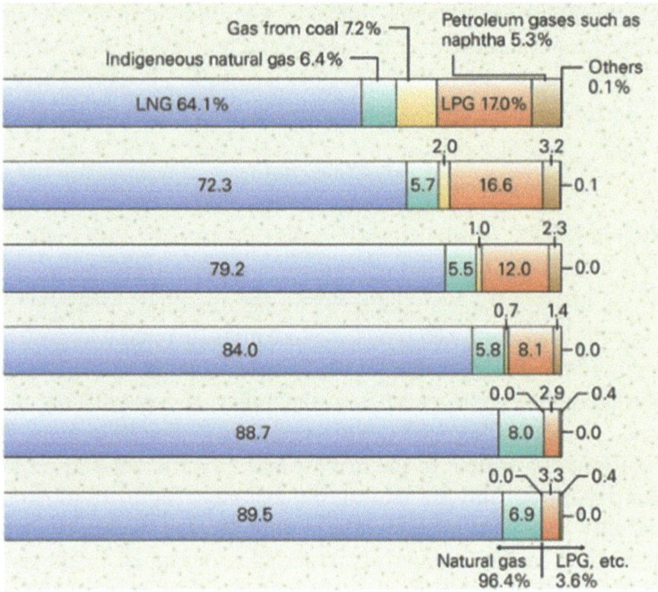

Fig. 2 Trends of resources of city gas [4]

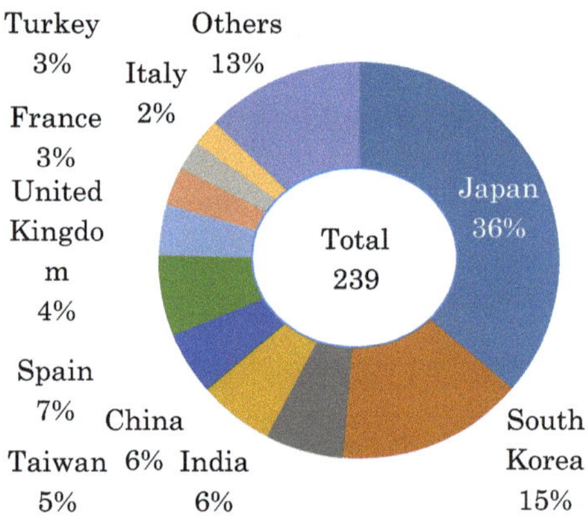

Fig. 3 LNG importers [5]

the total, are Asia-Pacific countries such as Australia, Malaysia, Russia, and Brunei. LNG production is dispersed more widely around the world, making it less exposed to geopolitical risk than oil.

Figure 4 shows the LNG caloric value of projects that began to supply LNG in years shown on the horizontal axis. Caloric values of nearly all projects used to be 43–46 MJ/m^3. Since around the early 2000s, however, the number of projects producing LNG with relatively small caloric value (~41 MJ/m^3) has steadily increased. As a result, the volume of LPG required to adjust caloric value rose in

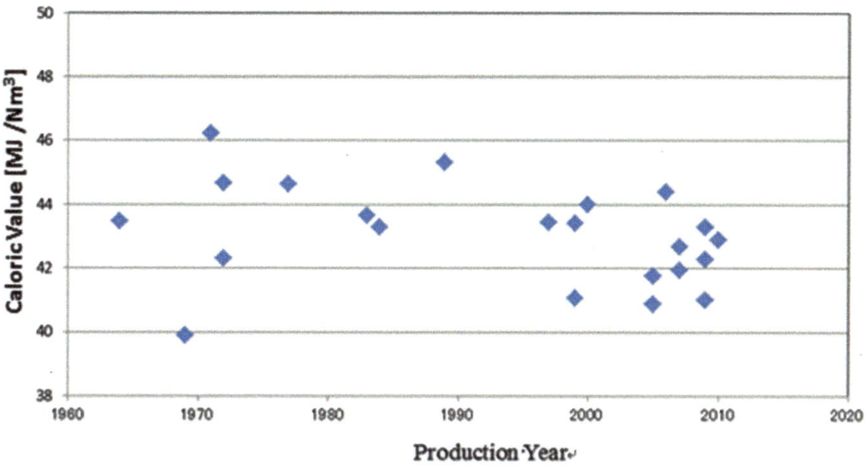

Fig. 4 LNG caloric value

Japan, so that several city gas suppliers lowered the standard caloric value of city gas to reduce the volume of LPG used. The caloric value of gas from unconventional sources now being developed (such as shale gas and coalbed methane) also tends to be low, from 40 MJ/m^3 or less to ~42 MJ/m^3.

The characteristics of LNG as a city gas resource are as follows. LNG is cooled and liquefied natural gas and under normal pressure has a very low temperature, approximately −160 °C. The main component of LNG is methane, and its caloric value varies with its percentage of ethane, propane, butane, and others. Because it has the smallest C/H ratio of any fossil fuel, it produces the least CO_2 per unit caloric value when combusted. Because impurities such as sulfur and nitrogen content are removed during liquefaction, LNG produces the least NOx and SOx emissions of any fossil fuel when combusted (Fig. 5). However, special receiving terminals and gas facilities are required to handle very low-temperature LNG. Given the importance of its low exposure to geopolitical risk and strong environmental friendliness, LNG is rated an "important energy source that will assume a growing role" in Japanese energy policy [2].

2.3.2 Facilities for Producing City Gas from LNG

2.3.2.1 LNG Terminals in Japan

Japan had 32 LNG import terminals in operation as of the end of March 2014 as shown in Fig. 6.

The main facilities of LNG import terminals are receiving, storage, and regasification (Fig. 7).

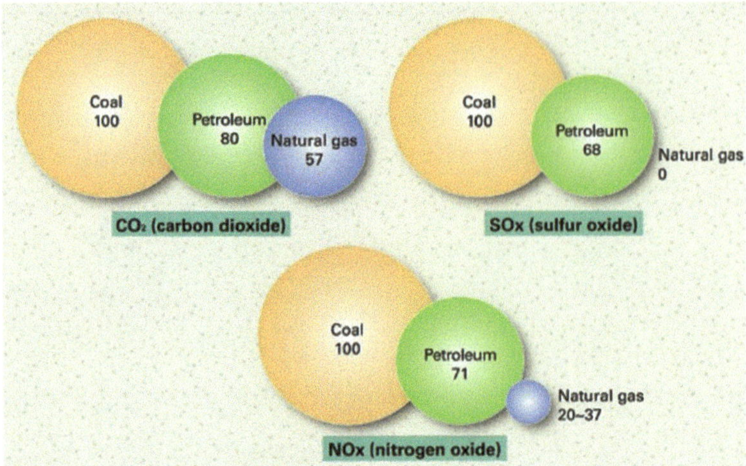

Fig. 5 Emissions of fossil fuel combustion (Coal = 100) [4]

Fig. 6 Regasification plant sites in Japan [4, 6]. There is one more LNG import terminal, "Yoshinoura-Okinawa Electric" on Okinawa Island, which is south of the area shown on this map

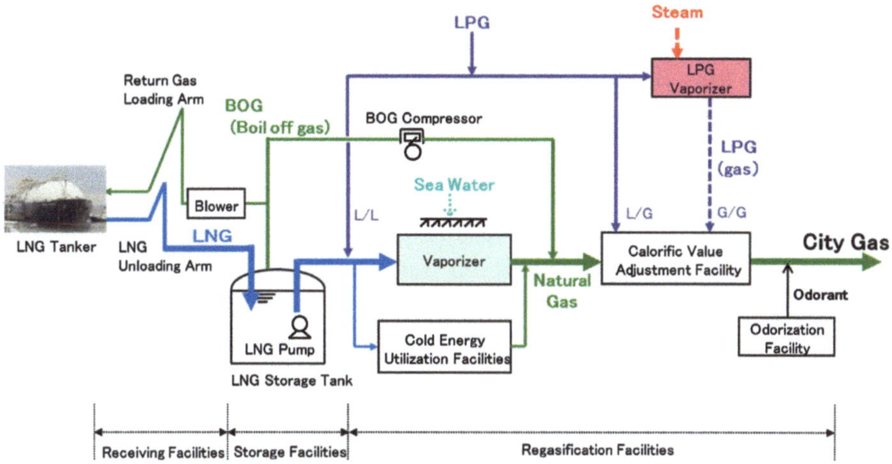

Fig. 7 Typical city gas production flowchart

LNG terminal facilities have high earthquake resistance in accordance with technological standards based on relevant laws and regulations and the industry's own engineering guidelines and are equipped to withstand a seven-class earthquake on the Japanese scale of seismic intensity. Measures are also taken to minimize as far as possible interruptions to production and supply after assuring prevention of secondary disasters caused by tsunamis, in accordance with the same guidelines.

To minimize damage to facilities and impacts beyond a terminal from a disaster or other incident, numerous safety systems are equipped to quickly detect incidents, minimize damage, and protect neighboring facilities. Smaller LNG terminals called "satellite" terminals are sometimes found at sites of demand located far from existing LNG terminals and city gas pipelines. Satellite terminals produce city gas from LNG transported to them by small LNG carriers such as lorries.

2.3.2.2 Receiving Facilities

These are facilities including jetties where large LNG carriers can berth to unload their LNG, and there are LNG unloading and return-gas loading arms connecting LNG carriers to the terminal, receiving pipes, and return-gas blowers for feeding gas back to the carriers. After berthing at a jetty, an LNG carrier is connected to the receiving terminal's facilities by LNG unloading and return-gas loading arms. LNG on the carrier is then pressurized by cargo pumps and discharged to the storage facilities on the terminal side. To maintain pressure in carrier cargo tanks during unloading, boil-off gas (BOG) is sent from the terminal by a return-gas blower. So that carriers can unberth as soon as possible in an emergency, LNG unloading and return-gas loading arms are equipped with emergency release mechanisms.

2.3.2.3 Storage Facilities

LNG received from jetties is sent to storage tanks via receiving pipes. The design and operating vapor pressure of high-capacity LNG storage tanks capable of handling large LNG carriers is normally around dozens of kPa-G, and LNG temperature in the tanks is around −160 °C. The BOG generated owing to heat input from the environment is maintained at ~0.1 wt.%/day storage capacity by a heat-insulating material. BOG is pressurized by a compressor and used as a city gas resource or fuel for nearby thermal power plants (some terminals are also equipped with "BOG reliquefaction systems," which reliquefy BOG using the cold energy of LNG and then inject it into LNG pipelines).

There are two main forms of LNG tanks, aboveground and in-ground (Fig. 8). As construction engineering has advanced in recent years, the tank capacity that can be constructed has increased. The largest capacities in 2014 were 230,000 kL (aboveground) and 250,000 kL (in-ground). LNG tanks built since the 1980s are equipped with liquid densimeters, top and bottom feed lines, and jet mixing nozzles that enable them to store LNG of dissimilar densities without causing rollover.

"Rollover" is a phenomenon that can occur when two types of LNG with dissimilar densities are introduced in the same storage tank. The heavier LNG moves to the bottom and the lighter type to the top. The density of the lower layer then decreases as its temperature rises because of heat from the outside, while the upper LNG density becomes heavier as BOG is generated. When the densities of the two approach each other, they mix rapidly, and a large quantity of BOG is generated. This should be avoided, because it can cause a sudden pressure surge in the tank.

2.3.2.4 Regasification Facilities

(a) Vaporization facilities

There are two main types of vaporization facilities, open rack vaporizers (ORVs) and submerged combustion vaporizers (SCVs) as shown in Figs. 9 and 10. ORVs vaporize LNG using water such as seawater as a heat source. Seawater flows outside of panels consisting of numerous finned heat transfer

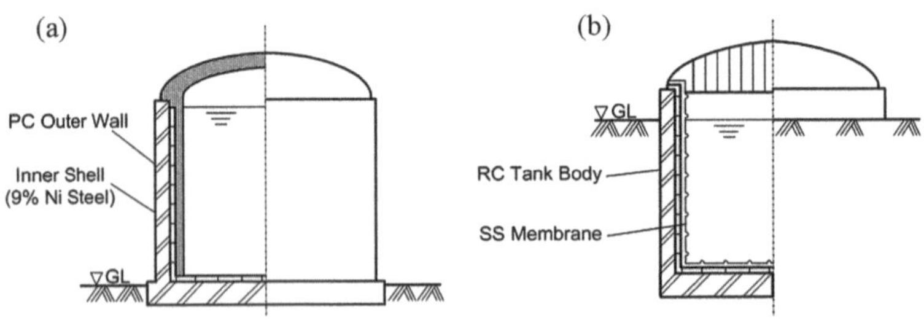

Fig. 8 LNG storage tank: (**a**) aboveground; (**b**) in-ground

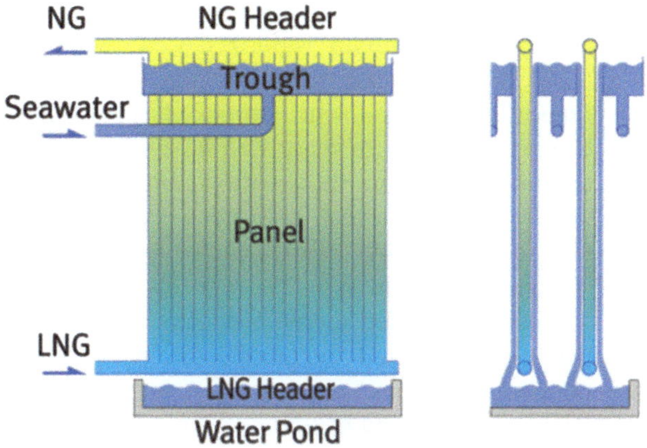

Fig. 9 Open rack vaporizer [7]

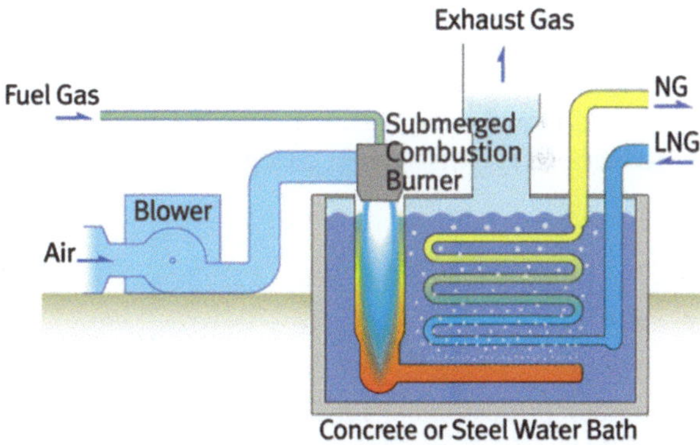

Fig. 10 Submerged combustion vaporizer [7]

tubes ("fin tubes"), and LNG is vaporized by heat exchange between the film of seawater outside the tubes and LNG inside the tubes. Because of their low-operating cost, these are widely used in Japan to meet baseload needs. However, their vaporization performance typically declines when the seawater temperature falls below ~8 °C. SCVs vaporize LNG using heat from submerged combustion of natural gas (fuel gas) as a heat source. Natural gas is combusted by a submersion burner in a concrete or steel water bath to make hot water in the bath. LNG is vaporized by heat exchange between the hot water in the bath and LNG in the heat exchanger tubes. Being less economical than ORVs, these are commonly used for peak saving and as emergency backups in Japan. Other vaporizers in addition to the above include air-fin types that use air-finned heat exchangers and hot water types that use shell and tube heat exchangers.

(b) Caloric value adjustment facilities

The caloric value of city gas is adjusted to the standard caloric value by mixing LPG or vaporized LPG with LNG or vaporized LNG. There are three ways of adding LPG to LNG to adjust the caloric value: the gas-gas (G/G) method, which mixes vaporized LNG and vaporized LPG; the liquid-gas (L/G) method, whereby LPG is mixed with vaporized LNG; and the liquid-liquid (L/L) method, which mixes LNG and LPG. The L/L method is the most economical, because vaporization facilities using this method can also be used to adjust the caloric value without any other equipment. The next cheapest is the L/G method, which does not require LPG vaporization, followed by the G/G method, which requires such vaporization.

(c) Odorization facilities

Odorant is added so that city gas can be quickly detected in the event of a leak. It is injected by pumps or drip into city gas pipelines. In Japan, odorization is a statutory requirement imposed on the city gas industry.

(d) Use of cold energy of LNG

About 840 kJ/kg of cold energy is normally contained in LNG, and effective use of such a vast amount of energy has long been studied. The main practical applications of cold energy are in BOG reliquefaction, air separation, and refrigerated warehousing, where it contributes to energy savings and cuts in CO_2 emissions. LNG is vaporized using cold energy, and the vaporized LNG is used to make city gas. Use of LNG cold energy thus also saves energy and reduces CO_2 emissions in city gas production.

2.4 *LP Gas*

The number of customers for LP gas is nearly the same as that for natural gas. Around 80 % of LP gas consumed in Japan is imported, and 20 % is generated domestically by refined crude oil. This gas in an underground gas layer exists as a mixture with other gas species, such as methane and ethane. Therefore it should be separated from the mixture and refined by removing impurities before its distribution as LP gas. LP gas dissolved in crude oil is separated and extracted from oil during petroleum refining.

2.5 *Transport*

LNG can be transported in several ways, such as tank lorries and tank containers, to satellite bases for distribution. Gas pipelines are the primary means of this distribution in Japan, where there are two types of pipelines. One is for the transport of natural gas produced in domestic gas fields. The other is for the transport of city gas produced in LNG terminals to areas of demand. The latter pipelines are primarily operated by city gas utilities. Table 1 lists the extent of these gas pipelines according to the Gas Industry Handbook published in 2013 [4]. Highly

Table 1 Total length of gas pipelines installed by city gas business entities [8]

	High	Medium	Low
Pressure level	≥1 MPa	<1 MPa, ≥0.1 MPa	<0.1 MPa
Total length	2220 km	32,644 km	250,649 km
Main role	Transport	Transport and distribution	Distribution
Material	Steel	Steel, cast iron	Steel, cast iron, polyethylene

earthquake-resistant polyethylene pipes are used for ~40 % of low-pressure (≤0.1 MPa) gas pipelines.

Historically, gas pipelines managed by city gas utilities were constructed over long periods, mainly driven by potential demand and private-sector businesses. As a result, the pipeline network has become extensive and based on LNG terminals, and LNG has become an important source of energy for households, consumers, and industrial applications in urban areas. However, it cannot be said that the development of these pipelines is progressing well.

Figure 11 shows pipeline construction conditions for high-pressure (≥1 MPa) gas pipelines in Japan that were primarily built for transport [5]. Although some regions have high-pressure gas pipelines connecting metropolitan areas, pipeline networks of each gas operator have been established separately, and standardization or unification of their specification (such as diameter) has not been achieved. Construction of gas pipelines to connect LNG terminals and main metropolitan areas has not progressed. Therefore, it cannot be said that present gas pipeline networks have sufficient robustness to secure gas supplies for large-scale demand areas in case of emergency.

Gas pipelines in Europe differ greatly from those in Japan in terms of density and network maturity. Those pipelines, connecting gas fields and consumption areas, have been established to secure a stable supply of natural resources. These have not been built in Japan. Even though a pipeline has been envisaged to transport natural gas produced at Sakhalin to Japan, such a project has not materialized.

2.5.1 Case of LP Gas

Imported and domestically produced LP gas is transported by coastal tanker ships and tank lorries from primary to secondary facilities, which are operated by import traders and wholesalers, respectively.

For residential use, LP gas is filled into compressed gas cylinders and distributed by other retail sellers, so this gas is used as a fuel in areas without gas pipelines.

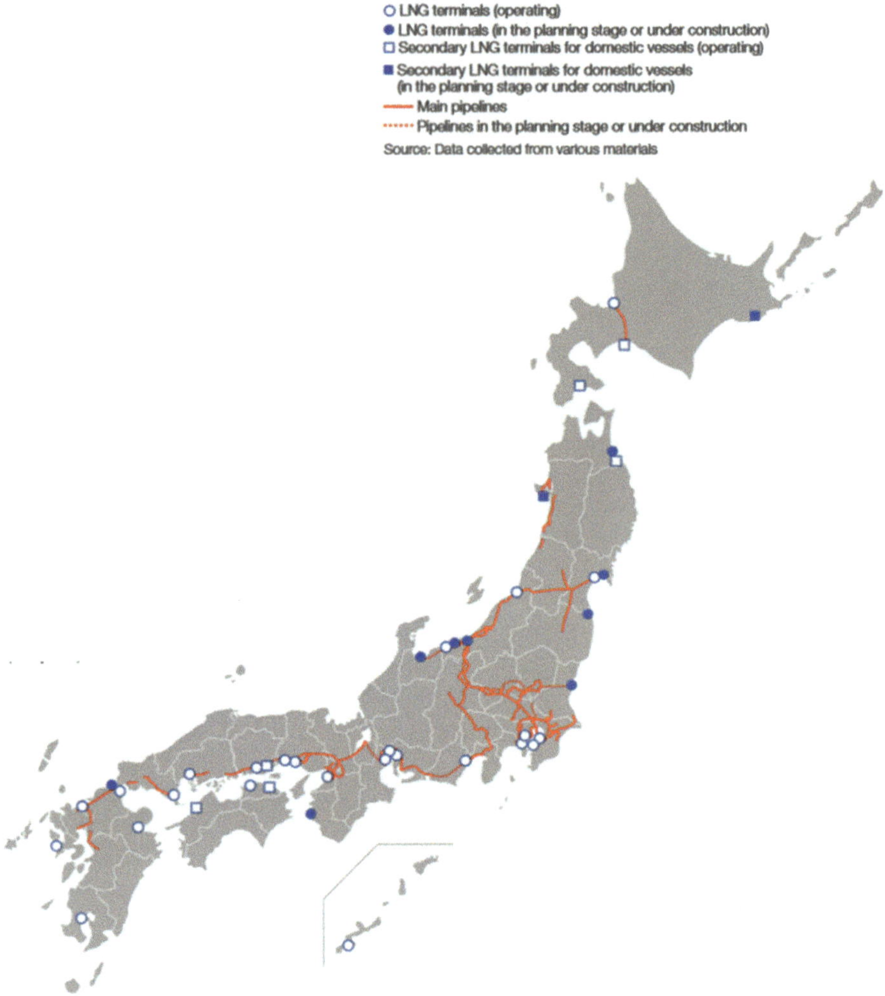

Fig. 11 Route map of high-pressure (≥1 MPa) gas pipelines in Japan [9]

2.6 *Gas Meters*

2.6.1 Overview

Japan imported all its gas meters until early in the Meiji period, 1868–1912. In 1904, the country developed its own gas meter (type A). Since then, this model has been modified and improved by manufacturers (changing the type from A through B and C and H to T). In 1969, when natural gas was first introduced in the country, the Japan Gas Association standardized the gas meter, and, accordingly, the type N meter was developed. In 1983, microcomputer-controlled gas meters called

"Micom Meters" for residential customers (see section below on their safety functionality) were developed to ensure safe gas operation. This meter was capable of detecting earthquake waves or abnormal gas flow rates; once detected, it blocks the flow of gas. Among these meters, types NB and NS (the latter adding telecommunication functionality) are the most common at present.

Manufacturers have furthered new developments in the gas measurement method of the diaphragm gas meter, and the ultrasonic gas meter entered use in 2005. This meter uses the difference in propagation times of ultrasonic waves in the gas flow to measure its rate. At locations such as industrial plants where the gas consumption rate is high, rotary, turbine, and vortex flowmeters are used instead of diaphragm gas meters.

2.6.1.1 Safety Functionality of Micom Meters

Micom Meters have various built-in sensors and a shutoff valve. The gas flow is blocked automatically during the situations listed below. When the gas flow is blocked, the user must manually restore the meter, and the gas can be used after the built-in microcomputer confirms that it is safe to operate. Gas is shut off in the event of a strong earthquake, with intensity equivalent to 5 or greater on the Japanese seismic scale.

1. The gas remains on for a long period, for reasons such as forgetting to turn off the equipment.
2. The gas hose detaches, or there is a problem in the piping, and a large amount of gas flows immediately.
3. An obstacle blocks the gas supply, reducing gas pressure.

2.6.2 Types of Gas Meters and Their Characteristics

2.6.2.1 Diaphragm Gas Meter (Fig. 12)

This meter measures the actual amount of gas used. Within the meter are chambers of a certain volume, partitioned by rubber diaphragms. After the chambers are filled with gas and discharged, a crank mechanism converts diaphragm motion into rotary motion, and the coupled mechanical counter counts the number of revolutions and displays the value. The meter is highly reliable and has been used not only in Japan but all over the world. The meter is used to measure small rates of gas flow for households, up to a maximum flow rate of 120 m^3/h.

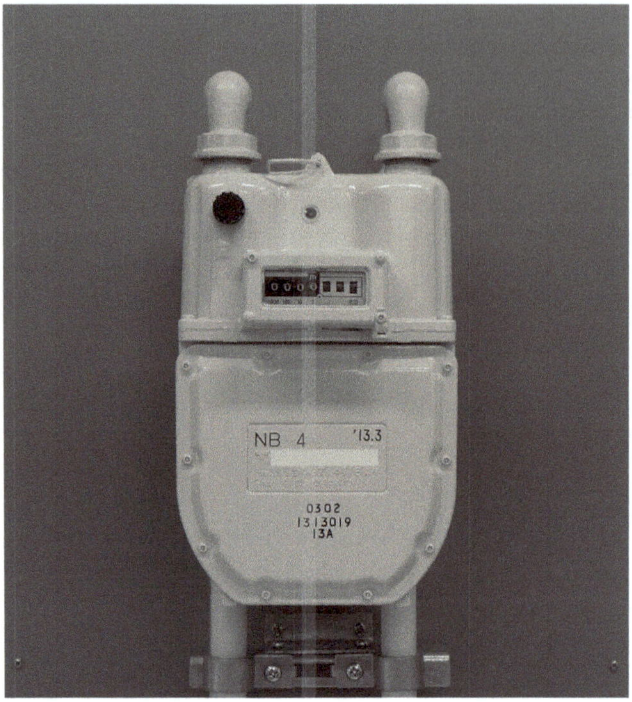

Fig. 12 Diaphragm meter (household use)

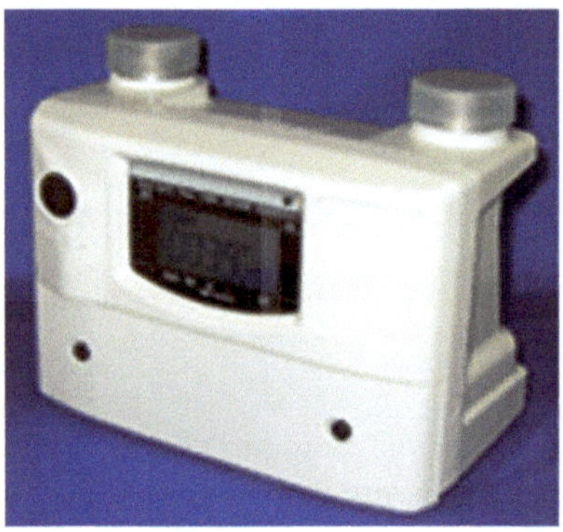

Fig. 13 Ultrasonic meter (household use)

2.6.2.2 Ultrasonic Gas Meter (Fig. 13)

This meter estimates the amount of gas used. It transmits ultrasonic waves through the gas flow channel and detects differences in propagation times of these waves to measure flow speed. It thereby computes the volume of gas used and displays the value. The compact size of this meter was achieved using the ultrasonic sensor for measurements. This also enables instantaneous measurement of flow rates. The meter is used to measure gas flow rates up to 6 m^3/h for household use and between 200 and 4000 m^3/h for industrial use (at medium gas pressure).

2.6.2.3 Rotary Meter

The rotary meter measures the actual amount of gas used. Precise measurement is possible because two rotors directly measure the volume of the gas flow. Instrumental error has not changed significantly over the years because the rotors and body of the meter do not make contact. The meter is capable of measuring gas flow rates up to 2000 m^3/h (at low and medium gas pressures).

2.6.2.4 Turbine Meter

This meter measures the volume of gas flow by counting the number of rotations of a bladed rotor within the gas flow channel. Gas volume is measured using only the bladed rotor, which makes the body of the meter small and light. This also allows precise measurement. The meter is capable of measuring flow rates up to 4000 m^3/h (at medium gas pressure).

2.6.2.5 Vortex Flowmeter

Theodore von Kármán discovered in 1911 that when a pillar is placed in a flowing medium, a repeating pattern of swirling vortices is formed in the downstream direction. The volume of gas flow in this meter is measured by counting these vortices. The simple structure of the meter allows a small and light body and precise measurement. The meter is capable of measuring flow rates up to 7500 m^3/h (at medium and high gas pressures).

3 Technology Road Map

We have classified the infrastructure of gas energy, one of the core energies in Japan, into the three categories of production, transport, and metering and described present and future states of the technology. Future states of gas energy will be tied to the advancement of many energy technologies. We hope that the following description of gas energy infrastructure improves the energy technology road map.

1. Response to trend toward diversification and lower caloric value of LNG
   The downward trend in LNG caloric value is being accelerated by rising output of LNG from shale gas, coalbed methane, and other unconventional natural gas projects. In the United States, numerous LNG projects (including shale gas) are being pursued, and Japan is set to begin importing from those in 2017.
2. Action on aging facilities
   Because some 45 years have elapsed since Japan began importing LNG, further investigation and implementation of steps to counter the aging of all types of LNG facilities will become necessary. More specifically, technologies to assess aging facilities, extend their useful lives, and upgrade large-scale plants must be developed.
3. Wide-area gas pipelines
   Considering construction costs and recovery of investment, wide-area gas pipelines clearly cannot be realized solely on the initiative of private-sector companies as before. Approaches to accelerate the construction of gas pipelines have been investigated for projects related to increasing the robustness of the industrial energy infrastructure [3, 10, 11]. As reported in the "List of Officials of Agency for Natural Resources and Energy (2014)" [12], deregulation has been broached as one measure for promoting robustness from an institutional aspect, via the construction of these pipelines. Support measures include the national government (a) exerting initiatives related to pipeline construction, (b) granting rights for gas pipelines to pass through agricultural land, and (c) introducing funding incentives like subsidies and interest waivers, which will be effective for constructing wide-area gas pipelines.
   Gas pipeline construction is more advanced overseas, so examples and lessons from other countries can be used to overcome technical challenges in Japan. Although high-strength steel pipes, undersea pipelines, and design factors local to construction sites are effective approaches for accelerating pipeline construction, all these have little or no proven record in Japan. Technical study is necessary to introduce these practices. In such a study, it is necessary to consider the unique characteristics of Japan, such as gas pipelines buried in earthquake-prone regions and densely populated cities.
4. Prevalence of smart gas meters
   There is increasing need for remote meter reading, because many residences are now using advanced home security systems. Thus, entering these houses and buildings for the inspection of gas meters has become challenging. For example, certain apartment buildings have automatic security lock systems, which require permission from a resident before entry. Furthermore, it is believed that demand will increase for services that emphasize security and safety or that permit monitoring of solitary or senior citizens through their gas use. Saving energy is another important consideration.
   To meet all these expectations, a smart gas meter is being developed to provide telecommunications functionality, remote switching systems, and detection systems that can acquire information on the amount of gas used per time zone.

4 Benefit and Future Vision

4.1 Production

Because LNG gas pricing for recent projects has been linked to the US Henry Hub price rather than JCC (Japanese Customs-Cleared Crude Oil), procurements from these projects are likely to increase because of the greater diversity of pricing mechanisms anticipated.

Advances must be made in LNG mixing technology to furnish the flexibility to receive the many types of LNG. Using simulations and actual testing on LNG tanks, the main challenges in the near future will be to clarify conditions required to mix LNG of variable densities in LNG tanks with various mixing capabilities and to enable mixing of LNG in LNG tanks without such capabilities.

4.2 Transport

Even in 2050 (the target in the third edition of the *Energy Roadmap*), natural gas is expected to be an important source of energy in Japan, and gas pipelines are anticipated to continue to be the primary method of natural gas transport. City gas utilities are actively pursuing construction of gas pipelines to meet the tremendous demand for natural gas and improve energy security [13]. In addition, as shown in the New Strategic Energy Plan of June 2014 [4], because a stable supply of energy is the foundation of energy policy, enhancing the robustness of the natural gas supply system is a major challenge. As shown in Fig. 14 and discussed in "Working for Gas Infrastructure Development" [14], it has been clearly demonstrated that wide-area gas pipelines that connect metropolitan areas and LNG bases effectively increase the robustness of the natural gas supply system.

4.3 Smart Gas Meters

To achieve the functionalities of smart gas meters, a telecommunications network is required that connects individual meters to a centralized control center, where information from each household is managed. The telephone line was once used for this purpose, but with new technological advancements such as cellular phones and the Internet, the telecommunications method for individual households has become much more diverse and liable to variation. Therefore, a method that does not depend on the telephone line has become necessary.

Consequently, infrastructure that uses radio communications is being developed to facilitate the use of smart gas meters (Fig. 15). Through this type of system, energy consumption data can be collected in detail. The collected data will be useful for

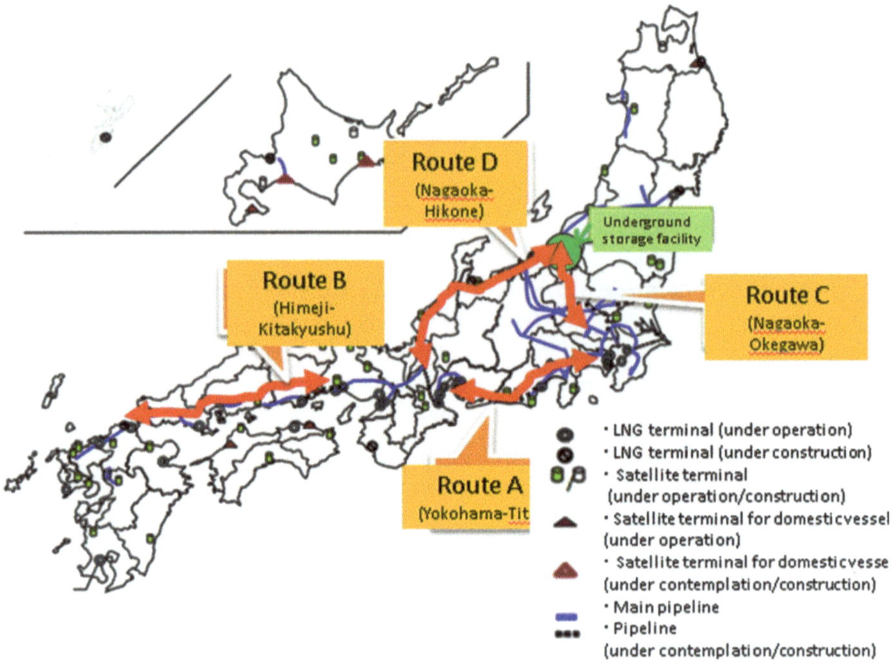

Fig. 14 Concept of wide-area transport gas pipelines to enhance energy security [12]

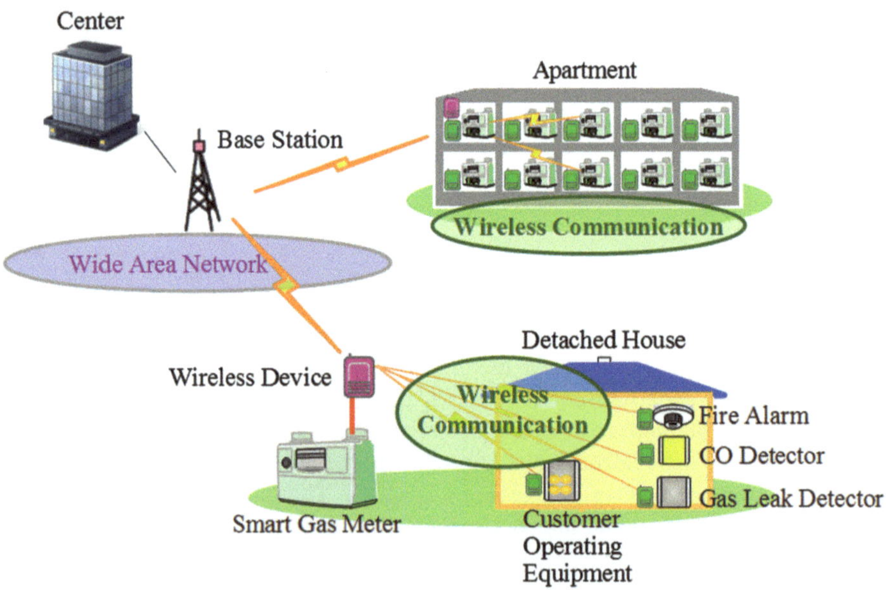

Fig. 15 Schematic illustration of remote meter-reading system

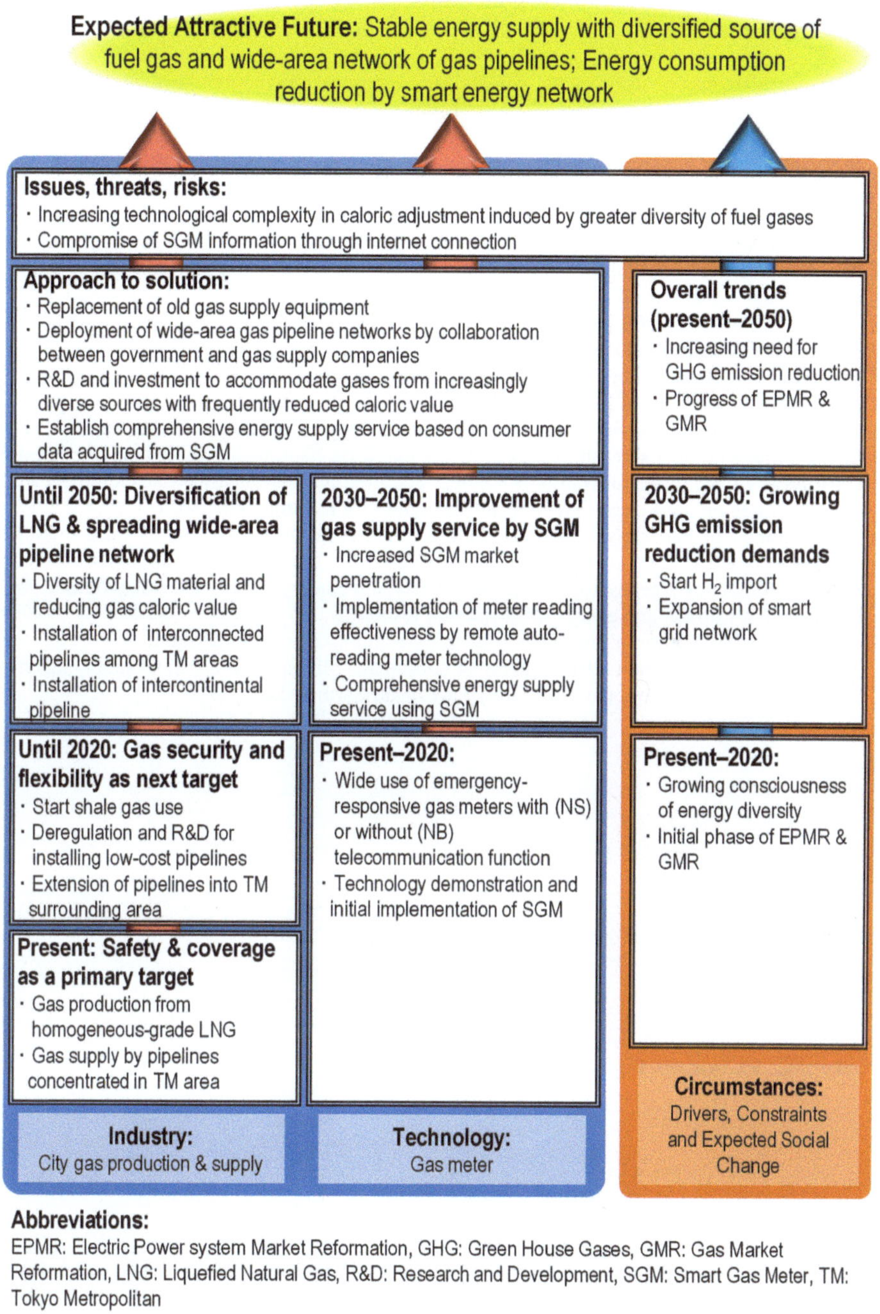

Fig. 16 Road map for gas supply infrastructure

multiple services, such as remote meter reading, remote control shutoff valve opening and closing, security, safety, monitoring services, and others.

References

1. Energy white paper, website by METI. http://www.enecho.meti.go.jp/about/whitepaper/2014pdf/
2. Energy Supply and Demand Policy Office, General Policy Division, Commissioner's Secretariat, Agency for Natural Resources and Energy (2014) the New Strategic Energy Plan
3. Working Group for Gas Infrastructure Development (2011) Report on Working for Gas Infrastructure Development
4. Japan Gas Association (2014) Gas facts in Japan 2013
5. British Petroleum (BP) Statistics 2013 (Data for fifth meeting of the Strategic Policy Committee of the Agency for Natural Resources and Energy) (2013)
6. GIIGNL (2014) The LNG Industry 2013
7. Sumitomo Precision Products Co., Ltd.
8. The Japan Gas Association (2013) the Gas Industry handbook
9. Tokyo Gas (2014) Annual Report of Tokyo Gas
10. The Advisory Committee for Natural Resources and Energy, Coordination Subcommittee, and Subcommittee on the Establishment of Infrastructure for a Natural Gas Shift (2012) Report on the Advisory Committee for Natural Resources and Energy, Coordination Subcommittee, and Subcommittee on the Establishment of Infrastructure for a Natural Gas Shift
11. Kanto Bureau of Economy, Trade and Industry (2014) Forum of for energy infrastructure resilience
12. List of Officials of Agency for Natural Resources and Energy (2014) Investigation into the enhancement of trade and energy infrastructure resilience
13. HP of Oval Corporation: https://www.oval.co.jp/techinfo/keisoku/vortex.html
14. The Japan Gas Association (2012) Report on Earthquake Mitigation Program of supplying city gas based on lessons from the 2011 Great East Japan Earthquake

Infrastructure for Next-Generation Vehicles

Seiichiro Kimura and Hiroshige Matsumoto

Abstract During and after the intense growth period of the economy in Japan around the 1960s, the number of fuel filling stations increased with the rapid spread of automobiles. However, two oil crises in the 1970s triggered the introduction of "next-generation vehicles." Examples include battery electric vehicles (BEVs), compressed natural gas vehicles (CNGVs), and hydrogen fuel cell electric vehicles (FCEVs). After the 1990s, CNGVs began to be introduced, and the development of BEVs and FCEVs accelerated. However, penetration of these next-generation vehicles was not fully successful, owing to their inferior performance (range, acceleration, durability, economic efficiency, and other factors) compared with conventional internal combustion engine vehicles (ICEVs) and a lack of infrastructure, e.g., insufficient CNG stations for CNGVs.

Since around 2010, the introduction of next-generation vehicles has progressed gradually. The higher price and shorter cruising range relative to ICEVs has been improved, and their infrastructure has expanded. FCEVs are scheduled to be on the market in 2015, and their hydrogen infrastructure is also being developed. This study discusses next-generation vehicles' fuel supply infrastructure, particularly its technical goals, challenges, and risks, and surveys Japan's past approaches and efforts and future prospects.

Keywords Fuel cell vehicle • Hydrogen station • Battery electric vehicle • Charger • Refueling station

S. Kimura (✉)
The Matsushita Institute of Government and Management, Kanagawa, Japan
e-mail: s-kimura@mskj.or.jp

H. Matsumoto
Kyushu University, Fukuoka, Japan
e-mail: matsumoto@i2cner.kyushu-u.ac.jp

Y. Kato et al. (eds.), *Energy Technology Roadmaps of Japan*,
DOI 10.1007/978-4-431-55951-1_13

1 Introduction

In 1886, Gottlieb Daimler and Karl Benz invented gasoline-fueled cars. After the Ford Motor Company produced the Model T in the USA, car prices began to decline, and ICEVs become one of the major transportation means. In Japan, motoring accelerated from the 1960s, and around 1980, the number of automobiles nearly equaled the number of households [1, 2]. The increase in the number of automobiles was rapid, which caused them to consume the most energy in transportation.

Japan's primary energy consumption in 2012 was 19.2 EJ through which various services including heat and transportation were facilitated. The transportation sector consumed approximately 17 % of this or 3.28 EJ. Eighty-nine percent of all energy consumed in this sector went to automobiles. Figure 1 shows the breakdown of energy consumption by vehicle type. Approximately 60 % of total vehicle energy consumption was for passenger cars, whereas taxis and buses consumed 2 % each. Nearly all buses use diesel fuel, and most taxis are operated by LPG fuel.

Ten percent of freight trucks have Otto engines, and 98 % of those run on gasoline. The energy consumption of diesel-fueled freight trucks constitutes 28 % of that of total vehicles, which is the second greatest after passenger cars. The number of registered passenger cars in Japan as of 31 March 2013 was 59.36 million, and the number of registered freight trucks was 14.85 million [1]. Energy consumption of freight trucks is one and a half times that of passenger cars.

Figures 2 and 3 show trends in primary energy consumption from 2000 to 2012. A consistent decrease of consumption in the transportation sector is clear in both figures. A decrease in passenger cars is particularly evident in Fig. 2. A closer look also reveals a drastic decrease of freight vehicles with diesel engines. Reasons for this are fuel efficiency improvement and a decline in the number of registered freight trucks. The fuel efficiency improvement was affected by the introduction of

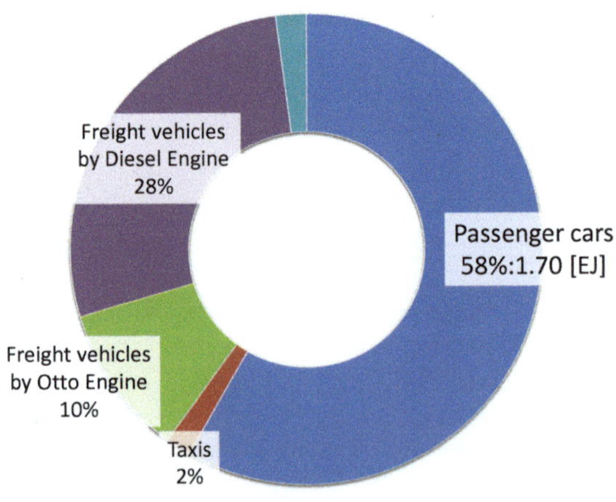

Fig. 1 Overall energy consumption in transportation sector during 2012 [3]

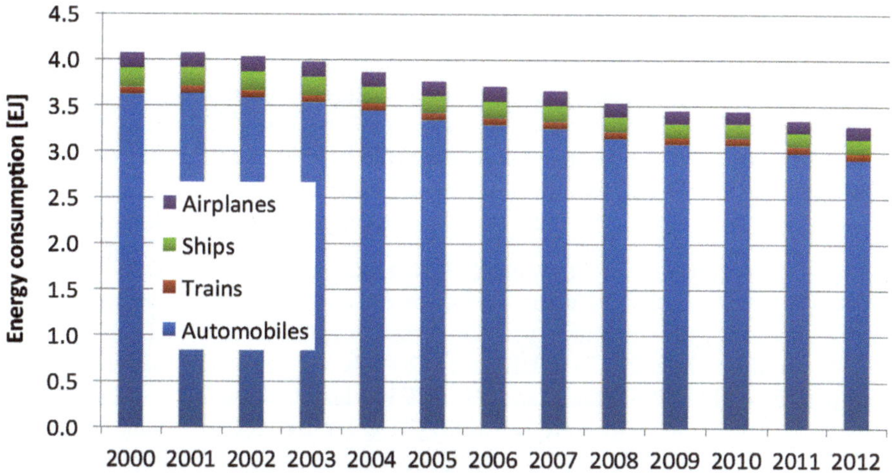

Fig. 2 Trend of primary energy consumption in transportation sector [3]

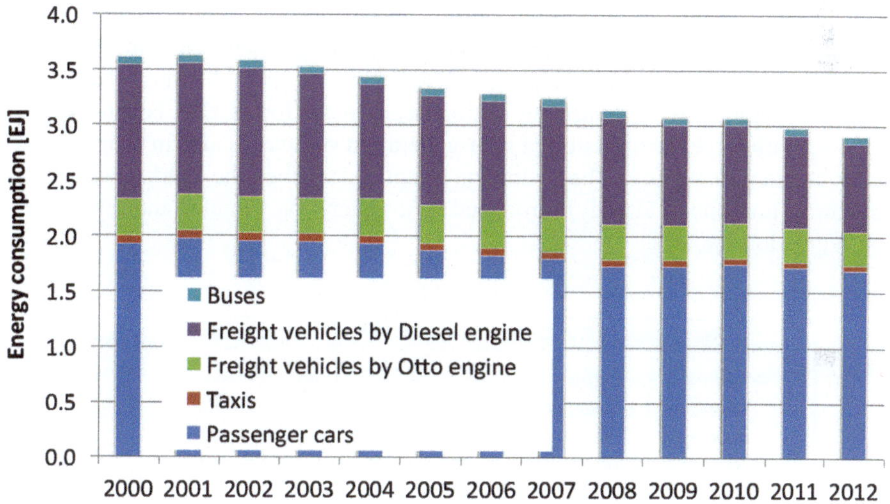

Fig. 3 Trend of primary energy consumption on type of automobile [3]

fuel-efficient vehicles such as hybrid electric vehicles (HEVs). However, these fuel-efficient vehicles still used conventional gasoline refueling stations. The number of registered freight vehicles peaked in 1991 at 21.15 million and then declined.

Taking such situations into consideration, in 2010, the Ministry of Economy, Trade and Industry (METI) specified trends and market needs of new-generation vehicles and developed a strategy in the *Next-Generation Vehicle Strategy 2010*. This strategy was planned by considering not only that next-generation vehicles are efficient in mitigating climate change caused by greenhouse gas (GHG) emissions but also the following three items:

1. Medium- to long-term energy restrictions such as continued high prices of crude oil
2. Maintaining Japanese companies' leading role in merger and acquisition activities in the field of environmental technology for automobiles
3. Economic development by industry related to next-generation vehicles

Next-Generation Vehicle Strategy 2010 also includes a concrete action plan and targets. Figure 4a shows diffusion projections with utmost private-sector efforts, and Fig. 4b shows government targets. Government goals are that total vehicle sales in 2020 should include 20–30 % HEVs, 15–20 % battery electric vehicles (BEVs)/ plug-in HEVs (PHEVs), less than 1 % FCEVs, and less than 5 % clean diesel vehicles. Targets for 2030 were set at 20–30 % for BEVs/PHEVs and ~3 % for FCEVs.

For the required infrastructure deployment, the targets are 2 million normal charging stations and 5000 quick charging stations for BEVs/PHEVs [4]. To achieve these targets, the Japanese government provided 100.5 billion JPY under its 2012 supplemental budget for charging infrastructure deployment. Based on the targets and action plan in this strategy and allocated budget, charging infrastructure for BEVs/PHEVs is rapidly progressing, as shown in the following chapter. In addition, preparation for hydrogen station deployment for FCEVs began in 2013 [5]. Preparation for the diffusion of next-generation vehicles is also in progress, but many challenges remain. The following chapter describes the market and infrastructure situations of already-introduced next-generation vehicles and related law and regulation updates.

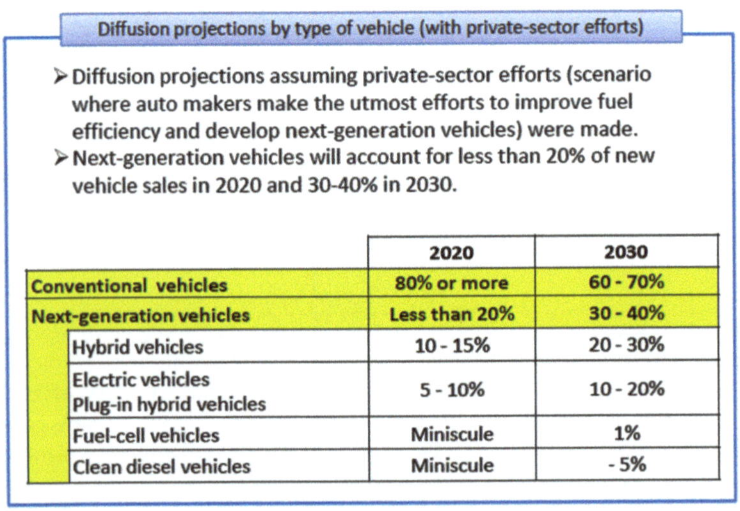

	2020	2030
Conventional vehicles	80% or more	60 - 70%
Next-generation vehicles	Less than 20%	30 - 40%
Hybrid vehicles	10 - 15%	20 - 30%
Electric vehicles Plug-in hybrid vehicles	5 - 10%	10 - 20%
Fuel-cell vehicles	Miniscule	1%
Clean diesel vehicles	Miniscule	- 5%

Fig. 4a METI next-generation vehicle plan 2010 with private-sector effort only [4]

Diffusion targets by type of vehicle (government targets)

- The government has set diffusion targets to pursue for each type of vehicle for accelerating the spread of next-generation vehicles.
- Next-generation vehicles should account for up to 50% of new vehicle sales in 2020.
- To achieve this target, the government should provide effective incentives.

	2020	2030
Conventional vehicles	50 - 80%	30 - 50%
Next-generation vehicles	20 - 50%	50 - 70%
Hybrid vehicles	20 - 30%	30 - 40%
Electric vehicles Plug-in hybrid vehicles	15 - 20%	20 - 30%
Fuel-cell vehicles	- 1%	- 3%
Clean diesel vehicles	- 5%	5 - 10%

Fig. 4b Government target on METI next-generation vehicle plan 2010 [4]

2 Present Status

To address the present status of infrastructure deployment for next-generation vehicles such as BEVs/PHEVs and FCEVs in the introduction phase, the current status of conventional ICEVs and gasoline refueling stations is first examined.

Figure 5 shows the number of vehicles and gasoline refueling stations in Japan after 1991. Vehicles here include those with more than three wheels, such as passenger cars and light and heavy duty vehicles, excluding two-wheeled vehicles. The figure shows that the number of gasoline refueling stations declined after a peak in 1993. In 2013, there were ~35,000 gasoline refueling stations, and the annual number of such stations closed was more than 1000.

The increase in the number of vehicles slowed temporarily after the Lehman Brothers bankruptcy in 2009 and Great East Japan Earthquake in 2011. The increase began again after 2012. Figure 6 shows that the number of HEVs owned has rapidly increased since 2009. Conventional gasoline refueling stations can also be used to fuel HEVs, and because they have a fuel efficiency double that of conventional ICEVs, HEVs have a longer range per liter so they need fewer visits to fuel stations. Considering that fuel efficiency of conventional ICEVs is improving and high-efficient vehicle also increases, the number of gasoline refueling stations is likely to continually decline in the future.

CNGVs were considered the potential winner among next-generation vehicles in the early 1990s, before HEVs were developed for the market. This is because CNGVs can use existing auto cycle engines, so their technical hurdles are not substantial. Moreover, CNGVs are more environmentally friendly because they emit less SOx, particulate matter, and CO_2. The numbers of CNGVs and CNG stations in Japan are shown in Fig. 7. Since around 1991, pilot CNG stations began

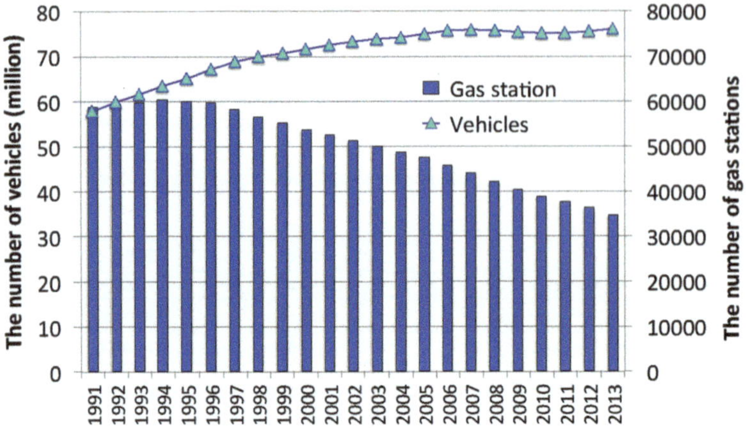

Fig. 5 Number of gasoline refueling stations and vehicles [1, 6]

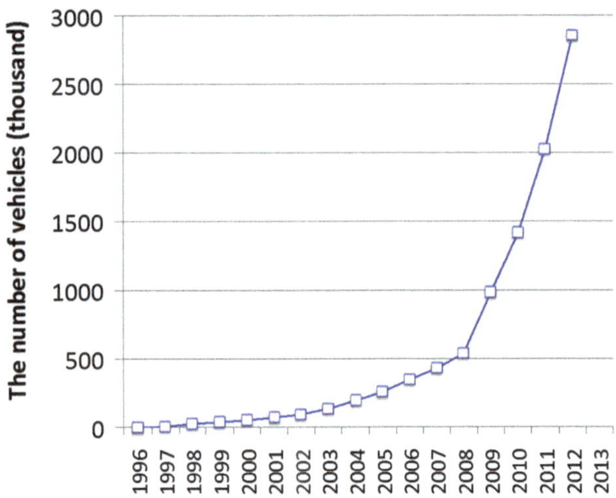

Fig. 6 Number of hybrid vehicles [7]

to be installed inside gas companies, and CNGVs were on the road for testing. CNGVs entered the market in 1996, and since then, the numbers of CNGVs and CNG stations have increased rapidly. In 2008, there were 344 CNG stations in the country. However, after the dissolution of "The Japan Eco-Service Stations Promotion Association" in 2007 [8, 9], CNG stations began to decrease, and the total number at the end of 2013 was less than 300. CNGVs, on the other hand, were adopted by cost-competitive delivery service providers and environmentally conscious transport companies. These vehicles reached 40,000 in 2010 and have slowly increased in number.

BEV diffusion was not successful in Japan until 2008, owing to their higher prices and shorter cruising ranges compared with ICEVs. Under the influence of the California state's Zero Emission Vehicle (ZEV) program, there were ~2500 BEVs

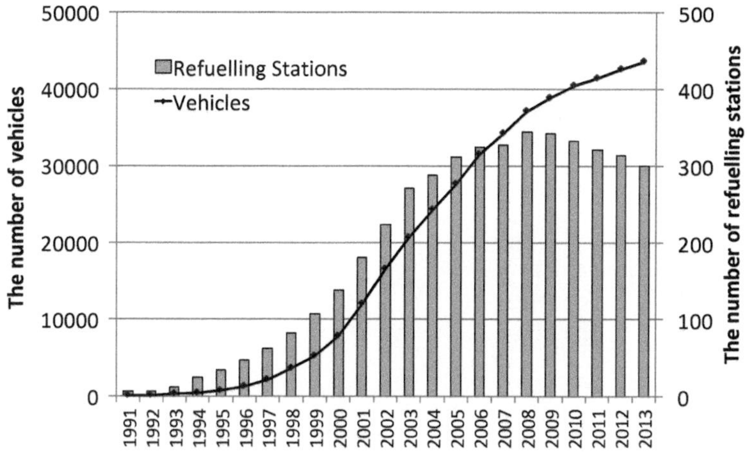

Fig. 7 Number of CNG stations and CNGVs [10, 11]

in 1996, but this decreased to 387 in 2008 [11]. After 2009, when development of the lithium-ion battery made great progress [12], automobile manufacturers began to bring BEVs and PHEVs to the market. Quick and normal charging stations have also been in development. Figure 8 (left, middle) shows quick charging station including TESLA supercharger, and Fig. 8 (right) shows a normal 200-V charger installed in a parking lot. Full charging takes about 30 min[1] for a BEV with a 24-kWh battery and CHAdeMO 50-kW quick charger (QC), whereas a 200-V normal charger takes 6–8 h. The total number of BEVs and PHEVs at the end of March 2012 was slightly less than 60,000 (Fig. 9).

FCEVs driven by fuel cells were introduced in 1994. DaimlerChrysler AG (currently Daimler AG) in Germany developed the New Electric Car (NECAR) with compressed hydrogen and a polymer electrolyte fuel cell (PEFC) for producing electricity onboard. Later, Toyota, Honda, GM, Ford, Nissan, and Matsuda followed with fuel cell development. Initially, gasoline or methanol was considered an FCEV fuel to generate hydrogen onboard because the early development stage of FCEVs was based on using existing gasoline refueling stations. Methanol was also considered for sale at modified gasoline refueling stations. However, because FCEVs with a gasoline/methanol reformer require at least several tens of seconds to start and additional fuel consumption by the reformer, FCEV development shifted to store pure hydrogen onboard [11].

The FCEV hydrogen storage method includes high pressure, metal hydride, and liquefied hydrogen, all of which were tested for verification. According to the results, 70-MPa compressed hydrogen gas has become mainstream for every automobile manufacturer [11]. It is pointed out, however, that quick filling at 70 MPa is costly in terms of pressure to added mass ratio. Therefore, cost reduction

[1] Using CHAdeMO, power 50 KW – for 24 kWh BEV.

Fig. 8 Examples of charging equipment for BEVs/PHEVs (*left*: TESLA supercharger, *middle*: quick charger, *Right*: 200-V normal charger)

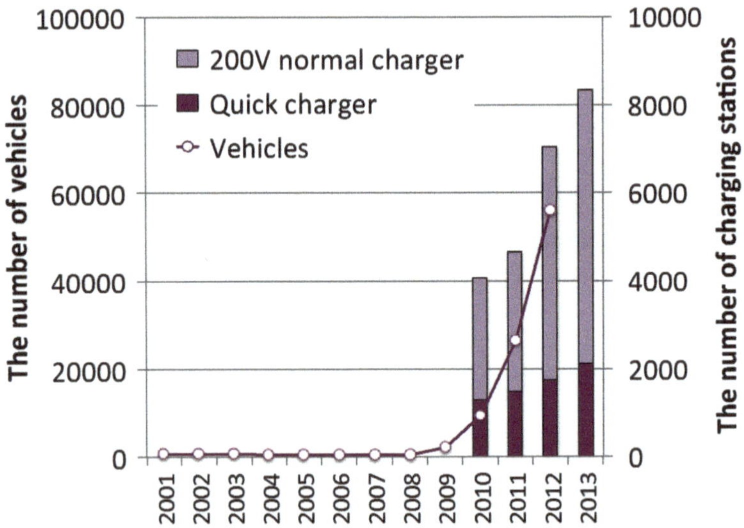

Fig. 9 Number of charging stations and BEVs/PHEVs [11, 13]

is necessary with the aid of technological development and revision of regulations [14].

The deployment of hydrogen refueling stations has also been discussed. In 2008, the Fuel Cell Commercialization Conference of Japan (FCCJ) presented a hydrogen station deployment scenario for 2015. The following year, the Council on Competitiveness-Nippon (COCN) released a proposal for FCEVs and hydrogen supply infrastructure deployment, in which hydrogen stations were to be deployed mainly in four metropolitan areas, Tokyo, Nagoya, Osaka, and Fukuoka. Later, the

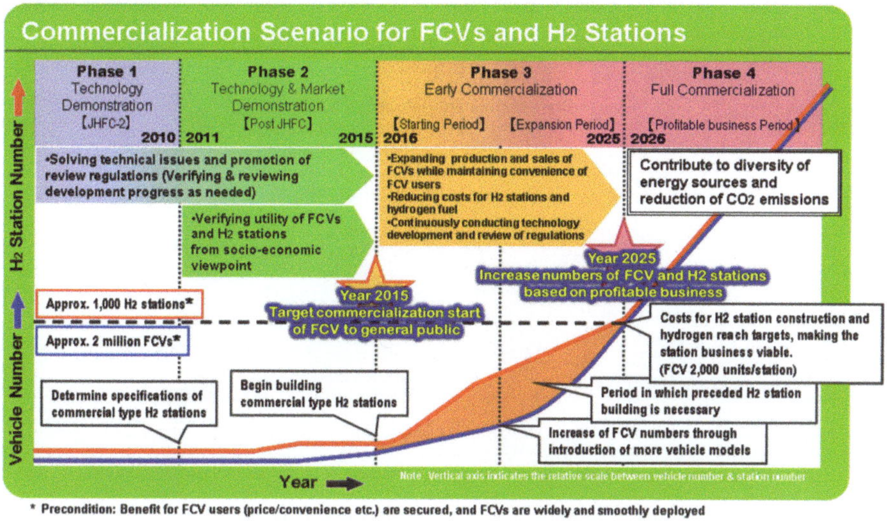

Fig. 10 Commercialization scenario for FCEVs and hydrogen stations [15]

FCCJ updated its scenario, which shows 1000 hydrogen stations by 2025 for two million FCEVs on the road (Fig. 10).

There are currently two types of hydrogen station, on-site and off-site. At an on-site station, hydrogen is generated on-site and compressed and supplied to FCEVs. It is common at such stations to generate hydrogen by natural gas or LPG reforming (Fig. 11). At an off-site hydrogen station, hydrogen is produced outside the station and delivered by trailer or pipelines and then supplied to FCEVs after compression (Fig. 12). An off-site station has less construction cost but has a hydrogen transportation cost.

The construction cost (excluding land) for one commercial on-site hydrogen station in Japan was estimated at ~500–600 million JPY [14], whereas that of an off-site station was ~400–500 million JPY [14]. This cost is greater than in Europe or the USA because of differences in materials of storage tanks and criteria and standards in the design and regulation of station construction, such as offset distance [16]. Discussions on deregulation for cost reduction are ongoing in the Japanese government. Issues under consideration include the introduction of liquid-hydrogen stations, installation of compound vessels for stations, increases in fueling pressure of hydrogen transportation trailers, and others [17]. Subsidies from METI to introduce hydrogen stations were 4.6 billion JPY in 2013 and 7.2 billion JPY in 2014, equivalent to 280 million JPY paid by the government per station [18]. Table 1 shows the distribution of 42 hydrogen stations, including those under construction. There are more than 50 stations expected by the end of 2015.

Companies have presented challenges for hydrogen station cost reduction. On 18 September 2014, Honda Motor Corporation and Iwatani Corporation announced the completion of a 20-ft container-sized “mini-hydrogen station” in the city of

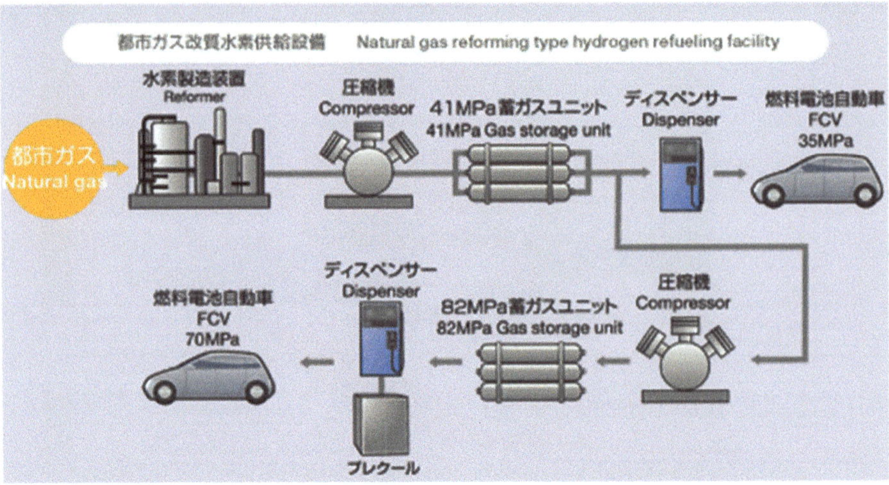

Fig. 11 On-site (natural gas reforming type) hydrogen station [14]

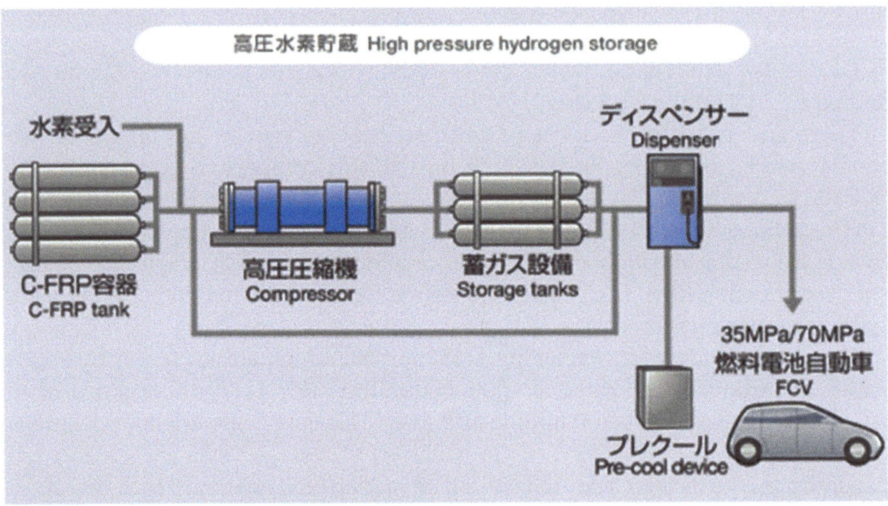

Fig. 12 Off-site hydrogen station [14]

Saitama that can produce 17-Nm^3 hydrogen per day and store 200 Nm^3 [20, 21] (Fig. 13). Figure 14 shows that via a high-pressure electrolyzer, a compressor is unnecessary at such mini-hydrogen stations. The mini-hydrogen station can be installed in 1 day, and the construction cost is much less than that of existing commercial hydrogen stations. The technology is considered appropriate in the early penetration phase with limited demand. However, because the technology did not see widespread use before October 2014, the High Pressure Gas Safety Act requires special approval by the minister of METI for its use. Therefore, legal

Table 1 Number of deployed hydrogen stations [18, 19]

Prefecture	Already deployed	Under construction
Saitama	0	5
Tokyo	4	4
Kanagawa	1	12
Chiba	1	3
Yamanashi	0	1
Aichi	3	10
Shiga	0	1
Osaka	2	2
Hyogo	0	1
Yamaguchi	0	1
Fukuoka	2	3
Saga	1	0
Total	14	42

Fig. 13 On-site mini-hydrogen station [20, 21]

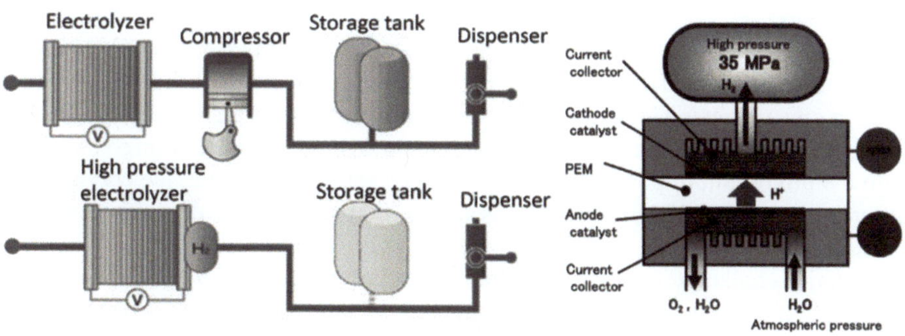

Fig. 14 Mechanism of direct-type, high-pressure water electrolysis [21]

deregulation is necessary for further cost reduction if this type of hydrogen station is introduced to facilitate FCEV market penetration [17].

3 Technology Road Map

Considering past diffusion trends of next-generation vehicles, BEVs/PHEVs and FCEVs are expected to be candidates for next-generation vehicles in the future, as described in the previous section. This section provides the technology road map of infrastructure for these vehicles as shown by Fig. 15.

There are five major challenges to be addressed in infrastructure deployment for the vehicles:

1. Appropriate locations for hydrogen and charging stations are limited, as are required spaces for station installation.
2. Lack of hydrogen demand because of the slow penetration of FCEVs.
3. Various energy supplies and coordination with energy storage systems.
4. Existing infrastructure replacement and necessity of introducing new infrastructure.
5. Cooperation of society and energy systems.

Efforts to tackle these challenges are described as follows. For charging stations, installation of public chargers for PEBs/PHEVs is in progress with active support from the government. The government target is 5000 quick chargers and 2 million 200-V normal chargers by 2020 [16]. Toyota, Honda, Nissan, and Mitsubishi motor corporations provide charging services at their local dealers and cover expenses for charging station installation. Their charging services and cost burden for installing these stations will continue for another 7 years, which is the depreciation and payout period [22].

It is expected that charging station deployment will be completed to a certain extent and that charging infrastructure will then play a different role, from mere charging to absorbing fluctuating supply energies from renewables. In other words, future charging stations may be used as a means of demand and response. It is also anticipated that BEVs/PHEVs will play a role in energy storage, including charging and discharging.

Supplier challenges include improvement of the quick charger when technology development makes it possible for BEVs/PHEVs to extend the driving range. Greater power demand is more likely in the future because of increased BEV/PHEV diffusion, so it is expected that power suppliers will respond accordingly, such as with the promotion of power system reform. To address this issue, the cooperation of three parties, BEV/PHEV users, charging station suppliers, and power suppliers, is important.

A major challenge for hydrogen stations is low usage of a station in the early introduction phase, caused by weak hydrogen demand. Building hydrogen supply chains deeply tied to a region is considered effective in solving this problem, which

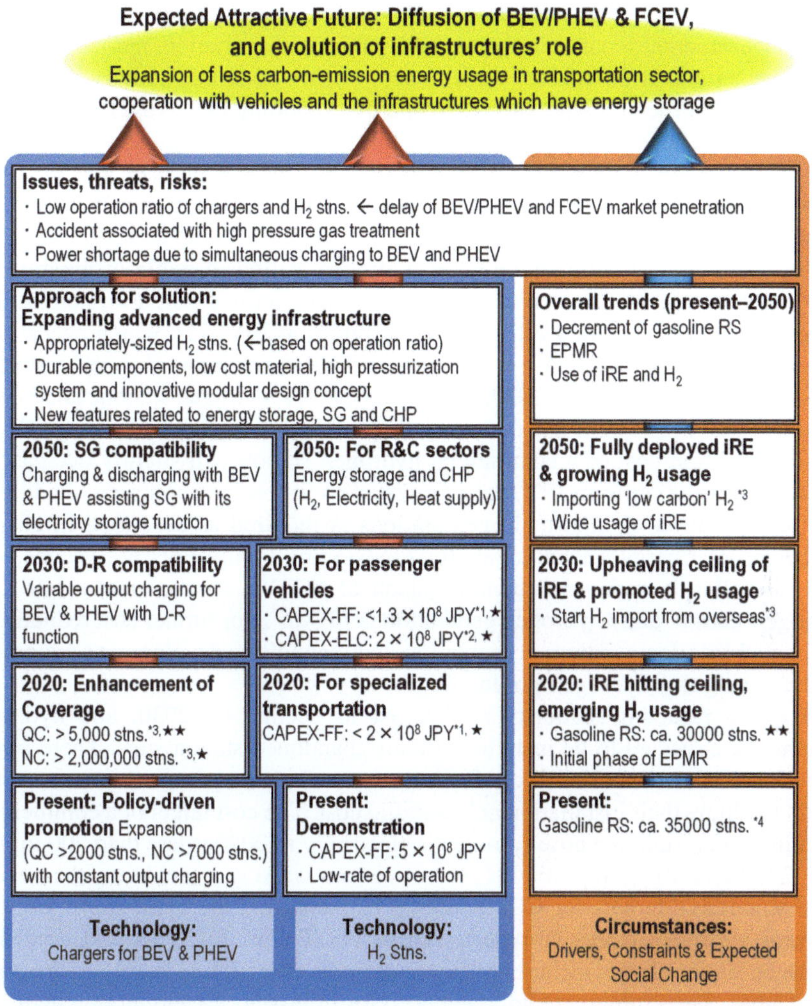

*1 NEDO target at 2020 [23]
*2 NEDO target at 2030 [23]
*3 Ministry of Economics, Trade and Industry (METI) roadmap at 2030 [4]
*4 Agency for Natural Resources and Energy, statistics at March 31st 2015 [6]

Abbreviations:

BEV: Battery Electric Vehicle
CAPEX: CAPital EXpenditure of commercial size H_2 station (Supply capacity > 300 Nm^3-H_2/h)
CAPEX-FF: CAPEX for H_2 stns. using Fossil Fuels as resource
CAPEX-ELC: CAPEX for H_2 stns. using ELeCtricity as resource
CHP: Combined Heat and Power
D-R: Demand-Response
EPMR: Electric Power system Market Reformation
FCEV: Fuel Cell Electric Vehicle
NC: Normal Charger (200V)
NEDO: New Energy and Industrial Technology Development Organization
iRE: intermittent Renewable Electricity
PHEV: Plug-in Hybrid Electric Vehicle
Refueling Station: RS
R&C: Residential and Commercial
stns.: stations
SG: Smart-Grid
QC: Quick Charger

Fig. 15 Road map of infrastructures for FCEV and BEV/PHEV

involves the introduction of hydrogen-fueled official cars, buses, and forklifts by local governments, local companies, and public transportation service providers [16]. The state is also expected to be actively involved in this activity by 2020 or thereabouts. Further, creating more demand by expanding FCEV uses is considered essential. The FCEV power supply capability is five times greater than an EV [16]. This feature is expected to be valuable in times of disaster and has great potential when applied to special-use vehicles such as power-source vehicles at construction sites.

Cost reduction of commercial-sized hydrogen stations capable of greater than 300-Nm^3/h supply is also a challenge. In particular, compressor, accumulator, and dispenser cost are currently estimated at 250 million JPY [11], which needs to be reduced by deregulation, mass production, and standardization of specifications.

It is expected that around 2020, there will be ~200 hydrogen stations and 30,000 gasoline refueling stations and that most on-road vehicles will be ICEVs and HEVs. The construction cost of a commercial hydrogen station will slightly decrease by that time, to ~300 million JPY, the same level as in foreign countries. In addition, legal deregulation of mini-hydrogen stations is required as described in chapter "Roadmap of Energy Technologies for Envisioning Future Energy Systems," for easier installation and consequent diffusion of FCEVs. At a current small-sized hydrogen station, daily hydrogen generation is 17 Nm^3. For further cost reduction at designated full depreciation, deregulation and technology development are crucial for increasing this hydrogen production.

Gasoline refueling stations are forecast to decline by 2030, and hydrogen stations are expected to be installed at their abandoned station sites. There would be no foundation cost or ancillary facilities for such new hydrogen stations, so this would facilitate their installation at less initial cost. The cost target of a commercial-sized hydrogen station whose resource is electricity in 2030 is less than 200 million JPY [23]. In addition, importing H_2 from overseas is likely, considering the trend of more aggressive targets for CO_2 emission cuts and that the hydrogen price from overseas hydrogen plants is expected to be 30 JPY/m^3 [6]. This will accelerate hydrogen station deployment, but it is expected to take time to supply hydrogen to wider areas. Therefore, there will be a need for mini-stations and mobile hydrogen stations, for which cooperation from both FCEV users and hydrogen suppliers is critical.

Around 2050, it is expected that construction costs of commercial hydrogen stations will be similar to those of conventional gasoline refueling stations in 2013. It is also projected that around 2050, CO_2-free imported hydrogen will be supplied, accelerating FCEV market penetration. The future vision also includes a network in which hydrogen stations supply hydrogen to homes and industrial buildings such as community gas businesses. Energy storage plus power and heat energy supplies using hydrogen as a medium are foreseen in this network.

Technical goals and challenges for 2050 include reliability improvement of equipment used at hydrogen stations, low-cost material and design applications, and cost reduction for hydrogen production.

If the infrastructure for BEVs, PHEVs, and FCEVs is not developed as described in the road map, both hydrogen and electric charging stations face the following challenges and risks: securing precious metals such as cobalt and platinum, procuring low-cost CO_2-free hydrogen and electricity, continued high prices for vehicles, and developing competitive technology such as hybrid vehicles. Hydrogen stations may have additional challenges and risks, such as social acceptance of hydrogen energy and stations and limited site conditions for the stations in the early stage of FCEV diffusion until ~2030. For electric charging stations, failure to secure sites on expressways could be one of the hurdles to infrastructure deployment.

4 Benefits and Attractive Future Vision

The benefits of infrastructure deployment for next-generation vehicles in Japan are listed as follows:

1. Contribution to the diffusion of next-generation vehicles, followed by reductions of energy consumption, GHG emissions, and energy procurement costs
2. Function and role as regional energy supply and load-change absorption facilities, such as smart grids
3. Ensuring international competitiveness in the area of infrastructure for next-generation vehicles
4. Enhancing social acceptance of hydrogen

Each benefit is detailed below.

4.1 Contribution to the Diffusion of Next-Generation Vehicles

Currently there are waiting times for BEVs/PHEVs at certain high-demand charging stations, such as those on expressways. Charging infrastructure is expected to progress with the aid of charging infrastructure deployment described in the preceding chapter. Therefore, BEVs and PHEVs will spread. Assuming BEV penetration, the benefit can be calculated as follows [11].

Table 2 shows the results based on the assumption of ten million BEVs and FCEVs. Ten million BEVs would be able to reduce 1.5 % (0.29 EJ) energy consumption in Japan (compared with 2010; the same comparison applies to the following). This may instead increase by ~0.5 % (0.09 EJ), depending on the means of electricity generation. Regarding CO_2 emission, electricity is supplied from natural gas, and as much as 1.4 % (18 million tons) of CO_2 can be reduced by BEVs. If electricity is supplied from renewables, 1.8 % (22 million tons) of CO_2 can

Table 2 Reductions in energy consumption, CO_2 emission, and procurement cost by introducing BEVs and FCEVs

	Energy consumption	CO_2 emission	Procurement cost
BEVs	−1.5 % to +0.5 %	−1.8 % to −0.1 %	−500 billion JPY[a]
FCEVs	−1.3 % to +0.7 %	−1.8 % to +0.1 %	−500 billion JPY[a]

[a]Based on the assumption that the average crude oil procurement cost is 1.8 JPY/MJ (2008 figure)

be reduced by BEVs. The energy self-sufficiency rate is related to prices of domestic electricity, imported oil, natural gas, and others. In the diffusion scenarios of BEVs, if fuels are generated only by domestic renewable energy, the annual contribution to energy procurement cost reduction would be 500 billion JPY by BEVs. This projection is based on the assumption of an average crude oil procurement cost of 1.8 JPY/MJ (the 2008 figure). An approximate 1 % increase in energy self-sufficiency rate is estimated from BEVs/PHEVs.

Applying the calculation above to FCEVs, assuming a widespread adoption of ten million, hydrogen stations are estimated at ~5000. This is based on the correlation between existing gasoline cars and gas refueling stations. There are now 35,000 gas stations and ~76 million vehicles (excluding two-wheel) on the road in Japan, indicating that one gasoline refueling station services more than 2000 vehicles.

By-product hydrogen in steel, oil, and other industries is estimated to be adequate to supply more than ten million FCEVs [24]. Assuming ten million of these vehicles on the road, the calculated benefit is 1.3 % (0.26 EJ) of the total energy consumption reduction in Japan [11]. This reduction is projected to increase by only 0.7 % (0.14 EJ), however, if FCEV fuel efficiency is poor and there are other energy consumption sources such as air conditioners during driving, and hydrogen is generated from fossil fuel. Regarding CO_2 emission, if all hydrogen and electricity are supplied from natural gas, as much as 1.4 % (17 million tons) of CO_2 can be reduced by FCEVs. If hydrogen is supplied by renewables, 1.8 % (22 million tons) of CO_2 can be reduced by those vehicles. In the FCEV diffusion scenarios, if fuels are generated only by domestic renewable energy, the contribution to energy cost reduction would be 500 billion JPY annually. This projection is based on the assumption of an average crude oil procurement cost of 1.8 JPY/MJ (the 2008 figure). An approximate 1 % increase in energy self-sufficiency rate by FCEVs is estimated.

There are potential increases in CO_2 emissions if fossil fuel is an energy source for next-generation vehicles, and this also depends on vehicle fuel efficiency. Therefore, it is desirable to use more renewables instead of fossil fuel for reducing CO_2 emissions.

4.2 Function and Role as Regional Energy Supply and Load-Change Absorption Facility

When hydrogen stations and refueling infrastructure are developed regionally, it is anticipated that hydrogen and refueling infrastructures linked to smart grids will function as energy supply and load-change absorption facilities.

There is already an example of a hydrogen station functioning as an energy supply facility. In the city of Kitakyushu, Fukuoka Prefecture, a pipeline and its core refueling station supply hydrogen to the region [25]. This hydrogen is then used for cogeneration at private residences and public facilities using fuel cells, and when the generated electricity supply cannot meet demand, measures such as increasing this supply can be taken according to the situation. Kitakyushu is also active in introducing FCEVs and developing hydrogen infrastructure [26].

Similarly, refueling infrastructure can help stabilize electricity supply by fueling FCEVs and BEVs/PHEVs and using fuels in various ways. An example at a commercial building demonstrates a reduction of peak demand for electricity and electricity costs by connecting the building with FCEVs [27, 28]. More effective use of energy is expected in the future from the merger of such chargers and the energy supply system.

4.3 Ensuring Infrastructure Investment Cost Reduction

In the development and diffusion phases of next-generation vehicles such as FCEVs and BEVs, infrastructure deployment is essential. The diffusion of such vehicles is greatly influenced by this deployment. Indeed, the deployment is inextricably associated with the diffusion of said vehicles. Progress in infrastructure deployment facilitates cost reduction and reliability improvement that will lead to infrastructure development. In the worldwide trend toward decreasing energy consumption and GHG emissions, the call for next-generation vehicles is expected to arise gradually from both developed and developing countries. Ahead of that time, if Japan can succeed in developing the infrastructure for those vehicles, it will ensure their diffusion across the world. Regarding energy storage, in some parts of Europe, surplus power from wind power generation is converted to hydrogen for storage and supply [29, 30]. One of the concerns of hydrogen infrastructure deployment is a lack of investment, owing to the weak demand of hydrogen for FCEVs. Therefore, if Japan succeeds in spreading FCEVs together with capacity-flexible mobile hydrogen stations, that would work even when hydrogen demand is weak. The impact would likely have an effect on Europe.

Japan's initiative for next-generation vehicle infrastructure deployment will therefore contribute to its overseas development in the future and reinforcement of the automobile industry's competitiveness and progress.

4.4 Enhancing Social Acceptance of Hydrogen

Hydrogen, which has not been widely used as an energy medium, is to be used for FCEVs. Generally, in Japan, not only is awareness of the safety of hydrogen energy sources poor but they have less technological reliability compared with gasoline, which has been used widely as vehicle fuel [31]. Therefore, more awareness and reliability of hydrogen is required among the public to introduce hydrogen infrastructure for further diffusion of the FCEV. Measures for this purpose can include advertisements and FCEV demonstration rides. Group interviews in Kitakyushu hydrogen town last year by the authors demonstrated the effectiveness of illustrating the safety of hydrogen stations and pipelines in use, toward broader and deeper understanding of hydrogen [32]. The more the public understands hydrogen, the greater the likelihood of FCEV purchase. FCEV diffusion and infrastructure development will then result in hydrogen's acceptance in society. The key to such acceptance is to introduce hydrogen infrastructure safely and reliably. The acceptance of hydrogen in one area may contribute to that in another area where understanding of hydrogen is more challenging, ultimately achieving acceptance in society as a whole. Each technology broached in this chapter has its own unique and important role. They are significant from a global perspective and can meet international standards. It is necessary to spread them appropriately to avoid specific optimization for domestic market, which is called as the "Galápagos syndrome" in Japan.

References

1. Automobile Inspection and Registration Information Association. Jidousya Hoyuu Daisuu no Suii. http://www.airia.or.jp/number/pdf/03_1.pdf. Accessed 17 Oct 2014
2. National Institute of Population and Social Security Research. Nihon no setaisuu no Syourai suikei (Zenkoku suikei) 2013nen 1 gatsu suikei. http://www.ipss.go.jp/pp-ajsetai/j/HPRJ2013/gaiyo_20130115.pdf. Accessed 17 Oct 2014
3. Agency for Natural Resources and Energy, Comprehensive Energy Statistics. http://www.enecho.meti.go.jp/statistics/total_energy/. Accessed 17 Oct 2014
4. Ministry of Economy, Trade and Industry (2010) Next-Generation Vehicle Plan 2010 http://www.meti.go.jp/english/press/data/pdf/N-G-V2.pdf
5. Next Generation Vehicle Promotion Center. FY2013, Subsidized project for hydrogen station deployment for fuel cell vehicles. http://www.cev-pc.or.jp/hojo/suiso_index_h25.html. Accessed 4 May 2015
6. Agency for Natural Resources and Energy. Kihatsuyu hanbaigyousyasuu oyobi kyuuyusyosuu no suii (touroku base). http://www.enecho.meti.go.jp/category/resources_and_fuel/distribution/hinnkakuhou/data/20140708_ss25fy.pdf. Accessed 17 Oct 2014
7. Next Generation Vehicle Promotion Center. Denki Jidousya tou hoyuu daisuu toukei (suiteichi). http://www.cev-pc.or.jp/tokei/hanbai.html. Accessed 17 Oct 2014
8. Tatsuzo Minami (2003–2004) Subsidy system for eco station (fuel filling facility). Clean Energy 12(4):16–18

9. House of representatives, The 166th Diet, Economy, Trade and Industry Committee, No. 5, April 4, 2007. http://www.shugiin.go.jp/internet/itdb_kaigiroku.nsf/html/kaigiroku/009816620070404005.htm. Accessed 3 May 2015
10. Organization for the Promotion of Low Emission Vehicles. Ogata ten-nen gasu torakku monita soukou ni tsuite. Conference material of the 59th Natural Gas Vehicles' Forum (2014.6.23)
11. Kimura S (2014) A study on the evaluation method of social impact by the introduction Next Generation Vehicles. Ph.D. thesis of Kyushu University
12. Nykvist B, Nilsson M (2015) Rapidly falling costs of battery packs for electric vehicles. Nat Clim Chang 5:329–332
13. Seiichiro Kimura, Material for EV・ PHV town symposium in Tokyo (2013.06.05)
14. Japan Hydrogen & Fuel Cell Demonstration Project. http://www.jari.or.jp/portals/0/jhfc/. Accessed 17 Oct 2014
15. Fuel Cell Commercialization Conference of Japan. Commercialization scenario for FCVs and H2 stations. http://fccj.jp/pdf/22_cse.pdf. Accessed 17 Oct 2014
16. Agency for Natural Resources and Energy (2014) The strategic road map for hydrogen and fuel cells. http://www.meti.go.jp/press/2014/06/20140624004/20140624004-2.pdf
17. Ministry of Economy, Trade and Industry. Saikin no nenryou denchi jidousya suiso sute-syon ni kansuu kisei minaoshi no sinchoku joukyou ni tsuite. http://www.meti.go.jp/committee/sankoushin/hoan/koatsu_gas/pdf/005_04_01.pdf. Accessed 17 Oct 2014
18. Next Generation Vehicle Promotion Center. Heisei 25 Nendo Nenryo denchi you suiso kyoukyu setsubi setti hojyo jigyo koufu kettei naiyo. http://www.cev-pc.or.jp/hojo/suiso_ketteinaiyou_h25.html. Accessed 17 Oct 2014
19. The Research Association of Hydrogen Supply/Utilization Technology. http://hysut.or.jp/en. Accessed 30 Sept 2014
20. Honda Motor Co., Ltd. (2014) News Release at 18th Sept., 2014, http://www.honda.co.jp/news/2014/4140918.html
21. Okabe M, Nakazawa K (2010) Effort to develop Honda Solar Hydrogen Station (SHS). Hydrogen Energy Syst 35(3):22–28
22. Nippon Charge Service, LLC. Charging infrastructure deployment project guide. http://www.nippon-juden.co.jp/wp/wp-content/uploads/2014/08/info_140801.pdf. Accessed 4 May 2015
23. New Energy and Industrial Technology Development Organization. Roadmap for fuel cell and hydrogen technology development. http://www.nedo.go.jp/library/battery_hydrogen.html. Accessed 6 May 2015
24. Development Bank of Japan. Fuel cell updates and challenges for its diffusion. http://www.dbj.jp/reportshift/report/research/pdf/83_s.pdf. Accessed 4 May 2015
25. Shibata T (2011) The Kitakyushu Smart Community Creation Project. Hydrogen Energy Syst 36(1):25–28
26. New Energy and Industrial Technology Development Organization (2015) http://www.nedo.go.jp/content/100567362.pdf
27. Nissan Motor Corporation (2013) News Release at 29th Nov., 2013. http://www.nissan-global.com/JP/NEWS/2013/_STORY/131129-01-j.html
28. Toyota Motor Corporation Develops Fuel-cell-bus Power Supply System. http://www2.toyota.co.jp/jp/news/12/08/nt12_0810.html. Accessed 4 May 2015
29. Gahleitner G (2013) Hydrogen from renewable electricity: an international review of power-to-gas pilot plants for stationary applications. Int J Hydrogen Energy 38:2039–2061
30. Germany Trade & Invest. Green hydrogen and power-to-gas technology. http://www.gtai.de/GTAI/Content/EN/Invest/_SharedDocs/Downloads/GTAI/Fact-sheets/Energy-environmental/fact-sheet-green-hydrogen-mass-energy-storage-for-future-en.pdf. Accessed 6 May 2015
31. Saito A (2014) Hydrogen social acceptance in Japan. Hydrogen Energy Syst 39(3):165-169
32. Itaoka K, Kimura S. A study on social acceptance of hydrogen technology in hydrogen utilization developed regions. Proceedings of the 30th conference of energy, economics and environment

Part V
Electric Power Generation and Its Backend Technology

Takao Nakagaki

Electricity is converted from various types of primary energy sources, such as fossil fuels, uranium and renewable. This part covers principal technologies for large-scale electric power generation systems connected directly to the power grid, including the backend of the nuclear fuel cycle. The share of nuclear power generation was about 30 % before the Great East Japan Earthquake. Thermal power generation uses coal, natural gas and petroleum equally as fuel, owing to energy security. Because of high mountains and a warm oceanic climate, hydropower can generate around 10 % of annual total electricity, and in the recent decades, pumped storage hydropower has increased for load-leveling use, especially in the summer peak season. The share of solar, wind and geothermal power generation is less than 1 %, but solar power generation has been dramatically increasing after enforcement of the Japanese "Feed-in-tariff".

Regarding the backend issue, reprocessing technologies of used nuclear fuel for recycling fast breeder reactors may be one solution, in addition to recycling back into MOX nuclear fuel for thermal reactors. To avoid CO_2 emission in tail gas, imported hydrogen is expected even for large-scale power generation in the future, in addition to CCS technology.

T. Nakagaki (✉)
Waseda University, Tokyo, Japan
e-mail: takao.nakagaki@waseda.jp

Thermal Power Generation

Takao Nakagaki

Abstract Thermal power plants will be a promising power source even in the 2050s, because they can generate a vast amount of electricity with low cost, high reliability, and stability. The plants will also have strong flexibility and controllability to compensate the gap between power demand and supply with coexistence of a certain amount of unstable renewable power sources. Coal, oil, and liquefied natural gas (LNG) are mainly used for thermal power generation in the electricity business and are evenly mixed because of energy security, the so-called best mix. After the Great East Japan Earthquake, thermal power generation of all electric companies drastically increased by 164 TWh (+33.9 %), to compensate their nuclear power generation.

Japan now has world-class, excellent thermal power technologies with heavy-duty steam turbines under the ultra-supercritical (USC) steam condition and high-efficiency 1600 °C class LNG-fired gas turbines. By 2030, if technical development projects of "advanced USC" and 1700 °C class gas turbines are completed successfully, 46 and 57 % net thermal efficiencies at the sending end (higher heating value, HHV) will be achieved in commercial power plants, respectively. By 2050, the integrated coal gasification fuel cell combined cycle (IGFC) is expected to appear, and CO_2 capture and storage (CCS) will move forward with full-scale implementation. However, introduction costs of these cutting-edge technologies are uncertain, and another technology for mitigation of operating restriction or life extension of aging plants may be preferable, depending on the situation after unbundling and electricity market liberalization by the "Electricity System Reform" in a few years.

Keywords Boiler turbine generator • Integrated coal gasification combined cycle • Gas turbine combined cycle • Ultra-supercritical • Turbine inlet temperature

---

T. Nakagaki (✉)
Waseda University, Tokyo, Japan
e-mail: takao.nakagaki@waseda.jp

Y. Kato et al. (eds.), *Energy Technology Roadmaps of Japan*,
DOI 10.1007/978-4-431-55951-1_14

1 Introduction

Japan is one of the world's major economic powers, where more than 120 million people live in prosperity, but it has a fundamental vulnerability because the country is strongly dependent on imports for most energy resources. After the Great East Japan Earthquake on 11 March 2011 and ensuing disaster, many energy utility customers have been diversifying their energy supplies, using combined heat and power, emergency power generation facilities, and gas-fired air conditioning systems for their business continuity plans. However, stable grid electric power is still the most important utility for the commercial and residential sectors, as well as many manufacturing industries that supply made-in-Japan products to domestic and global markets. Power line frequency differs between eastern (50 Hz) and western (60 Hz) Japan and the capacity of frequency converters between them is limited, but this is planned to increase from 1.2 to 2.1 GW [1]. Because the restart of nuclear power plants remains uncertain and market penetration of abundant renewable energy takes much time, thermal power generation is the main power source for the near- and midterm to secure GW-capacity bulk power sources and supplemental reserves near power-consuming areas.

Japan's electricity business including nuclear power stations is fully operated by private companies that bundle generation, transformation, and distribution. The thermal power generation business consists of regional monopoly general electric enterprises (ten companies), a wholesale trade electric enterprise (J-POWER), and joint or cooperative thermal power companies (7.6 % share in FY 2010) [2]. The Agency for Natural Resources and Energy (ANRE) publishes monthly electric power statistics [3] according to fuel type, including coal, biomass, heavy oil, crude oil, LNG, and others (naphtha, city gas, condensate, coke oven gas (COG), and liquefied petroleum gas (LPG)). However, all production of electricity is compiled into a numerical value of thermal power generation. There are gas turbines, steam turbines, and fuel cells as power sources. In fact, all coal and oil are used for boiler turbine generators (BTGs) with Rankine cycle, and LNG is used for a combined cycle with Brayton-Rankine cycles and BTGs. There is no large-scale fuel cell power plant for commercial use. The 250-MW Nakoso-plant No. 10 operated by Joban Joint Power Co., Ltd. is the only commercial integrated coal gasification combined cycle (IGCC) power station. Thermal power generation converts all chemical energy of fuel into heat energy of combustion gas, and subsequently, a thermal engine and generator set converts that heat energy into electrical energy via mechanical energy. With regard to these conversion pathways, the conversion efficiency from mechanical to electrical energy has already reached 99 % [4]. Therefore, thermal efficiency is mainly restricted by working fluid temperatures of heat engine cycles, i.e., main/reheat steam temperature and pressure for steam turbines and turbine inlet temperature (TIT) for gas turbines. A technology direction is improving heat resistance of machines to adapt to higher temperatures. Japan has world-class technologies for both steam and gas turbines.

2 Present Status: Target Market, Efficiency, Cost, Regulation

2.1 Overview of Thermal Power Generation Before and After Earthquake Disaster

After the earthquake disaster and subsequent nuclear power plant failure, the status of thermal power plants drastically changed. A summary of commercial thermal power generation operated by ten electric power companies and J-POWER in FY 2010 and 2014 is shown in Figs. 1a, b and 2a, b, respectively. All data are from published databases of ANRE's statistical survey of electric power [3]. The calculation procedure is as follows.

Fuels are categorized into three types: coal (coal/biomass), oil (heavy oil/crude oil/natural gas liquids), and LNG (LNG/gaseous fuels of NG, LPG, and COG). Installed capacity, generated energy, CO_2 emission, and thermal efficiency were calculated. CO_2 emission coefficients refer to Japan Gas Association data [5].

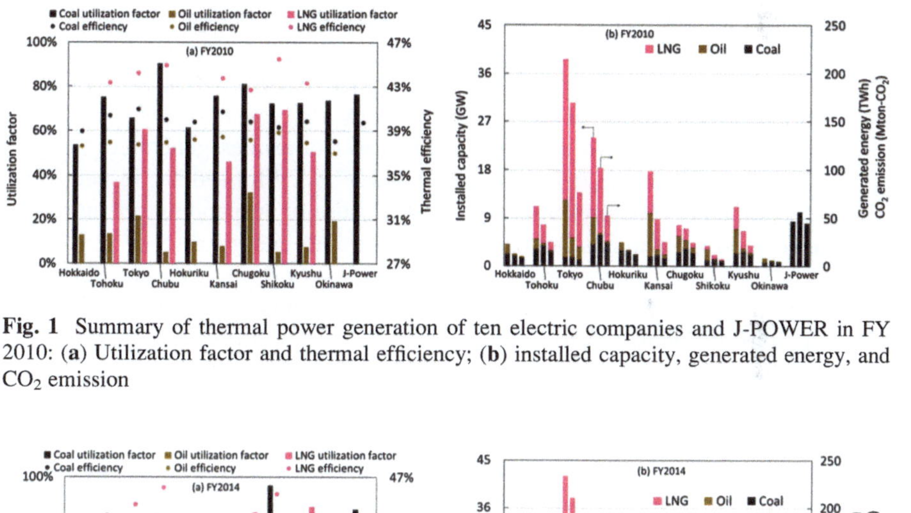

Fig. 1 Summary of thermal power generation of ten electric companies and J-POWER in FY 2010: (**a**) Utilization factor and thermal efficiency; (**b**) installed capacity, generated energy, and CO_2 emission

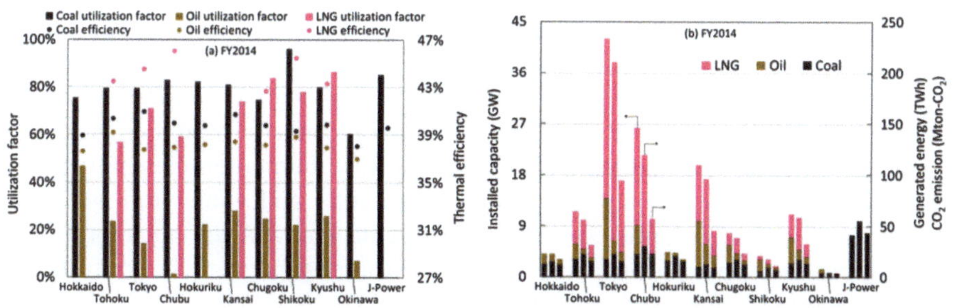

Fig. 2 Summary of thermal power generation of ten electric companies and J-POWER in FY 2014: (**a**) Utilization factor and thermal efficiency; (**b**) installed capacity, generated energy, and CO_2 emission

Referring to ANRE's statistical survey of electric power and the latest handbook of power generation facilities [6–8] published by the Japan Electric Association and the Thermal and Nuclear Power Engineering Society, all thermal power generation facilities were listed and categorized by types of fuel, types of cycle, and installed capacities. In addition, thermal efficiencies of BTGs (reheat and regenerative Rankine cycle) were estimated on the basis of temperature and pressure of main and reheat steam, and thermal efficiencies of gas turbine combined cycles (GTCCs) were from published data or TIT-based estimation. Eventually, representative thermal efficiencies of each electric power company were calculated as weighted-average values for three types of fuel, and utilization factors in each fiscal year were calculated using data of generated energy and installed capacity.

Note: *It is not known exactly which facility was used and how long it was operated, even by the combination of generated energy, fuel consumption, and estimated thermal efficiency. Thus, there is a discrepancy between thermal efficiency based on HHV and nominal best efficiency at the rated operating condition, because individual operations are unknown. The discrepancy is accentuated in BTGs because of the loss of start and shutdown, hot reserve, partial-load operation, and others. ANRE's published statistics had fuel-classified data of thermal efficiency and utilization factor, but these have not been published since FY 2012. The statistics of LNG does not distinguish BTGs and GTCCs, but their efficiencies are very different. All estimated thermal efficiencies agreed with the published data* [2] *within errors of 1 %, except for the oil of Tohoku and Hokuriku. All estimated utilization factors agreed with the published data within errors of 2 %.*

After the disaster, Tokyo and Kansai Electric Companies generated energies drastically increased, by 41.4 TWh (+24 %) and 47.3 TWh (+96 %), respectively. All other companies' energies also increased because of suspending operations of their nuclear power stations, and the total generated energy rose by 164 TWh (+33.9 %). Utilization factors increased by >15 % in oil-fired power stations of Hokkaido, Kansai, Shikoku, and Kyushu Electric Companies and also in LNG-fired power stations of Tohoku, Kansai, Chugoku, and Kyushu Electric Companies. CO_2 emissions increased by 73 % in Hokkaido, 82 % in Kansai, and 52 % in Kyushu. Thermal efficiencies of Chubu and Kansai Electric Companies' LNG-fired power stations increased by a 1.1 percentage point and a 3.9 percentage point because of newly commissioned stations and renewal of GTCC, respectively. After the second oil crisis of 1979, no new oil-fired thermal power plants were installed for 40 years in Japan, in accord with the International Energy Agency's (IEA) declaration of "Principles for IEA Action on Coal" [9]. Although Japan's oil-fired thermal power plants are generally old and have high costs and low thermal efficiency, they are frequently operated, especially so during the summer and winter following the earthquake disaster. Nevertheless, the total CO_2 emission was not much more than 24 %, in comparison with the 33.9 % increase in total generated energy from the enhancement in LNG utilization factor.

2.2 *New Installation, Extension, and Renewal of Thermal Power Plants*

Newly commissioned thermal power plants and near-future plans are listed in Table 1.

In principle, all electric companies announced that construction, expansion, and replacement of power sources would be consigned through bidding to enhance efficiency and transparency according to a METI guideline [10] published in 2012. Plans for thermal power purchase by bids are listed in Table 2.

It is generally considered that the design service life of power plants is ~40 years, but it can be extended by appropriate maintenance. It generally takes ~10 years from planning for power plant replacement to start of commercial operation. Table 3 shows capacities of Japanese thermal power plants that exceed 40 years from initiation of commercial operation by each decennial year from 2020 to 2050. More than 40 GW oil-fired power plants and 30 GW LNG-fired power plants will exceed 40 years by 2030, in contrast to many coal-fired power plants that are comparatively young.

2.3 *Best Available Technology (BAT) for Thermal Power Plants*

The Ministry of Economy, Trade and Industry, and Ministry of the Environment jointly published the best available technology (BAT), based on the best practice for thermal power plants in April 2013 to control GHG emission. The updated 2014 BAT [12] is summarized in Table 4. In principle, unless power producers meet this standard, no new installation or renewal plan will be approved.

3 Technology Road Map: Future Prospects, Problems, and Risks

3.1 *Overview*

Generated energy by general electric enterprises has declined, and this is expected to continue because of increasing disengagement of heavy users and electricity purchase from other independent power producers through electricity market liberalization. However, there is a strong possibility for further increase in energy generated by thermal power plants, including wholesale supply electric companies depending on resumed operation of nuclear power plants.

Table 1 Newly commissioned thermal power plants and near-future plans

Name of power plant	Fuel	Capacity (GW)	Specification	Efficiency (HHV %)	Start date of commercial operation
Hokkaido					
Ishikariwan-Shinko Unit 1–3	LNG	0.57 × 3 units	1600 °C MACC	57	Feb. 2019
					Dec. 2021
					Dec. 2028
Tohoku					
Hachinohe Unit 5	LNG	0.42	ACC	50	Jul. 2015
Noshiro Unit 3	Coal	0.60	600 °C USC	43	2020
Shin-Sendai Gr. 3	LNG	0.49 × 2 units	1500 °C MACC	53	Jul. 2016
Joetsu Gr. 1	LNG	0.60	MACC	>53	Early 2020s
Tokyo					
Kashima Gr. 7	City gas	0.42 × 3 units	1300 °C ACC	52	Jul. 2014
Hirono Unit 6	Coal	0.60	600 °C USC	43	Dec. 2013
Chiba Gr. 3	LNG	0.50 × 3 units	1500 °C MACC	53	Jul. 2014
Hitachinaka Unit 2	Coal	1.00	600 °C USC	43	Dec. 2013
Kawasaki Gr. 2	LNG	0.50 × 3 units	2-1	54	Feb. 2013
			1500 °C MACC	56	Jul. 2016 and Jul. 2017
			2-2 & 2-3 1600 °C MACC		
Goi Gr. 1	LNG	0.71 × 3 units	1600 °C MACC	56	After 2023
Nakoso	Coal	0.54	IGCC	48	Early 2020s
Hirono	Coal	0.54	IGCC	48	Early 2020s
Kashima Power[a]	Coal	0.65	USC	Unspecified	2020
Hitachinaka Generation[b]	Coal	0.65	USC	Unspecified	2020
Yokosuka[c]	Coal	1.00	IGCC	Unspecified	2020
Ichihara[d]	Coal	2.00	Unspecified	Unspecified	2020
Chiba[e]	Coal	1.00	Unspecified	Unspecified	2020
Chubu					
Joetsu	LNG	0.60 × 4 units	1300 °C MACC	54	May 2014
Nishi-Nagoya Gr.7	LNG	1.16 × 2 units	1600 °C MACC	57	Sep. 2017
					March 2018
Taketoyo	Coal	1.00	Unspecified	Unspecified	2021
Shimidzu[f]	LNG	2.00	Unspecified	Unspecified	2021

(continued)

Table 1 (continued)

Name of power plant	Fuel	Capacity (GW)	Specification	Efficiency (HHV %)	Start date of commercial operation
Kansai					
Himeji No. 2 Plant	LNG	0.49 × 6 units	1600 °C MACC	56	March 2015
Kobe[g]	Coal	1.40	USC	Unspecified	2022
Takasago[h]	Coal	0.60 × 2 units	USC	Unspecified	2021, 2027
Chugoku					
Takehara New No. 1	Coal	0.60	USC	43	2020
Misumi No. 2	Coal	1.00	USC	Unspecified	2027
Osaki[i]	Coal	0.17	O_2-blown IGCC	40.5	March 2017
Ube[j]	Coal	1.20	Unspecified	Unspecified	Early 2020s
Kyushu					
Matsuura no. 2	Coal	1.00	USC	Unspecified	June 2021
Hibikino[k]	LNG	1.60	Unspecified	Unspecified	2020

Over 150 MW only. Date last verified 15 March 2015

ACC Advanced combined cycle ~1350 °C, *MACC* more advanced combined cycle ~1500 °C

[a]Joint operation by J-POWER and Nippon Steel & Sumitomo Metal

[b]Joint operation by Tokyo Electric Company and Chubu Electric Company

[c]Joint operation by J-POWER and Tokyo Electric Company

[d]Joint operation by Kyushu Electric Company, Idemitsu Kosan and Tokyo Gas

[e]Joint operation by Chugoku Electric, JFE Steel, and Tokyo Gas

[f]Operated by Tonen General Sekiyu

[g]Operated by Kobe Steel

[h]Joint operation by J-POWER and Kansai Electric Company

[i]Joint operation by J-POWER and Chugoku Electric Company

[j]Joint operation by J-POWER, Ube Industries, and Osaka Gas

[k]Joint operation by Saibu Gas and Osaka Gas

Table 2 Plans for thermal power purchase by bids for each electric company

Company	Tohoku	Tokyo	Chubu	Kansai	Kyusyu
Capacity (GW)	1.20	6.00	1.00	1.50	1.00

Table 3 Decennial capacity of thermal power plants over the age of 40 years [11]

(GW)	2020	2030	2040	2050
Coal	3.90	6.59	13.71	13.61
Oil	28.12	13.10	1.37	0.02
LNG	21.29	8.15	21.79	8.69

Table 4 BAT for thermal power plants (last updated April 2014)

Class (MW)	Category	η_e (net HHV)
700–1100	Pulverized coal-fired power plant (PCPP) (USC)	40
500–600	PCPP (USC, SC)	39.5–39.5
200	PCPP (sub-C)	38
	IGCC (1200 °C)	40.5
800 (50 Hz)	GTCC (1450 °C, multi-shaft)	49
600 (60 Hz)	GTCC (1300 °C, multi-shaft)	51
500 (50 Hz)	GTCC (1500 °C, single-shaft)	52
400 (50, 60 Hz)	GTCC (1400 °C, single-shaft)	51
200–300 (60 Hz)	GTCC (1200–1400 °C, single-shaft)	50–50.5

The "Electricity System Reform" [13] underway at METI will have a great impact on the cost of electricity and result in price competition. Power producers expect load leveling and want to decrease their power generation cost by increasing the utilization factor of their facilities. However, in large-scale thermal power plants, coal-fired BTGs and LNG-fired GTCCs are mainly used for baseload power sources and intermediate and peaking power sources, respectively, because of the start and shutdown time and resulting loss. Since power suppliers purchase cheaper electricity on the market, power producers will replace expensive, inefficient, and decrepit oil-fired power plants and maintain comparatively efficient installed facilities, regardless of fuel type. Coal-fired power plants will be steadily installed in the future, because the Ministry of Environment jointly approved the BAT for cost-competitive coal-fired power plants. LNG will become increasingly important for GHG emission control, and GTCCs will largely replace decrepit oil-fired power plants near their LNG base. No LNG terminals and main gas pipelines are on the Sea of Japan side, except for Niigata Prefecture (Fig. 3). Therefore, oil-fired power plants remain on that side, but they are not large scale and have a limited impact on GHG emission. Further deployment of LNG bases and main network gas pipelines are also important from the perspective of diversification of risk and resilience of energy supply systems against the next mega-quake and tsunami disaster expected on the Pacific side, learning from the fact that Niigata-Sendai backbone pipelines supported the Gas Bureau of the city of Sendai in the last tsunami disaster. For example, Kobelco's Moka Plant will install two 700-MW gas turbine power plants that will draw natural gas from the Hitachi LNG base through the Ibaraki-Tochigi main pipeline. Tokyo Gas Co., Ltd. will purchase all generated electricity, and this project is reasonable in terms of diversification of risk, despite the inland location. There is concern that Japan has less underground storage of natural gas relative to other countries, despite being an isolated island without an international pipeline network (Table 5).

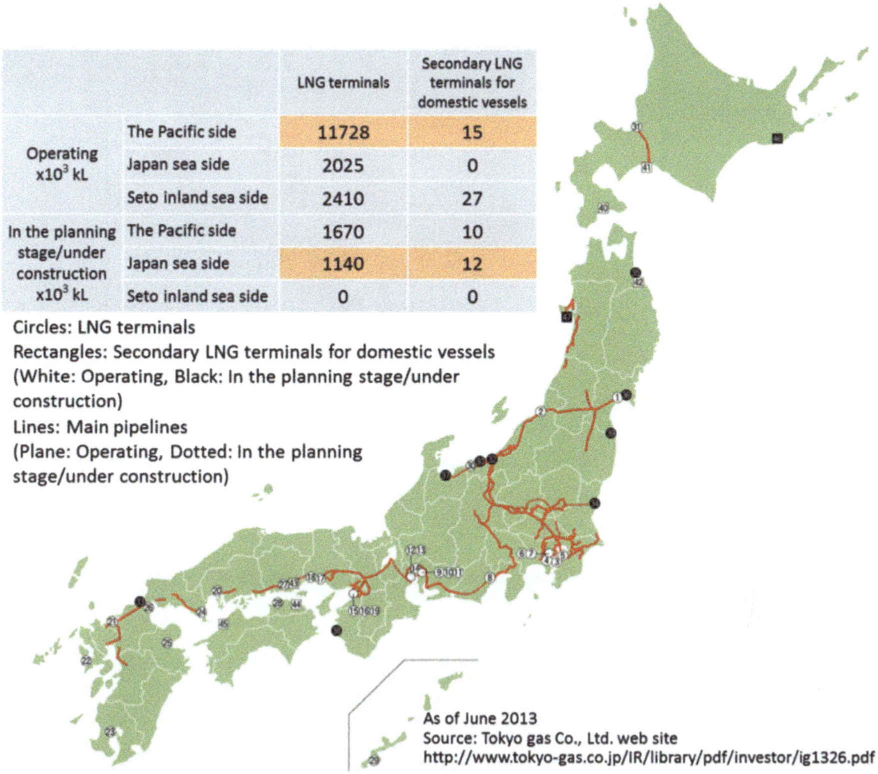

Fig. 3 LNG base and main pipelines

Table 5 Underground storage of natural gas [14]

	Canada	Italy[a]	UK	Japan
Demand of NG (billion m^3 per year)	84	86	91.6	77.1
Number of underground storage facilities and capacity of working gas (billion m^3)	49	10	49	5
	14.8	12.7	3.47	1.17

[a]In Italy, it is obligatory to store strategically 10 % of imported LNG from non-EU counties

3.2 Coal-Fired Power Plants

As mentioned in the introduction, higher working fluid temperatures permit greater efficiency of thermal power plants. Figure 4 shows a technology timeline for coal and LNG-fired thermal power plants, from 1970 to 2050. Coal-fired power plants historically overcame severe steam conditions incrementally by using heat-resistant technologies, especially via the evolution of material science of superalloys. The new No. 2 unit at Isogo Thermal Power Plant (J-POWER) achieved 43 % thermal efficiency (HHV) [15] by adapting the most severe ultra-supercritical (USC) steam

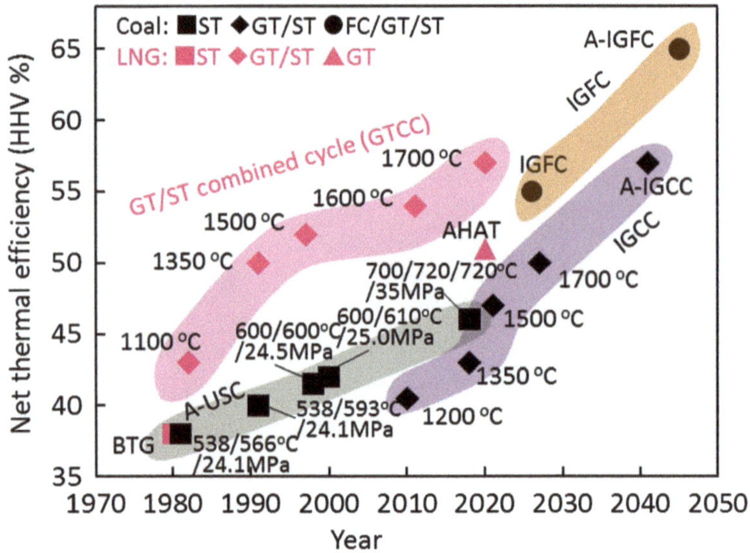

Fig. 4 Technology timeline for coal and LNG-fired thermal power plants [16]

condition of 25 MPa/610 °C main steam and 620 °C reheat steam in 2009. The METI national project will lead the way in the development of technology, adapting more advanced USC (A-USC) steam conditions of >700 °C and 35 MPa by the early 2020s. This project uses two-stage reheat cycles, and the performance of power plants is estimated in Fig. 5. The project targets 46 % thermal efficiency (HHV at sending end) by 720 °C and 35 MPa for the main steam condition and 720/720 °C for the two-stage reheat steam conditions; the 2014 BAT is ~42 %. Figure 6 depicts a process flow diagram (PFD) of the two-stage reheat cycle and development challenges of each component. To achieve the project target, many development subjects must be solved, and in particular, all components such as boilers, turbines, steam valves, and balance of plant (BOP) require heat-resistant superalloy materials at affordable costs.

Another option to boost thermal efficiency is coal gasification, especially the use of a low ash melting point coal like lignite. Japan's only IGCC used air-blown gasification technology with a two-stage, two-chamber entrained flow gasifier, achieving 42.9 % thermal efficiency [18] by eliminating air separation units, which consume a certain amount of electricity (Fig. 7). Tokyo Electric Company plans two new installations of 540 MW IGCC at Nakoso and Hirono power plants, which will use gas turbines with higher TIT than that of the first IGCC. Chugoku Electric Company and J-POWER are jointly installing an IGCC at Osaki Power Plant [19], which will use oxygen-blown gasifier technology. The plan of this project consists of three stages. The first will achieve 40.5 % thermal efficiency (HHV), envisioning a power generation cost comparable to conventional pulverized coal-fired power plants. The second stage adapts CCS technology, and the final stage will combine solid oxide fuel cells (SOFCs) with the IGCC plant. This

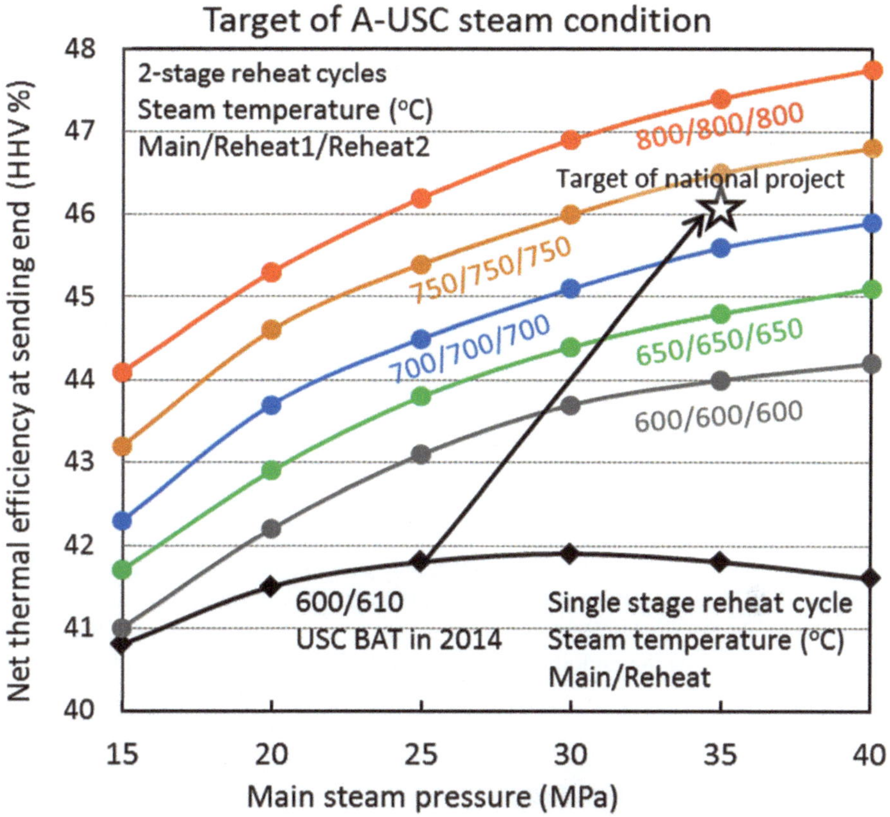

Fig. 5 Thermal efficiency for various USC steam conditions and target of national project [17]

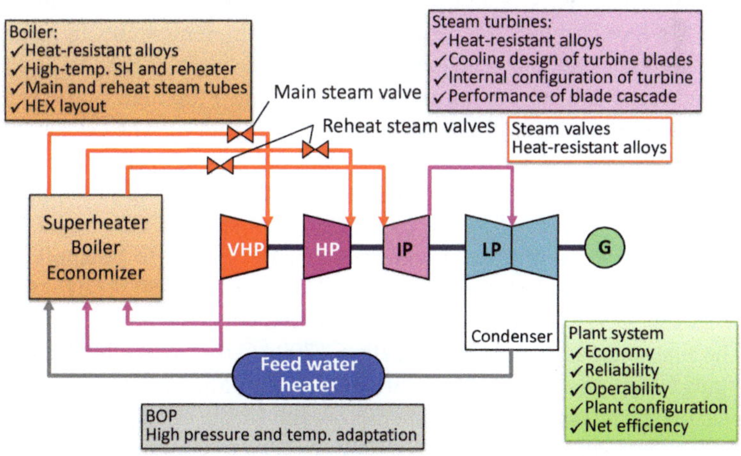

Fig. 6 PFD of two-stage reheat cycle and development challenges of each component [14]

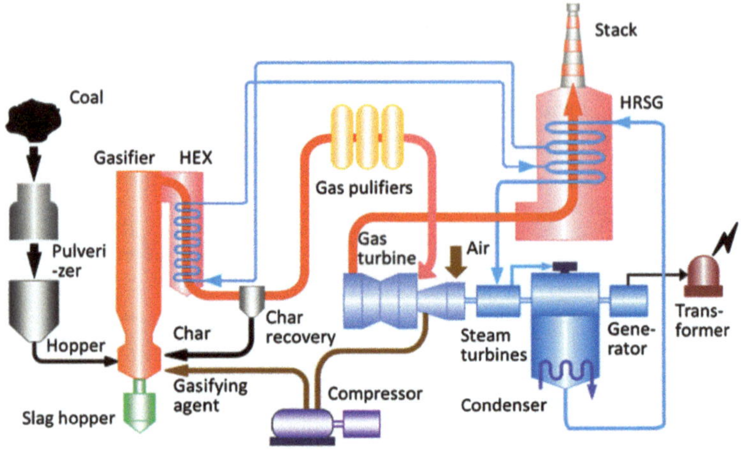

Fig. 7 Schematic of air-blown IGCC system

advanced combined system is the so-called IGFC, and the advanced IGFC (A-IGFC) [20] has also been proposed that targets 70 % thermal efficiency, using exergy recuperation technology and irreversibly generated heat in the SOFC from internal resistance as the endothermic reaction heat of coal gasification by steam.

3.3 LNG-Fired Power Plants

Because GTCCs can use LNG more efficiently than BTGs, almost all power producers will install GTCCs near LNG bases. As is the case for coal-fired power plants, GTCCs have broken the record for the TIT incrementally by using heat-resistant technologies and have achieved 1600 °C. The next stage is an approach to 1700 °C, and according to the national project for development of 1700 °C GTs [21], demonstration testing will be complete by 2020 (Fig. 4). The GT system plans to use a semi-closed cycle with 25–40 % exhaust gas recycling to reduce the production of thermal NOx, which increases exponentially at high temperature; it can be less than 50 ppm (at 15 % O_2) by reducing hot spots in the combustor. With regard to the turbine, an improved compressor is developed to be applicable to a high pressure ratio. Compressor efficiency can reach >89 % by controlling shockwaves and secondary flows in supersonic and sub-supersonic stages, respectively. The first-stage turbine blade targeting 1700 °C uses extended heat-resistant technologies based on 1600 °C class turbines. Technology components face three challenges, material (heat-resistant superalloys and single crystal), advanced shape film cooling (steam-air hybrid cooling system with 30 % reduction in flow rate), and advanced thermal barrier coatings (which realize a 20 % decrease in thermal conductivity by using Sa, Nb, and Pr, and a 20 % increase in durability). Moreover, three-dimensional turbine blades are designed using cutting-edge computational fluid dynamics, with turbine efficiency reaching >91 %. These technologies are also applicable to turbines for IGCCs.

The ultimate natural gas-fired power generation system is currently under development, aiming for realization in the 2020s. This power generation system, which has a composition similar to the IGFC, uses a solid oxide fuel cell as the topping cycle and cascadically combines a gas and steam turbine, targeting ~70 % [22] by 2050.

For all existing GTs, including mid- and small class, the intake air cooling system is known as an effective technology to compensate power reduction in summer. Advanced humid air turbines [23] represent a type of regenerative gas turbine cycle with highly humidified compressed air. The target of this technology is 51 % thermal efficiency (Fig. 4).

Mass import of hydrogen became a real possibility as a result of success in a demonstration of liquid organic chemical hydride. A full-scale (1 GW) hydrogen-fired power plant using a gas turbine combined cycle has been reviewed, and METI estimated that 2.37 billion Nm^3/year hydrogen [24] would be consumed under the condition of 51 % thermal efficiency and 49 % utilization factor. However, there is a technological problem in premix combustion, whereas diffusion combustion technology of hydrogen has been established. The co-firing ratio of hydrogen to natural gas, currently ~5 %, is difficult to increase because of the peculiar combustion characteristics of hydrogen, such as high flame speed or high adiabatic flame temperature.

3.4 CO_2 Emission

Figure 8 shows "life cycle CO_2 emission (g/kWh)" which previous fossil fuel-fired power plants emitted into the atmosphere over 40 years, from their construction to decommissioning, including fuel consumption. Coal-fired power plants, even A-USC, are estimated to emit nearly twice the CO_2 of 1700 °C natural gas-fired GTCCs. Carbon dioxide emissions of coal-fired power plants can be drastically reduced through CCS technology, under the assumption of amine processes with 90 % recovery rate. Simultaneously, however, net thermal efficiencies decline ~a 5 percentage point. This energy penalty is not negligible and so further technology development is necessary, for the example of typical amine processes, substantial reduction of regeneration heat of solvent in strippers or electricity burden in the CO_2 capture system. In addition, reducing wasteful fuel consumption associated with start and shutdown requires steady efforts in operation and control of all thermal power plants. Co-firing biomass with coal is another option for substantial CO_2 emission reduction, which has been demonstrated by adding a small percentage of wood biomass.

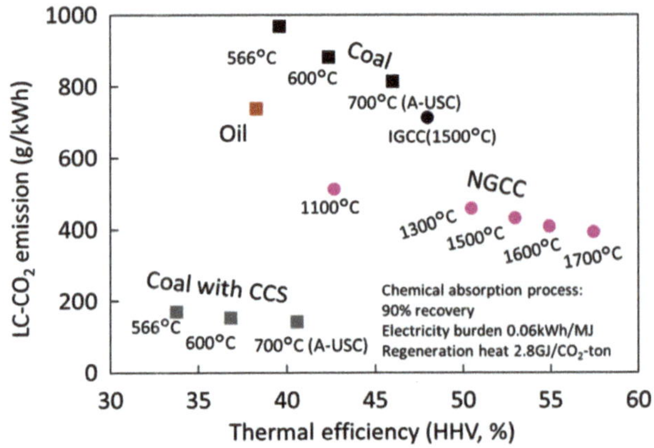

Fig. 8 Life cycle CO_2 emission for various thermal power technologies [25, 26]

3.5 *Possible Risk Factors*

One of the most plausible risk factors is increase in fuel procurement costs, owing to exchange fluctuations or the global scramble for natural resources. Mandatory implementation of CCS will elevate fuel consumption, while CO_2 emission will certainly decrease. Increasing unstable renewable energies, such as photovoltaics or wind turbines, may require more thermal power, which is used not only as bulk power management but also hot reserve and spinning reserve to stabilize grid frequency and power flow. Partial-load operations generally reduce thermal efficiency and may result in increased CO_2 emission per unit generated power. Frequent start and shutdown operations tend to cause mechanical troubles because of thermal fatigue degradation and require frequent diagnosis and repair of equipment. This may not only increase cost but also cause extensive power shortages in the worst case. To mitigate this problem, massive energy storage should be installed not only batteries but a combination of active use of variable-speed pumped hydroturbines and introduction of hydrogen fuel cell storage systems. As mentioned above, the Electricity System Reform underway at METI is aimed at market liberalization and consumers will welcome cheaper electricity. However, excessive price competition may decrease the number of engineers in the field and their morale and may eventually cause unexpected shutdowns and accidents.

4 Benefit and Future Vision

As a low-cost and stable power source in the world, by the 2050s, thermal power plants will still have a large share of power generation. Many of these plants are expected to be newly installed or replaced with earlier technologies equivalent to Japan's BAT, in both the domestic and global market. These would be 1500 °C or higher MACC fueled by LNG and reheat, and regenerative Rankine cycle BTGs with >25 MPa and 600/600 °C USC steam conditions or IGCCs fueled by coal, hopefully all CCS-ready (Fig. 9).

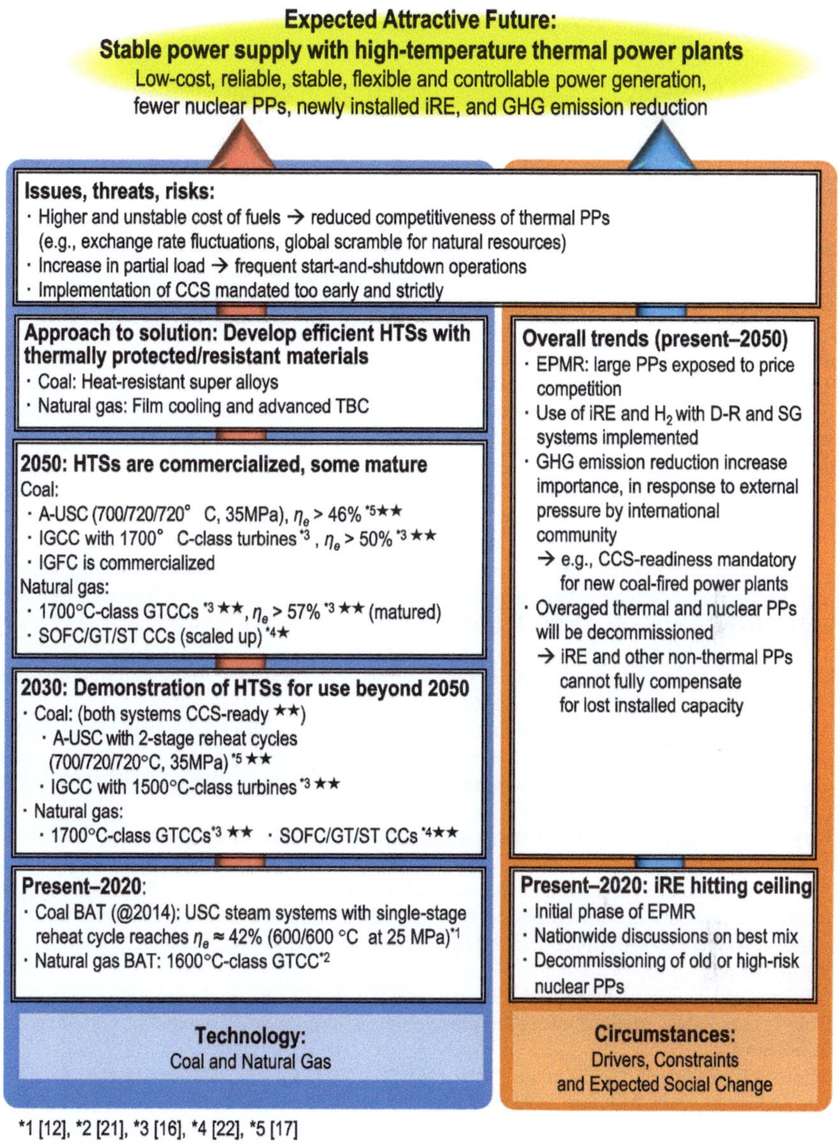

Abbreviations:
A-USC: Advanced USC, BAT: Best Available Technologies, CC: Combined Cycle, CCS: Carbon Capture and Storage, D-R: Demand-Response, EPMR: Electric Power system Market Reformation, FC: Fuel Cell, GT: Gas Turbine, GTCC: Gas Turbine CC, HTS: High Temperature Systems, HHV: Higher Heating Value, IGCC: Integrated coal Gasification CC, IGFC: Integrated coal Gasification Fuel Cell, iRE: intermittent Renewable Electricity, NEDO: New Energy and Industrial Technology Development Organization, PP: Power Plant, SG: Smart Grid, SOFC: Solid Oxide FC, ST: Steam Turbine, TBC: Thermal Barrier Coating, USC: Ultra-SuperCritical, η_e : Thermal efficiency (HHV%)

Fig. 9 Technology and circumstances perspectives of thermal power plants

References

1. Ministry of Economy, Trade and Industry (METI) (2012) Interim report on research council for the master plan of intensifying inter-regional tie-lines (Chiikikan rennkeisen tou no kyoka ni kansuru master plan kenkyukai chukan hokoku). METI, Tokyo, http://www.meti.go.jp/committee/sougouenergy/sougou/chiikikanrenkeisen/report01.html
2. Ministry of Economy, Trade and Industry (METI) (2012) Summary of power supply and demand in FY 2011 (Heisei 23 nenndo denryoku jukyu no gaiyo). METI, Tokyo
3. Agency for Natural Resources and Energy (ANRE), Ministry of Economy, Trade and Industry (METI). Open database for statistical survey of electric power (Denryoku chosa toukei). http://www.enecho.meti.go.jp/statistics/electric_power/ep002/ METI, Tokyo.
4. Yamakawa M, Kakiuchi M, Katayama J (2006) 50 Hz large-capacity water-cooled generator and hydrogen-cooled generator. Toshiba Rev 61(4):56–59
5. The Japan gas association (JGA) (2013) CO_2 emission coefficient of electric energy using evaluation for CO_2 reduction countermeasures (CO_2 sakugen taisaku no hyoka ni mochiiru denki no CO_2 haisyutu keisuu ni tsuite). JGA, Tokyo, http://www.gas.or.jp/kankyo/co2taisaku_20130221.pdf
6. The Federation of Electric Power Companies of Japan (FEPC) (2010) Handbook of electric power industry (FY2010). The Japan Electric Association, ISBN: 978-4889482317
7. The Federation of Electric Power Companies of Japan (FEPC) (2013) Hand book of electric power industry (FY2013). The Japan Electric Association, ISBN: 978-4889482706
8. The Thermal and Nuclear Power Engineering Society (TENPES) (2011) Directory of thermal and nuclear power plants (FY 2011) (Karyoku genshiryoku hatudensho setsubi yoran 23 nendo kaitei ban). TEMPES, Tokyo
9. International Energy Agency (IEA) (1998) International coal trade: the evolution of a global market. IEA, Paris. ISBN 978-9264156678
10. Ministry of Economy, Trade and Industry (METI) (2012) New guidelines for practice of thermal power purchase by bids (Atarashii karyoku denngen nyusatsu no unyou ni kakaru shishin). METI, Tokyo, http://www.meti.go.jp/press/2012/09/20120918003/20120918003-2.pdf
11. Koyama M et al (2014) Present status and points of discussion for future energy systems in Japan from the aspects of technology options. JCEJ 47(7):499–513
12. Ministry of Economy, Trade and Industry (METI) (2014) Reference table of best available technologies for thermal power plants (BAT no sankohyo). METI, Tokyo, http://www.meti.go.jp/policy/safety_security/industrial_safety/sangyo/electric/files/bat_sankouhyou/bat_20140501.pdf
13. Ministry of Economy, Trade and Industry (METI) (2014) Electricity system reform. METI, Tokyo, http://www.meti.go.jp/english/policy/energy_environment/electricity_system_reform/index.html
14. Ministry of Economy, Trade and Industry (METI) (2014) Present status and issues for natural gas and supply infrastructure in Japan (Wagakuni no tennen gas oyobi sono kyokyukiban no genjo to kadai). METI, Tokyo, http://www.meti.go.jp/committee/sougouenergy/sougou/kiban_seibi/001_06_00.pdf
15. Kimura H et al (2011) Development of technologies for highly-efficient coal-fired thermal power plants. Hitachi Hyoron 93(8):526–531
16. Ministry of Economy, Trade and Industry (METI) (2013) Promotion of introducing high efficiency thermal power plants (Kokoritu karyoku hatsuden no donyu suishin ni tsuite). METI, Tokyo, http://www.enecho.meti.go.jp/committee/council/basic_policy_subcommittee/past/002/pdf/002_009.pdf
17. Ministry of Economy, Trade and Industry (METI) (2010) Summary of element technology development for A-USC thermal power plants (Senshin cho-cho rinkaiatsu karyoku hatsuden jitsuyoka yousogijutsu kaihatsu no gaiyo ni tsuite). METI, Tokyo, http://www.meti.go.jp/committee/summary/0001640/034_05_09.pdf

18. Nunokawa M (2013) Progress in NAKOSO 250 MW air-blown IGCC demonstration project. Proceedings of ICOPE-2013, JSME
19. Fujii J (2013) The summary of the OSAKI CoolGen project and a future plan. J Jpn Inst Energy 92:408–412
20. New Energy and Industrial Techology Development Organization (NEDO) (2006) 2B5 next-generation, high-efficiency integrated coal gasification electric power generating process (A-IGCC/A-IGFC) clean coal technologies in Japan 32. NEDO, Kawasaki
21. Ito E et al (2011) Key technologies for ultra high temperature gas turbine. Mitsubishi Juko Gihou 48(3):2–7
22. Kobayashi Y et al (2011) Extremely high efficiency power system – SOFC triple combined cycle system. Mitsubishi Juko Gihou 48(3):16–21
23. Goto J et al (2011) Highly-efficient and flexible gas turbine system AHAT. Hitachi Hyoron 93 (8):532–535
24. Ministry of Economy, Trade and Industry (METI) (2014) Hydrogen-firing power generation (Suiso hatsuden ni tsuite). METI, Tokyo, http://www.meti.go.jp/committee/kenkyukai/energy/suiso_nenryodenchi/suiso_nenryodenchi_wg/pdf/004_02_00.pdf
25. Imamura E, Nagano K (2010) Evaluation of life cycle CO_2 emissions of power generation technologies -update for state-of-the-art plants-, CRIEPI Report No. Y09027
26. Matsukata M, Koyama M (2012) Sweeping review of energy system in Japan (Zero kara minaosu energy) Maruzen, Japan ISBN: 978-4621085134

Nuclear Power Generation

Hiroshi Sekimoto

Abstract Nuclear is a dense energy without CO_2 emission. It can be used for more than 1000 years using fast reactors and for more than 100,000 years using fast reactors with uranium from the sea. However, it raises difficult problems associated with severe accidents, spent fuel waste, and nuclear threats, which should be solved with acceptable costs. The Fukushima Daiichi accident seriously affected the Japanese atomic energy program. Before the accident, nuclear was considered one of the main baseload energies and was positively promoted. Shortly after the accident, closing and decommissioning of all nuclear reactors were seriously considered. Now, after 4 years, several nuclear reactors have a plan to renew operation. However, it will be difficult to construct new reactors in the near future. China and certain emerging countries have aggressive future plans for nuclear energy. Some innovative reactors have attracted interest, and many designs have been proposed for small reactors. These reactors are considered much safer than conventional large reactors and have fewer technical obstructions. Breed-and-burn reactors have high potential to solve all inherent problems for peaceful use of nuclear energy. However, they have some technical problems with materials. In Japan, even if nuclear energy is used, its contribution to mitigating global warming is very slight. However, if we contribute to developing innovative nuclear energy systems, the results will be global and the contributions considerable. The roadmap for large reactors in Japan is not a technological issue but a political and sociological one, while the roadmap for innovative reactors is technological. A roadmap for innovative reactors in Japan is presented herein.

Keywords Nuclear energy • Innovative reactor • Small reactor • Breed-and-burn reactor • Roadmap

H. Sekimoto (✉)
Tokyo Institute of Technology, Tokyo, Japan
e-mail: hsekimot@gmail.com

Y. Kato et al. (eds.), *Energy Technology Roadmaps of Japan*,
DOI 10.1007/978-4-431-55951-1_15

1 Introduction

Global warming is an increasingly urgent problem. Nuclear power is one of the few technologies that can mitigate climate change by reducing CO_2 emission [1]. The number of operating nuclear reactors was increasing in Japan, helping reduce CO_2 emissions, until the Fukushima Daiichi nuclear reactor accident in 2011 (Fukushima accident hereafter) [2].

The Fukushima accident greatly influenced the Japanese atomic energy program. Shortly after the accident, closing and decommissioning of all nuclear reactors were seriously considered. Currently, all reactors in the country have ceased operation, but there are plans for several to renew operation. However, it will be difficult to construct new reactors in the near future.

Nuclear energy not only has a safety problem but also several difficult problems caused by generated radioactive materials and technologies in common with nuclear weapon production. However, it has unique merits such as extremely high energy density, abundant resources, and zero CO_2 emissions.

According to a nuclear energy roadmap from international agencies [3], the capacity of nuclear energy will not change from the present through 2050 in OECD countries, but will grow rapidly in developing countries. Developed countries should help developing ones successfully use nuclear energy. Safe, simple, and easy reactors are required for developing countries.

Some small reactors (SRs) have these features, although their economy is a difficult challenge to overcome. Breed-and-burn reactors (B&BRs) have high potential to overcome all the above problems, but several technical problems await solution.

Even if nuclear energy is used in Japan, its contribution to mitigating global warming is very slight. However, if we develop innovative nuclear energy systems, their consequences would be global and their contribution considerable. The roadmap for large reactors in Japan does not depend so much on technological issues but political and sociological ones. However, the roadmap of innovative reactors mostly depends on technological issues. We present herein a roadmap of innovative reactors for Japan.

The following section describes fundamental characteristics of nuclear energy. A brief history and present status are presented in Sect. 3, and prediction of nuclear energy by international organizations is addressed in Sect. 4. Innovative nuclear energy is the topic in Sect. 5, where we address SRs and B&BRs. A technology roadmap is presented in Sect. 6, and benefits and future vision are addressed in Sect. 7

2 Fundamental Characteristics of Nuclear Energy

Human beings have been using renewable energy for a million years. The scale of this use and its environmental effects were small. Humans began to use fossil fuels on a large scale several hundred years ago and this made possible the industrial revolution and modern civilization. The important characteristic of fossil fuels is its energy storage. We call such energy concentrated or dense. Huge amounts of this energy can be used whenever required. Their consumption rate increased rapidly, and resource and environmental problems have become urgent issues.

Nuclear energy, which is also a dense energy, was discovered in the middle of the twentieth century. A list of energy density (specific energy) for several fuel materials is shown in Table 1 [4], where one can see a large difference (about a million times) of energy density between nuclear and fossil fuels. Compared with fossil fuels, nuclear energy has several merits in addition to the greater density, such as no CO_2 emission and no oxygen requirement.

The means for releasing nuclear energy were discovered during World War II. One of the main objectives of this war was to secure energy, especially oil. Nuclear energy became another energy resource. Figure 1 shows the resource amount and available period for each energy resource, estimated by the author

Table 1 Energy density [4]

Fuel material	Energy type	Specific energy (MJ/kg)	Direct uses
Uranium (in breeder)	Nuclear fission	80,620,000	Electric power plants (nuclear reactors)
Thorium (in breeder)	Nuclear fission	79,420,000	Electric power plants (nuclear reactors)
Hydrogen (compressed at 70 MPa)	Chemical	142	Rocket engines, experimental automotive engines
LPG (including propane/butane)	Chemical	46	Cooking, home heating, automotive engines
Jet fuel	Chemical	43	Aircraft
Fat (animal/vegetable)	Chemical	37	Human/animal nutrition
Coal	Chemical	24	Electric power plants, home heating
Carbohydrates (including sugars)	Chemical	17	Human/animal nutrition
Protein	Chemical	17	Human/animal nutrition
Wood	Chemical	16	Heating, outdoor cooking
TNT	Chemical	4.6	Explosives
Lithium battery (non-rechargeable)	Electrochemical	1.8	Portable electronic devices, flashlights
Alkaline battery	Electrochemical	0.67	Portable electronic devices, flashlights
Lead-acid battery	Electrochemical	0.17	Automotive engine ignition

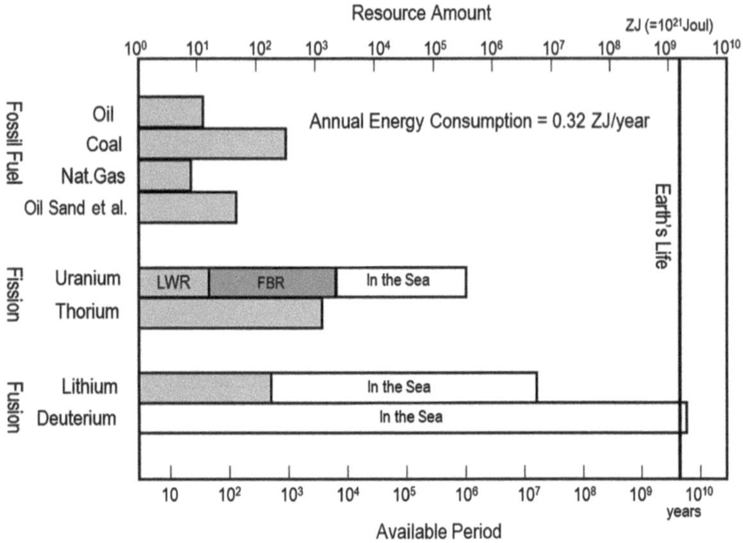

Fig. 1 Estimated resource amount and available period for individual energy resources [5]

more than 20 years ago [5]. Resource amounts for this graph were obtained from data available at that time. The values may be considerably different today, but the differences are sufficiently small for the present discussion. The available years were obtained by dividing resource amounts by total annual energy consumption in the future, 0.32 ZJ/year (1 ZJ $= 10^{21}$ J). Because several energy types are typically used in a mixed manner, actual available periods are considered to be longer than the values shown.

As already mentioned, there are dwindling supplies of fossil fuels. If only light water reactors (LWRs) with uranium fuel are used, nuclear energy sources can also provide only short-term supply. Nuclear reactors produce substantial radioactive materials during operation. This can cause devastating accidents during reactor operation and a problem with radioactive wastes, even after operation cessation. Another inherent problem of nuclear energy is that it uses materials and technologies that are closely related to atomic bomb production. Problems associated with these bombs require implementation of measures for safeguards, terrorist threats, and nuclear proliferation. In this work, we simply call these problems nuclear threats [6]. Cost-effectiveness is always an important requirement for energy. Thus, for nuclear energy to be used as a primary energy, all problems associated with (a) limited resources, (b) safety, (c) waste disposal, (d) nuclear threats, and (e) cost must be solved (Fig. 2) [7].

For (a), the resource problem, the use of fast breeder reactors (FBRs) can extend the available period to several thousands of years (Fig. 1). Even after this period, uranium in seawater can support a very long period of energy production. For (b), the safety problem, after the Three Mile Island (TMI) accident, several SR concepts were proposed [8], which are addressed in detail in Sect. 5.1 For (c), the waste

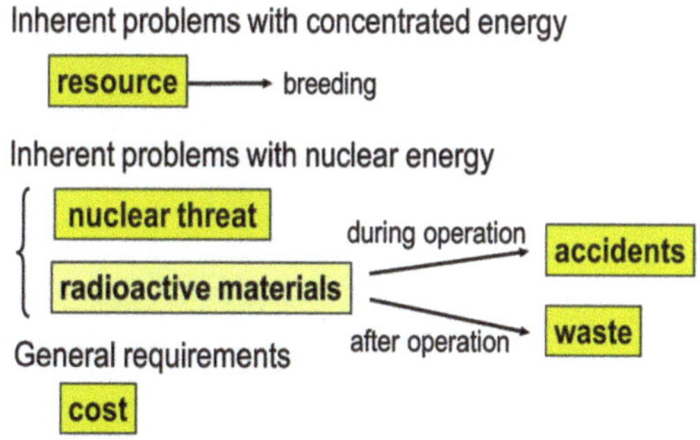

Fig. 2 Necessary and sufficient requirements for nuclear energy systems [7]

problem [9], at present only underground disposal is promoted in most countries with operational nuclear reactors. However, it is difficult to obtain public acceptance in many countries, and innovative technologies such as transmutation of long-life radioactive wastes are under study. Another creative solution to this problem is proposed in Sect. 7. The nuclear threat problem, (d) [6], is related to peaceful use of nuclear energy, but the relationship is not simple. Even if peaceful use ends, the risk of nuclear threats will persist. If care of nuclear materials and technologies is inadequate, the situation will become worse than the present peaceful use situation. We should develop a peaceful use system that strongly protects against this problem. Regarding (e), the cost problem is common to most issues related to daily life, and this is strongly tackled even at present.

3 Brief History and Present Status

Nuclear energy was discovered during study on basic physics in the twentieth century. In the beginning, release and utilization of this energy for human beings was considered impossible. However, fission was discovered in 1939, the chain reaction of neutron-induced fission was confirmed, and the criticality of CP1 was achieved in 1942. Since then, many nuclear reactors have been constructed and operated, in the beginning for military purposes and later for peaceful uses [10]. The development of nuclear power began with graphite-moderated water-cooled reactor ONPP constructed in 1951 in the USSR, followed by Shippingport PWR, constructed in 1954 in the USA.

The number of construction starts for nuclear reactors every year from 1955 to 2014 is shown in Fig. 3 [3]. The number generally increased from the beginning through around 1975, when the contribution of the USA was dominant, then decreased to nearly zero around 1995. Since the mid-2000s, the number of reactors constructed has been trending upward, largely because of rapid development in China. Years of the three major accidents (TMI, Chernobyl, and Fukushima) are also shown in Fig. 3. Despite the serious damage from the Fukushima accident, the recent upward trend appears sufficiently strong for increase in the future.

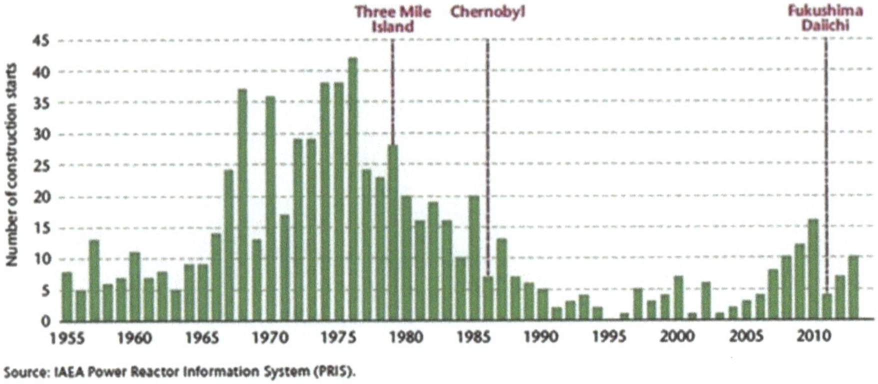

Fig. 3 Nuclear reactor construction starts 1955–2013 [3]

In Japan, the power generation capacity of nuclear reactors increased almost linearly from 1970 to 2000, and even afterward, it continued to increase until the Fukushima accident. The effect of that accident was so great in Japan that nuclear power generation ended and only a limited number of existing reactors could be operated after passing strict examinations and public acceptance [2].

4 Nuclear Energy Prediction by International Organizations

The Nuclear Energy Agency (NEA) and International Energy Agency (IEA) recently published "Technology Roadmap – Nuclear Energy" [3] in which they show electricity production by technology in the 6 °C and 2 °C scenarios (6DS and 2DS) (Fig. 4). Although the renewable and nuclear contributions are not substantial in the 6DS, they become so in the 2DS. The large share of variable renewables in 2DS significantly changes the nuclear operating environment. Nuclear power is traditionally operated to meet baseload demand. It is difficult for both PWR and BWR to execute rapid load following to meet the changes brought by renewables. New designs are required for this purpose.

The NEA and IEA roadmap also shows nuclear generation capacity variation from 2012 to 2050 in the 2DS by region (Fig. 5), during which the capacity of emerging countries, especially China, changes drastically. However, the capacity of OECD countries barely changes. Current reactors have been elaborated in developed countries for use in these countries. These countries should support developing countries in the use of nuclear energy. Safe, simple, and easy reactors are required for developing countries.

The Japanese contribution to meeting the 2DS target is small using nuclear energy only for the country, but it will grow if the country commits to helping developing countries and fabricating innovative nuclear reactors suitable for these countries.

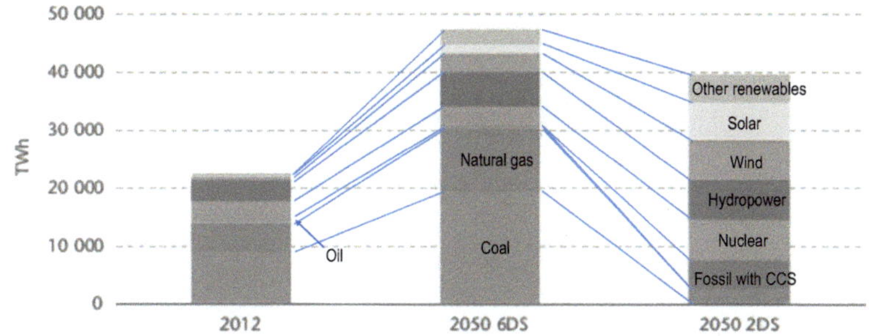

Fig. 4 Electricity production by technology in 6DS and 2DS [3]

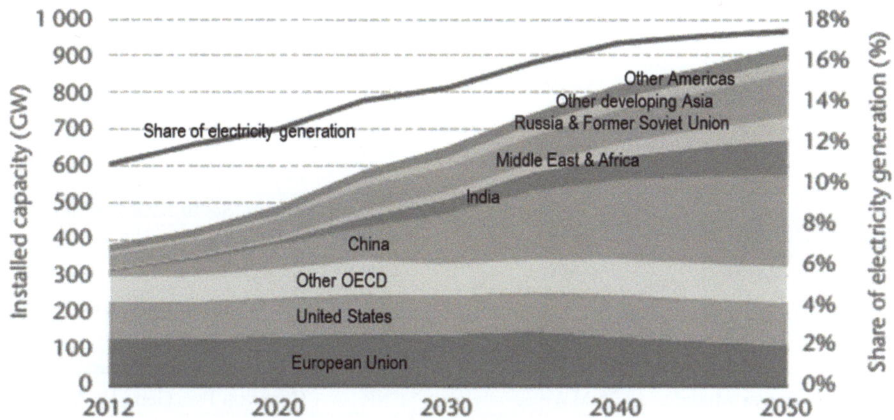

Fig. 5 Nuclear generation capacity in the 2DS by region [3]

5 Innovative Nuclear Energy

5.1 Small Reactors (SRs)

Soon after the TMI accident, many innovative reactor designs were proposed with inherent safety features, which depend not on the intervention of humans or electromechanical devices but instead on immutable and well-understood laws of physics and chemistry [11, 12]. These features are usually attained from favorable reactivity feedback for passive reactor shutdown, passive core cooling against decay heat, and superior confinement performance of radioactive materials. In general, SRs can perform these safety functions relatively easily [13].

Many SR designs have been proposed in Japan, and an international specialist meeting was held in 1991 [14]. The time for renewing old reactors has now arrived, and many SR designs have been suggested, especially in the USA. The numerous proposed SRs are shown in Table 2 [8, 15–17].

Table 2 Small reactors (SRs) [8, 15–17]

Name	Capacity	Type	Developer
CAREM	27–100 MWe	PWR	CNEA + INVAP, Argentina
MRX	30–100 MWe	PWR	JAERI, Japan
KLT-40S	35 MWe	PWR	OKBM, Russia
NuScale	45 MWe	PWR	NuScale Power + Fluor, USA
Flexblue	50–250 MWe	PWR	Areva TA, France
SMART	100 MWe	PWR	KAERI, South Korea
ACP100	100 MWe	PWR	CNNC + Guodian, China
NP-300	100–300 MWe	PWR	Areva TA, France
IRIS	100–335 MWe	PWR	Westinghouse-led, international
CAP-150	150 MWe	PWR	SNERDI, China
mPower	150–180 MWe	PWR	B&W + Bechtel, USA
SMR-160	160 MWe	PWR	Holtec, USA
Westinghouse SMR	225 MWe	PWR	Westinghouse, USA
VK-300	300 MWe	BWR	Atomenergoproekt, Russia
PHWR-220	220 Mwe	HWR	BARK, India
HTTR	30 MWt	HTR	JAEA, Japan
PBMR	165 MWe	HTR	Escom, South Africa, et al.
HTR-PM	2 × 100 MWe	HTR	INET + HSNPC, China
SC-HTGR (Antares)	250 MWe	HTR	Areva, France
GT-MHR	285 MWe	HTR	GA + Minatom, USA-Russia
4S	10–50 MWe	FNR	Toshiba, Japan
SVBR	10–100 MWe	FNR	AKME (Rosatom), Russia
Hyperion Power Module	25 MWe	FNR	Hyperion Pwr Gen., USA
LSPR, PBWFR	50–150 MWe	FNR	TokyoTech, Japan
ALFRED	120–600 MWe	FNR	Ansaldo, Italy
EM^2	240 MWe	FNR	GA, USA
BREST	300 MWe	FNR	RDIPE, Russia
S-PRISM	311 MWe	FNR	GE-Hitachi, USA
FUJI, miniFUJI	10, 100–200 MWe	MSR	IThEMS, Japan-Russia-USA
IMSR	45 MWe	MSR	Terrestrial Energy, USA
Mk1 PB-FHR	100 MWe	FGR[a]	MIT + UCB + UWM, USA
Leadir-PS100	36 MWe	LGR[b]	Northern Nuclear, Canada

[a]Flibe-cooled graphite-moderated reactor
[b]Lead-cooled graphite-moderated reactor

Table 2 lists many types of SRs, especially for PWR, high-temperature reactor (HTR) and fast neutron reactor (FNR). Categorizing these by their generation, PWR, BWR, and heavy water reactor (HWR) belong to Generation III and the others to Generation IV [18]. The Generation IV International Forum (GIF) was created in 2000. The GIF selected six systems as Generation IV technologies: gas-cooled fast reactor (GFR), lead-cooled fast reactor (LFR), molten salt reactor (MSR), sodium-cooled fast reactor (SFR), supercritical water-cooled reactor (SCWR), and very high-temperature reactor (VHTR). Using the type in Table 2,

VHTR belongs to HTR, and GFR, LFR, and SFR to FNR. According to the GIF 2014 Technology Roadmap Update for Generation IV Nuclear Energy Systems [19], VHTR moved from the viability to performance phase in 2010, and SFR and LFR should do so by the end of 2015.

Technical feasibility of the SRs of PWR, BWR, HWR, HTR, and some FNR (SFR and LFR) can be expected to be proven earlier. However, because of their scale disadvantages, the economy of these reactors is considered to be worse than conventional large reactors [19]. Means to overcome this obstruction are addressed in Sect. 6.

The US Department of Energy is supporting R&D on SRs [20]. The IAEA is supporting long-term international activities regarding SRs. Compared with these activities, the Japanese government is not active in supporting such R&D. In the future, if the Japanese public does not accept conventional large reactors and reduction of CO_2 emissions becomes a much more serious issue, or if developing countries ask developed countries to help with their nuclear energy programs, Japan should have the option of SRs.

5.2 *Breed-and-Burn Reactors (B&BRs)*

Safety and economy are only two of the five problems inherent to nuclear energy, as shown in Fig. 2. We should be concerned with the other three problems: resources, waste, and nuclear threats.

Another important reactor design concept for the future is the B&BR. The standing wave reactor (SWR), traveling wave reactor (TWR) [21], and CANDLE reactor [7] pertain to this concept. These reactors use only natural uranium, depleted uranium, and/or thorium for their fresh fuel and do not require reprocessing. Therefore, they are free from the nuclear threat problem. The burnup of these reactors is very high, i.e., 20 % for SWR and 40 % for TWR and CANDLE. Therefore, they can use fissile materials very efficiently, and the amount of spent fuel per energy produced is very small. These performances provide excellent solutions to resource and waste problems.

B&BRs have the excellent features above but also difficult technical problems. One is the material problem caused by their very high burnup, and the other is the difficulty of criticality performance. R&D on these reactors began recently in the USA and basic studies have initiated in several emerging countries. However, in Japan there is very little activity, even though the CANDLE concept originated there.

6 Technology Roadmap

The Japanese nuclear fleet is huge and has substantial numbers of excellent personnel, knowledge, and facilities. Given this, the fleet will overcome the present difficulties by promoting the conventional nuclear energy system with its great capability. The future of this energy depends on the economy, politics, and public acceptance, not on technology. However, the future of innovative nuclear energy, whose importance was cited above, depends mostly on technology. This section discusses the technology roadmap of such innovative energy.

A technology roadmap of Japanese SRs and B&BRs is shown in Fig. 6, together with predictions of conventional nuclear energy in Japan and forecasts of innovative nuclear energy in foreign countries (OECD countries, China, Russia, and certain emerging countries).

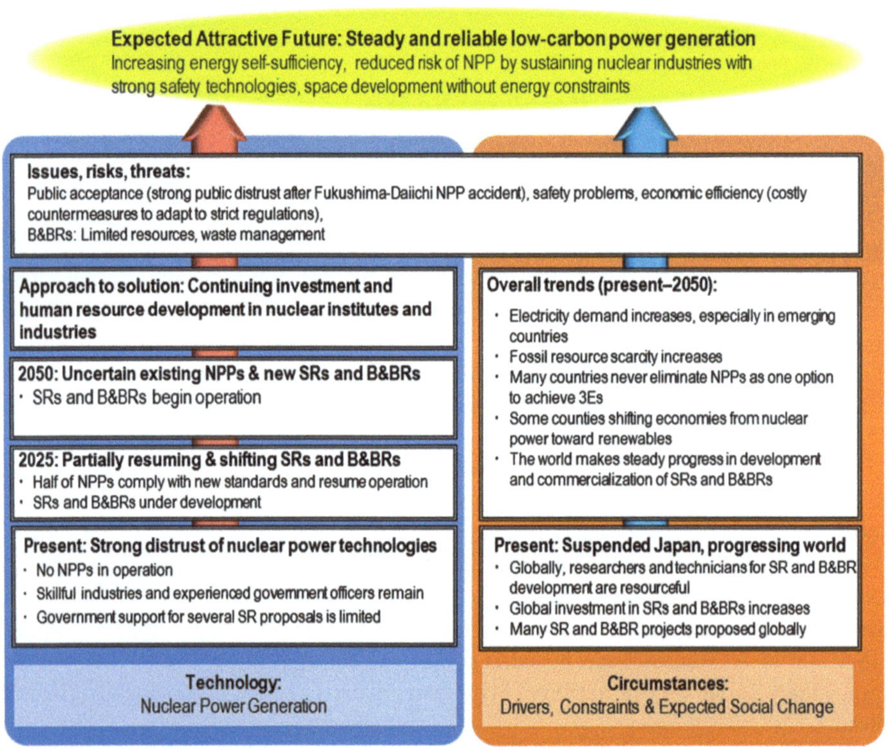

B&BR: Breed-and-Burn Reactor, NPP: Nuclear Power Plant, SR: Small Reactor, 3Es: Energy security, Economic growth, and Environmental conservation

Fig. 6 Technology roadmap of Japanese SRs and B&BRs, with predictions for conventional nuclear energy in Japan and forecasts of innovative nuclear energy in foreign countries (OECD countries, China, Russia, and certain emerging countries)

The greatest obstruction to SRs is economics. However, we have many ideas to overcome this problem, including factory mass production [13]. SRs have many potential markets not accessible to large reactors. Further, they can be used in conventional large energy grids as modular reactors, demonstrating substantial advantages.

The major obstacle to B&BRs is technical and is related to material development. It may take a long time to overcome this obstacle. The roadmap for Japanese B&BRs may be too optimistic, as the national government is currently disposed against R&D for such innovative reactors. However, it is not technically impossible to follow this roadmap.

7 Benefit and Future Vision

Nuclear energy can considerably reduce greenhouse gas emissions and other environmental pollutants with acceptable cost. Daily control and maintenance of nuclear reactors are easily accomplished. However, as mentioned in Sect. 2, current nuclear energy has inherent and difficult problems to be solved at acceptable cost, i.e., limited resources, safety, waste disposal, and nuclear threats. The Fukushima accident has slowed nuclear reactor deployment, particularly in Japan. However, China [22] is aggressively promoting nuclear energy development. In many countries, these inherent problems are being studied more than before.

SRs will substantially solve the safety problem. B&BRs have great potential to solve all the inherent problems. SRs can realize special usages such as desalination, district heating, and certain industrial types. They can be used by ships and submarines and are also expected to furnish undersea energy for investigation and mining. Many of these uses may be realized by the middle of this century.

One additional important characteristic of nuclear energy is its very dense nature, as mentioned in Sect. 2. It needs no additional material like oxygen for fossil fuel. Therefore, it is almost the only useful energy for deep space, where in the far future, humans will attempt to penetrate. At that time, they will need nuclear energy. Nuclear reactors are currently producing considerable spent fuel waste. Most of this waste can be used as new nuclear fuel. Fissile concentration in the spent fuel of B&BRs is particularly high. Although presently only underground disposal is considered for nuclear waste, controlled storage should also be evaluated. In the future, transport to space may be much safer, cheaper, and more reliable, if innovative techniques such as the space elevator [23] become available. At that time, storage of these wastes in artificial structures on/in the moon, asteroids, and outer space will become attractive options.

Nuclear energy will be important not only through 2050 but also beyond. In fact, it may become much more important than at present. Concerning the present, even if we continue using nuclear energy in Japan, its contribution to mitigating global warming is very small. However, if we develop innovative nuclear energy systems, their results would be global and their contributions considerable.

References

1. International Atomic Energy Agency (2014) Climate change and nuclear power 2014. http://www-pub.iaea.org/MTCD/Publications/PDF/ccanp2014web-14869824.pdf
2. Federation of Electric Power Companies of Japan (2014) Electricity review Japan. http://www.fepc.or.jp/library/pamphlet/pdf/2014ERJ_full.pdf
3. Nuclear Energy Agency, International Energy Agency (2015) Technology roadmap – nuclear energy. http://www.iea.org/publications/freepublications/publication/TechnologyRoadmap NuclearEnergy.pdf
4. Wikipedia (2014) Energy density. http://en.wikipedia.org/wiki/Energy_density
5. Sekimoto H (1990) Nuclear reactor theory, COE-INES textbook. Tokyo Institute of Technology, Tokyo
6. Nuclear Threat Initiative (2014) Understanding nuclear threats. http://www.nti.org/threats/nuclear/
7. Sekimoto H (2010) Light a CANDLE, 2nd edn. CRINES, Tokyo Tech, ISBN:978-4-905205-00-5 C3053 E, Japan
8. Sekimoto H (2014) A brief history of design studies on innovative nuclear reactors. In: Proceedings of the 4th International Conference on Advances in Nuclear Science and Engineering, (ICANSE), Bali, Indonesia, 16–19 September 2013, AIP Proc., New York
9. Wikipedia (2014) Radioactive waste. http://en.wikipedia.org/wiki/Radioactive_waste
10. Marcus GH (2010) Nuclear firsts: milestones on the road to nuclear power development. American Nuclear Society, Illinois
11. Lilienthal D (1980) Atomic energy, a new start. Harper and Row, New York
12. Weinberg AM, Spiewak I (1984) Inherently safe reactors and a second nuclear era. Science 224:1398–1402
13. Sekimoto H (1992) Several features and applications of small reactors, potential of small nuclear reactors for future clean and safe energy sources. Elsevier, Amsterdam, pp 23–32
14. Sekimoto H (ed) (1992) Potential of small nuclear reactors for future clean and safe energy sources. Elsevier, Amsterdam
15. World Nuclear Association (2014) Small nuclear power reactors. http://www.world-nuclear.org/info/Nuclear-Fuel-Cycle/Power-Reactors/Small-Nuclear-Power-Reactors
16. International Atomic Energy Agency (2012) Status of small and medium sized reactor designs. http://www.iaea.org/books
17. International Atomic Energy Agency (2014) Advances in small modular reactor technology developments. http://www.iaea.org/books
18. Wikipedia (2014) Generation IV reactor. http://en.wikipedia.org/wiki/Generation_IV_reactor
19. International Atomic Energy Agency (2014) Technology roadmap update for generation IV nuclear energy systems. www.gen-4.org/gif/upload/docs/application/pdf/2014-03/gif-tru2014.pdf
20. U.S. Department of Energy (2014) FY 2014 consolidated innovative nuclear research FOA. http://www.energy.gov/ne/downloads/fy-2014-consolidated-innovative-nuclear-research-foa-0
21. Gilleland J (2009) On the traveling-wave reactor. Nuclear News, September, 30–32
22. World Nuclear Association (2014) Nuclear power in China. http://www.world-nuclear.org/info/Country-Profiles/Countries-A-F/China--Nuclear-Power/
23. Wikipedia (2014) Space elevator. http://en.wikipedia.org/wiki/Space_elevator

Nuclear Waste and Power Generation

Norihiko Handa

Abstract Nuclear waste produced in light water-cooled reactors contains long half-life, high-level radioactive waste. Although this waste can be safely stored in underground repositories, the public remains anxious because they consider repository sites too close to their living space. Therefore, separation and transmutation of this radioactive waste is important. In this chapter, transuranic isotope separation and transmutation technologies using fast reactors are discussed, and a roadmap to achieve the realization of this process is proposed. This chapter introduces ways to maintain the use of nuclear power without producing additional waste because continued nuclear power is important to reduce carbon dioxide emissions.

Keywords Nuclear waste • TRU separation and transmutation • Fast reactor • The Japanese Strategic Energy Plan • Second reprocessing plant

1 Introduction

Uranium-238, a major component of uranium fuel for light water-cooled reactors (LWRs), is converted to heavier nuclides such as Pu (plutonium), Np (neptunium), Am (americium), and Cm (curium) by absorbing thermal neutrons in the reactor. The half-life of these nuclides is thousands to tens of thousands of years and they can be toxic to humans. This is called long half-life, high-level radioactive waste. These nuclides continue to be a serious concern with the use of nuclear energy.

As of 2014, the amount of spent LWR fuel reached 17,000 tons in Japan. The amount of long half-life, high-level radioactive waste of this spent fuel is 170 tons. If existing LWRs are used for the next 60 years, this waste will reach 496 tons [1].

If reliance on LWR is reduced, the emission of carbon dioxide will increase. Because there is no other reliable solution for generating power without emitting carbon dioxide, it is difficult to avoid using nuclear power. The problem of high-level radioactive waste must be resolved for newly built nuclear reactors.

N. Handa (✉)
E2M Inc., Chiba, Japan
e-mail: norihiko.handa@e2m.mygbiz.com

Y. Kato et al. (eds.), *Energy Technology Roadmaps of Japan*,
DOI 10.1007/978-4-431-55951-1_16

In this chapter, several methods for separating the long half-life, high-level radioactive waste from spent LWR fuel are discussed, as well as the potential of transmutation using fast reactors, in which the nuclides are converted to other nuclides with a shorter half-life. Also discussed is a strategy introducing fast reactors, which are expected to resolve these problems.

2 Present Status

2.1 Current Strategy: Waste Disposal and Transmutation

Currently, several countries are looking into the possibility of using underground repositories for disposing of high-level radioactive waste. Although no countries are actually operating such repositories, there are a few that have selected sites. There are five other countries (including Japan) that have begun studying the geologic conditions of potential sites. In underground repositories, high-level radioactive waste is stored as vitrified waste and covered in steel. It has been shown theoretically that this can confine the waste for at least 70,000 years. After that, the waste may begin to leak, but based on analyses, the radioactive hazard to local residents through drinking water is less than individual exposure from natural radioactivity [2]. Despite this, the public remains anxious. Although the waste is stored underground, the public considers repository sites too close to their living space. Therefore, a few countries, including Japan and France, have stressed in their policy statements that separation and transmutation technology studies will be conducted.

In Japan, the Japanese Strategic Energy Plan [3] of 2014 states, "Specifically, development of technologies for decreasing the radiation dose remaining in radioactive waste over a long period of time and enhancing the safety of processing and disposal of radioactive waste, including nuclear transmutation technology using fast reactors and accelerators, will be promoted by utilizing global networks for cooperation." Nuclear transmutation technology using fast reactors must be further developed.

2.2 Current Strategy: Ratio of Electricity Generated by Nuclear Plants

The Japanese Strategic Energy Plan states that reduced dependency on nuclear power in electricity generation is planned, "Dependency on nuclear power generation will be lowered to the extent possible by energy saving and introducing renewable energy as well as improving the efficiency of thermal power generation, etc."

If carbon emission levels are maintained within the pledged range, it will be difficult to reduce dependence on nuclear power. However, producing more nuclear waste is not accepted by the public. Therefore, introducing fast reactors, which can solve the high-level radioactive waste problem, is important. In fast reactors, Pu, Np, Am, and Cm are converted to other nuclides with shorter half-life fission products at high probability through nuclear fission reaction.

Fuel for fast reactors is fabricated from the major nuclides of high-level radioactive waste from LWRs. Thus, introducing fast reactors is a win-win situation because it reduces this waste. Establishing a well-balanced ratio of electricity generated by nuclear plants is possible.

2.3 *Current Status of Technology: Separation and Its Efficiency*

Radioactive waste treatment covers a variety of technologies. Among them, the following two are important:

- Technology to separate the long half-life, high-level radioactive waste from LWR spent fuel with 99.9 % efficiency
- Technology to shorten the half-life by transmutation in fast reactors

With waste separation technology, it is important to prioritize the nuclides. The long half-life nuclides fall into two categories, fission products and the transuranic (TRU) isotope. TRU is composed of Pu, Np, Am, and Cm, which is produced with uranium-238 from the LWR fuel. Water-soluble fission products may leak from disposal sites after the degradation of vitrified waste, causing exposure to humans. However, the dose caused by radioactivity of leaked fission products is well below the exposure dose from natural radioactivity [2].

Several countries have focused on the transmutation of TRU because of its serious hazard to humans, although TRU stored in vitrified waste does not leak. TRU is a major concern, and its separation efficiency from spent LWR fuels is a high priority. A separation efficiency of 99.9 % and transmutation of separated TRU lead to a storage period of 300 years, much shorter than tens of thousands years [4]. The public may be willing to accept this period.

There are two methods to separate TRU from spent LWR fuels. One is pyro-reprocessing using an electrochemical process and the other is aqueous reprocessing, which is based on the principle of extraction of materials with a solvent. The high separation efficiency of 99.9 % has been achieved for Pu with the conventional aqueous method, and reprocessing plants using this method are already in commercial operation in several countries. R&D is still needed for the separation of Np, Am, and Cm, because of different chemical behaviors of Pu in an aqueous process. New solvents should be developed. According to recent French studies, 99.9 % separation efficiency has been achieved for Am and Cm using a newly developed solvent [5]. The pyroprocess can theoretically recover almost 100 % of TRU using electrodes in electric field vessels. Because TRU not recovered in the first treatment remains in the electric field vessel, a larger decontamination factor or collection rate can be ensured by repeating this treatment. This process has achieved a recovery rate of 99.5 % or higher through the treatment of 4,400 kg of spent fuel from fast experimental reactor EBR II in the United States [6].

In Japan, R&D is still in the experimental phase. Because it is not easy to conduct demonstration tests on the separation of TRU using actual spent LWR fuels in the country, it is preferable to participate in overseas studies and share information.

2.4 Current Status of Technology: Transmutation in Fast Reactors

There are three options for transmuting TRU using fast reactors. The first two concern only the use of fast reactors and the third the use of an accelerator and fast reactor. With the first two options, there is a choice of fast reactor cores, one with a breeding ratio of one or one with a ratio of zero. The third option uses an accelerator to accelerate heavy particles into a target. High-energy neutrons generated by spallation in the target maintain the high power level, while the reactor core is subcritical. This is called an accelerator-driven reactor system (ADS).

Because ADS raises electricity generation costs as much as 25 % compared with present costs [6], it is not a high priority in this chapter. The fast reactor core with zero breeding ratio is not a high priority either, because it causes new safety issues [7]. Further, neither ADS nor such fast reactors can meet future electricity demands. They are suitable to transmute TRU but do not generate sufficient electricity.

Upon examining the development of the fast reactors with breeding ratio of one or larger, we see that this technology is ready for commercial application. One such fast reactor is the sodium-cooled type, which has a particularly long history. Argonne National Laboratories in the United States built the first such reactor for experimentation, EBR I, which began generating electricity in 1946. EBR II has been successfully operated for a long time. Other experimental or prototype fast reactors have been safely operated in the world. In France, Super Phoenix with 1200 MWe electrical output reached critical condition in 1985. It was closed in 1997 because of political debates regarding the use of plutonium. Recently, China has expressed the desire to import Russian fast reactors in the near future. In 2014, the Russian fast reactor 800-MWe BN800 reached critical condition in June 2014. The 500-MWe Indian PFBR is scheduled to reach critical condition in 2015. Both Russian and Indian fast reactors are intended to produce electricity only, and transmutation of TRU is not within their scope. However, France aims to start operating ASTRID, a 600-MWe fast reactor in 2025. The main purpose is to reduce the amount of TRU generated from LWRs.

Although there are still some safety issues to be resolved with sodium-cooled fast reactors, solutions may have been found. These issues are fragility of thin-walled structures against earthquakes, difficulty of visual inspection inside sodium, rapid accident progression because of positive coolant reactivity feedback upon mismatch between power and flow rate or gas entrainment into the core, pressure buildup caused by sodium water reactions in steam generators, and potential explosions caused by re-criticality during severe accidents. Several countries have conducted studies to overcome these difficulties, and solutions to improve safety are being developed [4].

Reduced construction costs are required before commercialization is possible. Simplified structures and components are being adapted to recent fast reactors, reducing cost.

In Japan, prototype reactor Monju has been inoperative for a long time because of political reasons, and the government has requested to move toward a restart. This will take some time, so international cooperation is a necessary step toward commercialization of fast reactors in Japan.

3 Technology Roadmap

3.1 Technological Goals and Solutions

As of 2014, the amount of spent LWR fuel was 17,000 tons, with 170 tons TRU. Based on discussions to develop the Japanese Strategic Energy Plan, existing LWRs were assumed to operate until 2069 if the life extension of LWRs is approved by the regulatory authority. The amount of spent LWR fuel would reach 49,600 tons, with 496 tons TRU.

3.1.1 Separation of Np, Am, and Cm from Highly Radioactive Liquid Waste

The Japanese reprocessing plant Rokkasho can reprocess spent LWR fuels with an annual capacity of 800 tons. The plant is planned to be in operation for 40 years, beginning in 2014 or 2015. Because this plant is only designed to separate Pu, other

TRU such as Np, Am, and Cm are left in highly radioactive liquid waste. To separate Np, Am, and Cm, an additional chemical process is needed downstream of the main process at the plant. In this process, Np, Am, and Cu can be separated from the highly radioactive liquid waste with 99.9 % efficiency. Development of this technology is a high priority.

3.1.2 Separation of TRU in a Second Reprocessing Plant

The Rokkasho plant will be closed in 2054 after operating for 40 years. Because LWRs can operate until 2069, 49,600 tons of LWR spent fuel will accumulate, so 20,000 tons of that fuel will remain un-reprocessed by Rokkasho. Because all TRU in the spent fuels is required to be separated and transmuted, a second new reprocessing plant is necessary to reprocess the remaining spent fuel. More comprehensive technology is needed for that plant, including technology for treating mixed oxide fuel, MOX fuel, which is richer in Pu than conventional LWR fuel. In addition, to reduce reprocessing costs, a single process is necessary to separate all TRU with 99.9 % efficiency. Establishing this technology for the second reprocessing plant is also a priority.

3.1.3 Transmutation of TRU Using Commercially Developed Fast Reactors

Introducing safe and economical fast reactors is essential to transmute TRU separated from both the Rokkasho and second reprocessing plants. Safer fast reactors are being designed to comply with current safety guidelines in several countries. These guidelines focus on the enhancement of inherent safety features of reactor cores, steam generators, and seismic resistance. Several countries have developed post-Fukushima guidelines that require greater robustness and resilience against natural hazards and terrorist attacks. New safety features in recent fast reactors can meet these guidelines. To introduce more economical fast reactors, a reduction in the quantity of necessary material can be achieved using more simplified structures and components. Recent reactor designs in France and India are using these simplified design features. The Japanese reactor Monju will be used for studying efficient transmutation methods. It is important to participate in overseas projects to build demonstration reactors and in basic studies using Monju. Thus, developing safe and cost-effective commercial fast reactors should be a high priority. If commercial fast reactors are successfully introduced beginning in 2030, it will be possible to complete the transmutation of 496 tons TRU by 2069, providing 25 GWe using fast reactors [1]. This is half the power currently generated by existing Japanese LWRs. The TRU is converted to fission products by nuclear fission reaction with shorter lifetime.

Transmuting TRU and power generation using fast reactors is the goal of the Japanese roadmap in this chapter.

3.2 *Steps Toward the Goal*

The following steps are recommended to reach the goal:

- Start separating TRU from highly radioactive liquid waste in the Rokkasho plant in 2020.
- Start introducing commercial fast reactors in 2030.
- Start building the second reprocessing plant in 2050.
- Completion of TRU transmutation using fast reactors by 2070.

It is important to establish the technology for separating Np, Am, and Cm from the highly radioactive liquid waste in the Rokkasho plant. Several methods can be applied, and development of this technology is necessary in Japan.

International cooperation is needed to introduce commercial fast reactors in Japan. Demonstration fast reactors have been built in Russia and India, and France is aiming to operate a demonstration reactor in 2020. In Japan, experimental studies on TRU transmutation using Monju are of first priority. In addition, international cooperation with the above countries is necessary for introducing commercial fast reactors by 2030, without building a demonstration reactor in Japan.

Before constructing the second reprocessing plant, it is important to select a reprocessing method. Because two methods (aqueous or pyro) can be used in the plant, a comparison of construction costs between the two methods is critical to the selection. This selection should be made by 2040 to launch the plant by 2050.

3.3 *Problems and Risks*

Most of the public do not support a restart of existing LWRs because of the unresolved problem of nuclear waste. Repository sites have not been determined and the technology of TRU separation and transmutation has not been fully developed. Therefore, it is unlikely that the public would accept the roadmap described in this chapter.

There is a technological risk in introducing commercial fast reactors without building a demonstration reactor. Despite operational experiences with Monju and shared information of demonstration reactors overseas, a careful approach is needed for design, fabrication, and building of the first commercial reactor.

4 Benefit and Future Vision

According to the roadmap, increased emissions of carbon dioxide caused by reduced dependence on nuclear power generation from LWRs can be controlled. Dependence on existing LWRs is assumed to decrease and become zero by 2070, according to the rule that old plants must cease operation upon reaching their 40-year lifespan. Recently built plants can operate for 60 years. New LWR plants (except already planned ones) cannot be built after 2014.

If fast reactors are not introduced beginning in 2030, 110 million tons of carbon dioxide will be released by 2050, but this can be reduced by 57 million tons with the use of fast reactors. Power generated by LWRs is shown by the blue broken line in Fig. 1, and the power is 23 GWe in 2050. Another 13 GWe will be generated using fast reactors by 2050, assuming a required TRU of 19.5 ton/GWe for one fast reactor core and recycling facility.

By 2070, dependence on LWRs could be zero, and a power generation of 25 GWe could be achieved using fast reactors instead. In addition, long-life, high-level waste produced by LWRs could be completely eliminated. Figure 1 shows these benefits.

After 2070, power supply using fast reactors with higher breeding ratio could meet future electricity demands without producing long half-life, high-level radioactive waste (Fig. 2).

Fig. 1 Nuclear waste reduction and power generation by waste

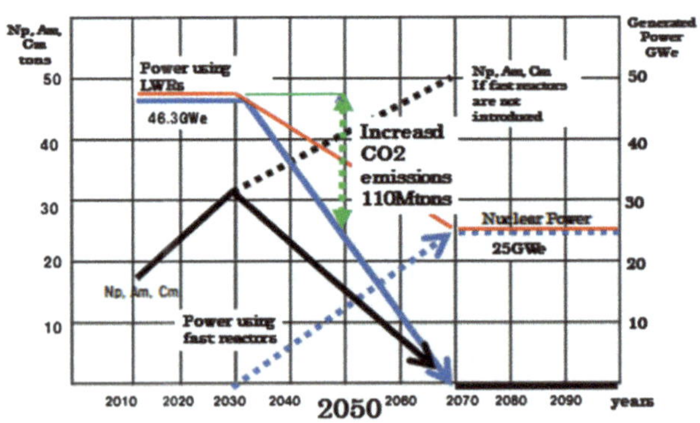

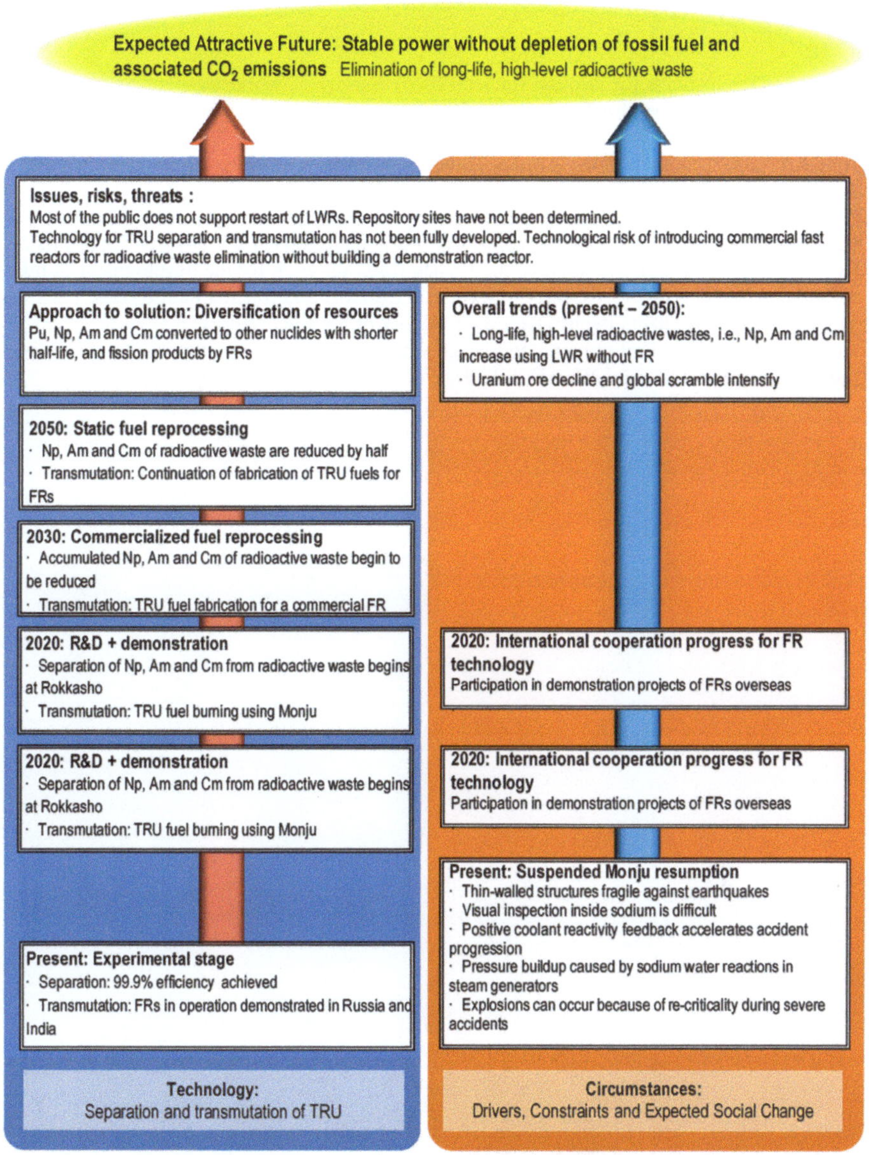

Rating	Technology parameters
★★★	Solid mechanistic estimates
★★	Extrapolation of trends
★	Controversial value (e.g., personal perspective, competition with existing technology)

FR: Fast reactor
LWR: Light Water Reactor
MA: Minor Actinides
TRU: Transuranic isotope

Fig. 2 Roadmap toward 2050

References

1. The CANON Institute for Global Studies (2015) Survey of world trend of technologies of separation and transmutation for long half-life high-level radioactive waste (in Japanese). CIGS, Tokyo, Japan
2. ANDRA (2005) Dossier 2005 Argile, Tome- Évaluation de sûreté du stockage géologiqu, ANDRA and JAEA(1999) JNC TN1400-99-020 (in Japanese)
3. Agency for Natural Resources and Energy (2014) The strategic energy plan. METI, Japan
4. CEA (2012) Report on sustainable radioactive waste management vol 1–vol 5. CEA, France
5. Rostaing C et al (2013) Overview of the French R&D program for the development of minor actinides separation processes, Int. Conf. on Fast Reactors and Related Fuel Cycles, Paris
6. Till CE, Chang YI (2011) Plentiful energy the story of the IFR. ANL, U.S.
7. Hoffman et al (2006) Preliminary core design studies for advanced burner reactor over a wide range of conversion ratio. ANL AFCI, U.S.

Hydropower Generation

Morihito Inagaki

Abstract There are two main types of hydropower, conventional and pumped storage. The former dates back more than a century. It developed with the growth of Japanese industries, although its share subsequently declined with the rise of fossil fuel-based thermal power plants. Several decades later, as a global warming countermeasure, hydropower has reemerged as an important solution, together with an increase in the use of renewable energy sources. R&D of pumped-storage hydropower has progressed with the growth of nuclear energy, and Japan came into possession of world-class advanced technology. However, despite being an essential power source to stabilize power systems (particularly with the coming full electricity deregulation), the importance of pumped-storage hydropower has yet to be affirmed, because of a lack of methods to assess its value. Nevertheless, the increased use of renewable energies, which are also unstable power sources, has underlined the importance of power system stabilization measures. This explains the significant expectations of pumped-storage hydropower features, such as its storage function and its ability to swiftly absorb supply and demand imbalances. Both types of hydropower are power sources that suit the culture and geography of Japan and will certainly contribute to the energy society on the basis of an energy storage system.

Keywords Pumped-storage hydropower • Rehabilitation and renovation • Energy storage • Grid stability

M. Inagaki (✉)
Tokyo Electric Power Company, Tokyo, Japan

JSE Co., Ltd., Tokyo, Japan
e-mail: inachan@jse-3e.com

Y. Kato et al. (eds.), *Energy Technology Roadmaps of Japan*,
DOI 10.1007/978-4-431-55951-1_17

1 Introduction

Hydraulic power originated in Japan. It has matured throughout the country's history and has provided lasting energy. Hydropower remains a significant infrastructure component, underpinning the foundations of society as a crucial energy storage system for industrial development and daily life.

There are two main types of hydropower: conventional and pumped-storage hydropower (PSH). The former is a type of sustainable energy that takes advantage of nature's circulation system, in which energy is collected and extracted from the hydrologic cycle. Moisture from seawater and the ground, evaporated by solar energy, turns into rain, flows from elevated areas into rivers, and finally returns to the ocean. Conversely, PSH is an anthropogenic circulation system that repeats the exchange between potential and kinetic energy via an electric power system.

1.1 Conventional Hydropower Stations

Conventional hydropower is suitable for the precipitous terrain of Japan and a technically well-established energy source, with more than a century of development. Currently, it covers ~10 % of the country's annual electric power consumption and ~20 % of total installed capacity. As shown in Table 1, as of 2010, there were 1888 conventional hydropower stations nationwide, with a total installed capacity of 21,852 MW. These stations operate as eco-friendly, cost-effective, highly sustainable, and semipermanent energy storage systems.

Although average annual precipitation in Japan is ~1700 mm, nearly twice the global average, the amount of precipitation varies drastically by region and season, owing to south-north mountain ranges spanning the center of the main island, Honshu. Kyushu island, southern Shikoku island, Tokai, Hokuriku, and other regions see annual precipitation exceeding 2000 mm. Conversely, annual precipitation in the Seto Inland Sea coastal area, Kanto (inland), Tohoku (Pacific Ocean side), and other regions is less than the national average. Precipitation peaks during summer on the Pacific Ocean side because of the rainy season and typhoons, and in

Table 1 Water power resources in Japan

	Developed power stations			Undeveloped potential sites		
Type of hydropower system	Sites	Installed capacity (MW)	Electric power generation (MW)	Sites	Installed capacity (MW)	Electric power generation (MW)
Run-of-river type	1,178	5,235	27,707	2,535	9,028	36,171
Pondage type	467	10,197	45,359	152	2,361	8,325
Reservoir type	243	6,420	18,929	57	1,489	3,415
Total	**1,888**	**21,852**	**91,995**	**2,744**	**12,878**	**47,911**

Fig. 1 Keage power station

winter on the Japan seaside, owing to substantial snow. Given these conditions, Japan's hydropower potential is mostly in the central area of Honshu and in the Japanese Alps.

Japan's first arc lamp was lit on March 25, 1878, known today as National Electricity Day. In 1887, the country's first electric company, Tokyo Dento, provided electricity for the first time, generated via a 25-kW-capacity coal-fired thermal power station. In 1891, to provide electricity for manufacturing Nishijin silk, Keage power station operated by the city of Kyoto became the first station to provide electricity from hydropower (Fig. 1). Water stored in Lake Biwa was supplied to nearby areas for agriculture and drinking water in Kyoto and used as an energy source to generate hydropower.

Around 1902, amid news from America concerning the success of long-distance transmission of up to 60,000 V of power, momentum increased for developing hydropower in the mountains. A series of development projects for large-scale hydropower stations was initiated, and, in 1911, the ratio of thermal power to hydropower in total electricity generation capacity of all Japanese electric businesses reversed, making hydropower the largest energy source, followed by thermal power. The total electricity generation capacity of electric businesses at the end of 1914 was 555,000 kW, and the capacity for hydropower reached 377,000 kW (69 %).

During the post-WWII recovery years, domestic industry activities continued to expand because of the procurement boom caused by the Korean War. In 1955, the most advanced oil thermal power generation technology at the time was imported from America to meet increased electricity demand, and the largest power source again reversed, from hydropower to thermal power. The portion of hydropower in total generated electricity of electric utilities was 78.7 % in FY 1955, declining to 46.1 % in FY 1962, reflecting that reversal.

Although hydropower was considered renewable energy by most other countries, it was not defined as such in Japan until around the year 2000, whereupon its importance began to decline.

1.2 Pumped-Storage Hydropower Stations

Since the 1960s, when variation of electricity consumption between day and night began to increase, technologies from a number of foreign countries were adopted and PSH stations began to be constructed.

At the time when PSH stations were first being built by Japanese companies with larger capacities and used to effectively exploit power generated by nuclear power stations overnight (economic pumping), similar to the construction of conventional hydropower stations, PSH stations were built along large rivers to ensure abundant water. Subsequently, given the lack of candidate sites and the need to fabricate large pumps, companies began striving to find new sites and obtain economies of scale. This spawned super high-head, large-capacity PSH stations in areas (rivers) with precipitous terrain but no potential for ensuring abundant water.

As shown in Fig. 2, the head, which was 100 m at the time when PSH was introduced, is now at 700 m, following the development of a super high-head pump turbine. This yields a power density more than seven times the original value with the same amount of water.

Fig. 2 Pump head of PSH

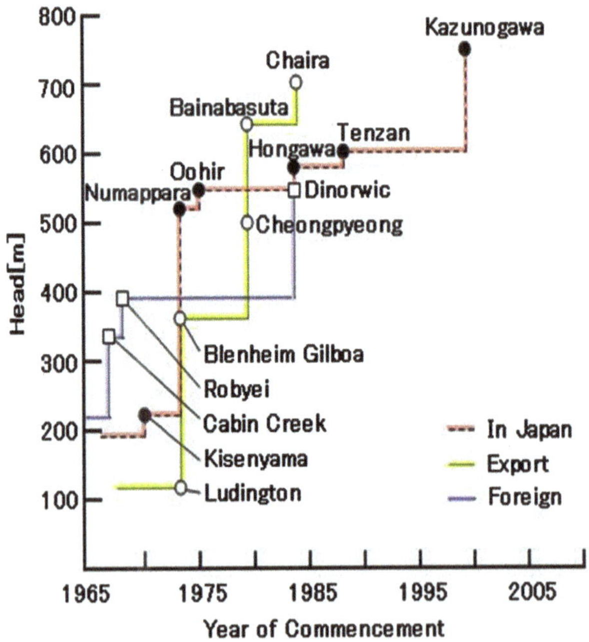

By the time the construction of new stations was underway, with capacity per plant exceeding 1 GW, the use of PSH was no longer limited to economic pumping for efficient nuclear energy use, and it began to function as an electricity storage system to adjust the supply and demand imbalance of power systems. Targeting further advancements, companies began to develop splitter runners with a wide adjustment range and variable-speed PSH systems enabling an ultrafast response.

However, a lack of quantitative PSH assessment in the ancillary services market during its transition from development and expansion based on the coexistence with nuclear power to its status as an adjustment power source for stabilizing electric power systems has caused its importance in the new market to go unnoticed.

2 Present Status

Although hydropower has been the victim of unfavorable circumstances, in recent years, general recognition has increased the importance of hydropower as a beneficial means for improving the global environment and as an energy source to support unstable renewable energy sources.

2.1 Conventional Hydropower Stations

Annual total output from conventional hydropower in Japan is ~90 TWh, and the country's hydropower potential is estimated at ~50 TWh (Table 1).

Conventional hydropower has three forms of water energy utilization, run-of-the-river, pondage, and reservoir. When the forms are reclassified under the definition of energy storage, though run-of-the-river is excluded because it uses energy in the form it is provided from the natural environment, the pondage and reservoir types can be defined as hydropower energy capable of storing electricity. Total annual output from these two types is ~64 TWh.

2.1.1 Economic Efficiency

According to a report by the National Policy Unit of Japan, the generating cost of conventional hydropower plants is 10.6 JPY/kWh on a 40-year average. Most hydropower plants, however, are operated for 50 years or more, so the actual average generating cost should be about half.

Regarding the cost breakdown of hydropower, the proportion of fixed expenses (mainly initial cost) is large and that of variable expenses (mainly operating costs) is small. Despite the low energy density and high initial cost per kW, power generation management with few business risks can be expected, even with a long payback period. This is because mechanically, station structure features an extended service life. Thanks to these economic characteristics and more than a century of development and operation history, generating costs at established hydropower stations in which most equipment repayment has been completed are somewhere between 4 and 6 yen/kWh.

Since hydropower involves lower variable expenses, once the challenge of fixed expenses in the initial station operation stage is overcome, subsequent business management will be stable. However, further development of hydropower is currently being hindered by a decline in candidate sites that enable cost-effective operation and the burden of fixed expenses during the initial stages of station operation.

2.1.2 Legal Systems

Following adoption of the Kyoto Protocol in December 1997, Japan enacted the renewable portfolio standard (RPS) law in June 2002 to revitalize hydropower development through market transactions.

However, the expected results were not achieved. Consequently, the Special Measures Concerning Renewable Energy Bill and the Renewable Energy Purchase Bill were submitted to the National Diet on April 5, 2011, and activities to promote the use of renewable energies began anew with the establishment of the feed-in tariff (FIT) in July 2012.

FIT is a system in which the national government guarantees that electricity generated by renewable energies will be purchased by relevant regional utilities. Expenses for purchasing renewable energies incurred by utilities are then covered by collecting special charges from electricity consumers. The system thereby promotes and supports the adoption of renewable energies.

Through FIT, the profitability of an electric generation business is ensured by the guaranteed purchase of electricity for a certain period at a certain price. Consequently, though the outcomes depend on the power sources involved, this system has begun to catalyze and increase the market share of renewable energies. Figure 3 shows the number of FIT-approved hydropower sites following system establishment in 2012. It is reasonable to conclude that the total has been steadily increasing, considering the relatively long period required for hydropower station candidate site surveys, design, and construction. Due to less solid business, solar and wind power may be eliminated after completion of the FIT. Conversely, hydropower will continue to improve the global environment in association with long-term and down-to-earth business.

2.1.3 Efficiency and Optimization

Although some operating hydropower stations have been renewed, many still feature old designs from several decades ago. By renewing these outdated stations with current technology, a 3–8 % increase in efficiency can be expected. Loss (disc

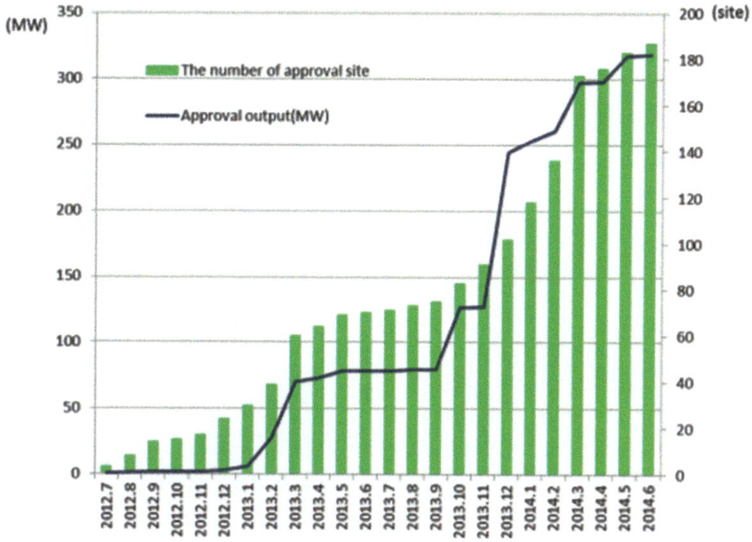

Fig. 3 FIT-certified hydropower plants

friction, secondary flow, and flow channel friction) is analyzed by making 3D computer-aided design (CAD) images in the form of rotary runners and static structures of flow channels, such as stay vanes and guide vanes, and then analyzing flow by computational fluid dynamics (CFD) using those images. Based on this analysis, an optimal solution is obtained, and by visualizing the flow, an optimal form is created by making fine adjustments to every detail of the runner form.

When hydropower was the largest source of electricity (followed by thermal power), hydropower stations built at the time were based on a design concept referred to as "kW-focused design." This concentrated on generating electricity in bulk, even if the amount could only be maintained for a limited period. Currently, kW-focused generation is rarely included in qualities required for conventional hydropower stations, and there has been a shift to a kWh-focused design, which focuses on generating as much electricity as possible.

By optimizing operation and technological innovations to use stored potential energy efficiently, an output boost is expected in addition to that from undeveloped potential hydropower.

2.2 *Pumped-Storage Hydropower Stations*

As shown in Fig. 4 and Table 2, located around Japan and mainly owned by electric companies, there are 45 PSH stations with a total output 27 GW. When the potential energy of water stored in dams is converted to electricity generation capacity, the total daily electricity storage capacity is 190 GWh. However, because PSH is an

Fig. 4 Schematic of PSH

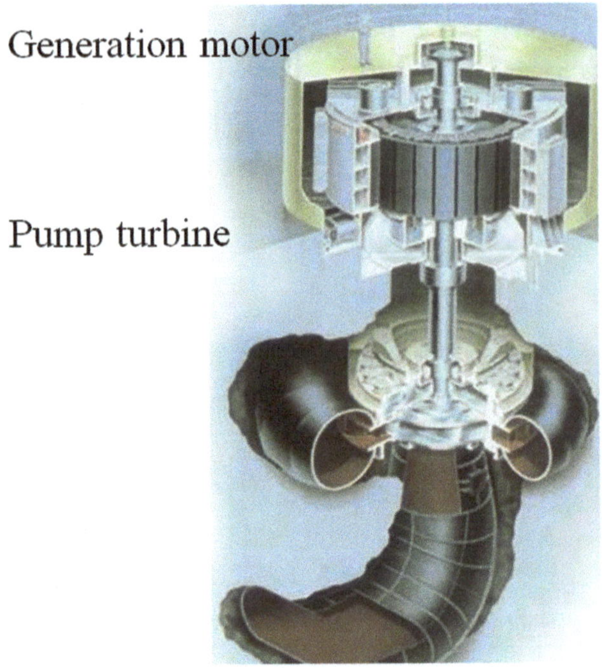

electricity storage (charge-discharge) system, hydro energy is stored at less than the amount of electricity generated (owing to hydrodynamic loss).

2.2.1 Economic Efficiency

The economic efficiency of PSH is increased by ensuring the required generating capacity at low cost. According to qualitative analysis, the portion of PSH in an optimal combination of power sources is 10–15 % of the total electricity generated. To date, the installation of PSH stations has been based on the installed capacity of nuclear and thermal power. Because these stations require power for pumping, determination of station installation is based on economic evaluation alone.

Although plans for PSH use in Japan have been postponed or shelved, there is a strong demand for PSH as a measure for managing unstable power sources and excess electricity in Europe, where the PSH market is booming. Amid the adoption of solar and wind power after implementing the FIT system, substantial unstable power sources began entering the electric system. The problem of power flow (static characteristics of electricity) is the only issue currently on the agenda. However, it is expected that with the current stability of the electric system, maintenance of supply and demand balance during sudden fluctuations of generated electricity caused by the short cycles of unstable renewable energies (dynamic

Table 2 Pumped-storage power station in Japan

Company	Power plant	Output (kW)	Commencing	Company	Power plant	Output (kW)	Year
Hokkaido Electric Power Company	Niikappu	200,000	1974	Kansai Electric Power Company	Kisenyama	466,000	1970
	Takami	200,000	1983		Okuyoshino	1,206,000	1978
	Shumarinai	1,120	2013		Ookawachi	1,280,000	1992
	Kyogoku	200,000	2014		Okutataragi	1,932,000	1974
		(200,000)	Schedule of 2015	Chugoku Electric Power Company	Matanogawa	1,200,000	1986
		(200,000)	From 2024				
Tohoku Electric Power Company	Daini Numazawa	460,000	1982		Shinnaruhagawa	303,000	1968
	Ikejirigawa	2,340	1934		Nabara	620,000	1976
Tokyo Electric Power Company	Yagisawa	240,000	1965	Shikoku Electric Power Company	Kagehira	46,650	1968
	Tanbara	1,200,000	1982		Ananaigawa	12,500	1964
	Kannagawa	940,000	2005		Oomorigawa	12,200	1959
		(1,880,000)	From 2024		Honkawa	615,000	1982
	Shiobara	900,000	1994	Kyushu Electric Power Company	Tenzan	600,000	1986
	Imaichi	1,050,000	1988		Oohira	500,000	1975
	Kazunogawa	1,200,000	2014		Omarugawa	1,200,000	2007
		(400,000)	From 2024	Electric Power Development Company	Shimogo	1,000,000	1988
	Azumi	623,000	1969		Okukiyotsu	1,000,000	1978
	Midono	245,000	1969		Okukiyotsu-Daini	600,000	1996
	Shin-Takasegawa	1,280,000	1979		Numappara	675,000	1973
Chubu Electric Power Company	Hatanagi-Daiichi	137,000	1962		Shintoyone	1,125,000	1972

(continued)

Table 2 (continued)

Company	Power plant	Output (kW)	Commencing	Company	Power plant	Output (kW)	Year
	Okuyahagi-Daiichi	323,000	1980		Nagano	220,000	1968
	Okuyahagi-Daini	780,000	1980		Ikehara	350,000	1964
	Takane-Daiichi	340,000	1969		Okinawa Yanbaru Seawater Pumped Storage Power Station	30,000	1999
	Hasegawa-Daiichi	288,000	1976				
	Okumino	1,500,000	1994	Kanagawa Prefecture	Shiroyama	250,000	1965
							(kW)
				Total	44 power plants	**27,352,810**	

electricity characteristics) will become an even more serious problem. Moreover, because PSH is capable of an ultrafast response, its importance is likely to increase.

2.2.2 Legal Systems

A power source for stabilizing the electric system must be capable not only of providing required power amounts but also doing so swiftly. As shown in Fig. 5, there is a menu of electric power exchange in Europe that specifies the time and amount of electricity required to maintain the supply and demand balance, based on which the electricity value is traded. Clearly, it is equally important for Japan to have a market transaction menu like this example, which must be established before power generation, transmission, and distribution are unbundled within the electricity supply system.

Classification in Europe	country	Classification of reserve	Time required
FRC:Frequenc Containment Reserves	Scandinavian countries	FNR(FCR N)	120s ～ 180s
	Scandinavian countries	FDR(FCR D)	30s
	Continental Europe	Primary Control Reserve	30s
FRR:Frequenc Restoration Reserves	Scandinavian countries	Regulation Power	15m
	Continental Europe	Secondary Control Reserve	Within 15 m
	Continental Europe	Direct Activated Tertiary Control Reserve	Within 15 m
RR:Replacemen Reserves	Scandinavian countries	Regulating Power	15m
	Continental Europe	Schedule Activated Tertiary Control Reserve	Different in each country
	Continental Europe	Direct activated Tertiary Control Reserve	Different in each country

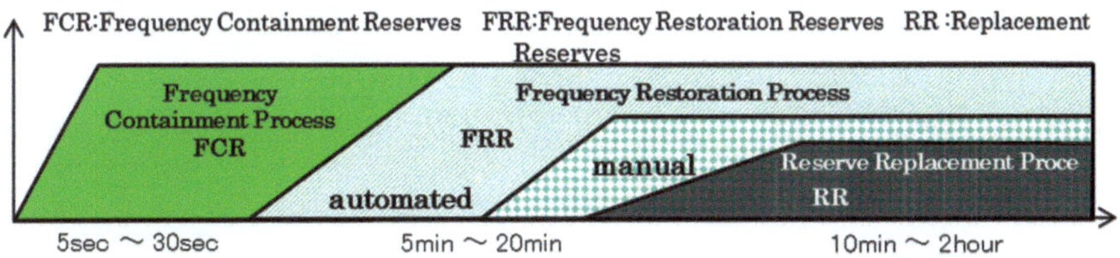

Fig. 5 Menu of electric power exchange in Europe

2.2.3 Efficiency and Optimization

In fluid machinery design, the use of CFD has enabled fluid mechanisms to be quantified and visualized, which has paved the way to efficiently create optimal designs.

Throughout a century of hydropower, it was considered taboo in the industry to increase the blade number (blade diversification) of the Francis pump-turbine runner, because it was believed that efficiency would decline because of greater friction loss with increased water contact area. By reexamining the issue from a different perspective and flow analysis technology, advantages and limitations of blade diversification were identified as shown in Fig. 6. In response, a pump-turbine (splitter) runner was developed, with a new form featuring splitter blades.

In addition, a variable-speed PSH system has been developed, allowing ultrafast input/output responses in both directions electric generation (supply side) and pumping (demand side). Figure 7 shows the operational results of this system on a given day in summer 2014. This system can respond at high amplitude with long cycles of 20–30 min at 120–130 MW (red line) and at a lower amplitude with shorter cycles of a few minutes and 20–30 MW (black line). Furthermore, though not shown explicitly by these data, the system can alter output within a few seconds and 2–3 MW, outperforming conventional, ultrahigh response hydropower systems in terms of response speed.

Since the system can follow and respond to sudden changes in the output speed of solar and other types of power, it could be useful in solving problems related to the adoption of renewable energies. It is expected that with the increase in the use of

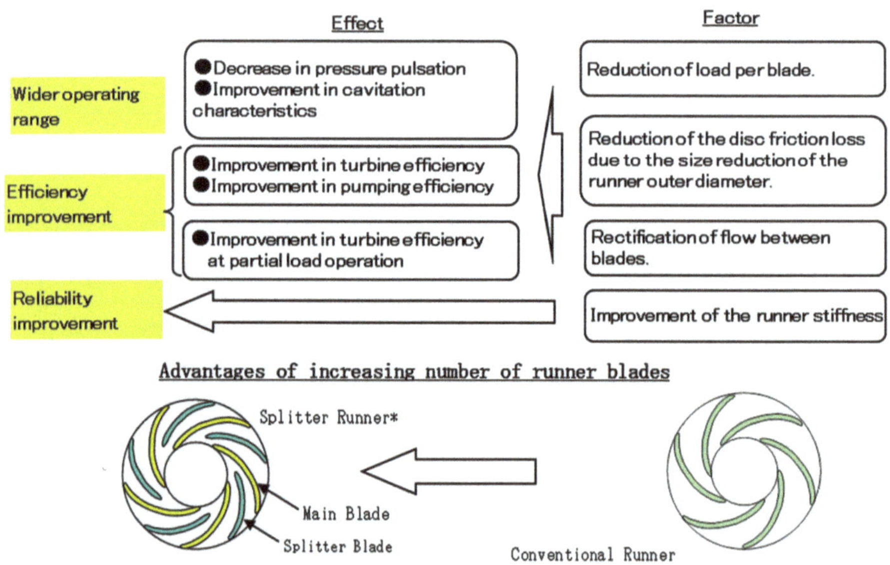

Fig. 6 Development of splitter blade runner

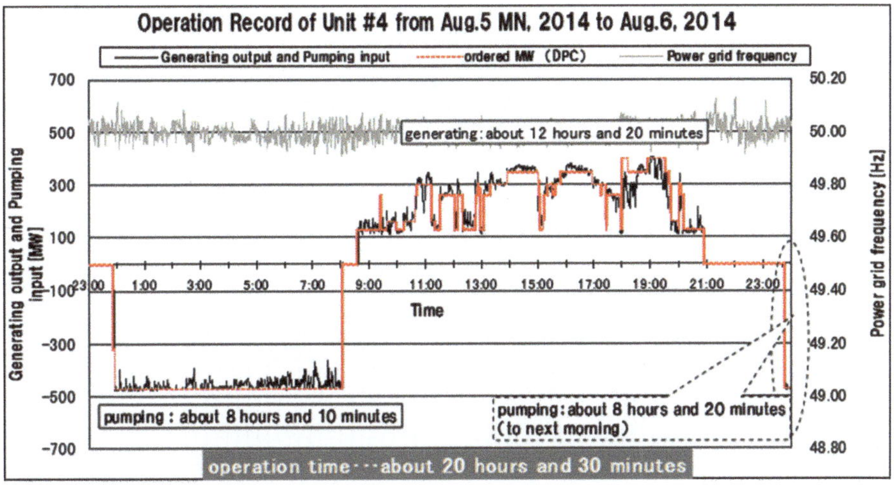

Fig. 7 Effect of pumped-storage power system

solar energy, the future total electricity generation capacity may exceed total demand. Based on this assumption, PSH pumping could be used to absorb excess electricity generated (Fig. 8). In future, conventional late-night PSH pumping may shift to daytime, and electricity may be generated by PSH when lights are turned on. In Europe, following large-scale adoption of wind power, PSH pumping during strong winds is frequent.

3 Technology Roadmap

The technological development for conventional hydropower shifted from a focus on increasing scale to cost reduction and high value-added features around 1970. With the decline of desirable installation sites and leveling off of positive economic effects from R&D in subsequent years, the importance of technological development has been declining since about 2000, along with a decrease in the number of new hydropower stations established. For PSH, research into systems with greater capacity and higher heads has begun, and a high-energy-density, super high-head pump turbine was developed.

Around 2005, progress in technological development projects spawned the creation of the splitter blade pump-turbine Francis runner and a variable-speed PSH system. By enhancing high value-added features that help stabilize the power system, such as a rapid response and spinning reserve, PSH has transformed into a major power source that functions as a power shifting system (energy storage) between day and night in the power system and enhances the overall power system.

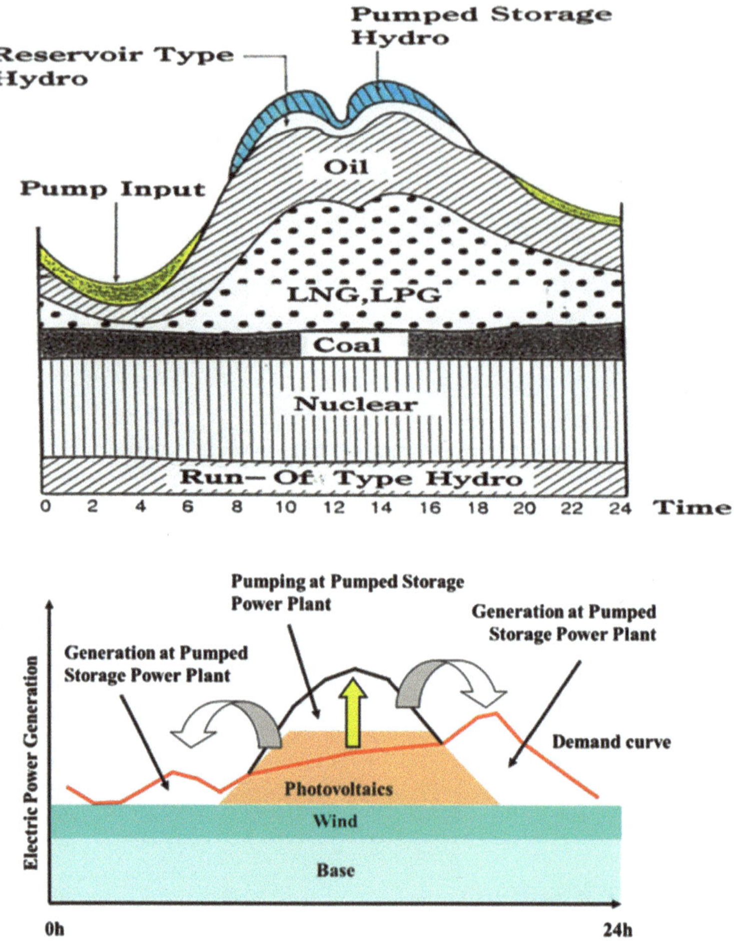

Fig. 8 Pattern of daily electricity usage

Because equipment/function-based R&D is now firmly established, the roadmap for future R&D should center on further enhancing potential high value-added features of hydropower and activities to build an optimal hydropower system, which can fulfill required power system qualities.

3.1 Conventional Hydropower

Following the decline of desirable candidate sites, there was a move toward constructing conventional hydropower stations in remote mountainous areas, away from power-consuming regions. Such stations were built with greater capacities and at greater distance to ensure economic efficiency.

When the law preceding the 2012 establishment of the FIT system (RPS law of 2002) was enacted, several methods were attempted to revive the use of hydropower in a manner suitable for the environment and culture of Japan. Consequently, the implementation of R&D and business models focusing on hydropower within the infrastructure prompted the introduction of a dependent-type hydropower (2005) using water and sewerage systems, agro-industrial water, and others.

Based on the experience of advanced PSH technology, the bare minimum specifications for a hydropower station were marginalized while ensuring safety, by thoroughly eradicating specifications that could be eliminated. Consequently, the economic efficiency of a hydropower station with capacity ~100 kW was ensured for the first time and its commercial use began, triggering widespread use of stations with this capacity. From around this time, hydropower overcame its crisis period and began a resurgence.

Following a period of decline, the construction of new hydropower stations resumed following implementation of the FIT system, bolstered by increased positive effects (on generating costs) of legal support.

In addition to the FIT, the behind-the-scenes catalyst for hydropower enhancement was the widespread use of cost-cutting methods beginning around 2000. Moreover, the market value of hydropower was positively reevaluated and its status acknowledged as a stable power source capable of playing a key role in improving the environment. For these reasons, many aging stations have been renovated in recent years. According to a report by the International Energy Agency, such renovation has been one of the activities implemented with a rise in export of electricity from northern European to continental European countries.

The number of new hydropower plants constructed following FIT introduction has been small, given the difficulties in expediting construction because of the comparatively long time required for surveys and procurement of materials and equipment. Nevertheless, it is assumed that the total will have steadily grown by 2020. At present, it is important to actively implement R&D and energy surveys for candidate sites (flow status observation).

In addition, large-scale hydropower stations built during the rapid economic growth of the 1970s will require renovation around 2030. When this is done, equipment designed for peak demand response as kW focused will be downsized, and the stations will be transformed to power sources capable of generating greater electricity.

3.2 Pumped-Storage Hydropower Stations

In addition to PSH operation requirements in combination with base power sources such as nuclear, with a greater need for PSH as a system stabilizing equipment in response to variable renewable energies such as wind and solar power, the qualities required of PSH are about to change dramatically.

Development activities have included a super high-head, pump-turbine runner facilitating heads of 700 m or higher with a single-stage pump turbine. This was based on R&D leveraging economies of scale to realize economical pumping. As measures to stabilize the power system, the development of a Francis pump-turbine runner with splitter blades, featuring a large output fluctuation range and variable-speed PSH system enabling rapid response to input/output during pumping and PSH output modes, must continue apace.

With the establishment of the FIT system, the share of renewable energies in electricity production has been changing dramatically. Under current circumstances, the locations and scale of equipment to be connected to the power system are not apparent, which hampers efforts to assume the required capacities and locations of PSH as a source to support those unstable renewable power sources. Furthermore, no clear roadmap can be defined, because PSH planning and management are influenced by the basic energy plan of the national government, which contains several variable elements. Until the characteristics required for PSH are clearly identified, it is assumed that the following activities will take place. While efforts are made to further reduce the cost of economic pumping and configure conditions for expanding function to stabilize the electricity supply, values required by the energy market (which are expected to change greatly) will materialize by 2020. By 2030, aging PSH stations designed for economic pumping will be renovated, and installation of PSH stations at sites near power-consuming regions to enhance electricity quality (i.e., deconcentrated-type urban PSH stations) will begin.

4 Benefit and Attractive Future Vision

Acknowledging that a storage system must provide required amounts to supply destinations and do so optimally, it is important to develop an overall optimal tuning technology in combination with existing hydropower technologies, thereby meeting requirements for such systems.

Subsequent development efforts for conventional hydropower will focus on reducing costs and increasing electricity generation capacity. For PSH, efforts will center on enhancing value-added functions to efficiently and quickly absorb electricity supply and demand imbalance (Fig. 9).

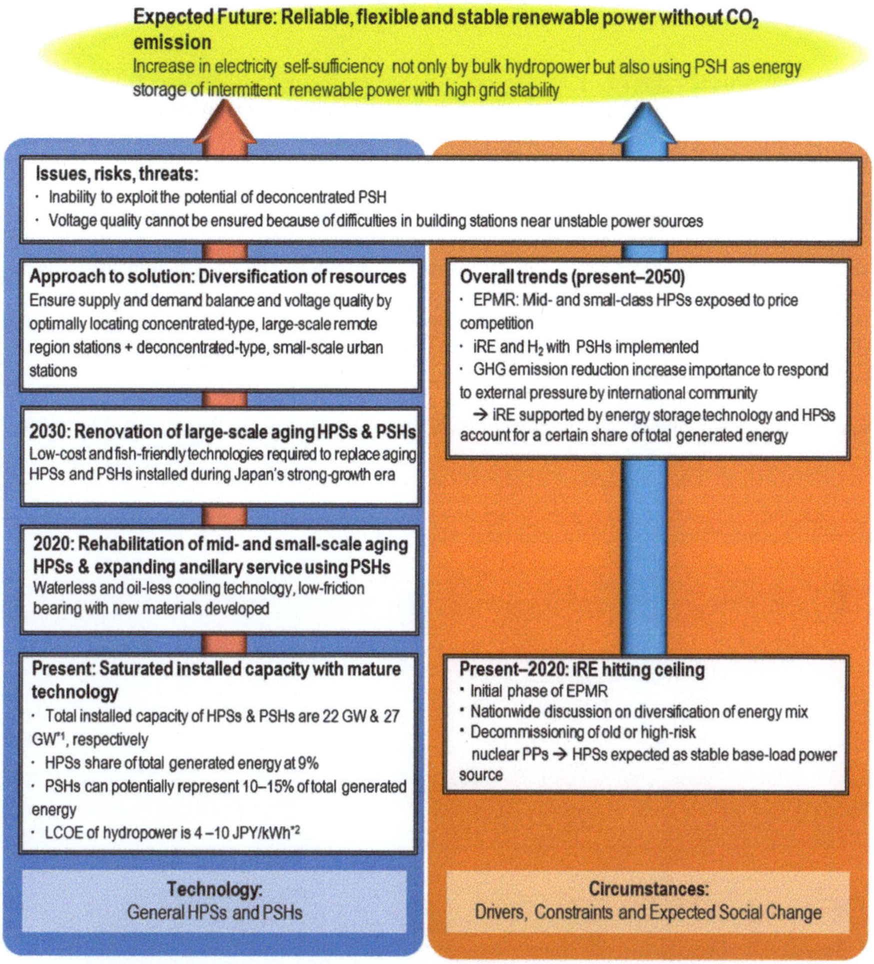

*1 The Federation of Electric Power Companies of Japan (FEPC), Hand book of electric power industry (FY2010), The Japan Electric Association (2010), ISBN: 978-4889482317
*2 Value based on internal information. IEEJ reported a similar number of LCOE in the following document: Matsuo, Y., Yamaguch, Y. and Murakami, T., LCOE transition of power sources evaluated by using asset securities reports (Yukashoken houkokusho wo mochiita hyokashuhou niyoru dengenbetsu chokihatsuden cost no suii) http://eneken.ieej.or.jp/data/5092.pdf, The Institute of Energy Economics, Japan (IEEJ), Tokyo, Japan (2013)
Energy and Environment Council of the Japanese Government published 10.6 JPY/kWh as LCOE of general HPS. https://www.env.go.jp/council/06earth/y060-100/mat02_3.pdf (2011, in Japanese).

HPS: Hydro Power Station, PSH: Pumped Storage Hydropower, LCOE: Levelized Cost of Electricity
Nuclear PP: Nuclear Power Plant, iRE: intermittent Renewable Electricity
EPMR: Electric Power system Market Reformation

Fig. 9 Technology and circumstance perspectives of hydropower generation

References

1. The Fourth Strategic Energy Plan, Agency for Natural Resources and Energy, Ministry of Economy, Trade and Industry (2014) 11 April. http://www.enecho.meti.go.jp/en/category/others/basic_plan/pdf/4th_strategic_energy_plan.pdf
2. Hydropower academy textbook (Suiryoku Academy Kyouhon). Hydropower Academy (Members-only available). March 2014
3. Inagaki M (2006) Technical issues of pumped storage hydropower stations and technology for improving its added value. Ph.D. thesis, Yokohama National University
4. Understanding renewable energy, agency for natural resources and energy, ministry of economy, trade and industry. http://www.enecho.meti.go.jp/category/saving_and_new/saiene/
5. Report by the Investigation Committee to Study Costs and Other Issues, rico Energy and Environment Council, National Policy Unit. http://www.cas.go.jp/jp/seisaku/npu/policy09/archive02_hokoku.html
6. Basic survey on promoting adoption of green energy etc. (Survey on Infrastructure Development for the Development and Adoption of Hydropower). New Energy Foundation. March 2012

Geothermal Power Generation

Keigo Matsuda

Abstract Geothermal heat is a huge amount of thermal energy obtained from the inside of the earth and emits a small amount of greenhouse effect gas (GHG) when used as energy. In addition, unlike other natural energies such as solar or wind, geothermal power is not affected by weather. Therefore, it is put to a variety of applications from power generation to heat downcycling as a clean energy source. Geothermal power generation is largely divided into two types: steam power generation based on a heat source of about 200 °C and the other binary generation based on a heat source of about 100 °C. Recently, more attention is drawn to binary power generation that uses a lower-temperature heat source. This chapter discusses the road map of geothermal-based power generation system and application of binary power generation technology.

Keywords Geothermal energy • Power plant • Rankine cycle • Binary cycle • Social receptivity

1 Introduction

Geothermal heat is the type of heat that is generated as a result of the decay of radioactive elements contained in rocks that constitute the crust. U^{238} and Th^{232} in particular generate a huge amount of thermal energy, 6000 °C, and the heat transfers from there to the ground surface for a distance of about 6000 km. Thus, generated thermal energy is used in power generation or in the form of heat, using water as a medium, depending on the place or location (on the earth, in other words, it is effective use of nuclear energy). Geothermal generation (mainly steam power) uses steam generated by geothermal heat to rotate steam turbines and generate electric power. This operational process emits less greenhouse gas (GHG) than a thermal power generation system and is thus less likely to cause depletion. It is capable of stably generating power regardless of the weather, season, or the time

K. Matsuda (✉)
Yamagata University, Yamagata, Japan
e-mail: matsuda@yz.yamagata-u.ac.jp

Y. Kato et al. (eds.), *Energy Technology Roadmaps of Japan*,
DOI 10.1007/978-4-431-55951-1_18

of day, unlike power generation by recyclable energy such as solar power, wind power, or water power. Operating ratio is a very high 70 %, and its GHG emission is said to be 1/30th of coal-fired power generation [1].

First tested in Larderello, Italy, in 1904, geothermal power generation has been used in a little less than 30 countries in the world, and the current total power output is over 10 million kW. In Japan, test operation was conducted at Beppu spring, in the city of Beppu in Oita Prefecture in 1925. In 1966, the Matsukawa geothermal power plant was constructed in the city of Hachimantai (formerly Matsuo village), Iwate Prefecture. Currently, about 20 geothermal power plants are operated to produce over 500,000 kW in total. Japanese geothermal resources are equivalent to over 20 million kW, the third largest after the USA and Indonesia, which this amount corresponds to over 20 nuclear power plants. However, no new geothermal power plants have been constructed in Japan for about 15 years, because of various causes including the time scale to introduction, cost, and restrictions. Geothermal power generation faces various problems in a wide range of fields, including technical, regulatory, institutional, and environmental problems and fears of earthquake inducement and fear of hot spring depletion. That is why it appears difficult to construct new geothermal power plants anytime soon. On the other hand, attention is being drawn to binary power generation, which can generate power by using thermal energy of existing hot springs without digging any new hot spring wells. Japan is said to have about 25,000 hot spring sources, and their energy reserves are estimated to be over 500 MW. Binary power generation can generate power by operating heat engines with a high heat source about 100 °C and about a few dozen °C low heat source. It can therefore make effective use of not only relatively low-temperature geothermal heat but also waste heat from the industrial fields [2].

It is logical to conclude that geothermal power generation technology with its peripheral technology is one of the tools to produce clean domestic energy that helps solve global warming problems and energy security issues.

2 Present Statuses

2.1 Geothermal Power Generation System

While the mechanism of geothermal power generation is basically the same as that of thermal power generation, the geothermal fluid taken from a geothermal reservoir deep underground is used for the former instead of steam generated by boilers used by the latter. The specific mechanism of power generation differs, depending on whether the fluid obtained from a geothermal well (called a production well) is superheated steam or multiphase flow. Figure 1 shows the current status of geothermal power generation technology. When superheated steam is obtained, there are not many cases of excellent conditions where the dry steam method is

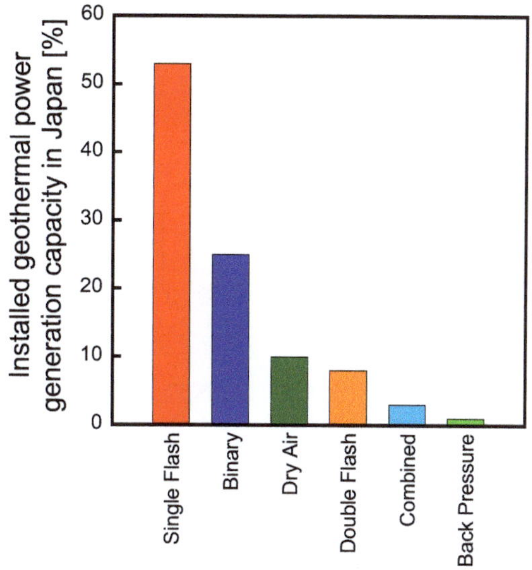

Fig. 1 Operational status of geothermal power generation system

appropriately employed. Therefore, the steam flash power generation system and hot spring binary power generation, both of which use the multiphase flow composed of geothermally produced steam and hot water as a heat source, are explained with a particular current status and problems.

2.2 *Steam Flash Power Generation*

The schematic diagram of the steam power generation system is illustrated in Fig. 2. The basic process of steam power generation consists of the following steps: to find a geothermal reservoir, excavate a well for steam tapping, suck a multiphase fluid composed of steam and hot water heated to temperatures of approx. 150–200 °C, separate steam from hot water (with hot water returned to underground), rotate turbines with a separated steam, and generate electric power. Used steam is cooled down by the cooling tower to become warm water, which is stored in the condenser for reuse as a coolant for the cooling tower.

Since the multiphase flow spouting out by itself from underground can contain noncondensable gas such as carbon dioxide or hydrogen sulfide, it may be sent through the flash drum and ejector to the cooling tower in some cases. When the temperature–pressure of the self-spouting multiphase flow is high, the double flash method, instead of single flash, may be used to generate secondary steam and ultimately enhance thermal use efficiency ($E = Q \cdot (T - T_0)/T$; here, E is exergy,

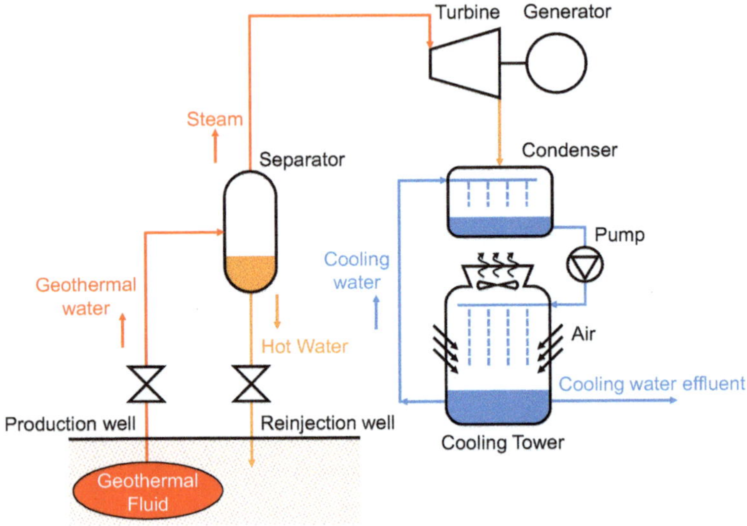

Fig. 2 Steam flash power generation system

an amount of heat, and T is temperature). However, when there is advanced use of heat conducted, solids composed of scale components may be separated from the return fluid from the recharge well (downhole pump) [1–4].

2.3 *Hot Spa Binary Power Generation*

Binary power generation, a.k.a. binary cycle power generation, is the system that only transmits heat to the secondary medium and generates electricity by the secondary medium as the process fluid, in case it is difficult to introduce the geothermal resource directly to the process side to rotate the turbines. Such a case is when the temperature of the geothermal resource (namely, steam or hot water) is less than 150 °C and no steam is generated, or self-spouting is difficult, or when the resource is strongly acidic. Secondary media used by binary power generation include fluids lower in boiling temperature than water, such as a hydrocarbon pure component system including pentane or isobutane, or hydrochlorofluorocarbon (HFC245fa) (ee cycle), or ammonia–water system (Kalina cycle). It therefore allows power generation from low-temperature heat sources. However, the power output (a few kW to a few MW per unit) is smaller than that of steam power generation (a few MW to tens of MW per unit). In other words, binary power generation technology is not the kind of technology that allows power generation

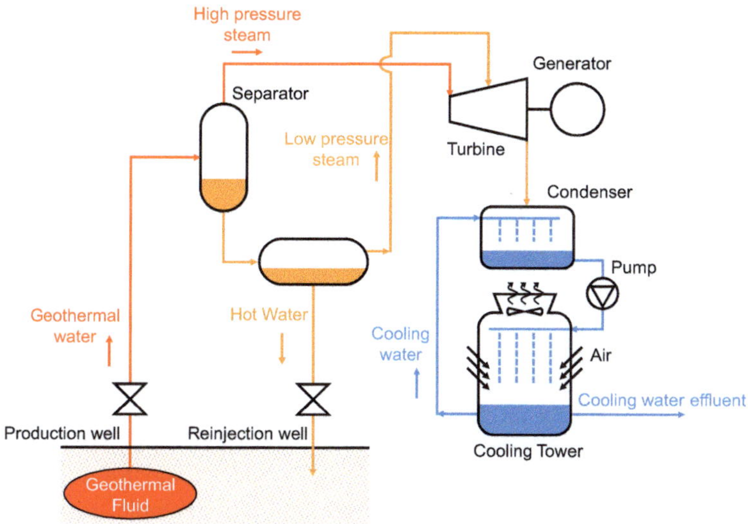

Fig. 3 Binary power generation method

only from geothermal resources but an energy-saving kind of technology that makes effective use of, for example, low-temperature waste heat.

There are a few ways of binary power generation that generates power using geothermal heat as a heat source. The following describes binary power generation using hot spring. Figure 3 shows a schematic illustration of the system. The hot spring binary power generation system consists of a heat exchanger, steam turbine, and pump. It receives hot spring heat from the heat exchanger, which evaporates the secondary medium and rotates the turbine to produce electric power. Since the spring source temperature of a hot spring ranges wide, it is difficult to generalize. However, if the source temperatures of approx. 80–100 °C is obtained, this system can generate electricity without digging a new hot spring. This is the major advantage of this system. Since the lowest temperatures can go down to approx. 60 °C because of the use of water for hot bathing, it is difficult to obtain a sufficiently wide temperature difference. It therefore makes it difficult for this system to obtain a sufficient efficiency ($\eta = W/Q$; here, η is efficiency and W is work) in conversion to electricity compared with the Rankine cycle (it can only obtain about 3–5 % if Carnot efficiency is possible). The binary power generation system also shares the scale problem attributable to hot springs and other problems, including sealing for prevention of secondary medium leakage and other equipment problems with the steam power generation system [5].

2.4 *Problems to Solve for Diffusion of Geothermal Power Generation*

As earlier mentioned, it is important to develop appropriate geothermal power generation technology at the initiative of the national government for energy security. Although Japan is rich in geothermal resources, it may be no exaggeration to say that the country cannot yet make effective use of them. According to Ehara [2, 4], the diffusion of geothermal (including binary power generation) power generation is hindered by three impediments that particularly Japan face, which are (1) the power generation cost problem (there are less national incentives to the introduction of this system than other recyclable energies and that has stagnated technical development); (2) the national park problem (it is not allowed to develop geothermal power generation in national parks; there is no coordination between the Ministry of the Environment that controls national parks and the Ministry of Economy, Trade and Industry (METI), because of their inflexibility); and (3) the hot spring problem (social receptivity related to, for example, the notion that geothermal power generation negatively affects spring sources) [2].

3 Technology Road Map

The construction of commercial geothermal power generation plants has stopped in Japan since the operational start-up of a geothermal power plant of Tokyo Electric Power Co. in Hachijo Island in 1999, except that proving testing and development of hot spring binary power plants has been conducted at a few places, including Matsunoyama hot spring in Niigata Prefecture and Obama hot spring in Nagasaki Prefecture since 2010. Although geothermal power generation has many problems, it has also the greatest potential for further development.

As the first step of geothermal power generation development includes various activities including environmental measurement and survey, it has too long a lead time compared with other kinds of power plants. By the way, since 2012, there are an increasing number of cases where the construction of geothermal power plants is approved in national parks. In fact, Yuzawa Geothermal Power Co., Ltd. (established in 2012 by joint funds of Electric Power Development Co., Ltd., Mitsubishi Materials Corp., and Mitsubishi Gas Chemical Company, Inc.), was submitted to the Ministry of Economy, Trade and Industry (METI), the Construction Plan of the Wasabizawa Geothermal Power Plant (tentative name), and the environmental impact assessment preparation report for construction of a 42 MW steam power plant in Yuzawa City, Akita Prefecture. The company also sent its copy to the Akita Prefectural governor and city mayor of Yuzawa, so as to realize operational start-up in 2020. A local hot spa cooperative union of Tsuchiyu hot spring, in Fukushima Prefecture, plays a central role in the development of a hot spring thermal binary power generation plant (a few hundreds of kW to MW class).

They have a fund for its construction, with the national government as a debt guarantor, through a local bank. In the Kokonoe town, Oita Prefecture, the town and a private entity started up the Sugawara binary geothermal power generation project (5,000 kW) and intend to start operation in 2015 with the national government as a debt guarantor. In addition, geothermal power generation projects in the stages of investigation or development are ongoing at dozens of locations, centering in Tohoku region and Hokkaido.

Considering that the target of power generation using geothermal energy until 2020 is over 1 million kW (target of the Sunshine Project (2010) is 2.8 million kW), it should be necessary to clarify the diffusion scenario of hot spring binary power generation, and the maps and an introduction scenario, based on underground environmental measurements and surveys for construction of geothermal power plants. A steam flash power generation plant is a single dynamic process whose the amount of thermal energy dynamically changes depending on the condition of the underground reservoir, and its resource is basically a dirty system. As a result, maintenance frequency of its process is higher than that of other types of power plants. It is therefore considered relatively difficult to maintain the power generation unit price (currently estimated to be ¥16/kWh) and operational stability. Since reduction in power generation unit price requires reduction in input equipment cost and operation costs (including maintenance cost), if we can accurately predict the properties of underground resources, we should be able to accelerate the future diffusion of geothermal power generation. When high-performance materials with corrosion resistance and soaking resistance (particularly against scales) are developed to allow the system to cope with underground resources of various properties, the performance of the geothermal generation system should be further improved. The hot spring binary power generation system is accelerating diffused when heat media featuring low boiling points, large latent heats, and high and substantial safety are developed and the performance of the expander is improved. Furthermore, post-generation hot water may be used in various ways, including subsequent application of hot spring water to hot bathing and downcycling thereafter, including the use of hot water effluent in snowmelting or as a heat source of district heating, such as for heating of greenhouses. In other words, geothermal resources can create large energy-saving effects in terms of power generation, even though they are not directly used to generate electricity.

Development of binary power generation using heat other than geothermal sources is actively being progressed. For example, K. Matsuda operates ammonia–water cycle power generation using waste heat of steam from the top of the distillation column at Sodegaura Refinery of Fuji Oil Co., Ltd. The heat source for this system is 116 °C steam generated at 130 ton per hour. This steam rotates a turbine power generator to stably generate electricity of 4,000 kW, according to Matsuda's report [6]. It is necessary in the future to discuss social receptivity of geothermal power generation, including applicability to social systems, through industry, government, and academia, and cooperation while considering the ways to realize how to make hot spring areas be the manufacturing centers.

4 Benefit and Future Vision

Japanese geothermal resources are estimated to be 20 million kW. Currently, the total power generation output is 500,000 kW. The authors have presented a road map for solutions to various problems facing geothermal power generation and for its proliferation. The maximum power output feasible by maximizing the current available relevant technological capability is at most a few million kW. In order to envisage an ideal scenario of generating over 10 million kW from geothermal energy by 2050, technologies that are expected to play the key roles include commercialization of hot dry rock (HDR) power generation (including an enhanced geothermal system or EGS power generation) and magma power generation. Conventional steam power generation or hot water binary power generation pumps steam or hot water in underground reservoirs formed by rainwater up to the ground as heat media and uses such media to rotate turbines and generate power. Therefore, no power generation can be done if not for underground reservoirs that contain heat media. However, almost no underground reservoirs are formed other outside of that in volcano front areas. Therefore, no effective use of geothermal heat can be made. Now the attention is drawn to power generation using thermal energy of hot dry rock mass which is a rock area of approx. 200 °C with no water present at a depth of over approx. 3 km underground (in which no faults exist and no rainwater permeates because of greater depth and high pressure). HDR power generation generates electricity using steam generated from an artificial underground reservoir, formed by water injected into the HDR area through a hole drilled from the ground to that depth. EGS is the type of HDR power generation using a hot rock area where a slight amount of water exists. The basic mechanism also applies to magma power generation. Heat exchangers are inserted into the areas in volcanic fronts where rocks of 800–1000 °C in temperatures are molten, and water is injected to obtain steam. Direct injection of water may also be conducted to obtain steam. Extracted steam is then used to rotate turbines and generate electricity [2]. Either method is feasible in principle. They, however, face various problems such as excavation cost, stability of artificial reservoirs, or material durability. Despite these problems, the resources have a great potential of generating large amounts of electricity. It is important that the number of researchers engaged in this field be increased to promote this genre of power generation. The use of geothermal resources is deeply to clarify earthquake prediction mechanisms in a collateral sense. Therefore, technical development in this field is also beneficial in energy security and disaster prevention (Fig. 4).

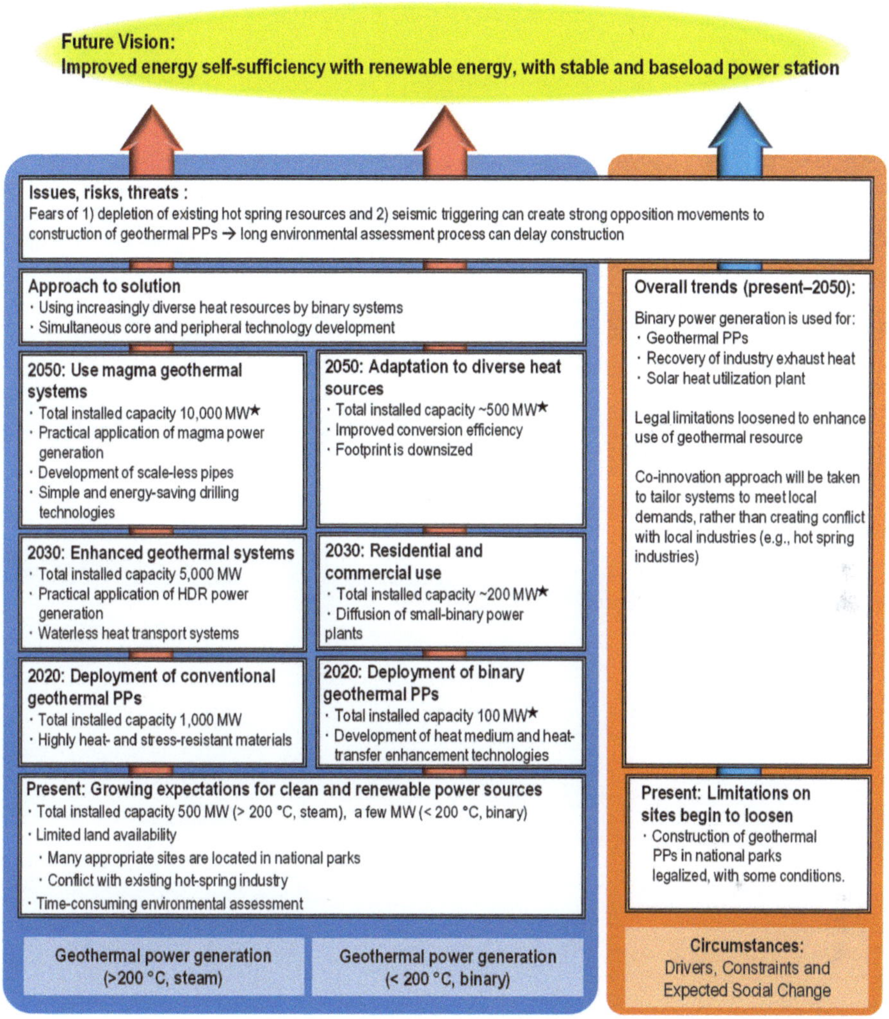

Rating	Technology parameters
★★★	Solid mechanistic estimates
★★	Extrapolation of trends
★	Controversial value, e.g., personal perspective, competition with existing technology

CC: Climate change
HDR: Hot dry rock
PP: Power plant
R&D: Research and development
UN: United Nations

Fig. 4 Road map

References

1. Heat Transfer Society of Japan (ed) (1996) Energy new technology (Kankyo to Syoenerugi notameno Enerugishingijyutsutaikei in Japanese). NTS, Tokyo
2. Ehara S (2014) Geothermal energy (Chinetsu Enerugi in Japanese). Ohmusya, Tokyo
3. Mori Y, Suyama J (1980) Geothermal energy (Chinetsu Enerugi Tokuhon in Japanese). Ohmusya, Tokyo
4. Ehara S (2008) Sustainable development of geothermal energy. J Jpn Inst Energy 87:792–800
5. Science and Technology Co. ltd (ed) (2014) The effective utilization of unharnessed waste heat from factory and their a practical applications (Miriyoukoujyohainetsu no Yukouriyogijyutsu to Jitsuyoutenkai in Japanese). Science and Technology Co. Ltd, Tokyo
6. Matsuda K (2014) Low heat power generation system. Appl Therm Eng 70:1056–1061

Wind Power Generation

Yosuke Nakanishi, Tetsuo Saito, and Ryuichi Yokoyama

Abstract Renewable energy, such as wind and photovoltaic power generation, is an attractive energy source to help solve global environmental issues surrounding the effects of greenhouse gases. Wind power ranks the largest among all renewable energy sources and offers the highest global market opportunity.

In this chapter, the status and potential of wind power energy in Japan are described for onshore, fixed offshore, and floating offshore wind power. We discuss roadmaps for increased wind power penetration in Japan using technical countermeasures.

Keywords Renewable energy • Wind power • Onshore • Offshore • Wind power penetration goal

1 Introduction

In Japan, the utilization of renewable energy sources is expected to impact not only energy security by increasing the self-sufficiency ratio of primary energy supply but also the development of new industries and creation of employment. However, large-scale installation of renewable energy sources, such as wind power, has yet to be achieved because of regulations originating from concerns about power system stability and others.

This chapter discusses roadmaps for increased wind power penetration in Japan through the year 2050, for which total electricity demand is estimated at 758–930 TWh per year [1]. The wind power penetration goal is intended to provide ~20 % of total electricity demand by 2050. Based on this goal, the benefits of CO_2 reduction and economic impact are also discussed.

Y. Nakanishi (✉) • R. Yokoyama
Waseda University, Tokyo, Japan
e-mail: nakanishi-yosuke@waseda.jp; yokoyama-ryuichi@waseda.jp

T. Saito
The University of Tokyo, Tokyo, Japan
e-mail: saitoute@iis.u-tokyo.ac.jp

Y. Kato et al. (eds.), *Energy Technology Roadmaps of Japan*,
DOI 10.1007/978-4-431-55951-1_19

2 Present Status

2.1 Comparison of Wind Power Penetration in Japan and Worldwide

According to the "Global Wind Report: Annual Market Update 2013" published by the Global Wind Energy Council [2], wind power generation capacity reached 318,105 MW worldwide in that year, with a 21 % annual increase rate. As shown in Table 1, China has the highest generation capacity at 91,460 MW, whereas Japan is ranked 18th with a capacity of 2670 MW. Table 1 also shows the penetration of wind power globally; wind power penetration in Japan remains as low as 0.5 %. In 1998, the Japanese government published the "General Outline for Introduction of New Energy," which set the goal to achieve wind power generation capacity of 30,000 MW by the year 2010. However, that goal was not met. Reasons behind the failure are:

1. Prolonged wind farm construction time caused by revision of the Building Standards Act
2. Elimination of subsidiaries for newly constructed wind farms in 2010
3. Addition of wind power generation facilities as subject to the Environmental Impact Assessment Law

Nevertheless, this situation is expected to change with implementation of the new feed-in tariff (FIT) scheme implemented in 2012.

Table 1 Installed capacity and share of wind generation in selected countries [3]

Country	Wind generation capacity [MW]	Ratio to gen. capacity [%]
Denmark	4747	33.2
Portugal	4557	27.0
Spain	22,637	20.9
Germany	34,468	11.7
UK	10,946	7.7
Sweden	4474	7.0
Netherlands	2714	4.8
Italy	8448	4.7
USA	61,292	4.1
France	8128	3.1
Canada	7813	3.0
China	91,460	2.6
Australia	3489	2.4
Japan	2670	0.5

2.2 *Wind Power Potential in Japan*

This section discusses the potential of wind power in Japan, i.e., the wind power generation capacity excluding unusable wind power resources (e.g., wind speeds <5.5 m/s, which are insufficient for commercial use) while taking limitations into account, such as onshore utilization and natural conditions.

Calculations were performed under the following assumptions:

- Diameter of a wind turbine is D.
- Wind turbines are separated by a distance of 3D laterally and 10D front to back.

Based on these assumptions, the wind power generation capacity per km^2 is calculated to be 10 MW.

2.2.1 Wind Power Potential of Onshore Wind Turbines

Assuming that wind turbines are placed under the following condition:

Wind speed at a ground clearance of 80 m is 6.0 m/s or higher.

Then the wind power potential of onshore wind generation facilities in Japan is 209,830 MW [4]. This is nearly equivalent to the entire power generation capacity of existing electrical utilities in Japan. Additionally, if the maximum wind power generation capacity of a region is limited to the power generation capacity of each existing power utility in that region, then the wind power potential of onshore wind turbines in Japan becomes 74,360 MW, which is equivalent to 36 % of the total power generation capacity of Japan's existing utilities (Fig. 1). For power security reasons, the idea is to use existing power generation facilities as backup generators

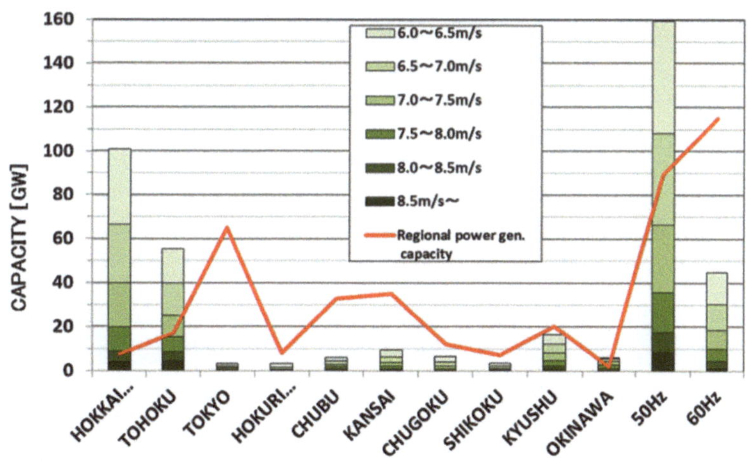

Fig. 1 Potential of onshore wind generation and generation capacity of utilities in 2010

in cases where no power is generated from wind turbines. In Fig. 1, the bar charts labeled 50 Hz and 60 Hz indicate the total capacity of the respective regions, because Japan's utility frequencies differ from region to region. The 50 Hz regional power utilities (such as in Hokkaido, Tohoku, and Tokyo) are located in Eastern Japan; the 60 Hz regional power utilities (such as Hokuriku, Chubu, Kansai, Shikoku, Kyushu, and Okinawa) are in Western Japan.

2.2.2 Wind Power Potential of Fixed Offshore Wind Turbines

Assuming that wind turbines are located under the following conditions:

- Wind turbines located within 30 km of the shore.
- Ocean depth <50 m.
- Wind speed at a ground clearance 80 m is 7.0 m/s or higher.

Then the wind power potential of fixed offshore wind turbines is calculated at 156,460 MW. This is equivalent to 75 % of the current power generation capacity in Japan. With regional upper limits, this potential becomes 61,650 MW, which is equivalent to 30 % of the entire current power generation capacity of Japan (Fig. 2).

2.2.3 Wind Power Potential of Floating Onshore Wind Turbines

Assuming that wind turbines are located under the following conditions:

- Wind turbines located within 30 km of the shore.
- Ocean depth greater than 50 m and less than 200 m.
- Wind speed at ground clearance 80 m is 7.5 m/s or higher.

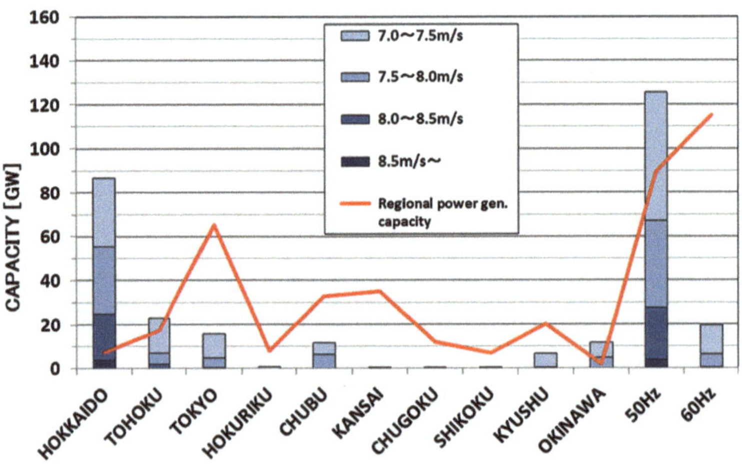

Fig. 2 Potential of fixed offshore wind generation and generation capacity of utilities in 2010

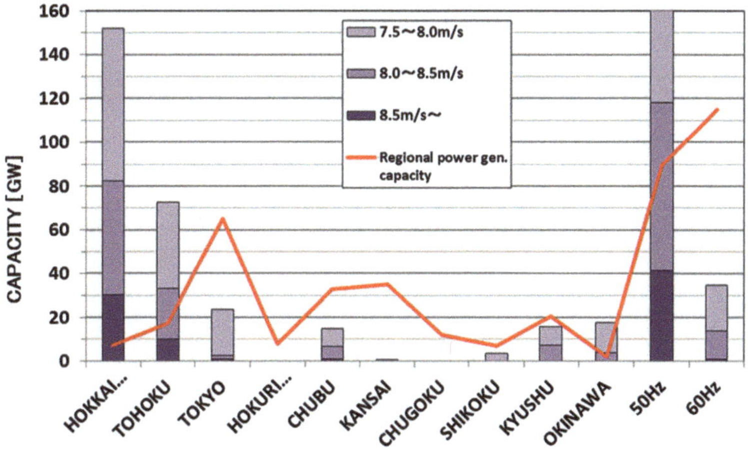

Fig. 3 Potential of floating offshore wind generation and generation capacity of utilities in 2010

Then the wind power potential of floating wind turbines is calculated at 300,460 MW. This is equivalent to 150 % of the current power generation capacity of Japan. With regional upper limits, this potential becomes 84,800 MW (Fig. 3).

2.2.4 Practical Wind Power Potential in Japan

Installation of wind turbines requires many preparation steps, such as a period of 1 year or longer for wind condition analysis at the planned site, environmental assessment, planning grid connection with utilities, and so on. Taking these matters into account, the practical potentials of onshore, fixed offshore, and floating wind turbines would be one-half, one-third, and one-fourth of the calculated capacities above, respectively, which amounts to 232,180 MW. This is about 110 % of the current power generation capacity in Japan. With regional upper limits, the wind power potential would be 76,720 MW or ~37 % of the current power generation capacity of Japan (Fig. 4).

2.3 Power Generation Costs of Wind Power in Japan

According to the report published by a cost examination committee, power generation costs for coal, nuclear, and hydroelectric power generation facilities in Japan for the fiscal year 2010 were 9.5 JPY/kWh, 8.9 JPY/kWh, and 10.6 JPY/kWh, respectively. Costs for onshore wind power facilities were 9.9–17.3 JPY/kWh (30 MW facility, 10.0 JPY/kWh; 6 MW facility, 14.0 JPY/kWh); offshore wind power generation facility costs were 9.4–23.1 JPY/kWh. Whereas the construction cost of offshore wind power facilities is considerably greater than that of existing

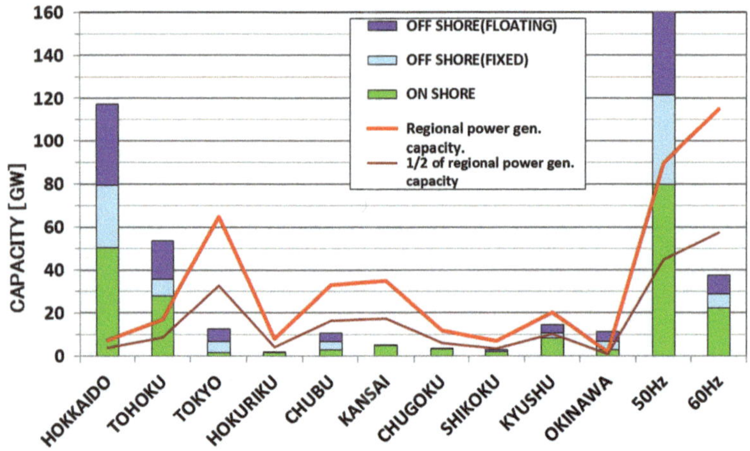

Fig. 4 Potential of wind generation and generation capacity of utilities in 2010

onshore wind power generation facilities, the power generation cost of the offshore wind power is expected to be lower than that of coal by the year 2030, with sufficient increase in wind power penetration coupled with proper cost reduction strategies. Realization of such cost reductions will require close cooperation between wind power generation facility owners and the Japanese government.

2.4 Connecting Wind Power to Existing Power Systems

Wind power is delivered from generation sites to consumers via transmission and distribution networks. Because existing power distribution and transmission systems are not designed with renewable energy sources in mind, many aspects must be considered when connecting large-scale wind power generation facilities to the power network. Elements that must be examined include the heat capacity of transmission lines, effect on power flow, and balancing between the power demand and supply to retain frequency stability. Additionally, short- and long-term capital investment planning will require consideration of the power output fluctuations of wind power.

3 Technology Roadmap

Major challenges to achieving higher wind power penetration in Japan are:

1. High generation cost (Sect. 3.3)
2. Connecting large-scale wind power generation facilities to existing power networks (Sect. 3.1).

To further improve the effectiveness and efficiency of energy utilization, energy management should encompass every energy source available within a region and should consider elements that are indirectly related to energy, such as transportation systems, lifestyles, and demographics of the region (known as the "smart community" concept).

3.1 Technical Issues

This section discusses functionalities that will be required for future wind power generation facilities and the breakthrough technologies required to achieve the installation capacity goal. The summarized countermeasures for both the wind farm and utility sides are shown in Table 2.

3.1.1 Wind Farm Side

1. Maximum generation and power ramp limitation

Table 2 Issues and countermeasures in wind generation connection to grids

Issues	Countermeasures in wind farms	Countermeasures in power systems
Insufficient regulation capacity for short period deviation	Restriction of wind output increase Use of batteries (group control)	Wide-area operation Re-dispatching of generators (types and numbers of generators) Economic load dispatching is sacrificed
Insufficient regulation capacity for long-period deviation (ramp deviation)	Restriction of output increase Restriction of maximum output	Re-dispatching and wide area operation Use of meteorological prediction
Insufficiency of downward output regulation	Restriction of maximum wind power Output restriction in case of frequency increase	Use of storage (pumped up hydro, battery) Use of batteries in substations Lower of the lowest output limit of thermal units Increase of output control speed of thermal units
Insufficient capacity of transmission lines	Restriction of maximum wind power	New inner-regional transmission lines New inter-regional transmission lines

Wind power generation should be controllable, so that the overall power generated does not exceed demand. This also applies to the power ramp so that the change in generation rate remains manageable by regulated power supplies.

2. Active power limitation to counter power system frequency increase
   This functionality is the expansion of functionalities presented in (1). Wind power generation should be controllable, to retain the power frequency of the power system.
3. Fault ride through
   This is a functionality that aids the recovery from brownouts without disconnecting from the power system.
4. Storm control
   Current wind turbines are equipped with a system to stop rotating when wind velocity exceeds a threshold (generally 25 m/s), to prevent damage to the wind turbine equipment. This functionality, however, leads to abrupt loss of generation capacity for the power system and may cause power system instability. A wind turbine is required with a velocity margin (e.g., 25–35 m/s) that can continue power generation while controlling the rotational speed and that does not damage wind turbine equipment.
5. Reactive power, power factor, and voltage control (for steady-state phenomena)
   The majority of newly installed wind turbines are equipped with the functionality to control reactive power for preventing the unnecessary transported reactive power over long distances of network. The combination of controlled reactive power operation and constant power factor operation will allow reduction of the capacity of static VAR compensators (SVC) or the need for SVCs installed in a network (utility) side. Some wind turbine models can provide reactive power even while not generating active power (the wind turbine is not rotating).
6. Supply source of reactive power in cases of voltage instability (for transient phenomena)
   The majority of newly installed wind turbines are equipped with functionality to control reactive power for supporting network voltages during a network fault such as voltage dip or voltage swell. This functionality will allow wind power generation facilities to be used as a reactive power supply, similarly to thermal and hydroelectric power plants. Herewith, it is possible to control the rapid voltage deviation such as monetary voltage drop.

3.1.2 Utility Side

1. Construction of regional interconnection/regional internal connection
   One problem with wind power is that some locations that are suitable for wind turbine placement (e.g., Hokkaido, Tohoku) have small local power consumption levels, resulting in insufficient capacity of transmission and distribution networks to transmit power generated at these locations to major consumption

sites. To allow such generated power to be transmitted to areas with large power consumption, both regional interconnection and internal network capacities must be increased. As shown in Fig. 4, the regional power generation capacity of Hokkaido is smaller than the potential capacity for wind generation. Thus, increasing the interconnection network capacity between Hokkaido and Tohoku is expected, to provide surplus wind power from Hokkaido to Tohoku.

2. Establishment of supply and load control and power network control protocols

   Conventional power grid operation protocols will not be suitable for full utilization of wind power. In addition to existing supply and demand balancing and grid operation control, a new system is required to integrate the wind power into the existing power system. To be specific, the following are required:

   - Regional interconnection/regional internal connection control system operated by the Organization for Nationwide Coordination of Transmission Operators
   - Wind power generation monitoring system with phasor measurement units (PMU) interconnected by wide-area situational awareness (WASA)
   - Demand side management system (DSM or DR)

3. Improved meteorological prediction

   By improving meteorological prediction, power supply planning can include prediction for wind and solar power generation in addition to demand prediction.

4. Installation of energy storage (batteries)

   Installation of batteries equivalent in cost to pumped-storage hydroelectric facilities but with shorter deployment times will facilitate interconnection of wind power to the power network.

3.2 Social Receptivity

We discuss the importance of social receptivity based on past cases and the current situation.

With respect to the construction of wind power facilities, plant owners have conducted independent environment assessments according to the "Guideline for environmental impact evaluation by wind power generation" issued by NEDO (New Energy and Industrial Technology Development Organization) or the "Procedure for environmental impact evaluation by wind power generation" issued by JWPA (Japan Wind Power Association).

However, those guidelines and procedures had no legal power and were interpreted differently because of their imprecise language. As a result, some local residents have filed complaints, such as about noise, after a facility began operation.

In November 2011, wind power plants generating more than 10 MW have become subject to similar treatment to thermal power plants of more than 150 MW, under Japanese environmental law. Hereafter, they are under obligation to undergo assessment of articles and methods, according to Japanese regulations [4].

Consequently, investigative commissions began convening between wind power plant owners and social organizations such as community residents, agriculture and fishery workers, and nature conservation groups. It is very important to obtain social acceptance by promoting community involvement and mutual dialogue with wind power plant owners when first planning construction.

3.3 Generation Cost Reduction

Reduction of power generation costs will require close cooperation between wind power facility owners and the Japanese government. The following are some suggestions for reducing the costs of wind power generation.

3.3.1 Wind Power Generation Facility: Owner Side Effort

1. Increased numbers and/or capacity of wind farms
   Constructing larger wind farms will reduce unit construction cost.
   - Onshore wind farms: 10 MW or greater
   - Offshore wind farms: 100 MW or greater
2. Installation of large-scale wind turbines to improve efficiency
   Installation of large-scale wind turbines will accommodate stronger winds at higher altitudes, which is crucial to improve the overall efficiency of wind turbines.
   - Onshore wind turbines: 0.5–1 MW/turbine $\Rightarrow$ 2–3 MW/turbine
   - Ocean wind turbines: 0.6–2 MW/turbine $\Rightarrow$ 5–10 MW/turbine
3. Utilization of accurate wind state simulation technologies
   Improved wind state simulation will allow wind farm operators to improve the efficiency of their wind farms and contribute to reducing the fault rate. Additionally, simulation data will provide resource information for wind farm location decision making and optimal placement of each wind turbine.
4. Development of smart maintenance systems
   Development of smart maintenance systems that provide precise wind turbine states will enable early fault warning, which will allow wind farm operators to plan optimal maintenance plans.

3.3.2 Government Support

1. Ambitious mid- and long-term wind penetration goal setting by the government

- Ambitious goals will encourage wind turbine vendors and utilities to invest more money into research and facilities.
- Mass manufacture will lower the cost of each unit.

2. Promotion of projects by the government and practical implementation

- Establishment and operation of the Organization for Nationwide Coordination of Transmission Operators
- Reinforce inter-region infrastructure
- Speed up environmental assessment process
- Elastic operation of regulations related to development, construction, and operations and maintenance
- Establishment of committees by local government and harbor administrators
- Technology to absorb power output variations (prediction, control, demand/supply balanced operation): NEDO
- Wind generation high-level implementation research and development project: NEDO
- Offshore wind generation technology research and development project: NEDO
- Government support for floating offshore wind power
- Zoning of floating offshore wind power and establishment of seaport infrastructure and construction/maintenance equipment (ships)

4 Benefit and Future Vision

4.1 Wind Power Penetration Goal

Assuming that the technological and legal breakthroughs presented in the previous section are realized, we have designed a wind power penetration roadmap to provide 20 % of the total electricity demand by the year 2050 [5] (Figs. 5 and 6; Table 3).

The installation target of wind generation by 2050 for each region, 50 Hz and 60 Hz areas, is shown in Fig. 5 with a consideration of the following.

As shown in Fig. 4, each potential of wind generation in Hokkaido, Tohoku, and Okinawa region is greater than each power demand which is equivalent to current power generation capacity owned by each utility.

Therefore the installation target of generation capacity was designed to be limited with following restrictions.

- Set to less than 50 % of the generation capacity of 50 Hz area with Hokkaido, Tohoku, and Tokyo.
- Set to 25 % of the generation capacity of Okinawa region.
- Set to no limitation of 60 Hz area without Okinawa for inshore.

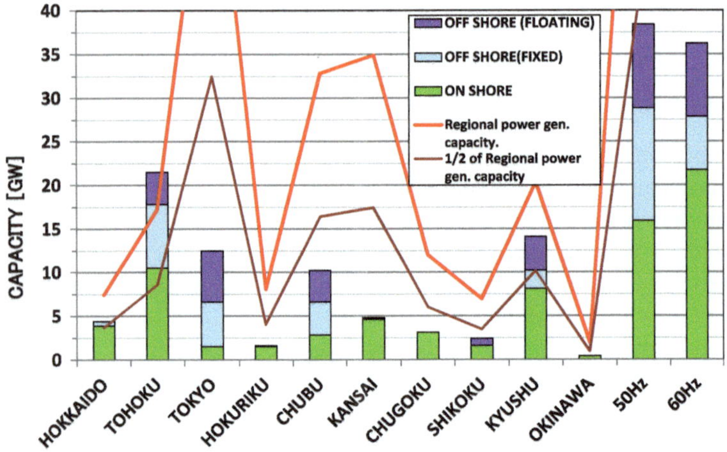

Fig. 5 Installation target of wind generation by 2050 and generation capacity of utilities in 2010

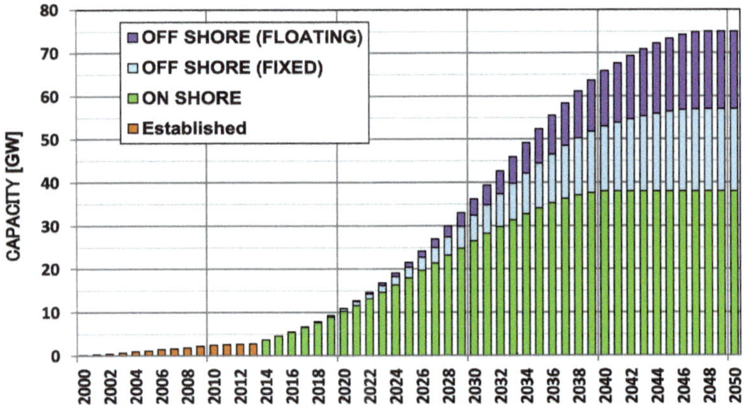

Fig. 6 Long-term vision: roadmap of wind generation installation

Table 3 Roadmap of large-scale wind generation

	Present capacity and installation target [GW]				Generated output [TWh]
Year	Sum	Onshore	Fixed	Floating	Sum
2010	2.5	2.5	0.0	0.0	4.3
2020	10.9	10.2	0.6	0.1	23.0
2030	36.2	26.6	5.8	3.8	84.0
2040	65.9	38.0	15.0	12.9	162.0
2050	75.0	38.0	19.0	18.0	188.0

The roadmap shown in Fig. 6 was designed using the S-characteristic, which means that penetration will show a gradual increase during early stages, followed by a stable increase during middle stages and saturation at end stages. And also in Table 3, the capacity factors of onshore, fixed offshore, and floating offshore used were 25 %, 30 %, and 35 %, respectively, in conversion from the capacity [GW] to the generated output [TWh].

4.2 Economic Impact

The economic impact on regional industries by wind power generation facilities has been discussed in the publications by Japan's Ministry of Economy, Trade, and Industry (METI), "New energy and industry vision" (2004), and "The Industrial Structure Vision 2010" [6]. The following benefits are expected for regional economies.

1. One-third of wind power generation facility construction costs, such as for a road, foundation, and building construction orders, will be expended in regional industries. For example, an economic impact of 500 million JPY on regional industries is expected in the case of construction of a 20 MW wind farm.
2. The majority of construction workers will be employed from the region. For example, construction of a 20 MW wind farm is expected to create new employment of 520 person-day. Additionally, the need for periodic replacement of facility equipment will create new and stable business and jobs.
3. Increases in installation capacity will lead to increased running and maintenance costs as well as insurance costs.

Table 4 shows the economic and job creation effects with construction costs, based on the wind power penetration goal. The estimated construction cost including operation, maintenance, and insurance costs is based on a European Commission roadmap and European Wind Energy Association records and roadmap [7] regarding the Japanese situation, described in the 29th meeting report (file 3-1) by the New and Renewable Energy Subcommittee, METI [8]. The market scale and cost differences between Europe and Japan are also considered. Additionally,

Table 4 The economic ripple effect and employment creation

Year	Items	Unit	Sum	Construction	Operation and others
2020	Construction cost	B.JPY	614	498	116
	Economic effect	B.JPY	1,133	898	235
	Employment	Employs	74,000	59,000	15,000
2030	Construction cost	B.JPY	1,635	1,009	626
	Economic effect	B.JPY	3,044	1,803	1,241
	Employment	Employs	197,000	121,000	76,000
2050	Construction cost	B.JPY	2,281	811	1,470
	Economic effect	B.JPY	4,484	1,452	3,032
	Employment	Employs	290,000	97,000	193,000

operation and maintenance costs are assumed to be constant. The insurance cost is based on a research report by the Japan Wind Power Association. As for the economic ripple effect and employment creation, we referred to the extended interindustry table issued by Professor Washizu of Waseda University [9].

Table 4 shows that the economic effect is roughly 3 trillion JPY and will create new employment of 200,000 persons by the year 2030. This effect will expand to 4.5 trillion JPY and 300,000 new jobs by 2050. Note that the construction cost is not directly proportional to the total generation capacity; construction cost includes replacement of wind turbines, whereas operation and maintenance costs are directly proportional to the total capacity.

4.3 Greenhouse Gas (GHG) Emissions Reduction

Based on the wind power penetration goal, the expected GHG emission reduction amount was calculated by multiplying the kWh by wind, multiplied by the emission factor conversion rate (kg-CO_2/kWh), as shown in Table 5.

During the first commitment period of the Kyoto Protocol (2008–2012), the average GHG emissions for Japan were 1.278 billion tons (1.343 billion tons in 2012). If the wind power penetration increases as planned, 44.13 million tons of GHG emissions can be reduced by the year 2030, which is equivalent to 3.5 % of the average GHG emissions during the first commitment period of the Kyoto Protocol. The GHG reduction will increase to 7.7 % by 2050 (Fig. 7).

Table 5 GHG emission reduction effect (%: reduction ratio to reference period)

Year	Items	Unit	Sum	Onshore	Fixed	Floating
2020	Capacity	GW	10.9	10.2	0.6	0.1
	Generated output	TWh	23.0	21.2	1.6	0.3
	GHG emission reduction	Mt-CO2	12.1	1.1	0.8	0.2
			1.0 %			
2030	Capacity	GW	36.2	2.7	5.8	3.8
	Generated output	TWh	84.0	57.1	15.2	11.7
	GHG emission reduction	Mt-CO2	44.1	30.0	8.0	6.1
			3.5 %			
2050	Capacity	GW	75.0	38.0	19.0	18.0
	Generated output	TWh	188.0	83.0	50.0	55.0
	GHG emission reduction	Mt-CO2	98.9	43.7	26.2	29.0
			7.7 %			

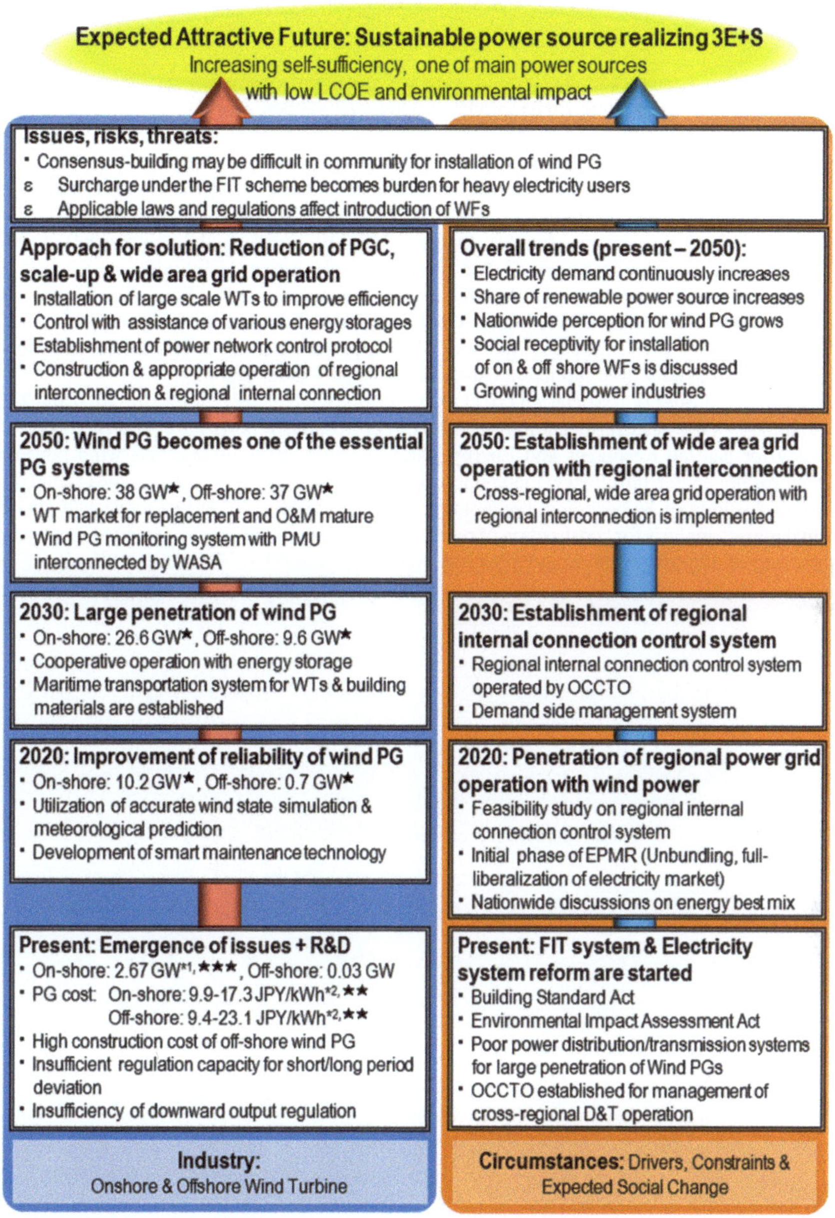

*1 Global wind report-annual market update 2013 [2]
*2 http://www.cas.go.jp/jp/seisaku/npu/policy09/pdf/20111221/hokoku_sankou2.pdf

Abbreviations:
FIT: Feed-in-Tariff, LCOE: Levelized Cost of Electricity, O&M: Operation and Maintenance, PG: Power Generation
OCCTO: Organization for Cross-regional Coordination of Transmission Operators, Japan
PMU: Phase Measurement Unit, WASA: Wide Area Situation Awareness, WF: Wind Farm, WT: Wind Turbine

Fig. 7 Roadmap for increased wind power penetration in Japan

References

1. Japan Wind Power Association (2008) Low carbon development strategy for Japan toward 2050 (in Japanese), February. http://jwpa.jp/pdf/50-02teigen2008.pdf
2. Global Wind Energy Council. Global wind report-annual market update 2013. http://www.gwec.net/wp-content/uploads/2014/04/GWEC-Global-Wind-Report_9-April-2014.pdf
3. Navigant Research. "World market update 2013", https://www.reportbuyer.com/product/2063494/World-Market-Update-2013.html
4. Ministry of the Environment. Act for Assessment of Environmental Impacts (in Japanese). http://www.env.go.jp/policy/assess//5-5advice/advice_h24_1/mat_1_3-2.pdf
5. Japan Wind Power Association (2014) Wind power penetration and Mid & Long term target ver. 4.3 (in Japanese), May 2014. http://jwpa.jp/pdf/2014-06dounyuumokuhyou.pdf
6. Ministry of Economy, Trade and Industry. The Industrial Structure Vision 2010 (outline). June 2010, http://www.meti.go.jp/english/policy/economy/industrial.html
7. European Wind Energy Association (2011) Pure power: wind energy targets for 2020 and 2030, July. http://www.ewea.org/fileadmin/files/library/publications/reports/Pure_Power_III.pdf
8. Agency for Natural Resources and Energy (2008) Current status of Japanese wind power (in Japanese), November. http://www.meti.go.jp/committee/materials2/downloadfiles/g81125a05j.pdf
9. National Institute of Science and Technology Policy (NISTEP) (2013) Economic and environmental impact analysis for construction of renewable energy power plants by extended input-output table, August. http://www.nistep.go.jp/wp/wp-content/uploads/NISTEP-DP096-FullJ.pdf

Photovoltaic Power Generation

Masakazu Sugiyama

Abstract Dissemination of photovoltaic (PV) electricity generation systems in Japan has been triggered by the start of the feed-in-tariff scheme in 2012, and the capacity of installed PV has increased almost linearly each year since then. Accordingly, the system cost of PV has decreased drastically such that the levelized cost of electricity (LCOE) of PV in Japan will reach 14 JPY/kWh in 2020, thereby creating a situation of so-called grid parity. Key issues for such a low LCOE are cost reduction of high-efficiency (>20 %) silicon PV modules and development of improved thin-film PV modules with slight material consumption and simple fabrication processes. The ceiling of PV installation capacity is likely determined by the acceptance of existing electrical grids; capacity is expected to be 70–165 GW by 2030. To make PV electricity generation compatible with existing grids, large-capacity electricity storage should be coupled with PV such that electricity generated by PV can be levelized for several days. Such storage systems include redox-flow batteries and hydrogen (H_2)-based energy storage combining water electrolyzers and fuel cells. Ultimately, large-capacity PV installations will be in countries of strong solar irradiance and combined with chemical energy transport. Concentrator PV modules with >35 % efficiency and abundant direct solar irradiation will cause LCOE to be much less than 7 JPY/kWh (NEDO target for 2030), and H_2 generated by such low-cost PV electricity will be transported to Japan using H_2 carrier technology that is currently under development.

Keywords Photovoltaic • Levelized cost of electricity (LCOE) • Installation capacity • Electricity management • Energy storage

M. Sugiyama (✉)
The University of Tokyo, Tokyo, Japan
e-mail: sugiyama@ee.t.u-tokyo.ac.jp

Y. Kato et al. (eds.), *Energy Technology Roadmaps of Japan*,
DOI 10.1007/978-4-431-55951-1_20

1 Introduction

Photovoltaic (PV) electricity generation is the most widely disseminated energy-harvesting technology from sunlight. Installation of PV equipment is simple compared with wind and other renewable power generations. The capacity of PV for electricity generation is very scalable, from a couple of kilowatt to sub-GW, making it suitable as both a small-scale distributed power source and large-scale electricity generation plant. Boosted by economic policies, the installation of PV first gained strong popularity in Europe. Subsequently, Japan introduced a feed-in-tariff (FIT) scheme in 2012, resulting in a rapid increase of installed PV capacity in the country. Such installation has greatly contributed to the reduction of module price, so the levelized cost of electricity (LCOE) of PV is steadily decreasing. The energy conversion efficiency of conventional silicon PV modules has reached 20 % and is still slightly increasing. However, technological breakthroughs for thin-film modules are ongoing, using new materials such as copper indium gallium selenide (CIGS) and perovskite, which will further reduce LCOE in the near future. Another high-efficiency technology, concentrator PV using epitaxial tandem cells, has become available, with module efficiency >35 %. Such technologies provide the option to install high-efficiency PV modules in countries with intense solar irradiance at GW scale.

The rapid increase in installed PV capacity has made it apparent that a large portion of PV electricity generation will never be compatible with a conventional large-scale electrical grid, because the temporal and seasonal fluctuations of PV power create unmanageable grid instability. It is therefore widely recognized that further installation of PV will not depend on the cost issue alone but also on the compatibility of PV electricity with the grid system. A genuine breakthrough is needed for large-capacity electricity storage and transport, to substantially increase the contribution of PV-based energy to total energy supply in the future.

The following sections provide a detailed description of the present status of PV dissemination, cost issues, and recent advances in PV technology, followed by a road map of PV through 2050 and future perspectives.

2 Present Status

2.1 Dissemination of PV Power Generation in Japan

2.1.1 Installed Power Generation Capacity

The installed PV power generation capacity in Japan increased almost linearly from the start of the FIT as shown in Fig. 1, with a slightly increasing slope, e.g., 7 GW/year around August 2013 and 10 GW/year around October 2014. In the FIT scheme, electricity suppliers must purchase PV-generated electricity at a much higher price

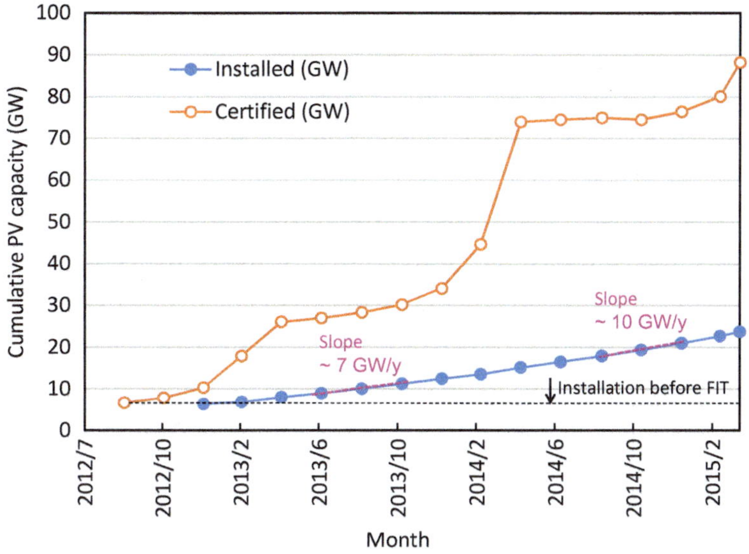

Fig. 1 Installed and certified photovoltaic power generation capacities from the onset of the feed-in-tariff (FIT) scheme in Japan [1]

than that of electricity purchased from a grid. Prices began at 42 JPY/kWh, and the price in 2015 is 27–35 JPY/kWh for small-scale (<10 kW) PV electricity generation and 27 JPY/kWh for large scale (≥10 kW) [1]. The initially higher purchase price boosted investment in large-scale solar power generation, so the capacity of industrial installations increased at a much higher rate than that of household installations. By the end of March 2015, installed capacity was 23.7 GW, of which 7.8 GW was small-scale (<10 kW) household installation. These values include capacities installed before the start of the FIT (4.7 GW for small scale and 0.9 GW for the rest). In addition, 82.6 GW is already certified for the existing electricity purchase price but has yet to be installed. Most of the latter capacity is for industrial installation. However, owing to the difficulty of connecting large PV capacity to an electrical grid, some of the already certified capacity will not be installed. Therefore, the certified capacity should be regarded as the potentially maximum installation capacity, which means that Japan will have the capacity for several tens of GW of PV electricity generation in a couple of years.

A problem with the FIT is that the price of purchasing PV electricity is uniform regardless of the site of electricity generation, whereas electricity demand varies significantly by region. As a result, most large-scale PV plants have been built in suburban areas where there is very limited electricity demand. This places a large burden on the electrical grid because time-dependent PV electricity generation cannot be managed when its capacity exceeds a certain fraction of the maximum power demand in a grid. Most electric power companies in Japan have therefore begun to refuse the purchase of PV electricity, especially from large-scale PV plants, despite the fact that much greater PV power generation capacity has been certified for installation in a couple of years.

Further installation of PV will require modification of electrical grid systems and/or energy storage systems coupled with PV that can levelize electricity from PV plants. Conventional electrical grid systems have been designed on the basis of unidirectional electricity flow from power generation plants to consumer sites. Once a large-capacity PV generation plant is installed near a consumer site, it is almost impossible to transfer the PV electricity from that site to other distant sites with greater electricity demand, owing to the inability of grid transformer stations to transmit electricity upstream. It is desirable to modify an electrical grid so it can cope with omnidirectional power transmission, but this is a strategy requiring a large budget and substantial time.

Once the issue of electricity transmission is solved, another topic of electricity management becomes problematic. The temporal fluctuation of PV electricity generation creates significant challenges for the balance between electricity demand and supply. Failure of this balance results in frequency fluctuation in alternating current (AC) electricity, the range of which is strictly regulated by each electric power company. If allowances for such fluctuation are expanded without affecting the operation of electrical generators and other equipments at consumer sites, management of the balance becomes easier, and more PV can be connected to electrical grids. It is also expected to extend the capacity of interconnection between electrical grid systems belonging to different electric power companies. The advantages of an extended electrical grid size are twofold: (1) increased possibility of connecting a large-capacity PV plant and a large-consumption site of electricity and (2) greater potential for averaging the fluctuation of electricity generation among distant PV sites.

Supply-demand balance management is most difficult when PV electricity output changes rapidly. Batteries can be used to reduce the ramping rate of this electricity. Predicting the temporal evolution of this electricity based on weather forecasts is another challenge mandatory for demand-supply balance management. Once the behavior of PV electricity output is known in advance, it is easier to prepare for rapid fluctuations in this electricity using other power generation mechanisms.

Another solution is combining PV with energy storage systems. For household purposes, batteries have received much attention for variations of PV electricity output between sunny and cloudy conditions. The capacity of such batteries is typically a maximum of a couple of hours of stored electricity. This is certainly effective for levelizing peaks and valleys of PV electricity. For PV to be a steady and reliable electricity source that can cover a significant fraction of energy demand in Japan, it is necessary to levelize PV power over a day or ideally over a couple of days. This necessitates an extremely large capacity of energy storage for the PV electricity. The primary existing technology for such large-capacity storage is the sodium sulfur (NaS) battery. As an example, on Miyakojima Island, a combination of 4 MW PV power generation and 4 MW, 28.8 MWh NaS batteries is in operation, permitting nominal PV electricity storage of 7 h. This is probably the most advanced combination of mega solar power generation with energy storage in

Japan, yet no trial has demonstrated the levelization of PV electricity over a couple of days.

Future installation of PV power generation facilities depends on such improvements in electrical grid systems, power management methods, and large-capacity energy storage.

2.1.2 Cost

The cost reduction of PV power generation is, of course, the strongest driving force for the dissemination of renewable electricity. Policies to boost the installation of PV power generation facilities, among which FIT is certainly the most effective, have made it possible to reduce the price of PV modules along with cumulative module production (Fig. 2). The cost benefit of PV, however, is dependent on LCOE over the lifetime of the PV modules, with units \$/kWh. LCOE is basically obtained by

$$\mathrm{LCOE}\left(\frac{\$}{\mathrm{kWh}}\right)$$
$$=\frac{\text{system cost}\left(\frac{\$}{\mathrm{kW}}\right)+\text{operation cost}\left(\frac{\$}{\mathrm{kW}}\right)}{\left(365\left(\frac{\mathrm{day}}{\mathrm{year}}\right)\right)\times(\text{lifetime (year)})\times\left(\text{effective operation hours}\left(\frac{\mathrm{h}}{\mathrm{day}}\right)\right)}. \quad (1)$$

Effective operation hours represent the effective period in a day during which solar irradiation is at standard intensity ($1\,\mathrm{kW/m^2}$) and is dependent on solar irradiance at each operation site:

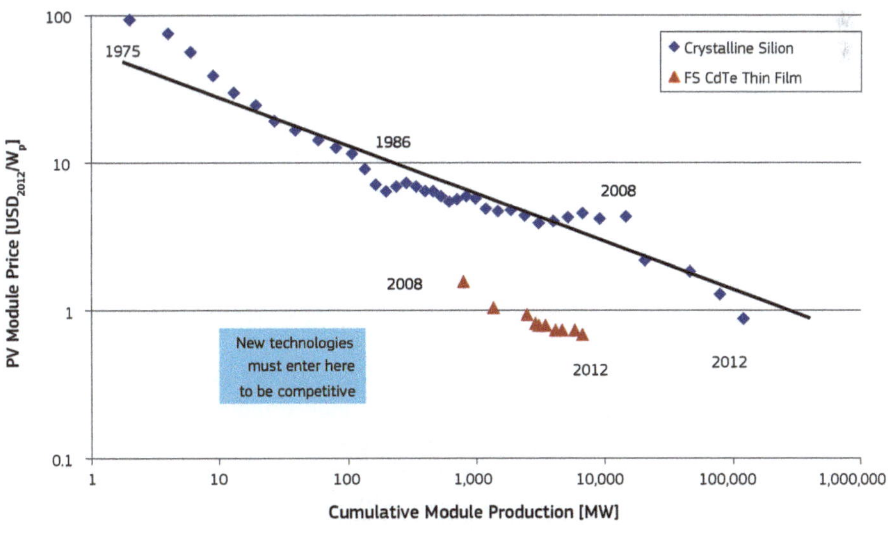

Fig. 2 Price of PV modules as a function of cumulative module production [2]

$$\text{Effective operation hours}\left(\frac{\text{h}}{\text{day}}\right) = \frac{\text{irradiance}\ (\text{kWh m}^{-2}\ \text{day}^{-1})}{1\ \text{kW m}^{-2}} \times \text{PR}\,, \qquad (2)$$

where PR is the performance ratio, which indicates the ratio of actual AC power generation per unit area under standard intensity to the DC power generation at the maximum power point, according to the module specification ($1\ \text{kW/m}^2 \times$ module efficiency). PR usually degrades because of several factors: (1) contamination and/or browning of the PV panel surface, reducing transmittance; (2) failure to track the maximum power point with a power conditioner; and (3) power loss in an inverter. The ratio of the effective operation hours per day to 24 h is often referred to as system utilization efficiency, the current value of which is typically 13 % for industrial installations in Japan [1]. Under typical irradiance at optimized tilt (3.6–3.9 kWh/m^2/day) in Japan, the utilization efficiency is 0.15–0.16 if there is no other mechanism of power loss. Both strong irradiance (geographic condition) and large PR (product development and good maintenance) are beneficial for LCOE reduction.

Sensitivity analysis [3] has clarified the key factors for reducing LCOE: (1) system utilization efficiency, (2) module efficiency, (3) system cost, and (4) operation lifetime. Module efficiency impacts system cost by reducing both the amount of raw materials and installation cost (owing to a reduction of module area) for a given power output. Enhancement of module efficiency is very important for LCOE reduction, and its impact on LCOE is discussed in detail in the following section. The typical value of system utilization efficiency in Japan, 13 %, is largely limited by solar irradiance, and there is little room for improvement. System lifetime can be extended to 25 or 30 years if packaging and wiring in the PV modules are free from significant degradation and have a sufficiently long lifetime of electric power parts (such as inverters and maximum power point trackers). Both of these are currently unsatisfactory, and intensive research and development is ongoing to improve the technology.

System cost has several components (Fig. 3). For household installations, the PV modules represent the largest fraction of cost, so system cost can be reduced by decreased module cost. For industrial installations, however, PV module cost is less than half the total system cost, which makes it difficult to reduce LCOE simply by reducing module cost alone.

Figure 4 shows annual price trends for PV systems and modules. Europe initiated a cost reduction of both systems and modules, and module cost is less than half the system cost. This trend in Japan has lagged (Fig. 4), but the FIT scheme initiated rapid cost reduction in 2012, reducing system cost to ~2.5 \$/W by the end of 2013 [1].

Figure 5 shows an estimation of LCOE for PV electricity as a function of solar irradiance and system cost. Low-cost PV systems indeed reduce LCOE.

Grid parity is a condition in which PV power generation cost is equivalent to that of electricity purchased from an electrical grid. This is the condition that drives installation of PV power generation, even without any political support. The

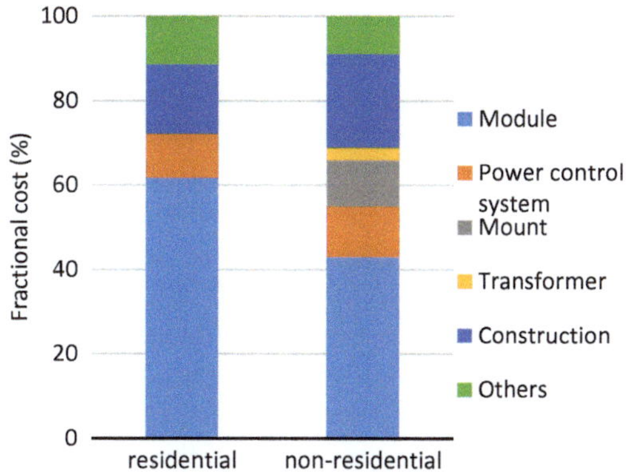

Fig. 3 Fractions of PV system cost [3]

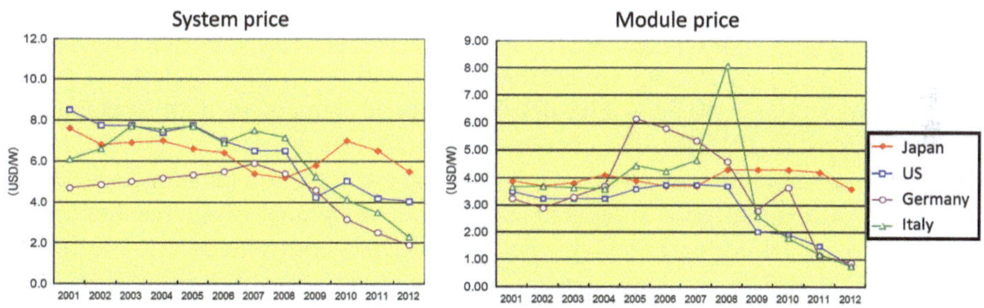

Fig. 4 Annual trend of PV system prices for industrial installations and module prices [3]

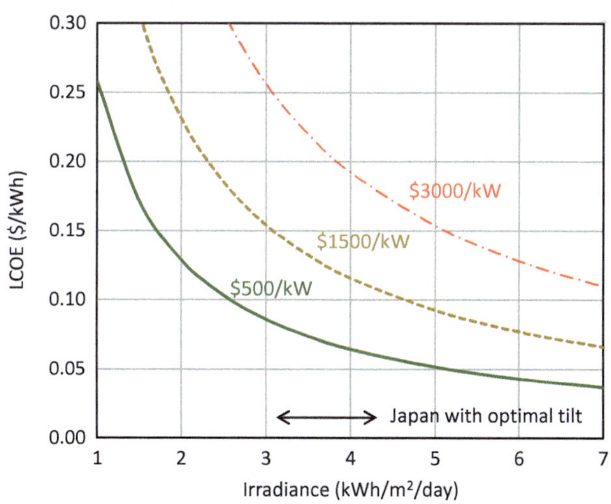

Fig. 5 Estimated LCOE of PV electricity for various system costs under the following assumptions: 20-year lifetime, 90 % performance ratio, 3 % interest rate, $50 /kW/year operation cost, and 20 $/kW grid connection cost. No profit is taken into account

average price of electricity in Japan as of 2012 was 22.3 JPY/kWh for residential and 15.7 JPY/kWh for industrial purposes, respectively, as estimated from the income of electric power companies and amount of electricity sold [3]. Taking typical irradiance in Japan, a system cost <3,000 $/kW can attain grid parity for household installation. The cost for parity in industrial usage is much less than that figure. NEDO, a Japanese funding agency for PV research and development, has set an LCOE target of 14 JPY/kWh (near grid parity for industrial usage) by 2020, and a more aggressive target of 7 JPY/kWh by 2030 will be pursued as an electricity generation cost competitive with existing generation mechanisms. The former figure (14 JPY/kWh) can be achieved with a system cost of 1,500 $/kW, which is within an achievable range following the trend of the cost-learning curve in Fig. 2, under an assumption that module cost is about half the system cost. A system cost of 500 $/kW is not easy to achieve based on conventional PV technology.

Precise estimation of system cost and a future vision are given in a report by the Center for Low Carbon Society Strategy of the Japanese Science and Technology (JST) agency [4]. Table 1 summarizes this estimation. Given the existing crystalline silicon (c-Si) cell, module cost is ~100 JPY/W. Considering the portion of module cost (~40 %) in Fig. 3 for nonresidential installation, the system price is 100 JPY/W × (100 %/40 %) = 250 JPY/W, in good agreement with average PV installation cost in 2013 [1]. The cost fraction in Table 1 is therefore reliable. More than 70 % of module cost is taken up by material cost for existing c-Si cells. Reduction in material cost is dependent on the development of new technologies to reduce material consumption. In the case of c-Si, both reduction of substrate thickness and efficiency improvement are necessary. The thin-film CIGS cell requires much fewer semiconductors than c-Si, and its cost reduction depends on production capacity (reduced fixed costs) and improved efficiency. According to estimations based on fractional cost, a system cost of 1,000 $/kW and LCOE of 11–14 JPY/kWh will be achievable. The vision in 2030 is an LCOE of 7 JPY/kWh, but its realization is mostly dependent on the availability of innovative technologies that simultaneously enhance energy conversion efficiency and reduce material consumption, accompanied by high-throughput production.

2.2 Enhancement in Energy Conversion Efficiency

As mentioned above, improved energy conversion efficiency of PV is a strong driver for LCOE reduction, and tremendous effort has been devoted to this improvement. Technologies currently available on the market are divided into three categories (Table 2). Naturally, crystalline silicon represents the majority of PV, and recent cost reductions in raw material silicon and fabrication technologies have dramatically reduced PV module cost, as described herein. Thin-film PV was originally intended as a low-cost technology because of the limited amount of material used in PV modules, i.e., there are active materials only on the surface of a glass substrate. Historically, amorphous silicon is representative of this

Table 1 Estimation of PV system costs and future vision by JST [4]

		Model case for existing technology		Vision 2020		Vision 2030	
Technology		c-Si 180 μm thick	CIGS	CIGS	New thin film	c-Si 50 μm thick	New CIGS tandem
Assumed efficiency		17 %	13 %	18 %	15 %	23 %	30 %
Assumed production capacity (GW/year)		1	1	5	1	5	5
Estimated module cost (A) (JPY/W)	Material	76	59	40	34	39	29
	Operation	5	2	1	2	1	1
	Equipment, labor	22	18	9	12	6	7
	Module sum	103	79	50	48	46	37
Estimated BOS (B) (JPY/W)	Construction	33	44	27	32	13	10
	Power lines	40	40	20	20	10	10
	BOS sum	73	84	47	52	23	20
Estimated system cost (A) + (B) (JPY/W)		176	163	97	100	69	57
Estimated CO_2 emission from production (g-CO_2/W)		1,210	590	410	–	550	260
CO_2 payback time (year)		2.5	1.2	0.9	–	1.1	0.5

Table 2 Current status of PV production and energy conversion efficiency as of 2013 [3]

Category	Material	World annual production (fraction)	Module efficiency
Crystalline	Silicon	27.8 GW (87 %)	~20 % (single crystalline)
			~18 % (polycrystalline)
Thin film	Amorphous Si	1.0 GW (5.0 %)	~9 %
	CdTe	2.0 GW (6.0 %)	~13 %
	CIGS	0.7 GW (2.0 %)	~14 %
	Perovskite	Not available	18 % (cell efficiency)
Multi-junction	III–V semiconductors	0.05 GW	~36 % (under sunlight concentration with tracker)

category, but recent developments have caused the efficiencies of both cadmium telluride (CdTe) and CIGS to overtake the value of amorphous silicon. Very recently, perovskite cells have exhibited remarkably high efficiency, close to the value of crystalline Si, despite their simple and low-cost fabrication method of spin coating a solution onto a substrate. These cells are, however, very susceptible to moisture in the atmosphere, so the development of water-resistant perovskite materials and encapsulation technology are required for the cells to be used in real-world applications.

Intensive research and development targeting higher efficiency is underway. Two examples of such activities are next-generation silicon cells and multi-junction cells using epitaxial III–V compound semiconductors. The former are for conventional flat-plate modules, and the latter use a light-concentration mechanism and sun-tracking systems in sunny regions with direct solar irradiation (little scattering by clouds or airborne dust).

2.2.1 Advanced Silicon Cells

As shown in Fig. 6, conventional cells use a p-type crystalline substrate with an n-type layer on the top formed by impurity diffusion, resulting in a p-n junction for separation of photo-generated electrons and holes. This configuration necessitates electrodes on the top side with light incidence and on the back side. Surface coverage of the top electrode must be as small as possible to avoid light shading, but too sparse electrodes on the top result in series resistance for lateral current conduction, because conductivity of the n-type region is not sufficient.

The back-contact design has been introduced to make the top surface electrode-free. Diffusion lengths of electrons and holes in crystalline silicon are long enough for these carriers to arrive at n- and p-type regions on the back side. The n-Si substrate behaves like a pool of electrons and holes that are generated by incident photons. The n- and p-type regions act as selective electrodes that collect either electrons or holes only. The absence of electrodes on the top surface, accompanied

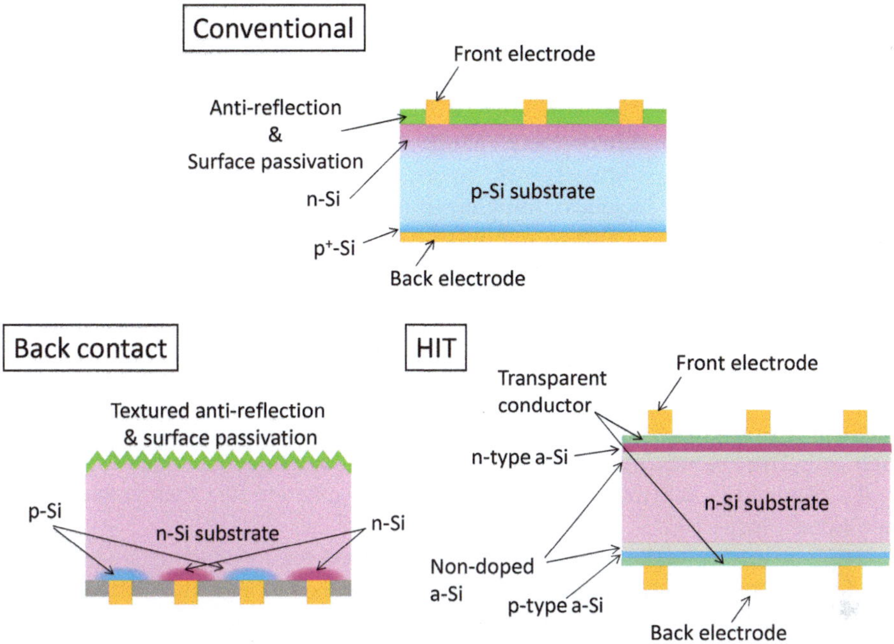

Fig. 6 Cross-sectional schematics for a conventional silicon cell, back-contact cell, and HIT cell

by antireflection texture and coating, maximizes the incorporation of photons into the n-Si substrate and thereby enhances energy conversion efficiency.

Heterojunction with intrinsic thin (HIT) layer structure uses an n-Si single-crystalline substrate, and both surfaces are coated with thin amorphous silicon (a-Si) layers. a-Si contains a small amount of H atoms, which terminates imperfect chemical (dangling) bonds at the surface of the silicon substrate and prevents electrons and holes from recombination at the surface. The a-Si layer on each side is doped to become n and p types, so that either electrons or holes are collected onto the electrodes on each side, respectively. The interface between the substrate and a-Si layer is not doped to prevent recombination between electrons in the substrate and holes in the p-type a-Si (and recombination between the holes in the substrate and electrons in the n-type a-Si). Such undoped layers, however, also block collection of the electrons in the substrate onto the n-doped a-Si (and collection of the holes in the substrate onto the p-doped a-Si). Therefore, there is an optimum thickness of the undoped a-Si to balance the two effects above. Both sides of a HIT cell can be transparent, and such bifacial cells sometimes benefit more efficient light collection. For example, bifacial cells set vertically on the ground covered with snow collect scatter light very efficiently.

The disadvantage of HIT cells is higher cost, mainly from the use of single-crystalline substrate, in contrast to low-cost silicon cells using polycrystalline substrates. Since HIT structure can perform at high efficiency even with a thinner substrate, a means of cost reduction is the use of thinner silicon substrates, which requires a more sophisticated wafer-cutting process. The HIT principle may be applicable to polycrystalline substrates, which is under investigation.

2.2.2 III–V Semiconductor-Based Multi-Junction Cells

A multi-junction cell includes multiple p-n junctions along the direction of light incidence, making it possible to minimize two fundamental energy losses in PV: (1) incapability of absorbing light with wavelengths longer than the bandgap of a semiconductor composing a p-n junction and (2) excess energy of photons in solar irradiation with respect to the semiconductor bandgap. As depicted in Fig. 7a, from the side with solar irradiation, a p-n junction with a wider bandgap absorbs the fraction of photons with greater energies, and the next p-n junction with smaller bandgap absorbs the remaining fraction of photons that are transmitted through the first cell. The low-energy fraction of photons with lesser energies than the second cell's bandgap is still transmitted to the third p-n junction and then converted to additional electricity.

High-quality epitaxial crystals of III–V compound semiconductors are normally used for this technology, because their bandgaps can be tailored with compositional variation. A tunnel junction, which is essentially a p-n junction with extremely high-concentration doping, functions as a transparent electrode between neighboring p-n junctions, making it possible to stack multiple such junctions in a single batch of crystal growth. The record conversion efficiency of such a multi-junction cell exceeds 46 %. Sunlight is concentrated on the surface of a cell through a

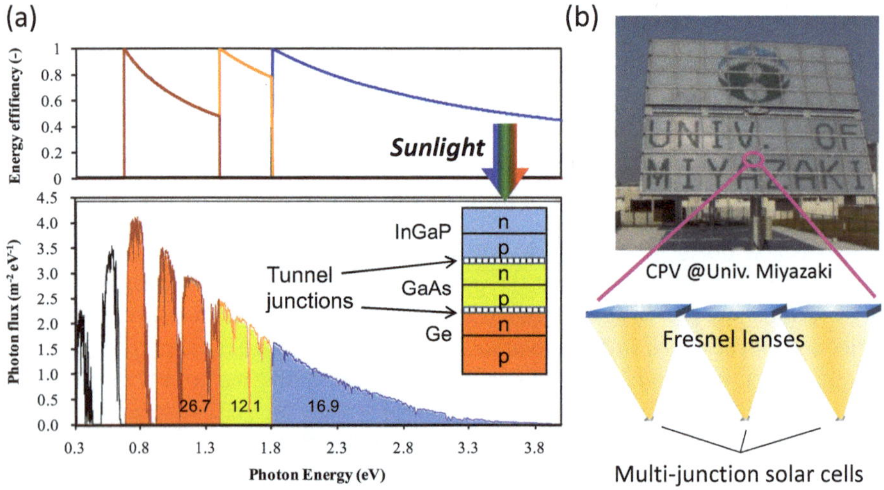

Fig. 7 A multi-junction PV cell. (a) Fractional photon absorption in each subcell, illustrated on a solar spectrum, and photon-energy utilization efficiency as a function of photon energy; (b) typical configuration of concentrator PV modules mounted on a sunlight tracker, with a schematic of sunlight concentration on the cell surface

lightweight lens that is always oriented toward the sun, because high-intensity concentration requires precise optical alignment, as in the PV module configuration shown in Fig. 7b. The reason for this sunlight concentration is twofold: (1) cost is reduced by minimizing the volumetric fraction of III–V semiconductor epitaxial crystals with respect to an entire module; (2) high-intensity light favors high-efficiency power generation by making parasitic loss mechanisms such as carrier recombination at crystal defects less relevant. Such a concentrator PV module with a sun tracker can facilitate much longer effective operation hours than a fixed flat-panel module, because the sun tracking allows the module to be irradiated at maximum intensity from sunrise to sunset. Since sun tracking demands direct irradiation sunlight without scattering by clouds, only limited regions, such as southern Europe, the central USA, and Australia, are suitable for concentrator PV. In these regions, concentrator PV can provide ultimately low-cost electricity from abundant solar irradiation.

3 Technology Road Map

3.1 Capacity of Future Installed PV Power Generation

The predicted potential for PV power generation is 13.8–193.6 GW for households and 25.3–359.5 GW for industrial installation, respectively, based on the area available for PV module installation [5]. However, dissemination of PV power generation is essentially driven by LCOE, considering subsidies. The current FIT scheme greatly accelerated PV installation, resulting in 23.1 GW already installed and 82.6 GW registered for installation in a couple of years. If we apply the rate of increase in installed PV capacity after FIT introduction (9.5 GW/year for the latter half of 2014) as a simplistic assumption, the capacity will be 172 GW in 2030 and 362 GW in 2050.

The other ceiling of PV installation capacity is the acceptance of PV electricity in the electrical grid, unless large-capacity energy storage is available. The electric power companies (excluding Tokyo, Chubu, and Kansai) declared PV electricity to be acceptable regarding manageability of their electrical grids, and the sum was 23.5 GW in 2014 [1]. The sum of electricity peak demand in 2013 forecast by those companies was 55.2 GW [5]. As a simplistic estimation, the ratio of acceptable PV capacity (23.5 GW) over peak electricity demand (55.2 GW) is multiplied by the forecast electricity peak demand (161.3 GW) for all electric power companies (including Tokyo, Chubu, and Kansai) [5]. This yields an acceptable PV capacity of 68.6 GW. This value is smaller than the PV installation capacity certified in the FIT scheme, 88.2 GW.

A more aggressive ceiling for PV electricity can be addressed on the basis of the expected power generation pattern during a day (Fig. 10 of Ref. [5]). According to this prediction, PV electricity generation capacity corresponding to 20 % of total annual electricity generation results in a situation in which PV electricity generation comprises almost 90 % of the power generation when PV power output is at the maximum. This is a very difficult situation from the standpoint of grid power supply stability and will be manageable only when the output power cut for PV can be accepted on request of grid power suppliers, and nationwide interchange of electricity is substantial. PV installation corresponding to such an aggressive situation is 165 GW, assuming PV module utilization efficiency of 13 %:

$$\begin{aligned}&(\text{Annual electricity generation in 2012: 941 TWh})\\&\times(\text{PV power generation capacity: 20\%})/(365\times 24\text{h/year})\\&\times(\text{PV module utilization efficiency: 13\%}) = 165\ \text{GW}\end{aligned}$$

The expected PV installation in 2030 based on the existing annual PV installation rate, 172 GW, almost matches this aggressive ceiling. This indicates that it may be possible to maintain the rate of annual PV installation through 2030 given the two prerequisites above, temporary output suppression from PV and a more robust electrical grid. Further installation of PV requires large-scale energy storage to manage the temporal fluctuation of PV electricity generation.

3.2 Cost of PV Power Generation

The NEDO research and development road map [3] targets LCOE reduction for nonresidential (industrial) installation, i.e., 14 JPY/kWh by 2020 and 7 JPY/kWh by 2030. An example cost analysis is presented by NEDO as in Table 3. According to the JST cost forecast in the preceding section, the cost of c-Si cells can be <1,000 $/kW including balance of system (BOS), and LCOE of 9–12 JPY/kWh will be realized according to the estimation in Fig. 5 with a lifetime of 20 years. As in the

Table 3 LCOE in 2013 and future perspective by NEDO [3]

	2013 (based on data published by METI)	2020 (perspective of NEDO)	2030 (perspective of NEDO)
System price (10^3 JPY/kW)	27.5	200	100
Operation years	20	25	30
Module efficiency (%)	16	22	25
Utilization efficiency (%)	13	15	15
LCOE (JPY/kWh)	23.1	13.2	6.9

NEDO perspective, an extended lifetime makes it easier to achieve an LCOE of 14 JPY/kWh with higher system price.

The LCOE of 7 JPY/kWh is somewhat aggressive and the reduction of production cost alone will not reach the target. Innovative technology for high efficiency and low cost is definitely required, such as ultrathin c-Si cells and high-efficiency thin-film cells, which are free from the constraint of material resources. As in Table 3, extension of lifetime is required to attain 7 JPY/kWh. The semiconductor itself will function for more than 30 years, but reliability associated with wiring and packaging is a key issue.

For residential installation, NEDO targets a system cost <20 JPY/kWh by 2030, including energy management and storage for levelizing the fluctuation of electricity generation, so that PV electricity cost would be competitive against the existing price of electricity purchase from an electric power company. Considering the tendency of cost reduction for PV modules, realization of that target cost is dependent not on the PV module cost but on that of installation and an energy storage system, especially the battery.

Ultrahigh-efficiency cells using III–V epitaxial layers with a sunlight concentration mechanism will facilitate a LCOE <3 JPY/kWh only in regions with strong direct irradiance, which unfortunately does not exist in Japan. This disadvantage of solar irradiance, accompanied by the aforementioned difficulty of grid management, urges the installation of GW-scale, ultrahigh-efficiency PV modules in sunny regions abroad, in combination with chemical energy storage and transportation technology such as water electrolysis using PV electricity to generate H_2 and its transport to Japan. This can be done with H_2 carrier technology such as chemical hydrides, CH_4 and NH_3, which are under development. It is expected that by 2050, Japan will import “solar chemical energy” from countries with strong solar irradiance, and the major PV installation sites will be in those countries.

4 Benefits and Future Vision

PV electricity can provide energy free of fossil fuels. Installation of 165 GW PV by 2030 with target utilization efficiency of 13 % would reduce CO_2 emissions by 0.98×10^8 t/year through substituting power generation using fossil fuels:

$$
\begin{aligned}
&\left(\text{PV capacity: } 165 \times 10^6\text{kW}\right) \times (24 \times 365\text{h/year}) \\
&\times (\text{PV utilization efficiency: } 13\%) \\
&\times \left(\text{CO}_2\text{emission by electricity generation in 2013: } 0.521 \times 10^{-3}\text{ton-CO}_2/\text{kWh [6]}\right) \\
&= 0.98 \times 10^8\text{ton-CO}_2/\text{year}
\end{aligned}
$$

Here, the value of CO_2 emission by electricity generation is affected by the lack of nuclear power generation. If we use a corresponding value before 2011, the CO_2 reduction potential is approximately 60–70 % of the above value.

Such substantial PV electricity installation requires breakthroughs in (a) the style of PV modules, (b) stationary energy storage to compensate for the instability of PV power supply, and (c) energy transport to Japan from countries with strong solar irradiance.

4.1 Novel Style of PV Modules

Further increases in PV installation will require not only conventional styles of installation, flat panels on rooftops and on flat land, but new styles such as bifacial panels standing on reflective ground; flexible PV sheets covering non-flat land, somewhat fragile factory roofs; and partially transparent panels on farmland. These new styles, of course, must meet the criteria of sufficiently low LCOE by low system module cost per area and high power conversion efficiency. For flexible modules, not only thin-film PV technology such as CIGS and perovskite but also crystalline silicon cells can be made flexible by thinning substrate thickness to ~50 μm, the technology for which is already in production. The installation on farmland is a typical example in which existing regulation prevents the dissemination of PV. Such installation, so-called solar sharing, needs the concurrence of agricultural affairs committees in each region, which is often very difficult. Therefore, a steady increase in PV installation requires not only careful control of FIT prices but also elimination of regulations that prevent this installation.

4.2 Stationary Energy Storage to Compensate for PV Power Supply Instability

The dissemination of PV is severely limited by temporal and seasonal fluctuation of electricity generation. It is therefore mandatory to combine PV with a large-capacity energy storage. Batteries are advantageous in terms of rapid temporal response and high efficiency between input and output electricity. Batteries, however, have disadvantages in extending their capacity to cope with the energy storage for a couple of days because their cost is proportional to energy capacity (kWh). Conventional electricity storage using batteries cannot levelize electricity for even a day, and it is not sufficient to compensate for the fluctuation of power output from PV.

Water pumping up is a reliable and ultra-large-capacity form of energy storage, but its geographical constraints make it suitable only for a large-scale electrical grid. PV electricity generation rather requires on-site energy storage attached to each generation site so that it can provide steady electricity either to consumption sites directly or to an electrical grid.

A technology to meet such a requirement is the redox-flow battery, in which redox reaction of ions in solution is used for energy storage. In this system, two solutions containing vanadium ions, one for the cathode and the other for the anode, are circulated from a liquid container to the vicinity of the electrodes, so storage capacity is limited by the capacity of the liquid tank. In addition to the large capacity, redox-flow batteries are advantageous for high-speed response to variation of power in/out. It has been demonstrated that a microgrid system containing PV panels, a wind turbine, and a model load can be managed using a redox-flow battery, so that no voltage fluctuation at the load was observed. This is a promising result, demonstrating that such batteries can manage the temporal fluctuation of PV power output. The drawback of these batteries is relatively low-energy capacity per volume, because the energy storage media, vanadium ions, are in solution. Another obstacle is their high cost, although this will decline with their dissemination.

Another promising method is energy storage using H_2, which is a system combining (1) H_2 generation by water electrolysis, (2) H_2 storage using either gas tanks or metal hydrides, and (3) electricity generation by H_2 fuel cells. The conversion efficiency between input/output electricity is likely $<50\,\%$, because it is limited by the energy conversion efficiencies of a water electrolyzer and fuel cell. In spite of the energy loss, H_2 energy storage has a definite advantage of large energy capacity. Unlike the battery, equipment capacity is determined by power (kW) of electricity for both an electrolyzer and fuel cell. Energy capacity (kWh) is easily increased by increasing storage tank size or the mass of metal hydrides, without a substantial increase in system cost. The H_2 energy storage therefore facilitates levelizing power from PV for a couple of days. Metal hydrides can reduce the volume of H_2 without pressurization, but they require heat to extract H_2. Utilization of waste heat from PV panels, electrolyzers, and fuel cells is therefore essential for enhancing energy efficiency when using metal hydrides for H_2 storage. For small-scale storage such as for a house or community, waste heat can be used for hot water supply and desiccant air conditioning, making it attractive to install H_2 energy storage at such a scale for maximizing total energy efficiency, including heat.

4.3 Energy Transport to Japan from Countries with Strong Solar Irradiance

Because of insufficient solar irradiation in Japan, energy transport and storage are necessary to use solar-based energy as a substantially large fraction of total national energy supply. Such long-distance transport of solar energy will be based on H_2 generation by water electrolysis using PV electricity. Transport of H_2 from countries with strong solar irradiance to Japan is possible using chemical H_2 carriers such as methylcyclohexane (MCH), NH_3, and CH_4. MCH is currently the most advanced technology and is described in another chapter. H_2 energy transport using

CH_4 is also an attractive technology. CO_2 and H_2 can be converted to CH_4 and H_2O via an exothermic reaction using a Ni-based catalyst. If there is substantial CO_2 emission, for example, reservoir CO_2 emission associated with natural gas mining, H_2 from PV electricity can convert this CO_2 to CH_4, which can be liquefied and transported to Japan in the exact manner as CH_4 from a natural gas well. The greatest advantage of this technology is that the product CH_4 can be consumed using existing infrastructure for natural gas. A typical natural gas well emits CO_2 at ~200–300 t/year, and a PV capacity ~20 GW is required to reduce all the emitted CO_2 to CH_4. This scheme will not be available when the mining of natural gas ends, but it would be in the distant future. Another drawback of this scheme is that CH_4 is used to carry the energy of solar hydrogen in only one way; once it is used, it is converted to CO_2 and emitted as long as existing natural gas infrastructure is used. Nevertheless, the compatibility with current infrastructure makes this scheme an option for introducing abundant solar-based renewable energy before new infrastructure for H_2 is available.

For realistic transport of solar H_2, the cost of H_2 production must be very low. The most essential factor of H_2 cost by water electrolysis is the cost of electricity. As shown in Fig. 8, the H_2 cost is nearly proportional to electricity cost, with an intercept corresponding to the equipment cost of electrolyzers. The US Department of Energy targets approximate H_2 costs of 2–4 $/kg as values for which fuel cell vehicles can be competitive with hybrid electric vehicles using gasoline [7]. With an electrolyzer cost of 500 $/kW, the target requires approximately 0.03–0.08 $/kWh for LCOE. Taking operation cost, interest, profit, and transportation cost into account, an even lower LCOE is desirable. The target LCOE, at which solar H_2 transported to Japan is economically feasible, is lower than the aforementioned NEDO target (7 JPY/kWh) by 2030. This is why such H_2 generation by PV electricity requires more abundant solar irradiation than what exists in Japan.

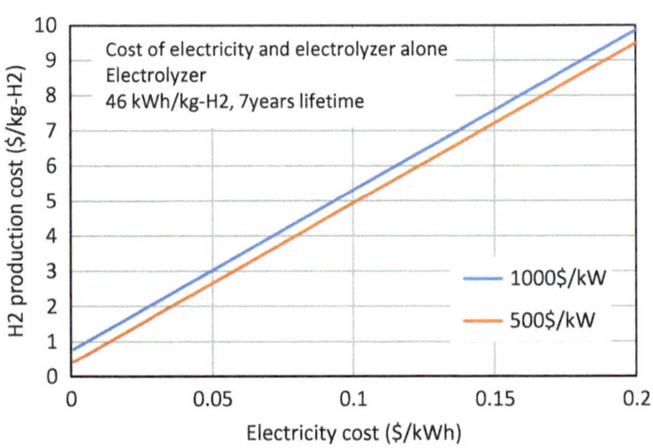

Fig. 8 Estimated cost of H_2 generated by water electrolysis. Only the costs of electricity and electrolyzers are considered; operation cost, interest, and profit are not considered

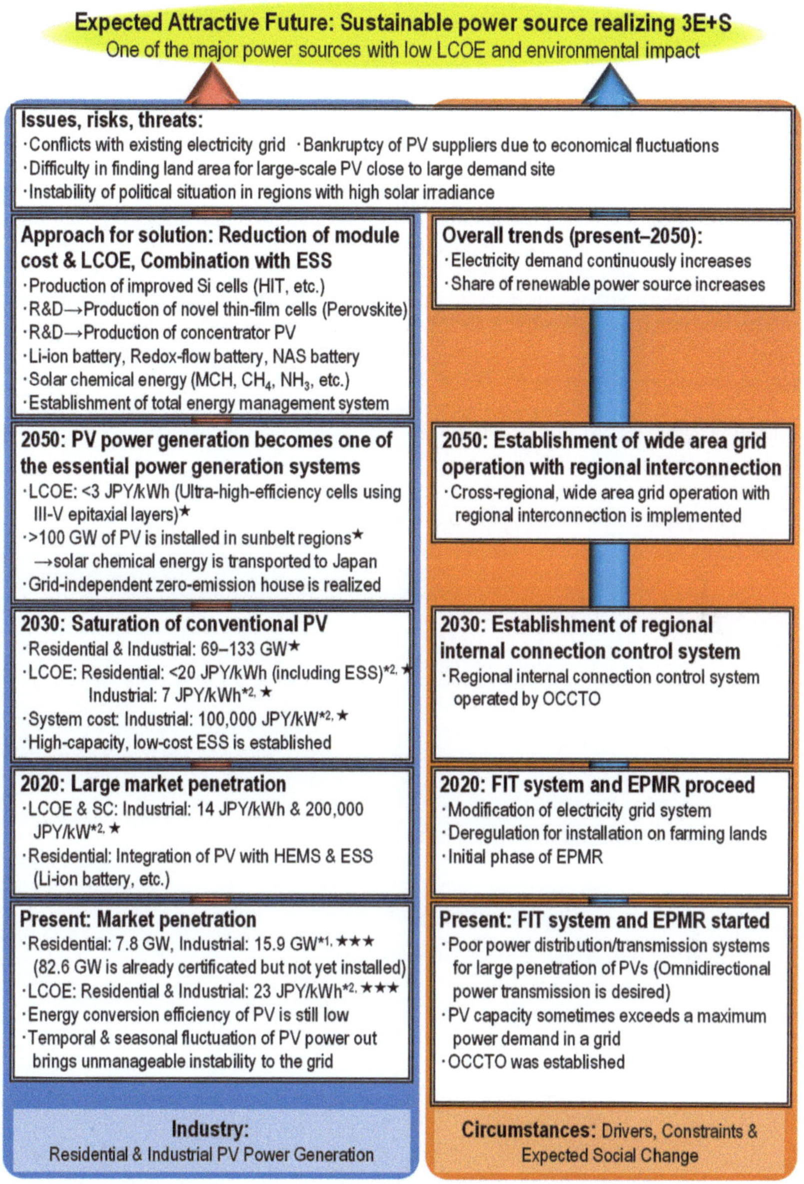

*1 Based on public data sources by The Ministry of Economy, Trade and Industry (METI)
*2 NEDO PV Challenges, New Energy and Industrial Technology Development Organization (2014)

Abbreviations:
FIT: feed-in-tariff, LCOE: levelized cost of electricity, PV: photovoltaic, ESS: energy storage system
OCCTO: Organization for Cross-regional Coordination of Transmission Operators, Japan
HEMS: home energy management system, MCH: methyl-cyclohexane, NAS: sodium (Na) sulfur (S)
HIT: heterojunction with intrinsic thin-layer, SC: system cost, EPMR: electric power system market reformation

Fig. 9 A technology road map of photovoltaic power generation

References

1. Based on public data sources by The Ministry of Economy, Trade and Industry (METI)
2. "PV status report 2013", JRC Scientific and Policy Report (2013)
3. "NEDO PV Challenges", New Energy and Industrial Technology Development Organization (2014)
4. "PV Power Systems: Quantitative Technology Scenarios, and Science and Technology Roadmap based on Elemental Technology Structure". Strategy for technology development, proposal paper for policy making and governmental action toward low carbon societies, Center for Low Carbon Society Strategy, Japan Science and Technology Agency (2014.3)
5. Koyama M, Kimura S, Kikuchi Y, Nakagaki T, Itaoka K (2014) Present status and points of discussion for future energy systems in Japan from the aspects of technology options. J Chem Eng Jpn 47:499–513
6. Based on electricity generation by Tokyo Electric Power Company in 2013, a press release: http://www.tepco.co.jp/cc/press/2014/1239772_5851.html
7. U.S. DOE program record 11007, "Hydrogen Threshold Cost Calculation". 25 Mar 2011

CO_2 Capture, Transportation, and Storage Technology

Ikuo Taniguchi and Kenshi Itaoka

Abstract Carbon dioxide (CO_2) capture and storage (CCS) can be deployed primarily at major point sources of CO_2, such as fossil fuel-fired power plants. For CO_2 capture, solution absorption as represented by liquid amine scrubbing is the most widely investigated. Membrane separation in pressurized flue gas, such as in an integrated gasification combined cycle (IGCC), is expected to reduce energy and cost. The majority of energy and cost during CCS is in the capture process. The roadmap of CO_2 capture targets a cost reduction from 4200 JPY (current references) to 1000 JPY by 2030.

For CO_2 transportation, although pipeline transportation is a mature technology, ship transportation is not widely used but is expected to be effective for offshore CCS in Japan. The roadmap of transportation has a 2050 infrastructure target connecting a cluster of substantial CO_2 sources to storage sites with a combination of pipelines and ship transportation.

There are three important factors for feasible implementation of CO_2 storage: capacity, injectivity, and security (containment). Japan has great potential to store CO_2 in offshore geological formations. CO_2 injection in geological reservoirs is also an established technology in oil mining via CO_2-enhanced oil recovery (EOR), which is a technology analogous to CCS. Stable long-term CO_2 containment in geologic formations is a scientific and technical agenda. The roadmap of storage technology has a target of monitoring cost reduction and technology development to enhance CO_2 trapping in those formations.

Keywords CCS • CO_2 capture • CO_2 transportation • CO_2 storage • Energy penalty

---

I. Taniguchi • K. Itaoka (✉)
Kyushu University, Fukuoka, Japan
e-mail: ikuot@i2cner.kyushu-u.ac.jp; k.itaoka@i2cner.kyushu-u.ac.jp

Y. Kato et al. (eds.), *Energy Technology Roadmaps of Japan*,
DOI 10.1007/978-4-431-55951-1_21

1 Introduction

Carbon dioxide (CO_2) capture and storage (CCS) is defined as "a process consisting of the separation of CO_2 from industrial and energy-related sources, transport to a storage location, and long-term isolation from the atmosphere" [1]. Operational CCS needs additional energy and cost compared with plants without CCS. CCS technology has substantially changed its international position in the last two decades regarding global warming countermeasures, from a mere technical feasibility to an important feasible energy solution to global warming [2]. According to the Intergovernmental Panel on Climate Change (IPCC) special report on "Carbon Dioxide Capture and Storage" published in 2006, CCS has the potential to reduce overall mitigation costs and increase flexibility in achieving greenhouse gas emission reductions. Since then, the G8 group agreed to launch 20 large-scale CCS demonstration projects by 2010, and the International Energy Agency (IEA) released a technology roadmap for CCS in 2011.

CCS is an integrated technology consisting of three stages: CO_2 capture, transportation, and storage, as shown in Fig. 1 [1, 3, 4]. In the post-combustion CO_2 capture process, CO_2 is separated from flue gases of fossil fuel. In the pre-combustion capture process, CO_2 is taken from syngas after a water-gas shift reaction. Widely studied CO_2 capture technologies include chemical (amine absorber) and physical (polyethylene glycol-based absorber) absorption and membrane separation. After capture, gaseous CO_2 is transported through a pipeline, or is liquefied to be transported by ship. CO_2 is then stored underground (geological storage) or under the sea (ocean storage) for isolation from the atmosphere for long periods. Geological and ocean storage involve different technologies, and the latter is currently prohibited by international laws such as the London Convention

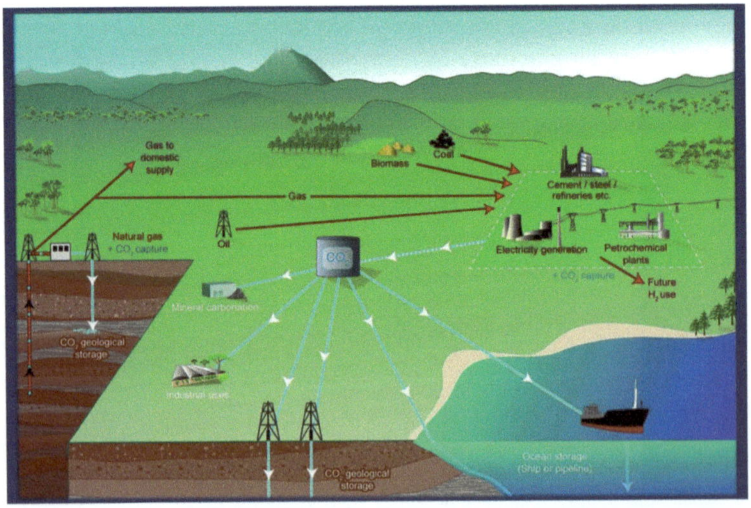

Fig. 1 Schematic drawing of CO_2 capture and storage from IPCC web page (http://www.ipcc.ch)

(International Maritime Organization 1996) and Conventions for the Protection of the Marine Environment of the North-East Atlantic and its Resources (OSPAR). Therefore, geological storage is the focus of this chapter.

CCS can be deployed primarily at large point sources of CO_2 (e.g., fossil fuel-fired power plants and steel plants) and medium-scale sources (e.g., fossil fuel-based hydrogen production plants, refineries, and cement plants). CCS could also be used at small-scale CO_2 sources such as hydrogen stations. This technology can reduce significant amounts (85–95 %) of CO_2 emission at the source [1].

CCS is recognized as an "end-of-pipe" technology, with its sole purpose to reduce CO_2 emission. Therefore, CCS facilities are regarded as additional, requiring supplementary cost and energy relative to conventional facilities producing original benefit without CCS. The energy used to operate CCS is called "energy penalty," i.e., a power plant with a CCS system needs roughly 10–40 % more energy than a plant of equivalent output without CCS [1]. Additional energy consumption entails additional CO_2 emissions as long as fossil fuel is used in the CCS. Therefore, the actual CO_2 emission reduction depends on the capture ratio of CO_2 and the energy penalty. Required energies for transportation and storage using pipelines are much less than those of capture. For example, when 1 t of CO_2 is transported 1 km in a pipeline with a 1.5-m head and a 1-m/s transportation rate, the required energy is calculated at 14.7 kJ/t-CO_2, using 1000 kg $\times$ 9.8 m/s^2 $\times$ 1.5 m. Thus, the required energy is 14.7 MJ/t-CO_2 if CO_2 is transported 100 km through a pipeline, which is ~0.37 % of the energy for capture. For CO_2 ship transportation, additional energy for CO_2 liquefaction is required but still not significant (~52 MJ/t-CO_2), because CO_2 has already been compressed and dried [5]. The energy required to compress CO_2 and inject it into a geological formation depends on depth but is typically ~3.2 MJ/t-CO_2 to pressurize from 10 to 15 MPa, when temperature and adiabatic efficiency are 313 K and 0.8, respectively [6]. However, costs of infrastructure including liquefaction facilities, buffer tanks, and ship transportation are high, so the technologies are compared not by energy required but by cost.

Currently, there have been less than ten CCS demonstrations worldwide [7]. The reason for such delay of CCS deployment is mainly the lack of economic incentive for implementation. Apart from mandatory development of legislation to provide such an incentive, cost reduction is one of the most important factors for CCS implementation. Cost depends strongly on the scale and operation site [1].

2 Present Status

2.1 Target of CO_2 Capture (CO_2 Sources)

CO_2 emission in Japan during 2012 was 1279 million tons, and the major emission sectors are energy industries (510 million t-CO_2, 39 %), manufacturing industries and construction (26 %, 333 million t-CO_2), and transport (17 %, 217 million t-CO_2) [8]. The mass of CO_2 in 2012 was emitted from major point sources

Table 1 Various CO_2 emission sources and gas separation conditions

Emission sources		Separation	CO_2 pressure (atm)
Thermal power plants	Post-combustion	CO_2/N_2	0.15
	Pre-combustion	CO_2/H_2	>10
	Oxy-combustion	O_2/N_2	–
Steel plants		CO_2/N_2	0.22
LNG purification		CO_2/CH_4	>10
H_2 generator after steam reforming		CO_2/H_2	0.4

such as fossil fuel-fired power plants (486 million t-CO_2) and steel plants (150 million t-CO_2) [9]. CO_2 capture should first be used at such major point sources.

In conventional thermal power stations such as coal-fired or LNG-fired plants, flue gas contains mostly N_2 and CO_2 (~14 % depending on the coal), so CO_2 is separated over N_2. In an oxygen-blown IGCC plant (EAGLE), syngas after the water-gas shift reaction contains H_2 (52 %) and CO_2 (40 %), with total pressure ~2.4 MPa. The CO_2 is separated over H_2. Separation conditions vary with the emission source. Table 1 lists major emission sources and target gas separations.

A report from the Japan Oil, Gas and Metals National Corporation (JOGMEC) states that 40 % of LNG deposits contain CO_2 at high concentration, >10 % [10]. CO_2 capture for LNG purification is a current topic. Crude LNG at a well is pressurized, and membrane separation would be a suitable separation technology. CO_2 removal in H_2 purification by natural gas and LPG reforming is also important for carbon-free H_2 production, which would contribute to early penetration of H_2 technology. Oxy-combustion is also a major technology for CO_2 capture [11]. Chemical looping combustion may also be an important alternative.

2.2 CO_2 Capture

The most widely investigated capture technology is solution absorption, such as liquid amine scrubbing [12], chilled ammonia [13], or Selexol [14, 15]. For example, in a thermal power station, flue gas is first passed through the solution at an absorber to separate CO_2 from other gaseous components by chemical or physical absorption. The CO_2-containing solution is then transferred to a desorber, and the CO_2 is recovered upon heating. CO_2 can be separated with high yield and purity by this method. However, the heating process is energy-intensive even though crude heat is used to generate steam for heating, which causes high-energy consumption and increases in capture cost. In the full chain of CCS, CO_2 capture cost is 58 % of the total [16], and cost reduction or development of effective capture technologies would be of prime importance for implementation of CCS. With monoethanolamine (MEA), the required energy at an advanced coal-fired plant to capture ≥90 % CO_2 with ≥95 % purity has been calculated at 4.0 GJ/t-CO_2 [17] (or 4200 JPY [16]), which is the current reference for developing novel absorbents. An amine sorbent recently developed by the Research Institute of Innovative Technology for the Earth (RITE) succeeded in reducing the required energy to 2.0 GJ/t-CO_2 at pilot scale (unpublished result). The Japanese Ministry of Economy, Trade and Industry proposed in its technical roadmap [18] that the capture cost should be 2000 JPY/t-CO_2 by 2020 and 1000 JPY/t-CO_2 by 2030. Further cost reduction would thus be needed, with the aid of novel capture technologies.

Other leading capture technologies are currently membrane separation and gas adsorption, which are also extensively investigated at pilot and demonstration scales. For membrane separation, cellulose acetate and polyimides were the first commercialized membranes for CO_2 capture in the natural gas industry. Crude natural gas at the well is highly pressurized, which is suitable for membrane separation because the difference of CO_2 partial pressure between feed and permeate sides drives the separation. This technology does not require any additional energy for the separation and is energetically very favorable in comparison to solution absorption, as described above. However, at thermal power stations or steel plants, the flue gas or blast furnace has an ambient pressure with low CO_2 concentrations. Thus, membrane separation might not be suitable for the post-combustion CO_2 capture process, where CO_2 is separated over N_2. Nonetheless, various polymeric membranes have been developed with moderate CO_2/N_2 selectivity (30–50) and high permeability, greater than 1000 GPU ($7.5 \times 10\,m^3$ (STP)/(m $^2 \cdot s \cdot Pa$) in SI units) [19–22]. For the pre-combustion CO_2 capture process, palladium-based membranes for H_2 separation have been examined [23]. In contrast, polymeric membranes by Ho's group show high CO_2 separation performance over H_2 and pilot-scale test opportunities are being sought [24]. This approach is preferable to CO_2 separation because pressurized H_2 is available, and energy required for further compression is considerably reduced.

Gas adsorption with thermal and pressure swing is also well established and has been used for gas dehydration. A number of new adsorbent materials have been developed [25, 26], and gas adsorption is promising for next-generation capture technology, in addition to membrane separation. However, such adsorption systems with sizes up to 10 MW have been studied, and further scale-up of facilities is needed. Chemical looping can be an important emerging technology and would be used in natural gas combined cycle (NGCC) or H_2 production [27]. Recently, bench-to-pilot-scale tests have been investigated. Conventional cryogenic separation is also available, but intensive energy loss limits its use in CO_2 capture. Table 2 summarizes various CO_2 capture technologies.

In the capture stage, two kinds of risks to human health should be indicated. One is the risk of aboveground CO_2 leakage during injection, which could impact humans and the environment. Although low CO_2 concentration is not harmful to human health, concentrations $>2\,\%$ have a strong effect on respiratory physiology. Concentrations 7–10 % can cause unconsciousness and death [1]. However, such high concentrations are believed to occur only in low-lying depressions with no wind, which is very unlikely.

The second risk to humans is from chemical solvent used to capture CO_2, particularly amine-based absorbent, which is one of the most efficient absorbents

Table 2 Various CO_2 capture technologies

Technologies	Characteristics	Challenges	Application
Solvent absorption	Well established	Reduce capture energy	Post-/pre-combustion
	Wide variety of absorbents	to <2.0 GJ/t-CO_2	Steel plants and
	High CO_2 purity (>99 %) and recovery		Cement refineries
	Energy penalty[a]: 11–22 % for natural gas combined cycle plants and 24–40 % for pulverized coal plants [1]		
Membrane	Established in LNG purification	Improve separation properties	Post-/pre-combustion
	Lower energy separation compared with solvent absorption	>1000 GPU with selectivity 100 in post-combustion	LNG purification
	Less infrastructure		H_2 purification
Solid adsorption	Well established in H_2 purification	Scale-up infrastructure size	Pre-combustion (PSA)
	Wide variety of adsorbents	to capture CO_2 >1×10^6 Nm^3/h	H_2 purification
	Energy penalty[a]: 14–25 % for integrated gasification combined cycle plants [1]		
Chemical looping	Less separation energy compared with solvent absorption	Develop novel oxygen carriers	NGCC
			H_2 production

[a]Additional energy consumption compared with a plant of equivalent output without CCS

for post-combustion CO_2 capture. Amines and degraded compounds (combination of oxidative and thermal degradation) can be released to the atmosphere in both vapor and droplet phases during capture [28]. Amine solvents such as MEA are basically hazardous [29]. Regarding degradation products, nitrosamines are carcinogens and nitramines are suspected to be carcinogenic [29]. There may be impacts to their health when humans are exposed to these substances at a certain concentration.

The aforementioned risks are mainly attributed to engineering, and their characteristics and magnitude can be compared with those in present industrial activities. Therefore, a risk management strategy should be developed at the same level as analogous industrial activities.

2.3 CO_2 Transport

Transportation of CO_2 is a mature technology [2]. The current method for CO_2 industrial uses is rationalized in light of required amounts. CO_2 for industrial use such as food production is transported mainly by tank trucks. CO_2 is transported by pipeline only for enhanced oil recovery (EOR), which needs a large amount of CO_2 in the USA. Because CCS is expected to be used primarily at large point sources and the amount of CO_2 to transport is much greater than that of general industrial use and equivalent to that of EOR activities, pipeline transportation is expected to play a major role in CO_2 transportation. The total CO_2 pipeline network length for EOR is 6000 km in the USA, and severe accidents in the network have not been reported [2]. Therefore, in CO_2 transportation for onshore geological storage, pipeline transport will be the only option, except for small-scale CCS experiments. In current CCS demonstration projects such as Weyburn and Snøhvit, pipeline transportation technology is used, for either onshore or offshore geological sites. The typical estimated cost of pipeline transportation is about 800 JPY/t-CO_2, accounting for 11 % of total CCS cost [16].

For far offshore CO_2 geological storage transportation, ships can also be an option. Compared with pipeline technology, CO_2 ship transportation is rarely used in commercial activities. There have been a few instances of such transport operationally [2]. The initial cost of CO_2 ship transport is considered to be more expensive than that of pipelines, because facilities for the latter, such as CO_2 buffer tanks and CO_2 tank ships, are costly [30]. However, the total cost of transportation would increase little with increases in transport distance compared with pipelines [1]. In addition, offshore pipelines would be very expensive for installation in the deep sea. Considering the geography of Japan, ship transportation would be a major option when storage sites are far offshore.

2.4 CO_2 Storage

CO_2 is injected into underground geological formations, such as depleted oil and gas fields, deep saline formations, and unmineable coal beds deeper than 800 m in a supercritical state. CO_2 storage in depleted oil and gas fields or deep saline formations is presently a practical option, whereas CO_2 storage in unmineable coal beds is still in the experimental stage. The typical estimated cost of CO_2 injection using extended reach drilling is about 2800 JPY/t-CO_2, constituting 38 % of total CCS cost [16].

Three important factors of CO_2 storage must be satisfied for feasible implementation: capacity, injectivity, and security (containment). In planning CCS, the capacity of a geological reservoir is one factor, because various facilities (including pipelines) would be connected from capture facilities to an injection well. A larger CO_2 reservoir is more favorable to store a large amount of CO_2 and for a long lifetime of facilities. Locations of potential reservoirs are shown in Fig. 2. It shows

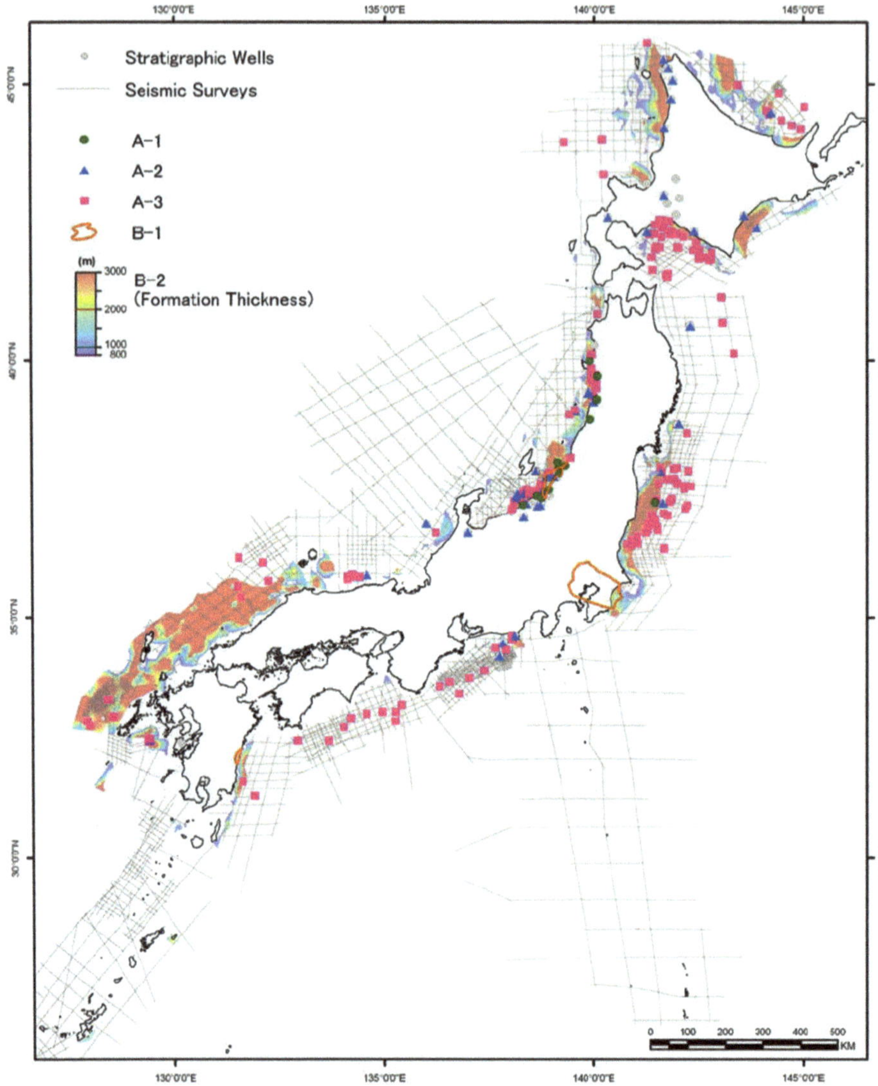

Fig. 2 Location of potential CO_2 reservoirs in Japan [31]

there are many potential storage sites in offshore surrounding Japan, but relatively less potential CO_2 reservoirs exist in regions on the Pacific coast from the west up to the Kanto region. The estimated capacity of geological reservoirs is summarized in Table 3, which shows Japan has a potential capacity of ~156 Gt-CO_2, equivalent to CO_2 emissions over more than a hundred years in Japan. The capacity of CO_2 storage for each geological structure in Table 3 was estimated based on geological data obtained by past oil and gas exploration activities and additional geological survey [32, 33]. The largest potential capacity exists in geological formation

Table 3 CO_2 storage potential in Japan

Data Source		Category A (Structural traps)	Category B (Gently dipping homoclinal structures and/or heterogeneous aquifers without trapping structures)
Oil & gas field	Data obtained during operation	A1: 3.5 Billion t-CO_2	B1: 27.5 Billion t-CO_2 (23.1 Billion t-CO_2)
Basic boring	Public domain data by seismic and drillhole	A2: 5.2 Billion t-CO_2 (3.4 Billion t-CO_2)	
Basic survey	Public domain data by seismic only	A3: 21.4 Billion t-CO_2 (13.5 Billion t-CO_2)	B2: 88.5 Billion t-CO_2 (69.4 Billion t-CO_2)
			B2*: 9.8 Billion t-CO_2 (6.2 Billion t-CO_2)
Scheme		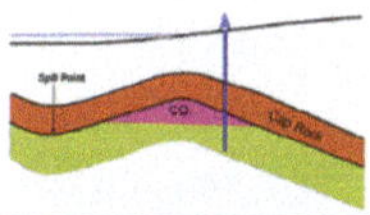	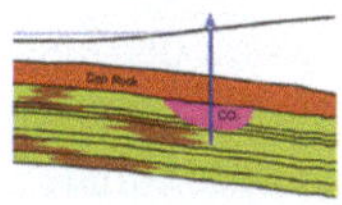
Sum		30.1 Billion t-CO_2 (20.4 Billion t-CO_2)	125.8 Billion t-CO_2 (98.7 Billion t-CO_2)
Total		155.9 Billion t-CO_2** (119.1 Billion t-CO_2***)	

Notes: All numbers in this table except for totals are cited from [31] and [32].
The figures are cited from [33].
B2* is an estimate of additional survey to supplement original B2 estimate.
Numbers without parenthesis: Distance from an active fault is not considered.
Numbers in parenthesis: Distance of 5km from an active fault is considered.
**=[Number of A1] + [Numbers without parenthesis of A2,A3,B1,B2,B2*]
***=[Number of A1] + [Numbers in parenthesis of A2,A3,B1,B2,B2*]

without trapping (confining) structure such as anticline structure. CO_2 storage capacity of potential reservoirs is shown in Table 3. Global estimated storage capacity is reported at ~2000 Gt-CO_2 [1].

Although capacity information is available for a particular storage reservoir, further data collection is always necessary for site characterization in CCS implementation, and this requires substantial investment in the CCS planning stage.

Currently, about ten projects, including small experimental ones, have been implemented worldwide. In this regard, CO_2 geological storage may be seen as an immature technology. However, if EOR with CO_2 is considered an analogous activity, CO_2 injection deep underground can be realized as an established technology. In North America, ~60 million tons of CO_2 are injected annually by CO_2-EOR [34]. Therefore, the basic physical processes and technical aspects of CO_2 injection into geological formations are well understood via CO_2-EOR experience, as well as via laboratory experiments, modeling, and CCS demonstration projects.

The non-established part of CO_2 storage is its long-term containment in geological formations, because CO_2 should stay in geological reservoirs for hundreds or thousands of years for climate mitigation purposes. This contrasts with the less-than-hundred-year operational record of CO_2-EOR. To ensure containment of CO_2 in reservoirs, a combination of simulation and monitoring of CO_2 behavior in those reservoirs is important. Reservoir simulation technologies that have been used in oil and gas mining are used in current CO_2 storage projects with adjustment per CO_2 properties. Among monitoring technologies, seismic technologies have been key in CCS. Seismic surveys are conducted in the planning stage of CCS before injection to characterize geological reservoirs and periodically during CO_2 injection. Moreover, after the completion of CO_2 injection, it is necessary to conduct seismic surveys periodically to monitor injected CO_2 migration in the reservoir. Seismic surveys are useful to communicate CO_2 containment in reservoirs, because they can visualize CO_2 migration. Simulation results of CO_2 migration should be periodically checked by seismic monitoring, and then simulation parameters should be adjusted to model long-term CO_2 migration [35].

The greatest problem of seismic surveys is the high cost of implementation, which is approximately 100,000,000 JPY per survey, depending on the geographical situation. Besides geological CO_2 storage, CO_2 utilization is an important issue. Effective CO_2 conversion to value-added products such as CH_3OH has been investigated all over the world, but current technologies emit more CO_2 relative to the case in which CO_2 is not captured. Agricultural use of CO_2 can be an option, and plant growth can be accelerated under higher CO_2 concentrations in greenhouses, but the potential demand is much less than the CO_2 emission from power stations. CO_2 fracturing in shale gas production and enhanced oil/gas recovery with CO_2 is more promising, and those technologies are categorized as one of the storage options [34].

In the storage stage, there is a risk of CO_2 leakage from storage reservoir to humans and the environment (including groundwater). This is attributed not only to the anthropogenic risk of reservoir engineering but also to natural risks caused by underground heterogeneity. Therefore, consideration of the intrinsic uncertainty of natural geology is necessary to manage such risks. Environmental risks of leakage depend on the characteristics of storage locations. The existence of the object of protection is a prerequisite to evaluate its risks. Nonresidential and low-ecology areas such as deserts therefore have advantages in reducing the risks.

The influence of CO_2 leakage from storage reservoir on shallow groundwater aquifers should be considered first. Particularly in the USA, groundwater is used for drinking and irrigation in many locales. Risks include effects on the quality of groundwater, caused by CO_2 leaked from storage reservoirs, which dissolves in water, decreasing its pH and promoting the elution of salt and metals (including toxic substances) into groundwater. Another possibility is that displacement of underground saltwater by CO_2 injection may cause increased saline concentration in shallow groundwater or spring water. In addition, the influence of acidified ground on vegetation and agriculture may also be considered. However, the possibility of such influences is believed to be slight, because they have not been observed in current activities such as acidic gas underground disposal. In addition, effects on the human body should be a concern when CO_2 is leaked into the atmosphere. There are several pathways through which stored CO_2 could migrate to the surface, including injection well failure, undiscovered active faults, and abandoned mining wells. Characteristics of the impact of aboveground CO_2 leakage are discussed in the capture section.

Because the probability and hazard of CO_2 leakage are scientifically estimated as low with consideration of analogous industrial and mining activities, appropriate social understanding of CCS would be an issue, requiring efforts in public communication and stakeholder engagement.

Finally, upon considering CCS projects as part of CO_2 management business, there are other types of risk. Obviously, the possibility of leakage is a source of risk, by reducing the efficiency of operation and usefulness toward the original objective. In addition, low injectivity of CO_2 into reservoirs could be a serious business risk. If the actual rate of CO_2 injection is much less than expected, a part of captured CO_2 cannot be stored in the reservoir, thereby decreasing the value of substantial investment in CCS facilities. Recent developments of reservoir engineering in oil and gas mining can help overcome the injectivity problem.

3 Technology Roadmap

3.1 CO_2 Capture

The roadmap of capture technology has targets of cost for capture processes. The current reference capture system using chemical solution specifies 4200 JPY to capture CO_2. This is expected to be 2000 JPY by 2020 and 1000 JPY by 2030 [18]. Given the aforementioned CO_2 capture technologies, both individual and combined technologies at each emission source are required. In addition, once novel separation materials are developed, impurity handling or tolerance, process design and heat integration, and environmental impact should be examined.

In solution absorption technology, research should focus on developing absorbents with low sensible and latent heats to reduce the energy penalty of CO_2 recovery at a desorber. Various membranes have been developed for CO_2 capture, and increasing CO_2 permeability should be a research target with high CO_2 selectivity. Recently, metal-organic frameworks have been popular as CO_2 adsorbents. Basic performance such as selectivity, capacity, kinetics, and stability will be examined at pilot scale. Thus, second- and third-generation technology development are needed for commercialization of CCS by 2030. Further diffusion of CCS at a global scale should be achieved by 2050.

3.2 CO_2 Transportation

Regarding transport, although a breakthrough is not likely in pipeline technology because it is already matured, ship transport can expect further progress. Around 2020, it is anticipated that the concept of CO_2 ship transportation for CCS will be established as an optimized chain for transporting CO_2 from capture and injection. This includes CO_2 liquefaction, temporary storage near a port, CO_2 ships, and additional temporary storage of CO_2 for further transport by pipeline or injection. Around 2030, further technological progress is expected to constitute the ship transportation chain, especially in liquefaction technology, and design of large vessels and tanks for temporary storage and material options. Meanwhile, CO_2 ship transportation systems connecting multiple sources and storage sites by shuttle ship (2000-t class) and direct injection from ships can also be established as part of the initial stage of future CO_2 transportation infrastructure [36]. By 2050, it is expected that CO_2 transportation infrastructure will be developed, connecting clusters of large CO_2 sources to storage sites by combined pipeline and ship transportation [37].

3.3 CO_2 Storage

It is a challenge in CO_2 storage to find appropriate storage sites with suitable geologic properties such as large capacity, favorable injectivity, and shielding, as well as feasible CO_2 transportation pathways from substantial CO_2 sources. In addition, while ensuring long-term containment should be a focus, geologic information databases of potential storage sites requiring substantial exploration activities should be developed to facilitate CCS project planning. To assure long-term containment, very accurate monitoring and modeling incorporating monitoring results are important. For monitoring technology, passive seismic monitoring, which does not need artificial seismic sources for surveying but uses surrounding noise, is expected to be developed and dramatically reduce the cost of seismic surveys by 2030. Another approach to ensuring long-term containment is enhancing trapping mechanisms. CO_2 trapping in reservoirs is expected to increase over time. However, this takes a very long time; thousands of years for capillary trapping and tens of thousands of years for mineral trapping [1]. Enhancing the trapping mechanism by injecting additional chemical substances with CO_2 would be developed by 2050. Also, injection technology using CO_2 microbubbles has been in development to enhance solubility trap mechanisms, which should be deployed by 2050.

4 Benefit and Future Vision

CCS is considered to be a large-potential and medium-cost technology in the marginal greenhouse gas (GHG) reduction cost curve [1] and an essential part of the GHG mitigation portfolio. The energy portfolio without CCS is believed to become very expensive relative to that with CCS. When CCS is fully developed, CCS will become a part of the energy infrastructure using fossil fuel. Most large CO_2 point sources would be equipped with CO_2 capture facilities and connected to transportation and storage infrastructure. The potential contribution of CCS technology to GHG emission reduction by 2050 can be calculated as follows.

CO_2 emissions reduction when CCS is deployed at all thermal power plants

$$= \mathrm{c} - (\mathrm{a} + 1)^{*}(1 - \mathrm{b})^{*}\mathrm{c} = 303\mathrm{Mt} - \mathrm{CO_2},$$

where:

(a) CCS energy penalty; 25 % (mean of 10–40 % [1])
(b) CO_2 capture ratio; 90 % (mean of 85–95 % [1])
(c) Annual CO_2 emissions at thermal power plants in Japan = 346 Mt-CO_2 (from electricity generation industry [38])

Among the above parameters, decreasing the energy penalty of CCS is one of the most important targets for R&D. Sensitivity of the energy penalty (10–40 %) to CO_2 emission reduction and the amount of CO_2 stored in geological formations are shown in Fig. 3. For an energy penalty increase from 10 to 40 %, the CO_2 emission reduction decreases by 3 % and the amount of CO_2 storage increases by 3 %. This indicates that reducing the energy penalty will contribute to saving energy and cost in CCS by reducing the amount of CO_2 that must be captured rather than decreasing atmospheric CO_2 emission (Fig. 4).

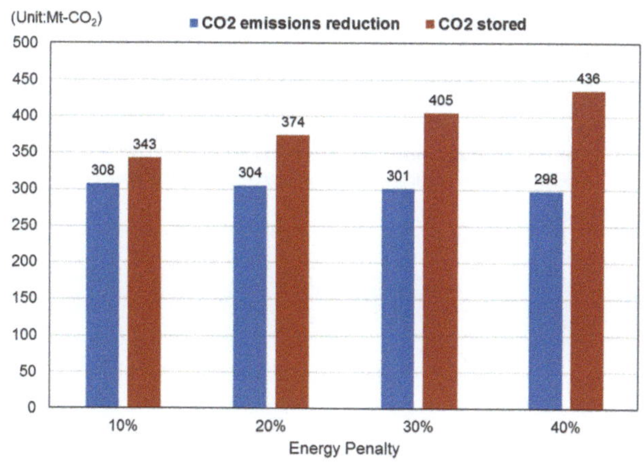

Fig. 3 Energy penalty of CCS and its impacts on CO_2 emission reduction. Note: Assuming all current fossil fuel-fired power plants in Japan are replaced with power plant CCS maintaining generation outputs the same as those of plants without CCS. All energy required to operate CCS is supplied by power plants with CCS at the same site

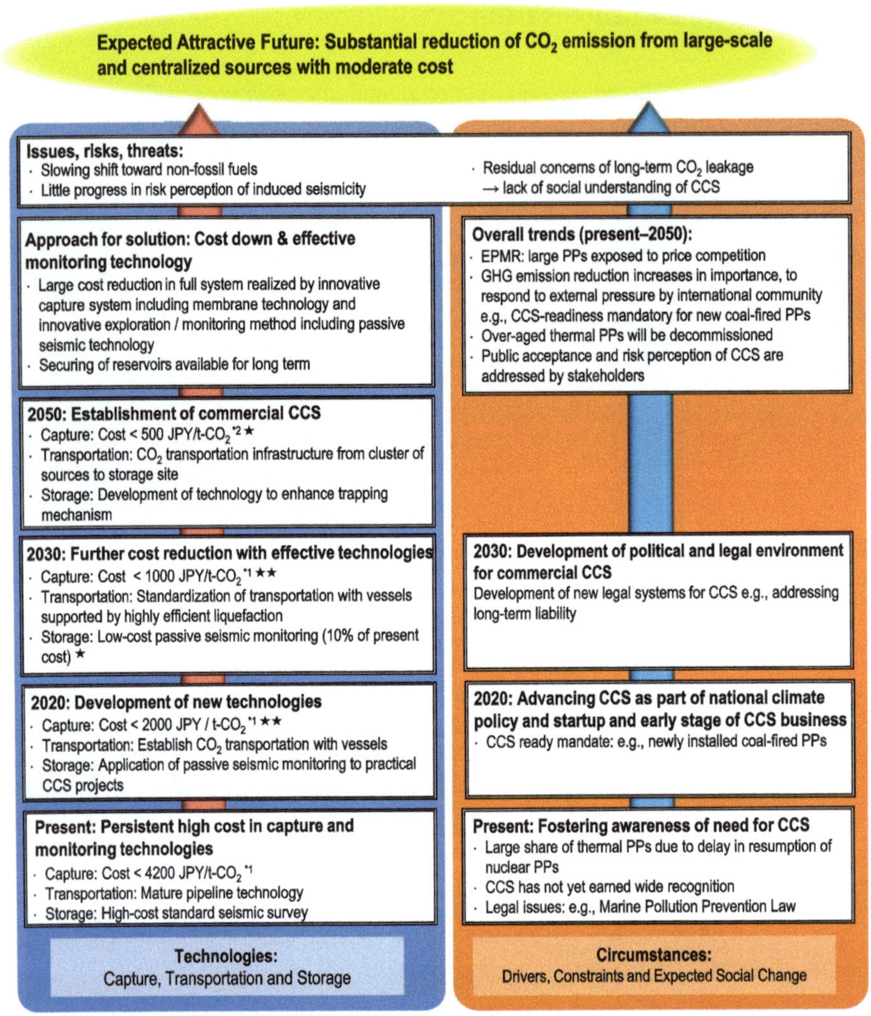

*1 [16], *2 NEDO, Advanced energy and novel environmental technology program, Project Target

GHG: Greenhouse effect Gas
CCS: CO_2 Capture and Storage
PP: Power Plant
EPMR: Electric Power system Market Reformation
NEDO: New Energy and Industrial Technology Development Organization

Fig. 4 Roadmap of CO_2 capture, transportation, and storage technology

References

1. IPCC (2005) IPCC special report on carbon dioxide capture and storage. Cambridge University Press, Cambridge
2. IEA (2013) Technology roadmap carbon capture and storage 2013 edition. OECD/IEA, Paris
3. Halmann MM, Steinberg M (1998) Greenhouse Gas carbon dioxide mitigation: science and technology. CRC, New York
4. Wilson E, Gerard D (2007) Carbon capture and sequestration integration technology, monitoring, regulation. Wiley, New York
5. IEA Greenhouse Gas R&D Programme (IEA GHG) (2004) Ship transport of CO_2, 2004/PH4/30. July 2004
6. The Society of Chemical Engineers, Japan (ed) (2011) Handbook of chemical engineering, 7th edn. Maruzen, Tokyo, p 202
7. Tanase D (2013) CO_2 geological storage on tomakomai large scale CCS demonstration project (in Japanese). Chem Eng Jpn 77:323–325
8. Center for Global Environmental Research (2014) National greenhouse Gas inventory report of Japan. National Institute for Environmental Studies, the Ministry of the Environment, Japan
9. Center for Global Environmental Research (2014) Rapid report of CO_2 emission. National Institute for Environmental Studies, the Ministry of the Environment, Japan
10. Fujii T (2009) UAE: short supply of LNG –current situation of sour gas development, JOGMEC report
11. Hadjipaschalis I, Kourtis G, Poullikkas A (2009) Assessment of oxyfuel power generation technologies. Renew Sustain Energy Rev 13:2637–2644
12. Rochelle GT (2009) Amine scrubbing for CO_2 capture. Science 325:1652–1654
13. Valenti G, Bonalumi D, Macchi E (2012) A parametric investigation of the Chilled Ammonia Process from energy and economic perspectives. Fuel 101:74–83
14. Henni A, Tontiwachwuthikul P, Chakma A (2005) Solubilities of carbon dioxide in polyethylene glycol ethers. Can J Chem Eng 83:358–361
15. Davis RA, Menendez RE, Sandall OC (1993) Physical thermodynamic, and transport-properties for carbon-dioxide and nitrous-oxide in solutions of diethanolamine or di-2-propanolamine in polyethylene-glycol. J Chem Eng Data 38:119–124
16. Research Institute of Innovative Technology for the Earth (2005) Effective CO_2 fixation and utilization research report
17. Mimura T, Honda A, Iijima M, Mitsuoka S (2001) Kagakukogaku Ronbunshu 27:335–378
18. The Japanese Ministry of Economy, Trade and Industry, Technology Roadmap (2009) in Japanese
19. Ho MT, Allinson GW, Wiley DE (2008) Reducing the cost of CO_2 capture from flue gases using membrane technology. Ind Eng Chem Res 47:1562–1568
20. Merkel TC, Lin H, Wei X, Bake R (2010) Power plant post-combustion carbon dioxide capture: an opportunity for membranes. J Membr Sci 359:126–139
21. Güiver MD, Lee YM (2013) Polymer rigidity improves microporous membranes. Science 339: 284–285
22. Park HB, Han SH, Jung CH, Lee YM, Hill AJ (2010) Thermally rearranged (TR) polymer membranes for CO_2 separation. J Membr Sci 359:11–24
23. Yun S, Oyama T (2011) Correlations in palladium membranes for hydrogen separation: a review. J Membr Sci 375:28–45
24. Bai H, Ho WSW (2011) Carbon dioxide-selective membranes for high-pressure synthesis gas purification. Ind Eng Chem Res 50:12152–12161
25. Samanta A, Zhao A, Shimizu GKH, Sarkar P, Gupta R (2012) Post-combustion CO_2 capture using solid sorbents: a review. Ind Eng Chem Res 51:1438–1463
26. Sjostrom S, Krutka H (2010) Evaluation of solid sorbents as a retrofit technology for CO_2 capture. Fuel 89:1298–1306

27. Adanez J, Abad A, Garcia-Labiano F, Gayan P, de Diego L (2012) Progress in chemical-looping combustion and reforming technologies. Prog Energy Combust Sci 38:215–282
28. Australian National Low Emissions Coal Research and Development (2012) Environmental impacts of amine-based CO_2 Post Combustion Capture (PCC) process deliverable 3.2 progress report
29. Norwegian Institute of Public Health (2011) Health effects of amines and derivatives associated with CO_2 capture
30. IEA Greenhouse Gas R&D Programme (2004) Ship transport of CO_2 report number PH4/30
31. Takahashi T, Ohsumi T, Nakayama K, Koide K, Miida H (2009) Estimation of CO_2 aquifer storage potential in Japan. Energy Procedia 1:2631–2638
32. Research Institute of Innovative Technology for the Earth (2013) Innovative zero-emission coal gasification power generation project, feasibility study on comprehensive system, from power generation to CCS: design of CO_2 geologic storage system and evaluation of storage potential (FY2008-FY2012) (in Japanese)
33. Ogawa T, Nakanishi S, Shidahara T, Okumura T, Hayashi E (2011) Saline-aquifer CO_2 sequestration in Japan-methodology of storage capacity assessment. Int J Greenh Gas Control 5:318–326
34. The Council on Competitiveness-Nippon report (2013) Development of CO_2 gas management technologies
35. IPCC (2006) 2006 IPCC Guidelines for National Greenhouse Gas Inventories, Prepared by the National Greenhouse Gas Inventories Programme, Eggleston H S, Buendia L, Miwa K, Ngara T and Tanabe K (eds). IGES, Japan
36. Chiyoda Corporation (2011) Preliminary feasibility study on CO_2 carrier for ship-based CCS. The Global CCS Institute, Canberra
37. Ozaki M (2013) CO_2 transportation by ship and injection to sub-seabed geological formation (in Japanese). Chemical Engineering of Japan 77:317–319
38. The Greenhouse Gas Inventory Office of Japan (GIO) (2014) National GHGs inventory report of JAPAN

Topic: Compressed Air Energy Storage (CAES)

Yoshiharu Toida

Abstract With the increasing share of fluctuating renewable energy sources, such as wind power and solar cells, demands for energy storage and load leveling in the electric grid are expanding. For this purpose, hydroelectric and thermal power generations are used. In Germany, however, for reducing greenhouse gases, development of second-generation compressed air energy storage (CAES) has been advanced to replace thermal power generation. Another type of CAES called advanced humid air gas turbine (AHAT), which uses moist air as compressed gas, is being researched in Japan. In 2013, 40-MW integral experimental equipment was constructed and technical data are now being collected.

Keywords Energy storage • Compressed air • Load leveling • Adiabatic

1 Introduction

In operating the electric grid, input and consumed amounts of electricity must be the same at all times. At present, hydroelectric and thermal power generations are used for load leveling in the grid. However, thermal power emits greenhouse gases, so other energy storage methods are desired to replace it.

In Germany, second-generation compressed air energy storage (CAES) has been advanced to replace thermal power generation. In this CAES system, energy is stored as compressed gases and sensible heat of solid substances.

A different type of CAES called an advanced humid air gas turbine (AHAT) has been developed in Japan. This system uses compressed gases and latent heat of water as energy storage.

Y. Toida (✉)
SCE-Net, Tokyo, Japan
e-mail: yo_toida@yahoo.co.jp

Y. Kato et al. (eds.), *Energy Technology Roadmaps of Japan*,
DOI 10.1007/978-4-431-55951-1_22

2 Technology Details

CAES is an energy storage system using air as a storage medium.

The system consists of:

- A compressor to compress the air and an air reservoir to store it
- A combustor to heat the air
- An expansion turbine to generate electricity

The air is compressed using surplus energy and stores the energy in the form of compressed air. When energy demand exceeds supply, the air is released and heated to drive an expansion turbine to generate electricity.

CAES systems in operation in Germany and the United States are both using salt domes with volumes of several 1 Mm^3 for the air reservoir. The New Energy Foundation of Japan used a 1,600-m^3 space in an abandoned mine as the air reservoir to build a CAES pilot plant of 2-MW output and duration 4 h [1].

However, this CAES has the following characteristics:

- Conversion efficiency is 20 % less than a battery and pumped hydropower.
- The construction cost of an air reservoir is high in Japan.

Reasons for the low efficiency of current CAES are:

- When the air is compressed by external work, it is heated and must be cooled to ambient temperature before storage in the reservoir.
- When this compressed air is expanded to drive the expansion turbine, it cools and must be reheated with natural gas prior to expansion.

Second-generation CAES in Germany introduces heat storage devices, which capture heat from the air during the compression and release this heat to the air during the expansion process. CAES's low conversion efficiency (50 %) could thereby be increased to 70 %, close to the value of a storage battery.

The AHAT system reduces compression power by the chilling effect of water sprayed on a compressor air intake, resulting in a calculated efficiency improvement. This improves the system's efficiency [2].

A 3-MW class pilot plant was built with subsidies from the Agency of Natural Resources and Energy in 2004. Based on the results, 40-MW integral experiment equipment was built in 2013. Engineering data of this system are currently being collected, and construction of a 200-MW system is a future goal.

There is no problem in Germany regarding construction costs of an air reservoir, because there is an extensive rock salt bed that can easily accommodate huge reservoirs. Unfortunately, there is no abandoned mine with a volume of several Mm^3 and no rock salt bed in Japan. Moreover, the cost is high to build the air reservoir in the ground, so placement on the seabed has been considered. However, there are fears of sea environmental destruction and problems of fishing rights, so only plans exist.

3 CAES Development in Other Countries

With the above technology, CAES of 2-GWh energy storage capacity is under construction in Germany [3]. CAES plants of a several 100-MW class are planned in the United States, with financial support from the Department of Energy. The University of Nottingham (UK) and Thin Red Line Aerospace have also co-developed CAES with a fabric balloon-like vessel as an air reservoir, which can be installed on the seabed for offshore wind power generation [4].

References

1. Moritsuka H, Morinaga M, Mimaki T (1993) Study on integrated compressed-air energy-storage advanced combined-cycle plant -thermal efficiency and operation. CRIEPI Research report, Nov 1993
2. Takahashi T, Koda E (2011) Study of compressed air energy storage generation system using humid air gas turbine. J Jpn Soc Energy Resour 32(6):19–27
3. http://www.rwe.com/web/cms/mediablob/en/391748/data/364260/1/rwe-power-ag/innovations/Brochure-ADELE.pdf RWE Corporate Website Jan 2010
4. de Jong M (2014) Commercial grid scaling of energy bags for underwater compressed air energy storage, Jul 2014. http://www.thin-red-line.com/140714_ThinRedLine_OSES2014.pdf

Topic: Distributed Cooperative Heat Supply System as a Measure Against Fluctuating Renewable Electricity Output

Kengo Suzuki

Abstract The distributed cooperative heat supply system (DCHSS), a cluster of heat supply appliances decentrally installed to the consumer side and centrally controlled by energy companies or aggregators, can be a means for absorbing the output fluctuation of renewable electricity. The DCHSS can contribute to both the penetration of renewable power plants and decarbonation of heat supply systems without district heating networks. The system appears especially well suited to Japan's north region, which has substantial wind energy potential and heat demand in winter.

Keywords Electricity and heat • Distributed system • Renewable electricity • CHP

1 Introduction

Heat supply systems can be a means for absorbing the output fluctuation of renewable electricity sources such as wind and solar power plants, considering that this renewable electricity covers a certain level of demand. Combination of heat supply technologies using electricity, such as heat pumps (HPs) and electric heaters and combined heat and power plants (CHPs), can enhance the flexibility of the electric grid. The surplus electricity can be used to produce heat when the outputs of renewable power plants are high, and the decreased electricity when these outputs are low can be covered by CHPs.

District heating (DH) networks combined with CHPs, HPs, and heat storage are regarded as a potent application of such flexible heat supply systems [1]. Through integrating regional heat demand by DHs, large-scale CHPs can be shared by a number of consumers. Then, the system becomes more flexible and reasonable compared with the case of installing small-scale discrete heating appliances.

However, DHs have not penetrated the market in Japan. Even in Hokkaido, the northernmost prefecture, the share of DH in total heat supply is merely 1 % [2]. The prime reason for this small share appears to be the high cost of pipeline installation.

K. Suzuki (✉)
Hokkaido University, Sapporo, Japan
e-mail: kengosuzuki@eng.hokudai.ac.jp

Y. Kato et al. (eds.), *Energy Technology Roadmaps of Japan*,
DOI 10.1007/978-4-431-55951-1_23

The joint trench for public infrastructure is not popular in the country, and heat pipelines should be buried under existing water and/or town gas pipelines, especially in urban areas. Thus, another type of flexible heat supply system without DHs should be sought to absorb the output fluctuation of renewable electricity in Japan.

2 System Details and Barriers

The authors' group has proposed the distributed cooperative CHP system (DCCS), a cluster of small-scale CHPs decentrally installed to the consumer side and centrally controlled by an energy company, to promote energy-efficient CHPs without DHs [3]. The CHPs and backup boilers are installed on the consumer side, and the controller monitors the electricity and heat demand of each consumer. Hot water tanks are also deployed on the consumer side to absorb the temporal mismatch between heat demand and supply. Surplus electricity from a CHP can be shared among consumers in the system through the existing grid, and electricity from outside the system is also available via the grid. The controller operates all the CHPs and boilers cooperatively to satisfy the demand. The DCCS can be cost-effective compared with large-scale DHs in Japan because existing well-developed town gas pipelines can be used by adopting fuel cell and gas engine-type CHPs, so additional investment in heat pipelines can be avoided.

The distributed cooperative heat supply system (DCHSS) proposed here is a developed DCCS concept. The consumer-side appliances are controlled not only to increase energy efficiency but also to absorb the output fluctuation of renewable electricity. In addition to CHPs and boilers, HPs and electric heaters are decentrally installed and centrally controlled. The controller can monitor the renewable electricity output and demand and can control all consumer-side appliances. The control algorithm is the same as the system with DHs. HPs and electric heaters are preferentially operated when the renewable electricity output is high and CHPs preferentially operated when that output is low. Such a system is rarely available at present, because there are few instruments to simultaneously monitor the electricity and heat demand of consumers. However, the Japanese government has shown direction toward disseminating energy management systems [4], and such a system will be technologically feasible before long.

However, there appear to be four types of barriers to the DCHSS. These are the seasonal variation of heat demand, the necessity of individual consumer flexibility, greater initial cost, and a regulated energy market. First, the DCHSS is less flexible in summer because of weak heat demand. Thus, the DCHSS should be positioned as a measure against winter season fluctuation of renewable electricity. In addition, the heat demand of every consumer must be individually satisfied, because the heat cannot be transported between consumers. For example, if a consumer has only an HP as a heat supply appliance, the controller must operate the HP even if the output of renewable electricity is low and system flexibility decreases. To ensure flexibility of the distributed cooperative system, heating appliances with multiple sources, such as an HP with a gas boiler and CHP with electric heating, are desirable for development and implementation. However, such multiple heat-source appliances would form another barrier, a greater initial cost than conventional appliances. The installation of energy management systems also raises the initial cost. Continuous efforts toward cost reduction are required to overcome the barriers. Regulations against emergence into the electricity and gas retail market and reverse power flow of electricity from CHPs also block the application of the system. However, it appears that these regulations will be relaxed in the near future. The full liberalization of the electricity retail market has already been decided, and relaxation of other regulations is also being discussed.

3 Future Vision

Japan's northern areas such as Tohoku and Hokkaido have substantial unused wind power potential [5], and wind power output and fluctuations there are greater in winter. If wind power achieves greater penetration and the winter output fluctuation emerges as a serious problem, the ample heat demand of this area in winter can be used to absorb the fluctuation by adopting the DCHSS. Because DCHSS flexibility is restricted by the amount and pattern of heat demand, combination with other types of infrastructure, such as dispatchable power plants, energy storage, and interconnection of electricity grids, is also important.

In spite of the various barriers, the DCHSS can boost the flexibility of the electricity grid without DH networks, and the heat production from renewable electricity and CHPs can contribute to decarbonizing the heat supply system. The deployment of multiple heat-source appliances can also augment the resilience of energy systems against natural disasters. Therefore, the DCHSS has the potential to be part of a sustainable energy system.

References

1. IEA (International Energy Agency) (2014) Linking heat and electricity systems. http://www.iea.org/publications/freepublications/publication/linking-heat-and-electricity-systems.html. Accessed on 23 Feb 2015
2. ANRE (Agencies for Natural Resources and Energy) (2014) Prefectural energy consumption statistics. http://www.enecho.meti.go.jp/statistics/energy_consumption/ec002/. Accessed on 23 Feb 2015
3. Akazawa M, Suzuki K, Tabe Y, Chikahisa T (2014) Model analysis of CO_2 emission reduction effect by introducing distributed cooperative CHP system in Hokkaido, Grand Renewable Energy 2014 Proceedings, Paper No. O-Po-2-3
4. ANRE (2014) Strategic energy plan (fourth plan). http://www.enecho.meti.go.jp/category/others/basic_plan/. Accessed on 23 Feb 2015
5. MOE (Ministry of Environment) (2011) Report of renewable energy potential survey in FY2010. http://www.env.go.jp/earth/report/h23-03/. Accessed on 23 Feb 2015. (in Japanese)

Part VI
Primary and Secondary Sectors of Industry

Yasuhiro Fukushima

There are three dimensions that each industry can explore: (1) Enhancement of efficiency, (2) increasing availability and use of renewable resources, and (3) cascade utilization of materials and energy beyond the borders of business entities, sectors, and locations. It is generally said that for (1), Japan has already been a top global performer in many of its industries, but for (2) and (3), there remains much room for exploration. Since the country is facing sustainability challenges in advance of most countries and regions in Asia, its experience of past and future strategies in adapting to changing socioeconomic and environmental circumstances furnishes a reference for many regions facing similar challenges in the near future.

Chapters in Part VI illustrate roadmaps to transform current industrial sectors other than electricity (Part V) and residential and commercial (Part VII) toward a sustainable future. These roadmaps highlight how new technologies can foster the co-prosperity of urban and rural communities, and where the opportunities lie in Japanese contexts.

Y. Fukushima (✉)
Tohoku University, Sendai, Japan

Chemical Industry

Tohru Setoyama

Abstract At the early twenty-first century, the chemical industry is facing a type of "climate change" regarding chemical resources. The shale gas revolution in North America and coal chemistry in China are becoming threats to conventional naphtha-based chemical processes. The lower production cost of chemicals derived from light alkanes, such as methane, ethane, and propane, especially abundant in wet shale gas, and chemicals derived from coal will make it difficult for naphtha-based chemical processes to be sustained, especially in East Asia and Japan. As a result, the chemical industry in Japan is focusing on value-added chemicals to maintain profit, and this trend will accelerate.

Because greater CO_2 emissions from conventional fossil resource, chemistry will exacerbate environmental climate change. Innovations in the chemical industry should be achieved by the concept of green sustainable chemistry (GSC) to realize both lower production cost and less CO_2 emission.

Keywords Climate change • Shale gas • LCA • CO_2 emission • Green sustainable chemistry

1 Introduction

While climate change such as global warming, mainly caused by anthropogenic CO_2 emissions, is becoming a very serious issue, the global chemical industry is facing another type of "climate change" regarding chemical resources. Both the shale gas revolution in North America and coal chemistry in China are changing the landscape of that industry.

Generally, lower alkanes such as ethane and propane are richer in shale gas than petroleum condensate [1]. Technological progress in hydraulic fracturing made it possible to extract them from shale gas in a cost-effective manner compared with the production of conventional oil [2]. Although both thermal cracking of ethane

T. Setoyama (✉)
Mitsubishi Chemical Group Science and Technology Research Center, Inc., Yokohama, Japan
e-mail: 1908182@cc.m-kagaku.co.jp

Y. Kato et al. (eds.), *Energy Technology Roadmaps of Japan*,
DOI 10.1007/978-4-431-55951-1_24

and catalytic dehydrogenation of propane are proven technologies, abundant methane is used as very cheap fuel or an energy resource [3]. Many projects are planning in North America to construct chemical complexes based on these technologies.

There is another threat to the naphtha-based chemical industry in East Asia. Growth of the chemical industry in China has been striking during the last decade and has been sustained by the increasing supply of chemical resources such as crude oil and coal. It has been recently said that the coal chemistry might occupy the major position in the chemical industry of China [4]. Coal is the cheapest chemical or energy resource among fossil resources. There are many projects in China to build coal-to-olefin (CTO) and coal-to-liquid (CTL) processes. Some of these will begin operations within a few years. Unit CO_2 emission of the conversion of coal into chemicals will be very great among various fossil resources, owing to the low H/C ratio of coal, an intrinsic property of that coal. Furthermore, emissions of environmentally harmful by-products such as particulate matter (PM) and acidic SO_x are problems associated with coal-based processes.

The gravity of climate change is argued in Intergovernmental Panel on Climate Change (IPCC) reports published in recent years [5]. They have indicated that it is necessary to deploy technologies on a huge scale that will facilitate zero or reduced CO_2 emissions by 2030. Further, the installation of such technologies should be expanded by the end of the twenty-first century to avoid climatic catastrophe. This means that one can expect an attractive business opportunity if any reliable innovative technology is proposed in the near future.

The two major CO_2 emitters, the USA and China, have been reluctant to mitigate climate change for a long period. However, the former country can reduce those emissions, mainly because of a drastic change in fuel resources for power generation, from coal to natural gas obtained from shale gas. China can no longer ignore environmental pollution. The two countries can change their policies toward reducing CO_2 emissions.

The Japanese government wishes to contribute to this issue by delivering Japanese low-CO_2 emission technologies and energy-effective processes to the world. These efforts will help mitigate global anthropogenic CO_2 and not just that of the domestic environment. Because the majority of chemical companies are using petroleum as feedstock, the impact of CO_2 reduction by changing resources from petroleum to superior fossil resources that emit less CO_2 is very straightforward and effective. Because the price of lower olefins as raw materials of chemicals is ca.1.5 times that of fuels, innovative technologies for lower olefin production would be more attractive than related technologies for fuels from an economic perspective.

In the past, the Japanese chemical industry demonstrated its technological excellence by providing numerous differentiated catalytic processes, especially in the late twentieth century [6]. Now, the industry is showing an excellent ability to provide many value-added chemicals and is maintaining its competitiveness in the global market. It is true that such chemicals are profitable, but their impact on CO_2 reduction is slight because the production scale is usually small compared with commodity petrochemicals. As a result, the domestic chemical industry has recently been declining production as an amount. There are many regions in the world where the chemical industry is regarded as a locomotive of national economic growth. Because Japan has a good quality culture of chemistry as demonstrated by the number of Nobel laureates in the field, we believe that it is possible to create innovative catalytic technologies that will facilitate zero or much-reduced CO_2 emissions, not limited to proposing value-added chemicals.

Considering the current situation, in which conventional chemical complexes based on naphtha crackers still have heavy infrastructure and their depreciation has nearly ended, newly proposed processes must be economically superior, even with the burden of depreciation.

Taking into account these subjects, we propose realistic strategies for the chemical industry as follows:

- Short-term strategy: Transformation of carbon resources from oil to lower alkanes and CO_2.

  Lower alkanes intrinsically emit less CO_2, owing to a higher H/C ratio than oil. In addition, incorporation of low-cost CO_2 can reduce the production cost of chemical products. Considering that we already have technologies using lower alkanes and CO_2 as chemical resources, the transformation of carbon as a fossil resource is desirable and realistic as a short-term strategy.
- Mid- or long-term strategy: Utilization of inedible biomass with high productivity.

  As sustainable chemical resources, biomass is attractive if its productivity is sufficiently high. If so, biomass should at least be economically competitive against fossil resources. Because biomass productivity based on natural photosynthesis is generally very low, the first priority among various challenges in biomass chemistry must be productivity enhancement. Preferably, processes to produce biomass-based chemicals should have sufficient access to current fossil resource-based processes.
- Long-term strategy: Solar hydrogen by water splitting and its use in chemical processes with CO_2.

  CO_2 emissions from fossil resources are inevitable. Further, there may be a shortage in the supply of fossil resources in the future. Therefore, utilization of solar energy to produce hydrocarbons is an attractive final goal for the sustainability of human society. It is possible to consume CO_2 as a chemical resource to provide hydrocarbons with a combination of solar hydrogen via water splitting. Among many methods for the conversion of solar energy, both solar cells and water splitting by semiconductor-type photocatalysts investigated in Japan occupy a cutting-edge position in natural science [7]. The installation of solar cells in Japanese society is not substantial, owing to the lack of an appropriate business model. If we can develop an effective business model based on innovations of the cutting-edge science, we can envision an attractive future. These innovations should be introduced in society together with the aforementioned technologies in short-term and midterm strategies with their full depreciation.

In any case, we should not treat these strategies as those in the domestic economy. It will be a great opportunity to extend them to the global market by means of innovative technologies from Japan.

2 Technology Details: Future Prospects, Problems, and Risks

2.1 Short-Term Strategy: Transformation of Carbon Resources from Oil to CO_2 and Lower Alkanes

Because methane is intrinsically chemically stable, a very high temperature is required to activate it for the production of syngas (a mixture of hydrogen and carbon monoxide), through methane reforming. Autothermal reforming is an optimal process in terms of energy efficiency, because it is based on a small exothermic reaction as shown below. This process has already been commercialized [8]:

$$CH_4 + 0.5\,O_2 \rightarrow CO + 2\,H_2 \quad \Delta H = -37\ \mathrm{kJ/mol} \tag{1}$$

Two as the H_2/CO ratio in this reaction fits well the Fischer-Tropsch (FT) reaction for fuel production [9]. A simplified equation of that reaction is as follows, in which (CH_2) denotes various hydrocarbons:

$$CO + 2H_2 \rightarrow (CH_2) + H_2O \quad \Delta H = -155\ \mathrm{kJ/mol} \tag{2}$$

By combining Eqs. 1 and 2, we obtain Eq. 3 below. This indicates that methane can theoretically provide hydrocarbons without emitting CO_2:

$$CH_4 + 0.5\,O_2 \rightarrow (CH_2) + H_2O \quad \Delta H = -192\ \mathrm{kJ/mol} \tag{3}$$

In the commercialization of this process, one of the key determinants of product cost is the price of methane. It is possible that this price will decline in the future. Currently, the price of LNG in Japan is twice that in the European Union, and this is called the "Japan premium."

In contrast, reforming methane using CO_2 has recently been commercialized (called dry reforming or DR) [10]. The theoretical reaction equation of DR is

$$0.75\,CH_4 + 0.25\,CO_2 + 0.5\,H_2O \rightarrow CO + 2\,H_2\ \Delta H{=}{+}163\,\mathrm{kJ/mol} \tag{4}$$

When we combine the FT reaction of Eq. 2 with that of Eq. 4, we obtain Eq. 5 as a reaction to form hydrocarbons:

$$0.75\,CH_4 + 0.25\,CO_2 + 0.5\,H_2O \rightarrow (CH_2) + H_2O \quad \Delta H{=}{+}8\ \mathrm{KJ/mol} \tag{5}$$

If the DR and FT reactions can be combined in a very energy-effective way without almost zero energy input from outside as shown in Eq. 5, the burden of energy management will be significantly lightened. Equation 5 shows a potential reduction of methane consumption through the utilization of CO_2 as one of the carbon resources. If the CO_2 price is low enough, we can also expect a significant reduction of the production cost of hydrocarbons.

As another olefin production process, methanol-to-olefin (MTO) is preferable to the FT process. Whereas MTO prepares olefins very selectively, the FT process requires substantial energy to separate lower olefin from the corresponding paraffin as a by-product. The MTO process via methanol synthesis is as follows:

$$CO + 2H_2 \rightarrow CH_3OH \quad \Delta H = -92\,KJ/mol \tag{6}$$

$$CH_3OH \rightarrow (CH_2) + H_2O \quad \Delta H = -63\,KJ/mol \tag{7}$$

Although methanol synthesis (Eq. 6) is regarded as an established technology, there is much room to improve its performance. For example, because the reaction is under the control of thermodynamic equilibrium, this synthesis generally requires high reaction pressure to achieve adequate conversion of raw materials (~50 %). This also means that a large amount of unreacted syngas must be recycled in the process. Regarding the olefin synthesis step (Eq. 7), only a few zeolite catalysts such as ZSM-5 and SAPO-34 are effective in satisfying the requirements of commercial processes. Therefore, there is only a narrow window to adjust the distribution of olefin products [11].

Table 1 compares the carbon footprint from mining to combustion and the estimated production cost of olefins in various processes, based on fossil resources [12]. It also shows the economic impact of CO_2 utilization as a resource. It is obvious that there is much room for innovation in C_1 chemistry, especially regarding the combination of fossil resources and CO_2.

Historically, the unit calorimetric energy price of LNG and coal vs. oil has been ~80 % and ~30 %, respectively [13a]. There has been a clear link between their prices and it is believed that this trend will continue in the future. The operational

Table 1 Comparison of CO_2 emission and production cost of olefins among various processes

Process	Material	CO_2 emission as a carbon footprint/MMton olefin	Consumption	Price ¥/kg	Material cost energy cost ¥/kg	Production cost ¥/kg
Cracker	Naphtha	584 MMton			105	120
					15	
Autothermal	CH_4	360 MMton	1.14	80	105	
	O_2		1.14	12	8	
Dry reforming	CH_4	318 MMton	0.86	80	72	
	CO_2		0.69	3	8	
	H_2O		0.28	2		

cost of CO_2 capture and storage (CCS) in the near future is estimated at ~3 yen/kg-CO_2 [13b]. When CO_2 is used as a chemical resource, energy for storage is unnecessary and, accordingly, CO_2 prices will be low. Based on these characteristics, we believe that the combination of CH_4 and CO_2 as chemical resources can realize relatively low olefin production costs.

However, ethane is also abundant in shale gas, and newly built chemical complexes in North America will use ethane crackers as the root of the supply chain of various petrochemicals [14]. Because ethane cracking produces only ethylene, there is the possibility of a supply imbalance between ethylene and propylene in the future.

Catalytic dehydrogenation of propane (PDH) is also an established technology to produce propylene, but it still has many disadvantages. For example, PDH needs a moving-bed reactor system that requires external heat supply to compensate the heat necessary for the endothermic reaction. In addition, rapid deactivation of the catalyst demands frequent regeneration. Further, the conversion of propane is low because the reaction is under equilibrium [13]. Although propane is easily extracted from shale gas, its price is not low as a chemical resource because it is a relatively expensive fuel as LPG. Therefore, the economic competitiveness of propylene production via PDH is not definitive. Considering the current situation, there are many candidates for innovative technologies occupying a certain position in the commercial production of propylene. Direct conversion of ethylene to propylene combined with an ethane cracker (ETP) [14] and methanol-to-propylene (MTP) are examples of such candidates.

Because the concentration of butane in shale gas is not high and BTX (benzene, toluene, and xylene) is not contained in it, the supply of butenes, butadiene, and aromatics may be a future bottleneck for chemical production that uses these as resources.

Considering that lower alkanes from shale gas will occupy an important position as chemical resources in the near future, innovative technologies such as the conversion of alkanes to olefins and interconversion between olefins are strongly likely to replace conventional technologies [15].

From a GSC perspective, coal chemistry will have a negative impact. This chemistry is an issue specific to China, and it will be very difficult for other countries to enact this strategy. The Chinese government wants to adopt more environmentally benign technologies such as solar cells and wind turbines, with the recognition that environmental pollution in the country is very serious and most of it is caused by coal-based technologies. Because both CTO and CTL are analogs of MTO and GTL, respectively, they will also be improved analogously to innovations related to shale gas, somewhat reducing CO_2 emissions.

2.2 *Mid- or Long-Term Strategy: Utilization of Inedible Biomass with High Productivity*

Chemicals derived from biomass are becoming more popular by the year. Ethanol, lactic acid, succinic acid, glucose, sorbitol, and others are provided based on biomass on a commercial scale. Biomass is assumed carbon neutral and will ably contribute to CO_2 reduction if used as chemical resources on a large scale. Currently, most biomass-derived chemicals are created by enzymatic processes of edible biomass such as corn, sugar, and others. Therefore, edible biomass is in competition with food for humans and resources for chemicals, which caused many problems in the last decade. The use of inedible biomass such as sugarcane, bagasse, wood, energy crops, and others is recommended for the chemical production.

Although shale gas application is not realistic in Japan, biomass appears to have some possibility there for use as a chemical resource. One candidate is wood. Wood for paper manufacturing is relatively abundant in Japan. Woody biomass is basically composed of three portions: cellulose, hemicellulose, and lignin. They are typically converted to various chemicals by very complicated and difficult processes, making it difficult to reduce production cost. For this reason, Japanese chemical companies are oriented toward value-added chemicals rather than commodity chemicals. This means that the production of biomass-derived chemicals in Japan is not suitable for significant reduction of CO_2 emissions. This is the intrinsic dilemma of such chemicals.

The productivity of biomass is basically controlled by the performance of natural photosynthesis.

Table 2 compares the productivity of biomass per unit area annually.

For biomass-derived chemicals to be competitive with fossil resource-based chemicals, enhancement of productivity of biomass will be the most important issue. Without any innovations for biomass to give enough productivity, it is difficult to recommend biomass as the candidate resources to mitigate the climate change in the twenty-first century [16]. When enough productivity is realized, biomass-derived chemical will contribute to CO_2 reduction even in Japan.

From this point of view, we can propose energy crop as another candidate. Because of the poor agricultural policy of Japan, huge areas of rice farmland are not used, and there may be a possibility to use these for energy crops.

For harmonization with the conventional infrastructures, biomass should be converted to the basic chemicals currently produced from fossil resources. From this aspect, both ethanol by the fermentation of glucose and syngas by reforming biomass would be good candidates.

Table 2 Comparison of biomass productivity

Biomass	Productivity liter oil/year hectare	Note
Corn	140	As bioethanol
Soybean	450	
Sunflower	960	
Palm oil	6,000	As biodiesel oil
Microalgae	14,500	Experimentally obtained
Microalgae	45,000–140,000	Theoretically expected

2.3 Long-Term Strategy: Solar Hydrogen by Water Splitting and Its Use in Chemical Processes with CO_2

The amount of anthropogenic CO_2 emissions is too large and is still increasing. We must do something to lower atmospheric CO_2 concentrations. Given this, the production of hydrocarbons using solar hydrogen and CO_2 would be one of the final goals in GSC. Solar hydrogen is defined as hydrogen produced by photocatalytic water splitting or by the electrolysis of water, with electricity generated by renewable energy such as solar cells or wind turbines. The production cost of chemicals derived from CO_2 and solar hydrogen must be nearly equal to or less than that of fossil-based chemicals. Because we can expect cheap CO_2 by means of CCS in the future, the cost of solar hydrogen is critical in determining the production cost of chemicals.

Although the material cost of solar hydrogen from water is very low, the depreciation cost of a production facility for solar hydrogen will be high, because it requires a large area for the reactors. Therefore, high photo conversion and low fabrication cost will be important to reduce the burden of depreciation cost. These are not easy tasks, in both the science and the technology. A long-term effort will be required to solve these problems for actualizing artificial photosynthesis.

Nevertheless, the combination of solar cells and electrolysis is regarded as commercially proven technology, mainly through the collaborative effort of academia and industry in Japan during the twentieth century. The domestic market for solar cells is not substantial, because of the lack of an attractive business model. It will be necessary to differentiate this market through innovations. It is clear that the established technology for water electrolysis is not suitable to scale up. If water electrolysis can be realized at a large scale in an economically feasible manner, it will be very attractive. A 10-year artificial photosynthesis project funded by the New Energy Development Organization has been ongoing since 2012 [17]. It will take a long time to implement it at large scale, but it is attractive because it will be a differentiated technology based on cutting-edge science.

From an economic standpoint, the following approach may be more realistic. First, we actualize innovative processes that will use CO_2 and fossil resources. Then, after depreciation completion of such processes, we install the innovations based on biomass and/or solar hydrogen, because solar radiation is limited by weather and daytime. This approach would be useful to solve the aforementioned problems.

3 Benefits and Prospective Future Vision

To achieve the transformation of conventional chemical processes and their infrastructures into innovative chemical complexes, the following tasks will be very important:

- Diversification of chemical feedstock following the GSC concept
- Application of new technologies to conventional ones, in which it is important to combine them harmoniously
- Economical superiority of new technologies over conventional ones, even considering depreciation

Figure 1 illustrates the concept of advanced chemical complexes from the combination of conventional and innovative processes. If solar or biomass-derived hydrogen can be supplied at low cost, such new complexes will have tremendous economic competitiveness. Furthermore, such complexes will not emit CO_2 but will consume it as a resource. It will be possible to reach the chemical complex goal in the twenty-first century through successive transformations and innovative technologies. The figure compares carbon footprints of various processes of olefin production. Although chemical complexes in Japan cannot realize the full features shown in Fig. 1, innovative technologies from Japanese industries can contribute to this, forming a new business model for those industries. Considering the scientific reliability of the IPCC reports, we must make our best efforts to transform the conventional technologies into innovative ones based on the GSC concept as soon as possible.

A road map of chemical industry is shown in Fig. 2. At present, it is not easy to postulate a practical road map, but we can propose an attractive business model based on innovations of the chemical industry. This is not an answer to a domestic problem but a proposal to create new business, which can be extended to the world market.

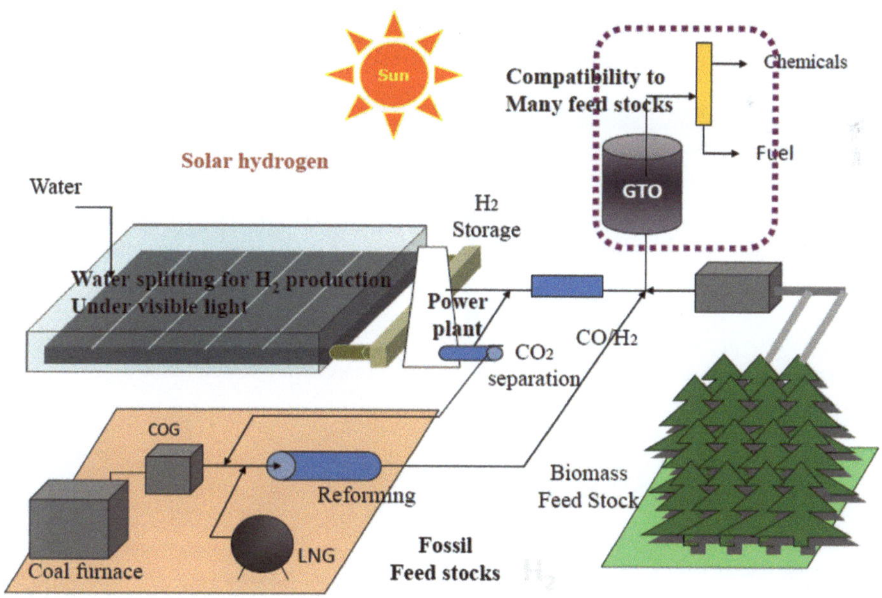

Fig. 1 Combination with conventional process with innovative technologies

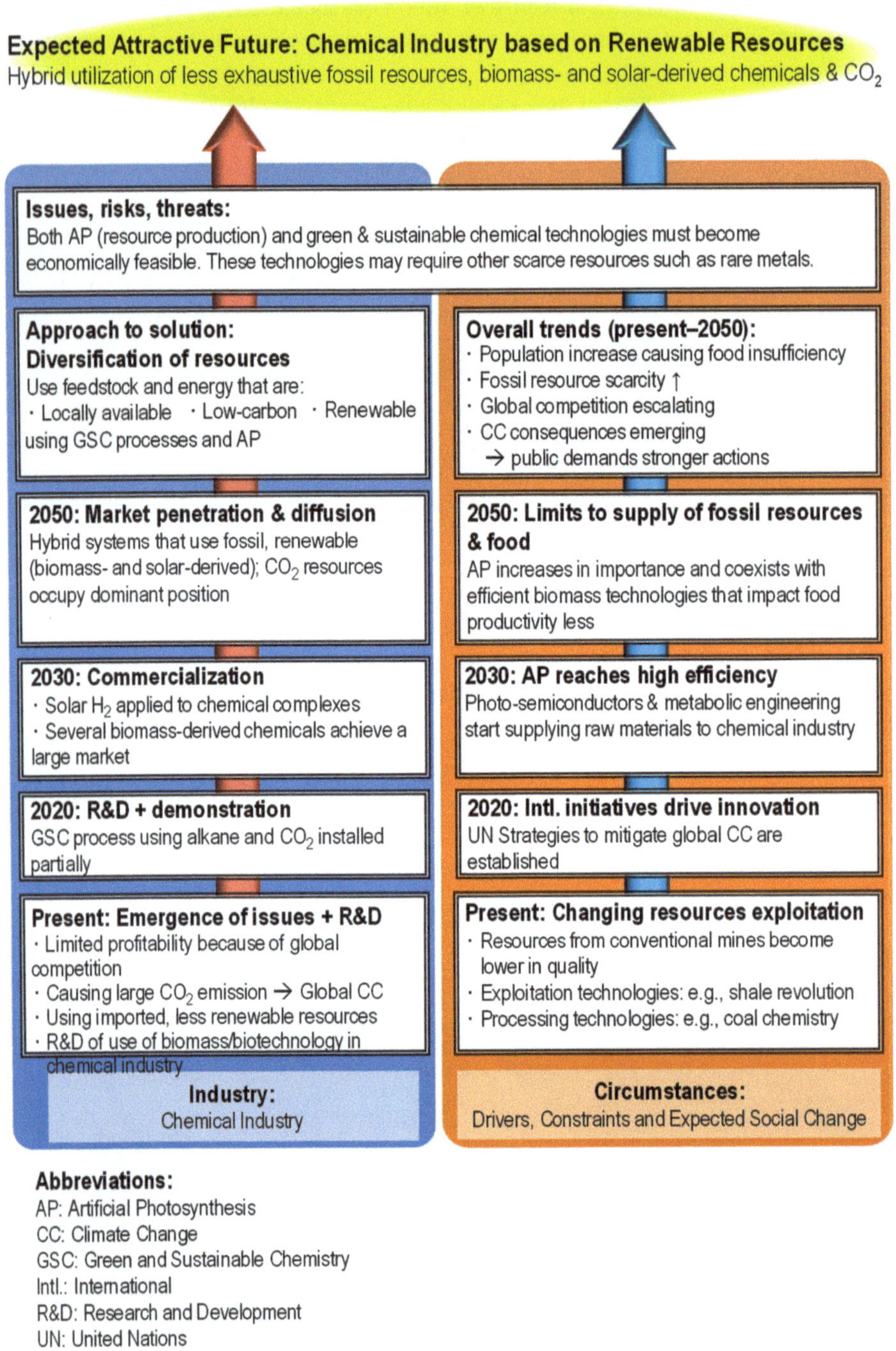

Fig. 2 Road map of chemical industry

References

1. http://chemistrytoenergy.com/states/chemistrytoenergy.com/
2. http://energy.gov/fe/science-innovation/oil-gas-research/shale-gas-rd
3. http://www.essentialchemicalindustry.org/processes/cracking-isomerization-and-reforming.html
4. Xie K, Li W, Zao W (2010) Energy 35:4349–4355
5. http://report.mitigation2014.org/spm/1pccwg3ar5summaryforpolicymakerapprovedpdf
6. a) Misono M, Nojiri N (1990) Appl Catal 64:1–30. b) Misonoo M, Inui T (1998) Catal Today 51(304):369–375
7. a) Fujishima A, Honda K (1972) Nature 238:37–38. B) Maeda K, Takata T, Domen K (2005) J Am Chem Soc. 127:8286–8287
8. www.pecj.or.jp/japanese/report/reserch/report-pdf/H10. . ./98cho9.pdf
9. Shulz H (1999) Appl Catal Gen 186(1–2):3–12
10. Sutton D, Kelleher B, Ross JRH (2001) Fuel Process Technol 73:155–173
11. Lefevere J, Mullens S, Meynen V, NOyen JV (2014) Chem Pap 68(9):1143
12. See the reference attached with Figure
13. a), b). World Energy Outlook (2012) International Energy Agency (Chap. 2–4)
14. Setoyama T (2014) Catal Surv Asia 18(2–3) ISSN 1571–1013
15. Sanfillipo D (2006) Catalysis Today 111:133
16. a) http://www.jie.or.jp/biomas/AsiaBiomass-handbook/Japanese/Part-2j.pdf. b) MJ Griffiths, STL Harrison (2009) J Appl Phycol 21:493–507
17. www.meti.go.jp/policy/tech_promotion/kenkyuu/saishin/46.pdf

Area-Wide Energy Saving in Heavy Chemical Complexes Using Area-Wide Pinch Technology

Kazuo Matsuda

Abstract It has been common to adopt a single-site approach to identify the energy-saving potential within individual industrial sites in heavy chemical complexes, and measures have been studied to optimize conditions. It has often been believed that by using this approach, all possible energy-saving measures are considered and implemented, and there is no additional energy-saving potential in a complex. In a challenge to conventional thinking and to achieve further economies, a new concept (area-wide approach) of area-wide energy saving is developed here. In this concept, utility systems for multiple sites in a heavy chemical complex are fully integrated, and surplus heat across multiple sites is used. For the area-wide approach, area-wide pinch technology is an excellent analytical methodology that can be used for multiple sites in a complex, in which all individual sites are viewed together and treated as a single entity for analytical purpose. This technology, consisting of a study procedure, R-curve analysis and total site profile (TSP) analysis, is applied to major heavy chemical complexes in both Japan and Thailand to investigate their energy-saving potential. Despite very high efficiency of the individual sites, the use of area-wide pinch technology confirmed that there is great energy-saving potential, and mid- and long-term plans are developed to achieve further economies of energy consumption within the heavy chemical complexes.

Keywords Area-wide • Energy saving • Multiple sites • Industrial area • Pinch technology

1 Introduction

Previous energy-saving studies for a heavy chemical complex have been carried out based on the concept of a single-site approach, which improves energy efficiency within the site itself. Klemes et al. [1] developed a total site approach, whereby pinch technology is used to study energy saving in utility systems. However, this was still a

K. Matsuda (✉)
Chiyoda Corporation, Yokohama, Japan
e-mail: ghg.matsudakazuo@gmail.com

Y. Kato et al. (eds.), *Energy Technology Roadmaps of Japan*,
DOI 10.1007/978-4-431-55951-1_25

single-site approach, and the limit has been reached for energy saving at single sites. A breakthrough was required, which came in the form of an area-wide approach for multiple sites. This approach has two aspects in energy saving, area-wide energy efficiency improvement and area-wide heat utilization. The former took the approach that if many utility systems in an entire industrial area were fully integrated, energy efficiency could be significantly improved by area-wide optimization in design and operation. Area-wide heat utilization makes use of heat across a large industrial area by accessing surplus heat from one or more sites to provide it to other sites. Area-wide pinch technology was developed to identify the area-wide energy-saving potential of an entire complex from the standpoint of the two aforementioned aspects, based on the concept of agglomerating multiple sites by treating them as a single entity. The consequence is that area-wide energy-saving projects can now be developed using the results of area-wide pinch technology studies.

2 Present Status

2.1 Area-Wide Pinch Technology

Area-wide pinch technology consists of three integral parts: a study procedure, R-curve analysis and total site profile (TSP) analysis. The study procedure shows how to perform the study based on a concept in which multiple sites are considered as one virtual site. R-curve analysis evaluates improvement in area-wide energy efficiency; the application of which is used to evaluate the feasibility of introducing a large-scale gas turbine system in an area-wide energy-saving project to attain energy saving. TSP analysis appraises area-wide utilization of low-grade heat and determines how the heat-sharing project will utilize surplus low-grade heat across the sites.

2.1.1 Study Procedure

Figure 1 shows a retrofitting procedure for the full-range program for existing sites to implement an area-wide energy-saving project, from the conceptual design to final

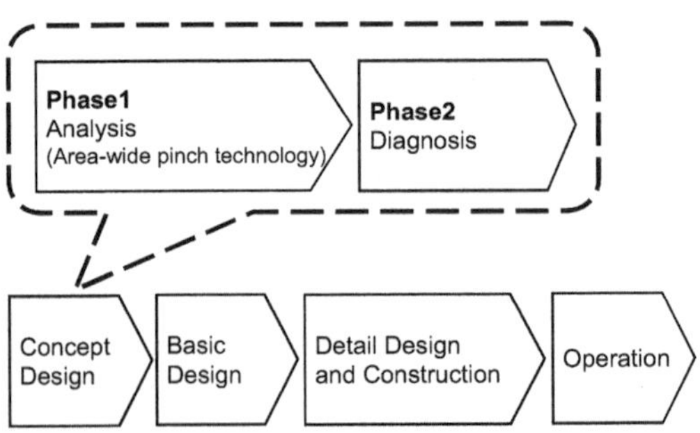

Fig. 1 Overall execution procedure

operation. Area-wide pinch technology study is used in the conceptual design stage, which consists of analysis (Phase 1) and diagnosis (Phase 2). In Phase 1, collected data are combined to become those of a virtual site. Such combined data are then used as input to the area-wide pinch technology, which evaluates area-wide energy efficiency improvement via R-curve analysis and utilization of low-grade heat via TSP analysis. In Phase 2, project ideas are identified and developed, such as area-wide energy efficiency improvement and area-wide heat utilization projects. Selected project ideas are then used to execute the basic design phase and continuing work.

2.1.2 R-Curve Analysis

The original concept of R-curve analysis was developed by Kenney [2], which was fuel utilization curve analysis (evaluation chart of total site cogeneration potential), but its scope and concept were limited to evaluate only fuel utilization efficiency for an in-house cogeneration system. To assess area-wide energy efficiency improvement, it was necessary to expand the scope and concept to include the power company, which led to development of the concept of "integrated energy efficiency" in R-curve analysis. The definition of the integrated energy efficiency is described by Eq. 1 as the ratio of available energy, which is heat (Q_{heat}) and power (W) over integrated energy consumption (Q_{fuel}). Equation 2 describes the R-ratio (the ratio of power-to-heat demand) in the process under site operating conditions.

$$\text{Integrated energy efficiency} = (W + Q_{heat})/Q_{fuel} \tag{1}$$

$$\text{R-ratio(power-to-heat ratio)} = W/Q_{heat} \tag{2}$$

Theoretical limit lines for two energy systems are shown in Fig. 2, the "gas turbine combined system" and "boiler and turbine conventional system." The

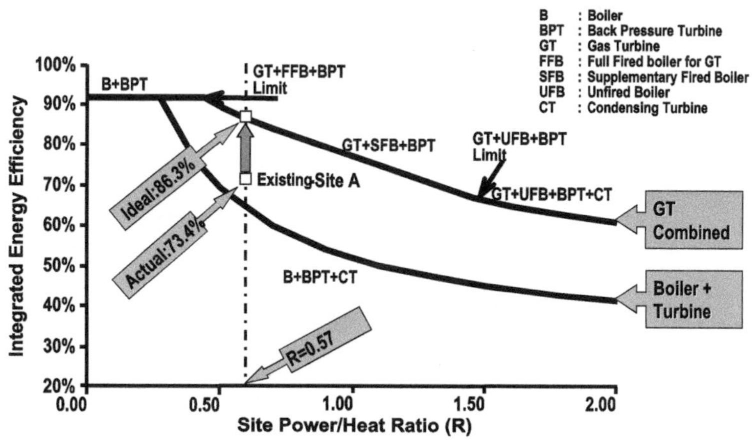

Fig. 2 R-curve analysis

maximum achievable efficiency can be identified in the R-curve for a given site R-ratio. The difference between existing and maximum efficiency shows the potential for improvement.

2.1.3 Total Site Profile Analysis

A graphical method, so-called total site profile (TSP) analysis, was first introduced by Dhole and Linnhoff [3] and later extended by Raissi [4]. Klemes et al. [1] considerably extended this method to encompass site-wide applications. TSP analysis is constructed from composite curves, which combines heat supply and demand based on heat exchangers of process-utility interfaces (e.g., heaters, coolers, and steam generators). TSP analysis is shown in Fig. 3. The right side of this figure shows composite curves of heaters, such as steam heaters and reboilers. The cold composite curve (heating demand) that consists of many cold process streams is heated by the heating media composite curve, consisting of flue gas, high-pressure steam, middle-pressure steam, and low-pressure steam. The left side of the figure shows composite curves of coolers, such as steam generators, cooling water/air fin coolers, and condensers. The hot composite curve consists of numerous hot process streams cooled by the cooling media, consisting of multi-pressure steam and cooling water. The dotted line in Fig. 3 shows actual operating conditions, and the solid line is the target condition for energy saving. The region at less than 130 °C of the hot composite curve on the left side of the figure has a large temperature space for the cooling media. The very low-pressure steam and hot water can be recovered instead of using the existing utility (cooling water), and they can be supplied to the right side to reduce low-pressure steam consumption for heating the cold composite curve. TSP analysis is used to integrate individual

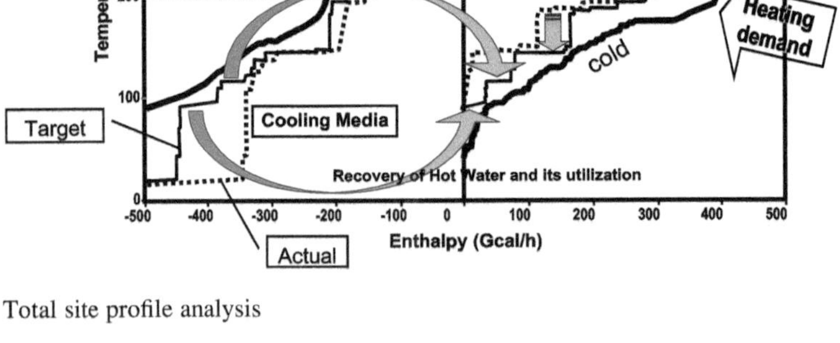

Fig. 3 Total site profile analysis

heating and cooling demands of the various processes across the entire site. Using a common utility system, excess heat from one process can be used as a heat source for another.

2.2 *Heavy Chemical Complexes*

Area-wide pinch technology was used to study three heavy chemical complexes (Chiba and Mizushima in Japan, Map Ta Phut in Thailand). Twenty-three sites agreed to cooperate in the Chiba complex study, which is on the northeast coast of Tokyo Bay. The complex was originally developed during the 1960s, and by the mid-1970s, the production of heavy metals and chemicals became the greatest in Japan. The Mizushima complex faces Mizushima Bay in Okayama Prefecture of western Japan and has 35 sites. Construction in this area began in the 1950s, and it rapidly developed into one of the nation's leading complexes by the 1960s. Map Ta Phut complex, 190 km southeast of Bangkok, was established in 1990. Its study covered 15 sites, each within 5 km of each other in the center of the complex. Sites included refineries, chemical plants, gas separation facilities, and a utility company.

2.3 *Energy-Saving Potential and Mid- and Long-Term Plan*

2.3.1 Energy-Saving Potential

As shown in Table 1, two complexes, Chiba (Matsuda [5]) and Mizushima in Japan, were studied for energy saving with area-wide pinch technology. There was tremendous energy-saving potential found for the heavy chemical complexes. Table 1 shows that Chiba has 23 sites and its integrated fuel consumption, which is the sum of fuel and electric power consumptions in the energy system, was 2.88

Table 1 Energy-saving potential

[Unit: annual crude oil rate equivalent, mil. kL = million kiloliter]	Chiba, Japan	Mizushima, Japan	Map Ta Phut, Thailand
1) No. of sites	**23**	**35**	**14**
2) Integrated fuel consumption (fuel + power)	2.88 mil. kL	3.79 mil. kL	3.18 mil. kL
3) Theoretical energy-saving potential by R-curve analysis	0.51 mil. kL	1.00 mil. kL	0.70 mil. kL
4) Theoretical energy-saving potential by TSP analysis	0.13 mil. kL	0.21 mil. kL	0.18 mil. kL
Total: 3 + 4	0.64 mil. kL	1.21 mil. kL	0.88 mil. kL

Note: 1 kL of crude oil is equivalent to 38.2 GJ

million kL/year (annual crude oil rate equivalent). R-curve analysis determined that Chiba has a potential energy saving of 0.51 million kL/year, whereas TSP analysis showed a potential of 0.13 million kL/year. Chiba therefore has the potential to save a total of 0.64 million kL/year. Table 1 also shows the result for Mizushima. Its energy- saving potential was a total of 1.21 million kL/year. The annual energy-saving potential at Map Ta Phut totaled 0.88 million kL/year (Matsuda et al. [6]). It is understood that the total energy-saving potential is around 30 % of the integrated fuel consumption at each heavy chemical complex. This indicates a vast energy-saving potential. Furthermore, from the standpoint of cost performance (less than a 5–7 years payback period) of area-wide energy-saving projects, almost 30 % of the theoretical energy-saving potential can be expected. It should be noted that 1 kL of crude oil is equivalent to 38.2 GJ.

2.3.2 Mid- and Long-Term Plan

Owners of heavy chemical complexes must look toward reducing energy consumption to strengthen their international competitiveness. The present study showed many potential opportunities for energy saving in such complexes. Despite the very high efficiency of individual processes, energy saving is possible by collaboration among sites (area-wide energy-saving projects) to enhance efficiency via integrating utility systems and sharing surplus heat across the sites. Seventeen ideas were developed from the results of R-curve and TSP analyses for collaborative energy-saving projects at Map Ta Phut (Matsuda et al. [6]). Mid- and long-term energy-saving plans were then drawn up, which included a road map for implementing practical ideas in appropriate order and avoidance of dual investment. Nine of the ideas were ultimately selected for plans (Fig. 4). These ideas confirmed further

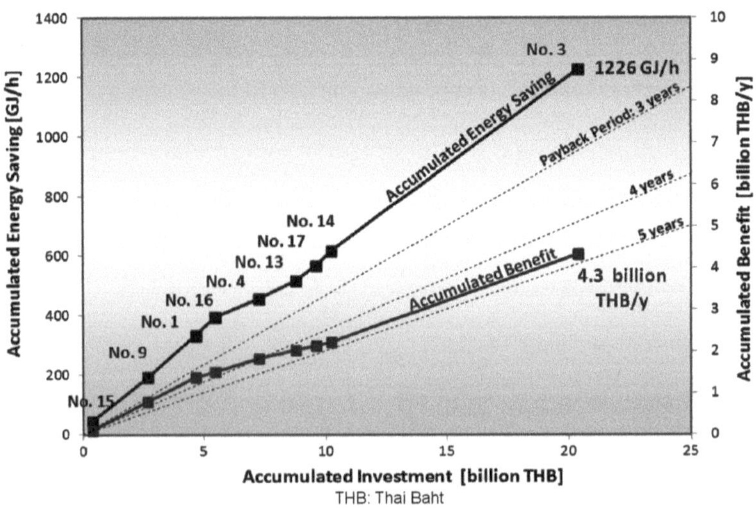

Fig. 4 Mid- and long-term plan

energy-saving potential of 1,226 GJ/h (0.26 million kL/year) in the entire complex (29 % of the theoretical energy-saving potential) and an overall payback period for the road map of less than 5 years. That period was calculated by dividing accumulated investment by accumulated benefit.

3 Technology Road Map

3.1 Area-Wide Pinch Technology

Area-wide pinch technology was developed and applied to two heavy chemical complexes in Japan and one in Thailand for area-wide energy saving. This examination clearly demonstrated tremendous energy-saving potential, despite the very high efficiency of individual sites in the complexes. In the near future (Fig. 5), area-wide pinch technology will be applied to foreign heavy chemical complexes, in areas such as the Middle East and Southeast Asia. After that, this technology will be used for new developments in developing countries to optimize the design of entire heavy chemical complexes. Eventually, the technology will be a de facto standard analytical methodology for evaluating and designing an optimal heavy chemical complex.

3.2 Area-Wide Energy-Saving Project

An area-wide energy-saving project has been demonstrated in Japan, and application of the area-wide pinch technology study of heavy chemical complexes in foreign countries will identify new such projects. These projects are collaborative across sites, and energy monitoring systems will be very important in maintaining effective operating conditions for all participants. These systems analyze collected data in real time, determine optimal operating conditions, and display optimized operating parameters to operators. It is recognized that the theoretical area-wide energy-saving potential is around 30 % of the integrated fuel consumption, based on the results of area-wide pinch technology studies. From the author's perspective, it is expected that 15 % of the aforementioned potential will be achieved around 2020, 30 % around 2030, 50 % around 2050, and 80 % at the end of the technological goal in Japan (Fig. 5).

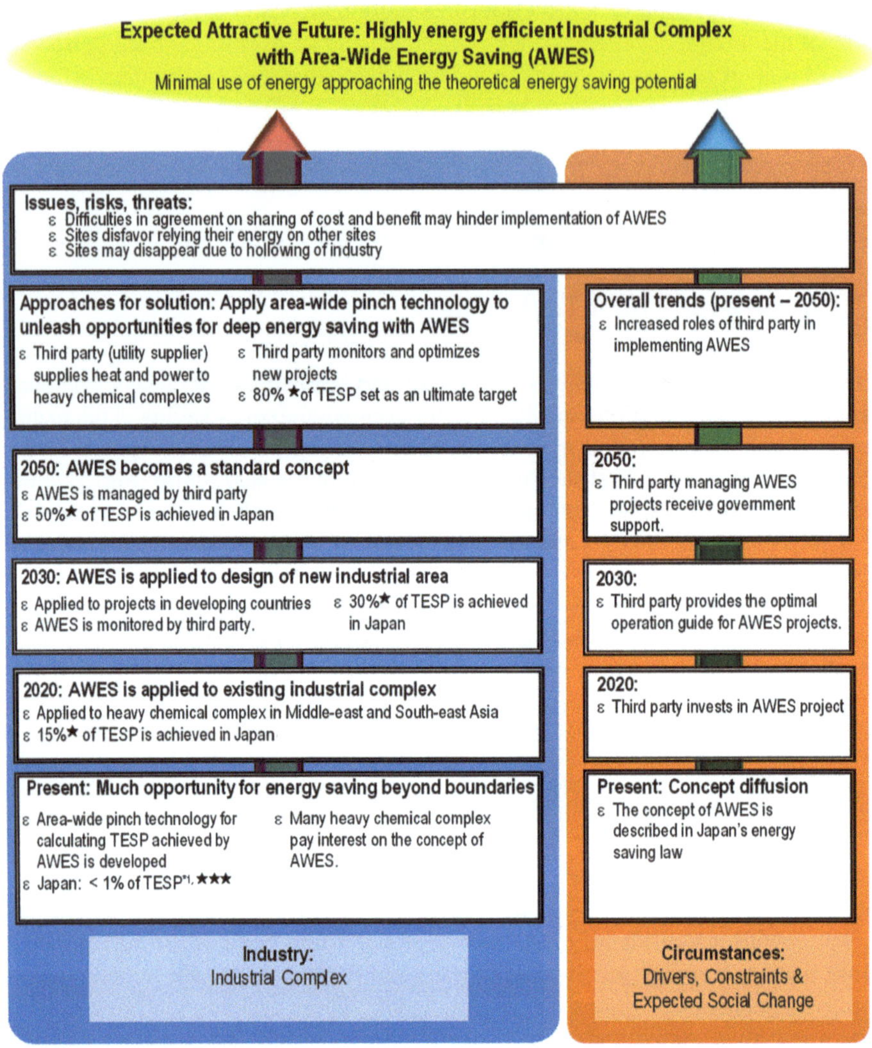

*1 NEDO project

Rating	Technology parameters
★★★	Solid mechanistic estimates
★★	Extrapolation of trends
★	Controversial value e.g.) personal perspective, competition with existing technology

AWES: Area-Wide Energy Saving
TESP: Theoretical Energy Saving Potential

Fig. 5 Road map

3.3 Social System, Regulation, and Other Considerations

The area-wide approach overcomes the limit for energy saving at single sites by treating multiple sites as a single virtual site. However, the legal system regulating Japanese energy saving has been slowly changing to follow the area-wide approach and still relies heavily on the traditional single-site approach. For example, piping regulation in the country has been strengthened to endure severe natural disasters. Once a site plans to install new piping connected with existing piping for energy saving and applies for a permit from the authorities, the latest regulation is applied to both piping, which stipulates replacement of the existing piping per that regulation. This extends the scope of work and increased project cost. Some of the latest regulations discourage implementing an energy-saving project. It has not been easy for sites to implement an area-wide energy-saving project, owing to unfavorable cost performance arising from the regulation and complicated contracting among sites as to how to allocate project profit among them. However, third parties have lately shown interest in investing in such collaborative projects, because the large scale of the projects will make them profitable for many years. Furthermore, the third parties will be able to maximize profit in area-wide energy-saving projects and readily calculate energy-saving amounts if they provide energy monitoring systems for optimal operation. The third parties are experts in solving such complicated contracts. In time, the government will provide support to promote area-wide energy-saving projects. Eventually, a strong network will be established to increase international competitiveness of heavy chemical sites in the complexes.

3.4 Issues

The installation of heat exchangers is necessary to achieve area-wide energy saving, and high-efficiency heat exchangers are critical to improve the economics. The safety design and operation of area-wide energy-saving projects are so important that deep mutual communication among sites is mandatory. There is a concern that some participating sites may withdraw from such projects, resulting in their failure. Any third party investing in and managing the projects will certainly need to consider and prepare countermeasures against such situations.

4 Benefits and Future Vision

The area-wide approach will overcome conventional barriers and achieve tremendous energy savings within heavy chemical complexes, as well as significantly reduce emissions of greenhouse gases. Techniques and know-how will be further developed by implementing area-wide energy-saving projects that will become a de facto standard in the world. Third parties will be able to invest in and manage these projects, and their support will optimize the operation of heavy chemical complexes. Owners of individual sites will be able to focus on investment in the process plants. Based on the success and profitability of energy-saving projects in large-scale complexes, the third parties will likely also turn their attention to optimization of clusters at mid-sized sites. Combination of a heavy chemical complex and other industry, such as steel, automobile, and others, will be expected to have much greater energy-saving potential. This is because the combination of sites in different types of business will be able to invite and identify new potential by sharing surplus heat of several grades, and compensate over- and short-capacity in utility systems of the sites.

References

1. Klemes J, Dhole VR, Raissi K, Perry SJ, Puigjaner L (1997) Targeting and design methodology for reduction of fuel, power and CO_2 on total sites. Appl Therm Eng 7:993–1003
2. Kenney WF (1984) Energy conservation in the process industry. Academic Press, London
3. Dhole VR, Linnhoff B (1992) Total site targets for fuel, cogeneration, emissions, and cooling. Comp Chem Eng 7(Suppl.1):s101–s109
4. Raissi K (1994) Total site integration. Ph.D. thesis, UMIST, UK
5. Matsuda K (2008) The development of energy sharing in industrial areas of Japan with pinch technology. J Chem Eng Jpn 41(10):992–996
6. Matsuda K, Hirochi Y, Kurosaki D, Kado Y (2015) Area-wide energy saving program in a large industrial area. Energy 90:89-94

Forestry and Wood Industry

Kazutake Oosawa, Yuichiro Kanematsu, and Yasunori Kikuchi

Abstract Excessively rapid socioeconomic changes in Japan have disrupted national policies related to the management of biomass resources. During the twentieth century, the Japanese forestry and wood industry transformed from producing wood for fuel to wood for construction to support national reconstruction in the postwar era. During this period, conifers were planted on most mountains. However, by the time sufficient conifer forests had grown to supply wood for construction, demand had declined because of the growth in imports. In the meantime, energy security and reduction of carbon emissions have become important issues, leading to the establishment of the "Biomass Nippon Strategy (2002)" and "Feed-in Tariff (FIT) Scheme for Renewable Energy (2012)," which have redirected interest back to the use of woody biomass for energy. However, the age class composition of wood resources in the forests is imbalanced because of the aforementioned transformation. Sustainable forest resource use demands a balanced structure and full use based on efficient collection of wood biomass generated in the supply chain, i.e., from forest to sawmill. It also demands the use of the best available technologies while maintaining other functions such as environmental conservation and watershed protection. In this study, we propose a road map to facilitate development of a sustainable forestry and wood industry that could satisfy demands for energy and construction.

Keywords Forest resource • Forestry • Energy • Construction wood • Harvested wood product • Sustainable

---

K. Oosawa (✉)
The University of Tokyo, Tokyo, Japan

Wakayama Prefectural Government, Wakayama, Japan
e-mail: oosawa_k0002@platinum.u-tokyo.ac.jp; oosawa_k0002@pref.wakayama.lg.jp

Y. Kanematsu • Y. Kikuchi
The University of Tokyo, Tokyo, Japan
e-mail: kanematsu@platinum.u-tokyo.ac.jp; kikuchi@platinum.u-tokyo.ac.jp

Y. Kato et al. (eds.), *Energy Technology Roadmaps of Japan*,
DOI 10.1007/978-4-431-55951-1_26

1 Introduction

Before 1950, domestic wood production in Japan was supplied for use as fuel more than for in construction. The fuelwood supply was about 30 million m^3/year, whereas that of construction wood was about 10 million m^3/year [1]. After the switch from wood fuel to fossil fuels and electrical energy consumption after 1950, the supply of fuelwood was limited to ordinary homes in rural mountain communities and for cooking in restaurants. Thus, the volume of fuelwood decreased to 6 million m^3/year by 1965. By contrast, the construction wood supply increased to 50 million m^3/year by around that year, because of strong national economic growth. However, this supply decreased to 16 million m^3/year because of large import volumes after abolition of the import surtax, strengthening of the Japanese yen, and a change in building construction from wood to concrete or steel. In the last few years, the demand for construction wood has slowly increased again, reaching 20 million m^3/year in 2012 after large sawmills began operation [2]. Moreover, the Japanese Ministry of Agriculture, Forestry, and Fisheries initiated the "Biomass Nippon Strategy" in December 2002, and public interest in woody biomass increased because of the high price of fossil fuel oil and promotion of forest sink thinning in accord with the Kyoto Protocol. This increased the demand for fuelwood after 2007, although the gross product from forestry was 143 billion yen/year in 2012, which meant that forestry only represented 0.03 % of the total national GDP.

In July 2012, the "Feed-in Tariff" (FIT) was initiated to promote renewable power generation in Japan [3]. Woody biomass was identified as one of the main options for renewable energy, and should be supplied to the power network over the next 20 years. Thus, this new demand for fuelwood is expected to grow, particularly for wood pellets and firewood for heating public facilities and apartment houses. The national production of wood pellets increased from 4,000 ton/year in 2003 to 98,000 ton/year in 2012 [4], and the potential of more heat from this source has been studied from existing cases in Europe. However, construction wood is considered a high-value-added product in the wood industry relative to fuelwood, even for power generation, in accord with the FIT. Thus, to enhance the feasibility of using large amounts of woody biomass, the added value in the wood industry should be kept at a reasonable and appropriate level. That industry has developed new products for construction, such as cross-laminated timber (CLT), the production of which is regarded as a key technology for increasing the demand for high-value-added wood. In May 2010, the Japanese government enacted a policy that prioritizes the use of wood in the construction of public facilities, i.e., the "Act for Promotion of Use of Wood in Public Buildings." As mentioned above, the utility of wood has increased because of potential wood power generation per the FIT, even though it is inefficient for this generation. Indeed, it should only be used for heating boilers, because its direct use in heating is more efficient. It is also important to increase the high-value-added use of wood as a construction material and in other applications.

In this chapter, we examine the major issues related to the forestry and wood industry in Japan. We discuss the functions of forestry in the country and the benefits of using wood as a construction material and fuel. We review the future sustainability of the forestry and wood industry in the form of a road map. Woody biomass can be described using various units, such as log-based or chip-based volume and weight. Some of these units are explained in the following sections.

2 Present Status

Some forests have a greater capacity for carbon fixation than others, and this depends mostly on the age of the trees. The carbon fixation capacity peaks in forest age class of 4–10[a], when the trees enter a period of rapid growth. If a forest is not used after this period, the growth rate declines with age. As a result, net annual greenhouse gas (GHG) emissions from the forest become positive when those emissions exceed carbon fixation at a certain point. The dynamic role of forests in carbon fixation has been increasingly incorporated into recent GHG accounting methods. For example, the 17th Conference of the Parties at the Intergovernmental Panel on Climate Change concluded that all changes in pools of harvested wood products should be considered in audits.

2.1 Forest Resources

Originally, forests had multiple functions, such as wood resource production, water resource conservation, landslide prevention, and soil conservation. Japanese forests have also been important in disaster prevention. Thus, wood resource production is only one role of the forests. Forest states should be changed from single storied to multistoried to ensure their continuing contribution to the public. Forests should also be maintained to ensure the production of high-value-added wood resources. Figure 1 shows how forest balance must be considered carefully.

In Japan, forests cover 25.10 million ha, or 68 % of the total land area (37.80 million ha). Forests are mainly in steep mountain areas with heavy precipitation. There are various forest zones because of the country's great north-south extent and complex topography. Planted forests are used mainly in forestry. The development of plantations throughout Japan peaked around 1960, and the area of planted forest comprised 41 % of the total forested area in 2014, or 10.30 million ha. Over half of these forests have entered the period of harvesting and use, because the plantations were established over 50 years ago. The Basic Plan for Forest and Forestry (July 2011) was established by the Japan Forestry Agency to determine desired future forms of the forests. This program aims to transform the age class composition of planted forests from the biased uniform state at present to a more heterogeneous distribution in 100 years, i.e., by 2110 (Fig. 2). In natural-regenerated forest, the

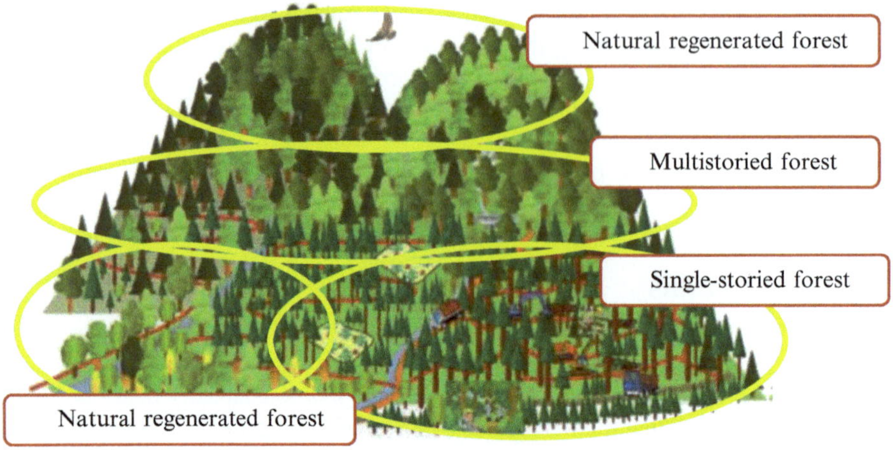

Fig. 1 Overview of forests over period of 100 years [4]

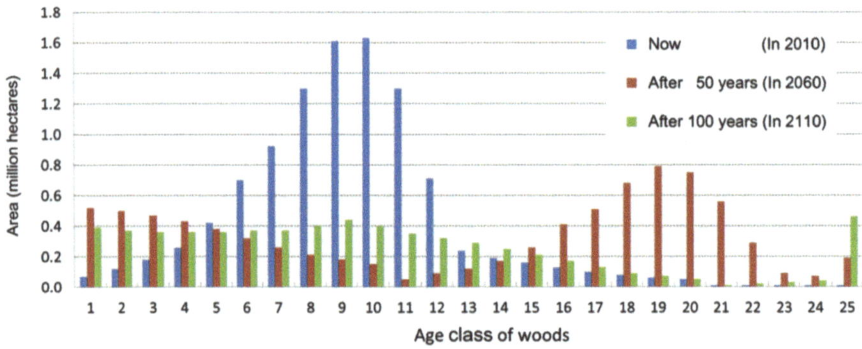

Note : The summed forest area is not given at each time, because a part of the single-storied forest will have been restructured as multistoried or natural regeneration forest. To avoid clear cutting, the forests should become multistoried, and forests of hard-regulated cutting should become natural regeneration forest.

Fig. 2 Age class composition of single-storied forest and the future of age class composition according to Annual Report on Forest and Forestry (goal state) [4]. Note: The summed forest area is not given at each time, because a part of the single-storied forest will have been restructured as multistoried or natural regeneration forest. To avoid clear-cutting, the forests should become multistoried, and forests of hard-regulated cutting should become natural regeneration forest

peak age class composition of planted forests is in the class of 12 [5], and this forest will be in the class of 19 by 2050. It is unlikely that regeneration will occur naturally, so the use of forests has become an urgent issue in the country.

As mentioned above, appropriate management of forests has improved the sustainability of forest resources. Those resources are living ones that change from year to year. Thus, a clear vision that considers appropriate forest management is required to ensure that wood biomass is used as part of the Japanese energy system.

Table 1 Forest state and goal forest states [5]

		Forest state				
		2010	2015	2025	2035	Goal forest state
Forest area	Million ha					
Single-storied forest		10.30	10.30	10.20	10.00	6.60
Multistoried forest		1.00	1.20	1.40	2.00	6.80
Natural-regenerated forest		13.80	13.60	13.50	13.10	11.70
Sum		25.10	25.10	25.10	25.10	25.10
Gross volume of wood	Million m^3	4,690	4,930	5,200	5,380	5,450
Volume per ha	m^3/ha	186	196	207	214	217
Gross growth of wood	Million m^3	74	68	61	55	54
Growth per ha	m^3/ha	2.9	2.7	2.4	2.2	2.1

Note: Single-storied forest: Forests used to supply timber are harvested by clear-cutting and grown artificially with a single-storied tree crown. In Japan, these forests are of Japanese cedar (*Cryptomeria japonica*), Hinoki cypress (*Chamaecyparis obtusa*), Japanese larch (*Larix kaempferi*), and other species

Multistoried forest: Forests used to supply timber are harvested by selective cutting and grown artificially with a multistoried tree crown. In Japan, these forests are of needleleaf and broadleaf trees that tolerate shade

Natural-regenerated forest: These forests are maintained by natural processes such as seeding and sprouting. Compositions of these forests vary with location, e.g., chinquapin (*Castanopsis*), oak (*Quercus*), beech (*Fagus crenata*), northern Japanese hemlock (*Tsuga diversifolia*), veitch fir (*Abies veitchii*), and Yezo spruce (*Picea jezoensis*)

To regulate forests, the Forest Agency has specified targets for forest management through 2110 (Table 1) in the Basic Plan for Forest and Forestry. As stated above, this program describes goals for forests over the next century. Table 1 gives a breakdown of the various forest states over time, including production levels. Appropriate forest management and forest resource utilization planning are required to ensure sustainable development of the forestry and wood industry. Energy production could be based on waste biomass, which is inevitably generated as an inexpensive by-product of sustainable forests, and this could enhance forestry in local areas. Depending on the demand for wood and energy, the scale of biomass energy systems should be designed appropriately, including, for example, power plants with FIT or local combined heating and power (CHP) generation.

2.2 *Wood Industry*

In Japan, the total volume of domestic and imported wood consumed was 71 million m^3/year[b] in 2012, of which self-supplied wood comprised about 28 %. This volume of wood included 20 million m^3/year[b] of domestic logs including forestland brushwood and 51 million m^3/year[b] imported logs, timber, and others. The number of new house starts, which constitutes the principal demand for timber, increased temporarily from 790,000/year in 2009 to 880,000/year in 2012 [4], but the number

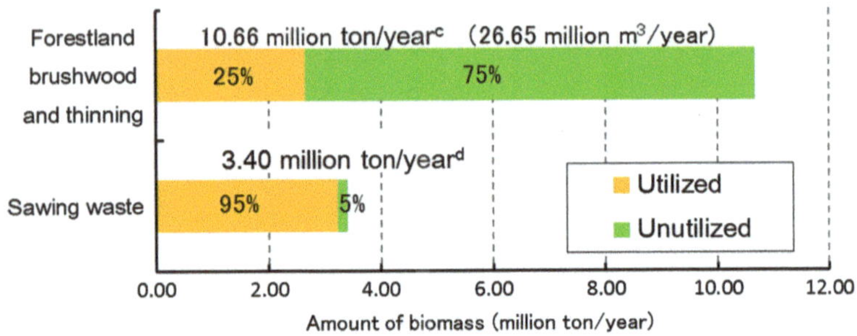

Fig. 3 Utilization of woody biomass in different states [4, 7]

of new wooden houses then decreased owing to an increase in the consumption tax rate in 2014. This explains the decline in the rate of new wooden housing, which currently makes up 55.1 % of all new housing. Areas producing normal-grade timber have increased in regions such as Kyushu, instead of in high-quality timber areas. In recent years, sawmills have been combined to establish large-scale sawmills.

The amount of forestland brushwood and unutilized thinned wood produced by forest management processes such as thinning has been estimated at 11 million ton/year[c], i.e., 27 million m^3/year in terms of log volume or 10 million GJ/year in terms of heat content as the lower heating value (LHV) (Fig. 3) [4]. The unutilized thinned wood is expected to be an inexpensive resource. If that wood were to be used instead of forestry wood production in the current situation, the cost of producing wood chips from thinning, including forestry, transport, and chipping, has been estimated at 17,526 yen/wet-ton [6]. Forestry wood production cost is about 7,000–11,300 yen/wet-ton [6][c]. The circulation price of domestic needleleaf wood chips for pulp is 12,200 yen/dry-ton [3], which is high compared with wood chips from thinning. Thus, it is presently necessary to exploit the unutilized thinned wood. Details on waste from sawing are considered later.

In the following, wood weight data address moisture content. In most cases, that content is on a dry basis, but we generally use the wet basis in this discussion.

2.3 *Energy Production from Woody Biomass*

Conventional utilization of wood involves heating in a wood dryer at sawmills or hot pressing to obtain laminated timber, among other uses. In 2014, the volume of sawmill waste such as shaving dust, sawdust, sawing remnants, and bark was estimated at about 3.40 million ton/year (Fig. 3) [4][d], or 33.32 million GJ of heat content as LHV. Sawmills and factories use sawing waste in woody-biomass boilers as a substitute for fossil fuels, thereby reducing GHG emissions and fossil fuel use. This system of heat utilization in sawmills and factories has become indispensable in the efficient utilization of resources, and for reducing production costs in medium-to-large sawmills that have sufficient funds for investment. In 2011, about 28.9 % [8] of sawmill waste was used in sawmills, and the remainder was supplied to the paper industry as wood chips.

In addition to its use in sawmills, woody biomass is used to heat direct-fired hot water boilers in public baths and agricultural heating, as well as for power generation in woody-biomass steam boilers and turbines. In Japan, some small power plants (tens to a few hundred kW) with gas engines have been examined in research studies, including wood gasification in CHP systems, which have an appropriate scale for the wood supply in local areas.

Large sawmills use woody biomass in steam boilers and turbines to generate electricity per the FIT scheme and supply heat to their own wood dryers and hot steam to other factories. In April 2014, there were plans for 81 woody-biomass power plants for power generation only, with total output 1,000,000 kW. Five power plants with outputs of 48,000 kW were in operation, excluding the plants built before FIT implementation. By 2016, it is expected that 52 power plants with outputs of 360,000 kW will begin power generation. Thus, it is expected that fuelwood demand will greatly increase. At present, the demand for fuelwood is estimated at 4 million ton/year[e] throughout Japan, for which it is assumed that a standard power plant with output 5,000 kW consumes up to 60,000 ton/year of fuelwood [9][e]. In addition, 38 % of the planned power plants have outputs 5,000 kW, with generation efficiencies below 20 %. However, the price of electricity produced by woody biomass is supported in the FIT scheme, so these power plants will still make a profit. In this scenario, the price of wood chip fuel from forests is estimated at 12,000 yen/ton [10][e]. This price will yield extra profits from chip production in sawmills, especially by using low-quality woods with curvature or worm damage. This chip price is too low to maintain chip production without the production of construction wood.

3 Technology Road Map

3.1 *Forest Management Activities*

The future status of forest resources is described in a road map for the age class composition in 2110 according to the Basic Plan for Forest and Forestry [5] and the Annual Report on Forest and Forestry (Fig. 2). In this road map, multistoried forests are expanded by replacing some single-storied forests, thereby providing multiple functions (including environmental conservation) by adjusting the biased age class composition in 6.80 million ha (Table 1). The remaining single-storied forests will be used to produce wood materials. In this basic framework, strategic and organized forest management actions are implemented in terms of planting and clear harvesting.

To adjust the age class composition, the supply of logs for sawing timber and making wood chips should be more than double than that of 2012, i.e., from 20 million m^3/year [2][b] to 39 million m^3/year[b] by 2020, and then to 50 million m^3/year[b] by 2030 [5]. Thus, the peak age class composition in 2050 will be in the class of 15–17 for the single-storied forests, which is the age of maturity for construction timber. In the future, with a shrinking or sustained population, ensuring that log demand matches supply must involve society-wide action to prioritize the use of domestic timber for houses, official buildings, furniture, interior material, and other uses. Therefore, the overall proportion of major-diameter wood will increase, and the supply chain will have to respond with adequate sawing machines and wood utilization measures. Challenges related to promoting timber use include constructing suitable infrastructure such as forest road systems, enhanced wood productivity by mechanization, recruiting and developing forest workers, and strengthening the capacity for sapling production.

3.2 New Technology for Wood Production

CLT is a new construction material used in medium-to-tall wooden buildings, which has generated new demand for wood. The Japanese Agricultural Standard for CLT was established in December 2013, and standards for buildings are being revised. CLT comprises large panels that are made by crossing and laminating wood plates at right angles (Fig. 4). CLT has favorable thickness and dimensional stability. About 5 million m^3/year[f] of CLT is manufactured worldwide, especially in Europe [11], where CLT is used in the construction of various buildings such as houses, apartment blocks, and high-rises (Fig. 5). The use of domestic wood may be influenced by construction activities in Tokyo in preparation for the 2020 Olympics, especially if CLT is used more widely. CLT has the potential to increase the demand for wood, which would enhance the forestry industry. In 2030, when log

Fig. 4 Cross-laminated timber [11]

Fig. 5 Apartments constructed from cross-laminated timber in Austria [11]

production will be at its maximum after its transformation by sustainable management, the wood demand should reach about 7 million $m^3/year^b$ in terms of plywood.

3.3 *New Technology for Wood Energy Production*

In fiscal year 2014, energy production from woody biomass was strongly dependent on support from the FIT scheme. At least 52 wood-biomass-based power plants will begin operation by 2016 [9]. Thus, 50 of 81 such plants will use domestic wood as fuel for their steam boilers and turbines [9], and fuelwood required for the plants will continue to be at least 4 million $ton/year^e$ for 20 years, throughout the period of the FIT scheme. Given these requirements and additional demand for newly planned power plants, the total demand for wood chips could reach about 15 million $m^3/year^b$, as well as about 10 million $m^3/year^b$ for fuelwood and 5 million $m^3/year^b$ for pulp wood in the FIT period. However, after this period, efforts will be required to maintain demand; otherwise, the supply chain for wood could collapse if demand decreases. The direct use of heat from biomass stoves or boilers could also increase fuelwood demand. However, exergy loss from the use of heat at low temperatures should be considered carefully. To increase exergy efficiency, gas engines or other types of cogeneration technology might be used. Based on these types of technology, local CHP generation systems have been introduced widely in advanced forestry countries throughout Europe, so these methods could maintain the demand for wood. One must anticipate a reduction in fuelwood demand upon termination of the FIT scheme, thereby ensuring a continuing demand. The Electricity Business Act and Heat Supply Business Act are the most important laws in this area, and they should be carefully examined to promote the diffusion of local CHP systems.

New types of firewood stoves represent another technology that could be used in Japanese households. The use of firewood, mainly from broadleaf trees in naturally regenerated and multistoried forests, can effectively reduce the use of fossil fuels in homes. It is important to promote regeneration by sprouting in naturally regenerated forests based on the use of firewood, thereby enhancing carbon fixation during natural regeneration. Most naturally regenerated forests are over 60 years old in Japan and have not been used since the shift in fuel use. Many of these forests have been damaged by insects or have failed to regenerate by sprouting. Therefore, it will be necessary to establish a system that involves moderate repeat harvesting, sprouting, and widespread firewood use in the country, based on well-balanced forest management.

4 Benefit and Prospective Future Vision

In Japan, most forest resources are near rural mountain communities where the population is declining. Revitalization of the forestry and wood industry is strongly related to conservation. Rural mountain communities cover 47 % of the total land area of the country [12]. Replacing local energy consumption based on fossil fuel with domestic wood could provide new employment opportunities and promote development of a new energy industry. In the past, energy expenditure has flowed out of the local economy, but this can be redirected to local industry and reduce local energy costs. By 2050, according to the road map, total energy utilization of domestic wood in the country will be 72 PJ/year [4, 5, 13]. If this wood energy could be converted into heavy oil A, the volume of that oil would be about 2 million L/year, with a value of about 160 million yen/year[g]. Thus, the energy industry and local economic circulation could be worth of about 160 million yen/year to the rural mountain communities.

Trees grow by photosynthesis using sunlight, i.e., carbon fixation. Thus, efforts to use domestic wood such as that sourced from timber forests and the use of wood in the construction of buildings and furniture can reduce GHG emissions. These efforts could also enhance carbon fixation levels throughout the world. For example, if a single house is built using wood, steel, or reinforced concrete, the amount of carbon fixed per wooden house is 6.0 tons. That for a steel-framed prefabricated house is 1.5 tons, and that for a reinforced concrete house is 1.6 tons, because of wood use in the interior or flooring materials. Therefore, a wooden house achieves four times the carbon fixation of other house construction types. Carbon emissions during the production of each material are 5.1 tons for a wooden house, 14.7 tons for a steel-framed prefabricated house, and 21.8 tons for a reinforced concrete house (Figs. 6 and 7). These values may vary depending on the calculation, but it is

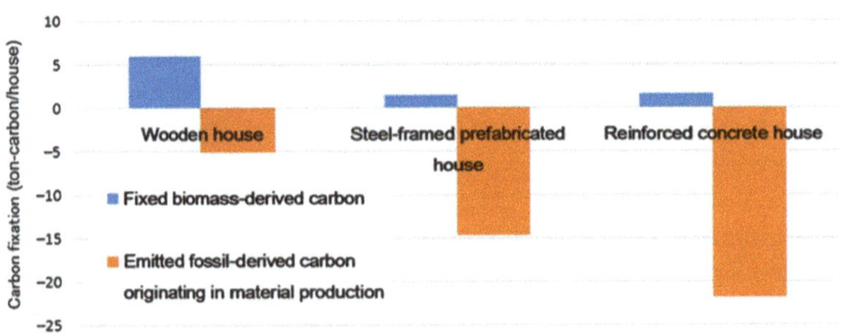

Fig. 6 Carbon fixation and carbon outputs in single house construction

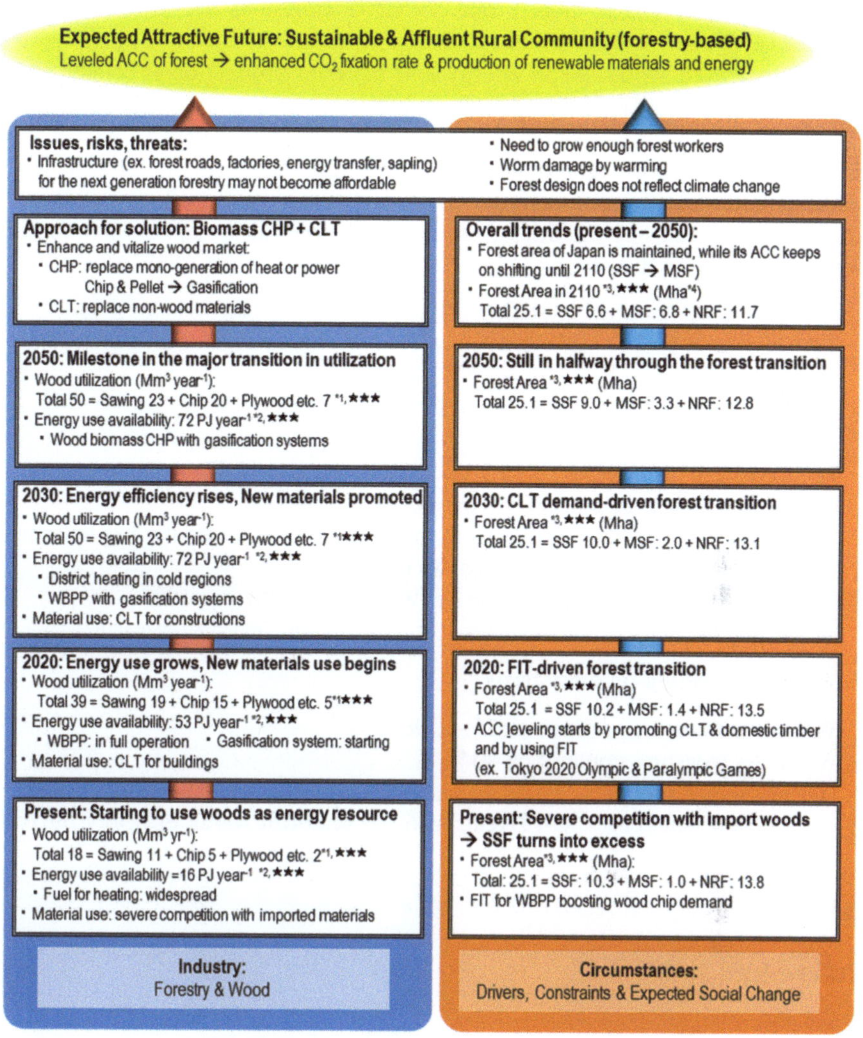

*1 Estimated from desired future state and supply presented in the "Basic Plan for Forest and Forestry [5]"
*2 Heat content of wood, calculated from volume of sawing waste (wet-base moisture content ca 15%) and wood chip (wet-base moisture content ca 40%) excluding those for paper industry [4, 5, 13]
*3 Estimated from desired future forms of forest in the "Basic Plan for Forest and Forestry [5]"
*4 Mega hectare, 10^4 km^2

Rating	Technology parameters
★★★	Solid mechanistic estimates
★★	Extrapolation of trends
★	Controversial value e.g.) personal perspective, competition with existing technology

ACC: Age Class Composition
CHP: Combined Heat and Power
CLT: Cross Laminated Timber
EC: Energy Conversion
FIT: Feed-In Tariff
MSF: Multistoried Forest
NRF: Natural regenerated Forest
SSF: Single-storied Forest
WBPP: Wood Biomass Power Plant

Fig. 7 Road map of forest, forestry, and wood industry

concluded that wood use greatly reduces GHG emissions and enhances carbon fixation [14]. Thus, it would be beneficial to promote the use of wood in many ways, because increased wood use in energy and material production is related to the regional economic cycle, reduced GHG emissions, and increased carbon fixation.

As explained in this chapter, increasing utilization of forest wood can be a driving force for the creation of new employment, greater energy self-sufficiency, and sustainable communities. Job creation may permit people who have moved to city areas for work opportunities to return to their rural mountain communities, thereby revitalizing them.

5 Notes

a. The forest age class represents a 5-year forest age period. Thus, the first age class is 1–5 years and the second 6–10 years.
b. Log-based volume transformed from the timber volume into the log volume
c. Wet basis moisture content rate: ca 40–60 %

$$\text{Wet basis moisture content rate} = (\mathrm{W}w - \mathrm{W}s)/\mathrm{W}w \times 100$$

$\mathrm{W}w$: Wood weight containing water
$\mathrm{W}s$: Wood weight on a dry basis

d. Wet basis moisture content: ca 15–20 %.
e. Wet basis moisture content: ca 40 %.
f. The volume of products
g. 160 million yen/year = 72 PJ year^{-1}/36 MJ (calorific value of heavy oil A) × 80 yen/L

References

1. Kato Y, Yasunaga Y, Kashiwagi T (eds) (2010) Jissoukano na Energy Gijyutsu de Kizuku Mirai – Honebuto no Energy Road Map 2 (Energy beyond '20 – Futures Built by Feasible Technologies)
2. Forestry Agency, Japan (2012) Mokuzai Jukyuuhyou (A balance sheet of woods) http://www.maff.go.jp/j/tokei/kouhyou/mokuzai_zyukyu/
3. Ministry of Economy, Trade and Industry, Japan (2012) Act on Special Measures Concerning Procurement of Electricity from Renewable Energy Sources by Electricity Utilities
4. Forestry Agency, Japan (ed) (2014) Shinrin Ringyo Hakusyo (Annual Report on Forest and Forestry in Japan) (in Japanese)
5. Forestry Agency, Japan (2012) Sinrin Ringyou Kihon Keikaku (Basic Plan for Forest and Forestry) (in Japanese)
6. Forestry and Forest Products Research Institute (ed) (2012) Kaitei Shinrin·Ringyo·Mokuzaisangyo no Syouraiyosoku (The prediction of forest, forestry and wood industry in future)
7. Forestry Agency, Japan (2014) Kanbatsu no Jisshijyoukyotou (A state of thinning) (in Japanese) http://www.rinya.maff.go.jp/j/kanbatu/suisin/zyokyou.html

8. Ministry of Agriculture, Forestry and Fisheries, Japan (2012) Mokuzai Ryuutuu Jyoukyou Tyousa Houkokusyo (Report of wood distribution structure 2011) (in Japanese) http://www.maff.go.jp/j/tokei/kouhyou/mokuryu/kouzou/#r
9. Ando N (2014) Miriyouzai no Kyoukyuubusoku ga Kenenn sareru Mokushitsu biomass Hatsududen – Tiikibetsusu Jyukyusuikeito Tenbou – (Wood biomass generation with worry of supply deficiency of unutilized wood, The estimate and perspective of supply and demand in local area) (in Japanese)
10. Ministry of Economy, Trade and Industry, Japan (2011) Zenryoukaitoriseido niokeru Mokusitsu Biomass Sigenbetsu Denryokutanka simulation Soukatsu (The summary of power cost simulation with each type of wood fuels in FIT/A paper of cost estimating committee for renewable power) (in Japanese) http://www.meti.go.jp/committee/chotatsu_kakaku/pdf/004_05_00.pdf
11. Japan Cross Laminated Timber Association (2014) What is CLT? http://clta.jp/clt/
12. Ministry of Agriculture, Forestry and Fisheries, Japan (2013) Sinkou Sanson no Genjyou (The state of rural villages) (in Japanese) http://www.maff.go.jp/j/nousin/tiiki/sanson/s_about/genzyo/index.html
13. A general incorporated association of improving and popularizing forestry Japan (ed) (2007) Mori no biomass energy (The forest biomass energy) (in Japanese)
14. Forestry Agency, Japan (ed) (2012) Shinrin Ringyo Hakusyo (Annual Report on Forest and Forestry in Japan) (in Japanese)

Agriculture

Yuichiro Kanematsu, Kazutake Oosawa, and Yasunori Kikuchi

Abstract The food self-sufficiency ratio in Japan based on caloric supply was 39 % in 2013, which is the lowest level among major developed countries. To achieve sustainable food supply in the future, state-of-the-art technologies should be combined to stabilize food production, increase productivity per resource, and ensure greater safety. This should be done even if it increases energy consumption. In a scenario of rising prices of fossil fuels, however, energy cost is one of the greatest risks for farmers. In addition to energy saving, energy production based on agricultural residue has the potential to revitalize agriculture. In this chapter, we review the current status of food production and energy consumption of agriculture. Based on this review, we suggest approaches to promote energy saving and production within agriculture that would stabilize and improve profitability for farmers. Energy saving would be effective and feasible in the field of protected horticulture by substituting waste heat and CO_2 from other industries and renewable resources for fossil fuels used in temperature control and CO_2 fertilization. To promote energy production by agriculture, a change of supply-chain management of resources from farmland is required for such a wide and thin distribution of resources. If agriculture can change to a more profitable and attractive business, new entries into agriculture would increase and food production would stabilize.

Keywords Combined food and energy production • Industrial symbiosis • Biomass • Bioenergy • Waste heat utilization

---

Y. Kanematsu (✉) • Y. Kikuchi
The University of Tokyo, Tokyo, Japan
e-mail: kanematsu@platinum.u-tokyo.ac.jp; kikuchi@platinum.u-tokyo.ac.jp

K. Oosawa
The University of Tokyo, Tokyo, Japan

Wakayama Prefectural Government, Wakayama, Japan
e-mail: oosawa_k0002@platinum.u-tokyo.ac.jp; oosawa_k0002@pref.wakayama.lg.jp

Y. Kato et al. (eds.), *Energy Technology Roadmaps of Japan*,
DOI 10.1007/978-4-431-55951-1_27

1 Introduction

As evidenced by traditional dietary cultures of Japanese "washoku" being designated as an intangible cultural heritage by UNESCO in 2013, Japanese traditional food culture has attractive features. However, recent dietary patterns in the country have been greatly Westernized since the 1950s, causing a discrepancy between food production and consumption. For example, meat consumption increased by a factor of 6 from 1960 (5.2 kg/year/capita) to 2010 (29.1 kg/year/capita) [1]. Domestic livestock production has also increased, but their feed largely relies on imported crops.

Such changes have reduced the food self-sufficiency ratio (SSR) in Japan. Moreover, the number of farmers in the country decreased from 2.9 million in 1990 to 1.9 million in 2010, and the proportion of an aged generation of farmers increased [2]. From a global standpoint, rapidly growing populations in developing countries will increase total food demand and can decrease the capacity of food imports to Japan and other countries. In such a situation, the food SSR of the country should be increased to improve the security and stability of the food supply.

From the energy aspect, energy saving and production by agriculture will be important in the near future. In Japan, agriculture and related industries such as starch plants are generally more energy consuming, and the price of energy resources is trending upward. In this chapter, we review the present status of food production and energy consumption of agriculture. Based on this review, we suggest approaches to promote energy saving and production within agriculture that would stabilize and improve profitability for farmers.

2 Present Status

2.1 Domestic Agriculture

The food SSR of Japan in FY 2013 was 65 % based on production value and 39 % based on caloric supply (Fig. 1). These ratios have persisted since about FY 2000 and are the lowest level among major industrial countries [3]. The Ministry of Agriculture, Forestry and Fisheries (MAFF) of Japan has set a target for the food SSR based on caloric supply to 50 % by FY 2020 [4] to improve the security and stability of the food supply. Achieving this target will require maximum effort on the part of all stakeholders and governmental support. Although monetary value is more common internationally for representing the food SSR, caloric supply is used in this chapter and in MAFF plans, because it reflects a more direct contribution to national health maintenance.

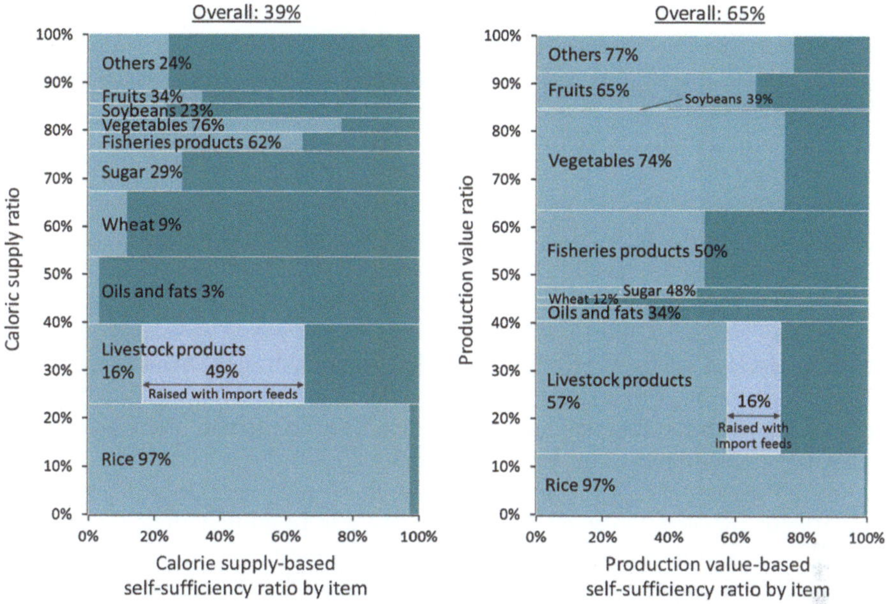

Fig. 1 Food self-sufficiency ratio in Japan (FY 2013), created based on data from [3]

The MAFF plan for improvement of food self-sufficiency is mainly aimed at increasing the production of wheat, soybean, and rice, which are expected to be grown by double cropping in existing paddy fields. The SSR of rice is already greater than 95 %, and the intent of its expansion is to replace imported wheat for bread or noodles with rice flour and livestock feed with forage rice.

2.2 *Energy Consumption and Production Within Agriculture*

Direct and indirect energy has been required for agriculture, in which technologies are used to stabilize food production, increase productivity per resource such as land area or materials (e.g., fertilizers), and maintain safety. However, in Japan, electricity cost has trended upward, caused by the cessation of nuclear power generation after the Great East Japan Earthquake in 2011. The country largely depends on imported fossil fuels. The price of them is unstable and has increased in the long term. Direct energy consumption by agriculture and forestry in the country was 131 PJ in FY 2007 [5], corresponding to 0.83 % of total national energy consumption. Direct consumption includes energy for operations on farms, e.g., fuel for farm machinery. Indirect consumption is associated with energy for the production and transportation of, for example, farm machinery, fertilizers, and pesticides. All energy consumption by agriculture, including direct and indirect, was estimated at 2–3 % of national total consumption [6], but these figures slightly vary with source data and calculation method. Although the impact of energy saving in the agriculture sector appears to be small relative to total national energy consumption, it would contribute substantially to the improvement of food production.

Agriculture can also be a supplier of energy via production from agriculture-derived waste and energy crops. Energy production from agriculture is only in the initial stages and is developing. Agriculture-derived waste mainly consists of livestock excreta and nonedible parts of crops.

Emission from livestock excreta in 2010 was 88 million tons/year. Although 90 % of this was used, most was used as compost and not as an energy resource. Power generation via biogas from methane fermentation has just begun experimentally on some farms. Transition from composting to methane fermentation keeping sufficient fertilization currently covered by compost is expected. The road map for achieving this is discussed in the next section.

Nonedible parts of crops such as rice straw, chaff, and pruned branches have not been effectively used. Emission from these parts was 14 million tons/year in 2010. Of this, 85 % of total amount was used: 30 % as compost, feed, livestock bedding, or fuel and 55 % as mixing in farms. The "Basic Plan to Promote Use of Biomass in Japan" (2010) [7] set a target to expand the utilization ratio to 90 % by 2020. This target includes increasing the ratio of conversion to energy.

3 Technology Road Map

3.1 Agricultural Production

Before investigating energy aspects in agriculture, forecasting of agricultural production is necessary, because the first priority of agriculture is to improve and stabilize food production. The first target is to achieve a food SSR of 50 % based on caloric supply in FY 2020, per the goal set by MAFF. After this target is achieved, the next one will be maintaining production amounts. Population trends should also be considered in forecasting food production. It is recognized as certain that the Japanese population will decrease to less than 100 million by 2050 [8]. If food production of the aforementioned 2020 target is achieved and maintained, the SSR will be increased by this population decrease. Estimates of population and food SSR based on these assumptions are shown in Table 1. The SSR will reach 64 % by 2050 in this estimation.

Table 1 Estimate of food self-sufficiency ratio based on its target and population trend in Japan

Year	2010	2020	2030	2040	2050
Population [million]	128.1	124.1	116.6	107.3	97.1
Food self-sufficiency (%)	39	50	53	58	64

To achieve these targets, certain measures must be taken. Achieving the SSR goal will require maximum effort from all stakeholders and governmental support as stated by the MAFF. To maintain the increased SSR against the decreasing number of farmers, one necessary measure is to increase that number, for example, by improving agriculture as a profitable and attractive business.

3.2 Energy Saving in Agriculture

Change or improvement in temperature control and CO_2 fertilization in protected horticulture has great potential to save on energy consumption. Energy consumption in agriculture can be separated into direct and indirect consumption, as mentioned above. Research has estimated energy consumption in wet-rice cultivation at 13.3 GJ/ha/year for direct consumption and 34.8 GJ/ha/year for indirect consumption [9]. Protected horticulture produces more products and consumes more energy than outdoor cultivation, mainly because of temperature control and CO_2 fertilization. Consumption rates differ greatly by cultivated species. Even for species with the least energy consumption, 500 GJ/ha/year is consumed directly and 540 GJ/ha/year indirectly. For species with the greatest energy consumption, 4,200 GJ/ha/year is consumed directly and 20,000 GJ/ha/year indirectly [10].

Oil-fired boilers are still widely used for heating but have been gradually replaced by heat pumps. Since the required temperature range of heat sources for greenhouse is less than 50 °C in heating with heat exchange and about 90 °C in cooling with absorption chiller, direct use of the combustion heat of fossil fuels involves substantial exergy loss. Utilization of low-temperature unused heat wasted from other industries can be an effective measure. Pipelines are generally used in such heat supply, but this requires rebuilding of infrastructure and geographical proximity between suppliers and consumers. If an area has difficulty in pipeline construction, heat storage and transport technology can be a future alternative. To realize penetration of heat storage technology in this purpose, improvement of heat storage density and simplification of heat storage/release operations are required.

CO_2 fertilization is currently sourced from the combustion of kerosene or heavy oil combined with air heating in most cases. A refined CO_2 cylinder is also used if necessary. Using CO_2 derived from agricultural waste toward reducing that waste or CO_2 from combustion process in other industries has the potential to reduce fossil fuel use and promote plant growth.

We propose the target for reduction of fossil resource consumption in temperature control and CO_2 fertilization as shown in the road map diagram at the end of this chapter. Such resource consumption can theoretically be reduced to zero, assuming that it can be replaced by unused excess heat and emitted CO_2 from other industries or by renewable energies. The 2050 target of 90 % reduction (over 2010) was set by assuming that required technologies are sufficiently established and with consideration of some sites that are inadequate to apply alternative energies or CO_2.

As an additional topic, plant factory (also called as indoor hydroponics), which has the potential to greatly affect future agriculture, has gradually become widespread in Japan. Although plant factories with artificial light consume significantly greater energy for the light source and temperature control and cultivable plant types are currently limited, there are many advantages, e.g., production stability against unstable weather, land use efficiency, and resource use efficiency. Water use will be 1/50 and fertilizer use will be 1/2 per unit production compared to conventional protected horticulture [11]. The aforementioned disadvantages regarding energy consumption will be reduced or eliminated by integrated use of waste heat and CO_2 from other industries and renewable resources.

3.3 Energy Production from Agriculture-Derived Resources

Livestock excreta can be transformed into a resource for methane fermentation. Most excreta is currently used as compost. By transferring them to methane fermentation, methane gas as an energy source and digested sludge that can be used as liquid fertilizer are simultaneously produced. Compatibility between compost and digested sludge should be demonstrated appropriately by a feasible application method of that sludge. Total-system design of biogas-based power generation considering the demand and usage of digested sludge should be addressed. If digested sludge can replace chemical fertilizer as well as compost, resources and energy for chemical fertilizer can be saved indirectly; in particular, phosphorus is a critical substance for food production [12].

Utilization of nonedible crop parts can be improved by solving the problem of resource collection from distributed sites. Good practices of resource collection have already been implemented, for example, the utilization of sugarcane bagasse in the sugar industry. Bagasse is a residue from sugarcane, which is generated in large quantities by the juice extraction process, and has been used as fuel for in-house cogeneration plants in sugar mills for self-supply of heat and power demand. In other words, fuels are carried and collected as part of producing the principal product (sugar) at sugar mills. Agricultural cooperatives, known as Japan Agricultural Cooperatives (JA), or wholesale markets can be new collection sites when similar means of transport are implemented for other crops. Although this involves an overall increase in transportation and requires consensus building with stakeholders to cooperate in this activity, there would be substantial benefits by enabling the collection of resources at centralized sites, without establishing new transportation routes.

For power generation use of dry nonedible crop parts, sharing the energy plant with forestry-derived biomass, which is cellulosic biomass similarly, is also worth considering. This can mitigate the influence of quantitative instability caused by seasonal variation or bad weather and can improve generation efficiency by scale-up of the power plant. This approach would be effective in the case where the amount of agriculture-derived biomass is not sufficient for use as fuel for the power plant. We should also consider the appropriate scale of the power plant for balancing the supply and demand of resources. Many projects failed in the past because of power plant overcapacity and shortage in procurement of fuels. It must be emphasized that maximizing energy production amount should not be the main object. Quantitative controllability of biomass-fired power generation should also be strongly valued in a local power grid. This characteristic is essential because photovoltaic (PV) and wind power generation are unsteady and depend on weather conditions. Some major Japanese power companies temporarily refused the purchase of electricity from renewables in 2014. This was because too many suppliers based on PV power rushed into the electricity market by the start of feed-in tariff (FIT) in 2012 and the existing grid reached overcapacity. In such a situation, power generation based on biomass will be important in a local power grid because of its controllability, especially after liberalization of the electricity retail market beginning in 2016.

The plan for biomass utilization should be based on activities of existing primary industry and management of its wastes and by-products, through reasonable energy recovery or material recycling. Since power generation efficiency is generally <20 % for steam turbines and <45 % for advanced gas engines, heat recovery with cogeneration is desirable to minimize energy loss. Recovered heat can be provided to greenhouses, plant factories, or drying processes of wood production, if facilities are located near the power plant. In this regard, location planning is important. Organization to integrate supply-chain and energy management is essential in both the establishment of new logistics of resources and collaboration of multiple industries.

The treatment of abandoned farmlands by considering various land conditions is also necessary as an agriculture-related matter. Abandoned farms increased from 135,000 ha in 1985 to 396,000 ha in 2010 [3]. Construction of PV generation panels on abandoned farmlands is a possible alternative. For lands that are still suitable for plant cultivation, crops that can grow in the shade can be raised under those panels. For lands that cannot be recovered for farming, simply constructing PV panels will be applied. Cultivation of energy crops on abandoned farmlands has also been considered for biofuel production.

In this article, we intentionally did not set a target for the amount of energy production from agriculture, because that production should be positioned as a by-product of food production. Our desired target for farmer participation rate in energy production is 90 % by 2050. Reconsideration and design of subsidy systems and legal restrictions are also important. For example, the FIT price setting in 2014 gives no advantage to heat utilization in biomass-derived power. Moreover, the obligation to supply registered amount of heat in Heat Supply Business Act in Japan has been a substantial barrier to starting a heat supply business based on biomass fuel.

4 Benefit and Attractive Future Vision

To achieve sustainable food supplies in the future, state-of-the-art technologies should be combined to stabilize food production, increase productivity per resource such as land area or materials (e.g., fertilizers), and ensure safety, even if this increases energy consumption. For achieving such changes in agriculture, simultaneous promotion of energy saving and production is strongly needed. Although energy consumption by agriculture and forestry is less than 3 % of the total consumption in Japan, energy saving and production in agriculture would enhance the profitability of food production and energy self-sufficiency of local communities. Additional income can be gained by selling the excess energy during seasons when energy consumption of farming activity is less than energy production. Such incomes can be a subsidiary financial resource against crop failure. The stabilization of local communities would revitalize agriculture and stabilize food production.

Full liberalization of the electricity retail market in Japan beginning in 2016 will boost such energy production based on agriculture and other primary industries. The new profit and energy can increase agricultural production by new investment. Using the produced energy for food processing such as cutting, converting, packing, refrigeration, and heating will add value to produced foods. Although agriculture is partly recognized as a low-profit business in the country, improvement of profitability by energy saving and production would increase the number of new farmers and might make it possible to achieve food self-sufficiency beyond current targets.

To promote energy saving and production in agriculture, the establishment of technologies and local management models is required. An example of such technology is heat storage and transport to recover excess low-quality heat from other industries. A CO_2 capture system with lower cost is desirable for recovered CO_2 fertilization. The establishment of local management models is needed for integrating local energy and supply chains including food products (Fig. 2).

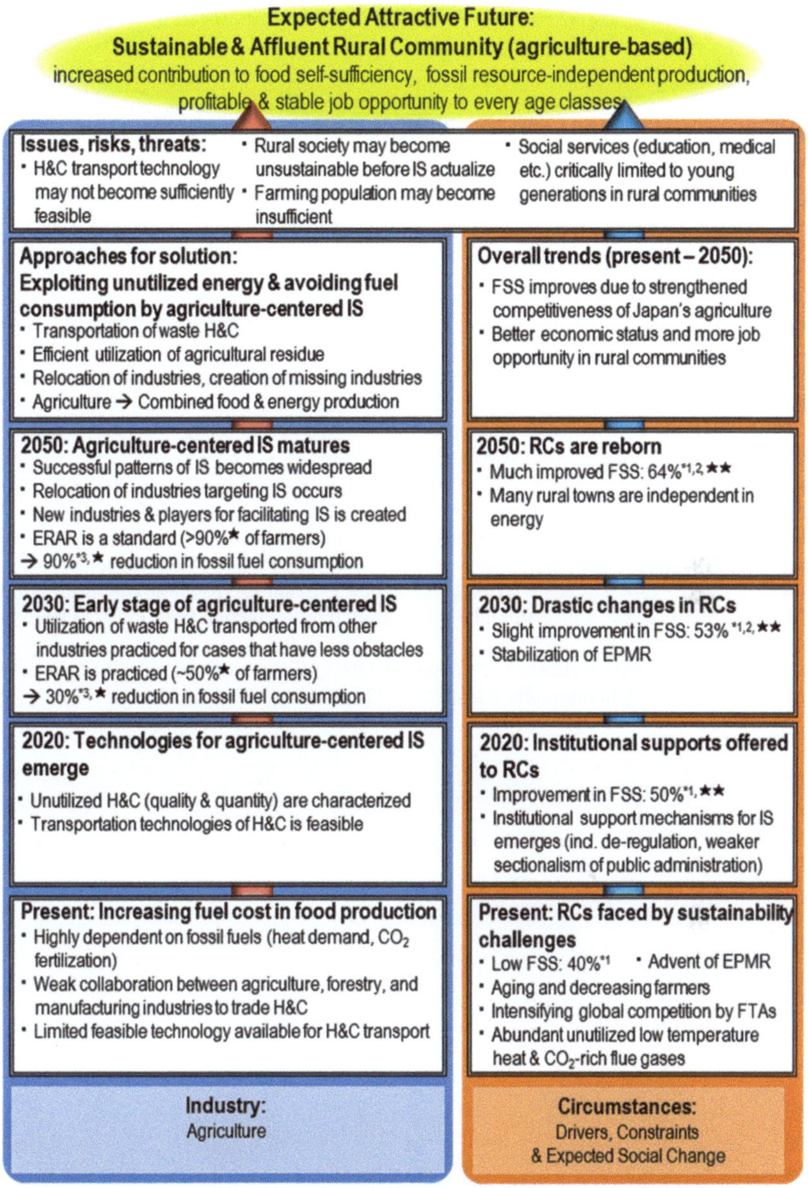

*1 Ministry of Agriculture, Forestry and Fishery (2012), Syokuryou Jukyuuhyou (Food Balance Sheet) (*in Japanese*), *2 National Institute of Population and Social Security Research (2012), Population Projections for Japan: 2011 to 2060, *3 On the basis of energy consumption for temperature control and CO_2 fertilization at present

Abbreviations: EPMR: Electric Power system Market Reformation, ERAR: Energy Recovery from Agricultural Residues, FSS: calorie based Food Self Sufficiency, FTAs: Free Trade Agreements, H&C: Heat & CO_2, IS: Industrial Symbiosis, RC: Rural Community

Fig. 2 Road map of agriculture

References

1. Ministry of Agriculture, Forestry and Fishery (2012) Syokuryou Jukyuuhyou (Food Balance Sheet) (in Japanese)
2. Ministry of Agriculture, Forestry and Fishery (2010) Nougyou Kouzou no Tembou (Outlook of structure of domestic agriculture) (in Japanese)
3. Ministry of Agriculture, Forestry and Fishery (2014) Annual Report on Food, Agriculture and Rural Areas in Japan FY2013
4. Ministry of Agriculture, Forestry and Fishery (2011) Annual Report on Food, Agriculture and Rural Areas in Japan FY 2010
5. Ministry of Economy, Trade and Industry (2008) Comprehensive energy statistics FY2007
6. Imou K (2009) Energy saving in agriculture. Mar Eng: J Jpn Inst Mar Eng. doi:10.5988/jime.44.719
7. Ministry of Agriculture, Forestry and Fishery (2012) Baiomasu Katsuyou Suishin Kihon Keikaku (Basic plan to promote use of biomass in Japan) (in Japanese)
8. National Institute of Population and Social Security Research (2012) Population projections for Japan (January 2012): 2011 to 2060
9. Saga K, Yokoyama S, Imou K (2007) Net energy analysis of bioethanol production system from rice cropping. J Jpn Soc Energy Resour 29(1):1–6
10. Inoue K (1998) Nougyou ni Okeru Enerugi Riyou no Genjou to Tembou (Present status and prospects of energy use in agriculture). JATAFF J 21(10):13–21 (in Japanese)
11. Kozai T (2012) Jinkoukougata Syokubutsukoujou – Sekai ni Hirogaru Nihon no Nougyou Kakumei (Plant factory with artificial light – agriculture innovation from Japan to the world -) (in Japanese)
12. Matsubae K, Kajiyama J, Hiraki T, Nagasaka T (2011) Virtual phosphorus ore requirement of Japanese economy. Chemosphere 84(6):767–772

Waste-Derived Energy

Ryo Moriyama

Abstract Waste is an inevitable by-product of human life and is generated by many sources, including residential, commercial, and industrial. In Japan, the amount of generated waste is decreasing through promotion of the 3R (reduce, reuse, recycle) policy and through population decline associated with aging. However, many issues remain in waste management, such as energy recovery from the incineration process, economic and environmental optimization for recycling systems, and residual life of final disposal sites. This article outlines the present status and issues of waste management in Japan, using various statistical data. Furthermore, a technology roadmap is developed based on Japanese waste management policy and development technologies, toward building a sound material-cycle society.

Keywords Waste management • Sound material-cycle society • Energy recovery • Waste power generation • Semi-aerobic landfills

1 Introduction

The first step in building a sound material-cycle society[1] is to understand material flow in the economic sector in terms of resources extracted, consumed, and disposed of.

Figure 1 shows all material flow in Japan in FY 2000 and FY 2011. The amount of natural resources (total domestic and imported) input declined to about two-thirds the FY 2000 amount, i.e., from 1,925 to 1,333 Mt. The amount of final disposal was reduced to about one-third, i.e., 56 Mt to 17 Mt. The proportion of final disposal waste to total materials input declined from 2.6 to 1.1 % as a result of 3R promotion. It can be argued that Japan is progressing toward a sound material-cycle society.

[1] A society in which the consumption of natural resources is minimized and the environmental load is reduced as much as possible

R. Moriyama (✉)
The Institute of Applied Energy, Tokyo, Japan
e-mail: rmoriyama@iae.or.jp

Y. Kato et al. (eds.), *Energy Technology Roadmaps of Japan*,
DOI 10.1007/978-4-431-55951-1_28

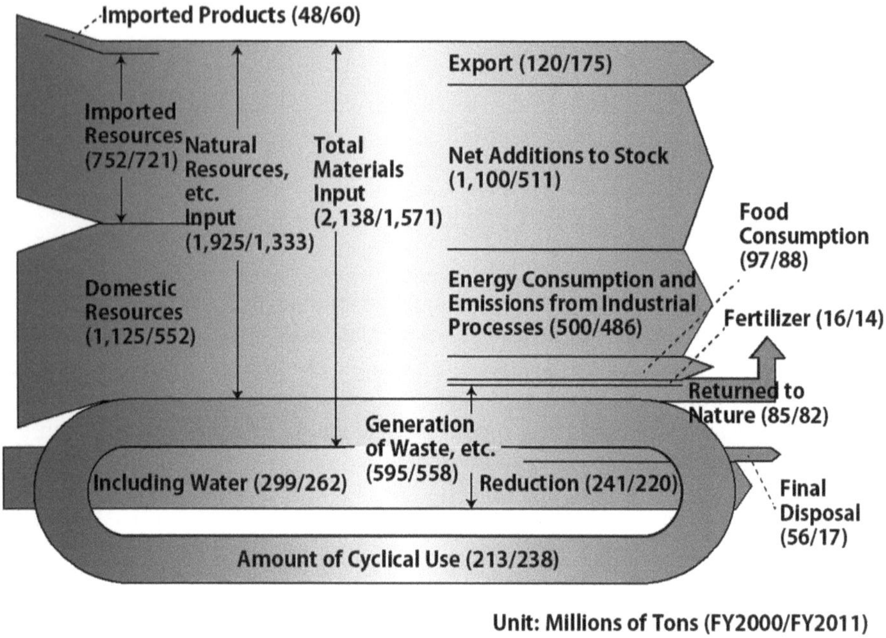

Fig. 1 Full material flow in Japan [1]

The amount of generated waste is expected to decrease in the future because of population decline associated with aging and business and lifestyle changes.

This article describes potential energy conservation and recovery in the waste management industry, providing an overview of that management in Japan. The first part outlines changes in waste management policy. The history of legal systems regarding the development of a sound material-cycle society (postwar period to present) is outlined.

In the second part, major trends and issues of waste management are described. Particular issues in Japan are energy recovery from the incineration process, optimization of recycling systems in terms of the environment and economy, and residual life of final disposal sites.

Finally, a technology roadmap is created with the concept of "green economy" and/or "green growth," which pursue sustainability from an environmental, economic, and social perspective.

2 Present Status

2.1 Waste Treatment and Recycling Policy

To understand the present status of waste management in Japan, it is necessary to describe the development of the legal system related to waste treatment and recycling, with its concepts and circumstances.

Modern system construction of postwar waste management began with the enactment of the *Public Cleansing Law* in 1954. The main purpose of this law was to improve public health by sanitary disposal of waste and cleaning the living environment. In the 1960s, the amount of generated waste increased considerably, particularly in urban areas, owing to population influx. Nearly 40 % of generated waste was disposed of in unlined landfills or dumped in mountainous areas, rather than in soil-covered landfills. Given this situation, the *Urgent Measures Law on Capacity Increasing of Waste Management Facilities* was promulgated in 1963. Based on this law, the introduction of waste incineration facilities was promoted.

The problem of household waste was not the only issue that arose from Japan's vigorous economic growth. Another was the problem of industrial waste generated by production activities of businesses entities. The *Public Cleansing Law* was completely revised in 1970 to become the *Waste Management and Public Cleansing Law*, which classified waste into "general" and "industrial." The law placed responsibilities for the treatment of waste from business activities on businesses and that from households on local governments.

Eventually, the Japanese social system transformed from mass production and consumption to recycling oriented after two oil shocks in the late 1980s. Emissions of harmful substances by incineration, especially dioxin,[2] have been a social problem because various products (including chemical and electrical) are mixed with household waste.

Since 1997, based on the *Guidelines for the Prevention of Dioxin Emissions from Waste Management*, *Air Pollution Control Act* and the revised *Waste Management Act*, the Japanese government has developed a variety of emission control measures, such as controlling dioxin emissions at incinerator stacks and improving waste incineration facilities.

In 1999, with a view toward preventing environmental pollution by dioxins and removing them from the environment, the Japanese government enacted the *Act on Special Measures against Dioxins*. Because of the development of emission control technologies and waste incineration facilities as well as a tightening of control regulations, dioxin emissions from waste incineration facilities were reduced in 2011 by ~99 % relative to 1997.

[2] Dioxins are substances (by-products) that are naturally generated by incineration processes. Among ~200 dioxins, 29 are regarded as toxic. Dioxins are currently generated by a variety of sources, such as electric furnaces for steel production, cigarette smoke, and automobile exhaust gas, but the major source is waste incineration.

- ***Basic Environment Law*** **(1994)**
  - ***Basic Act on Establishing a Sound Material-Cycle Society*** **(2001)**
    - ***Waste Management and Public Cleansing Law*** **(1971)**
    - ***Law for the Promotion of Effective Utilization of Resources*** **(1991)**
      - ***Container and Packaging Recycling Law*** **(1995)**
      - ***Home Appliance Recycling Law*** **(1998)**
      - ***Food Waste Recycling Law*** **(2001)**
      - ***Construction Material Recycling Law*** **(2002)**
      - ***End-of-Life Vehicle Recycling Law*** **(2005)**
      - ***Small Waste Electrical and Electronic Equipment Recycling Law*** **(2013)**

(Promulgated year)

Fig. 2 Legal system for establishing a recycling society

In the 1990s, the *Waste Management and Public Cleansing Law* was amended three times for stricter regulations on waste disposal facilities, creation of a certification system in accord with waste recycling, prohibition of open-air waste incineration, and others. As legislation toward the building of a sound material-cycle society, the *Law for the Promotion of Effective Utilization of Resources* was promulgated in 1991. Based on that, there are six laws pertaining to recycling individual waste (Fig. 2).

Furthermore, to promote the establishment of a sound material-cycle society designed to ensure implementation of 3R and proper waste management, the government issued the *Basic Act for Establishing a Sound Material-Cycle Society* in 2001. This law provides a clear vision for a sound material-cycle society, which has the following order of priority for resource recycling and waste management: (1) reduce waste generation control, (2) reuse, (3) recycle materials reclamation, (4) thermal recovery, and (5) appropriate disposal (Fig. 3). This means that the highest priority action is to avoid waste generation.

The following section describes the current situation of waste treatment and recycling in Japan, which is progressing toward a sound material-cycle society.

2.2 Current Situation of Waste Treatment and Recycling

Waste is classified as general and industrial, as described above. General waste is composed of night soil, household waste, and business waste from eateries and offices. Industrial waste is generated by business activities and is of 20 types, including cinders, sludge, and waste oil, as defined by the *Waste Management and Public Cleansing Law* (Fig. 4). This article covers general and industrial waste, except night soil and specially controlled wastes.

Figure 5a, b shows changes in the amount of waste treatment for general and industrial waste, respectively. In Fig. 5a, the amount of final disposal decreases from 11 to 5 Mt as a result of decreasing the amount of waste generation and

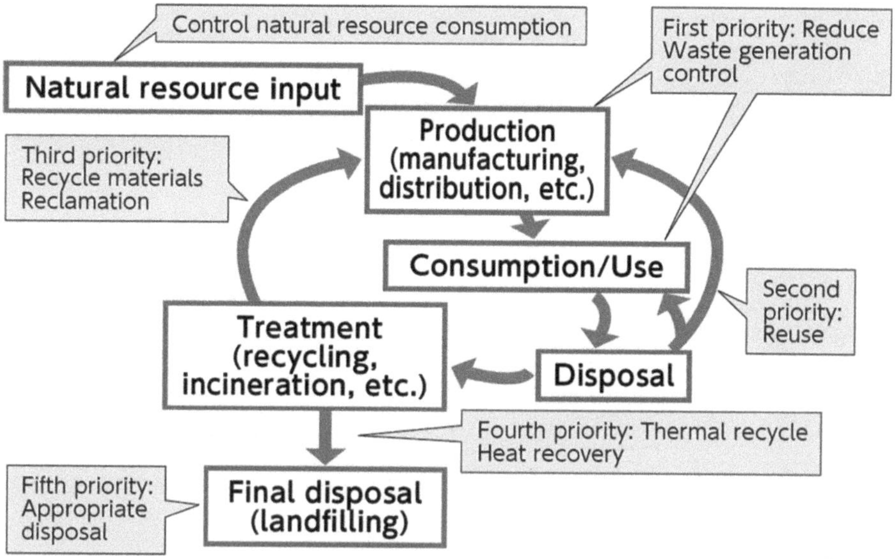

Fig. 3 3R concept in a sound material-cycle society [2]

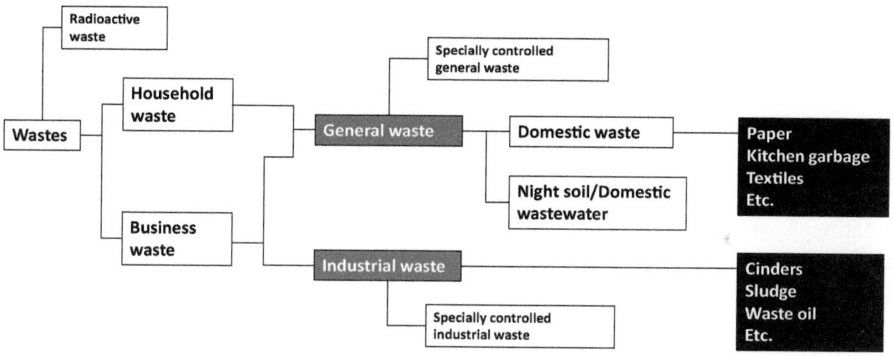

Fig. 4 Classification of wastes

increasing that of recycling. In Fig. 5b, the amount of final disposal also decreases, from 45 to 12 Mt, as a result of increasing recycling.

Although recycling has progressed, the major method of waste treatment is incineration for its volume reduction. In the incineration process, greenhouse gas (GHG) is generated along with thermal energy from the waste. Recovery of that energy is important for efficient energy use.

Regarding GHG emissions from the waste treatment industry, CO_2 emissions by incineration were 12 Mt in FY 2011, which is ~1 % of the national total (Fig. 6). These emissions were mainly from combustion of petroleum-derived waste plastics and waste oil. Furthermore, 4.8 Mt-CO_2eq (23 % of national total emissions) of

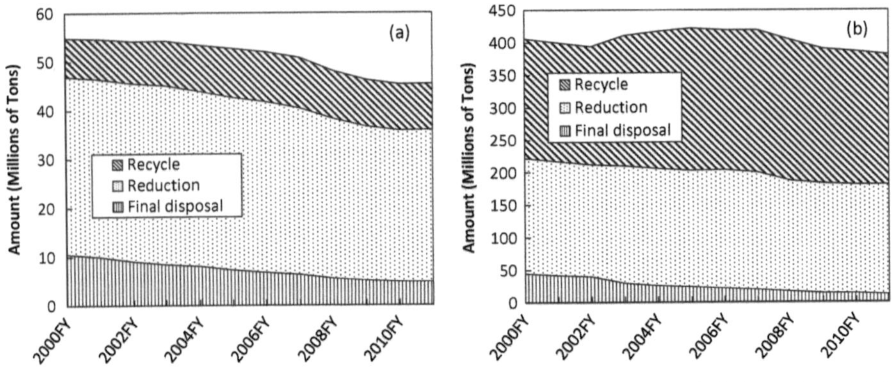

Fig. 5 Changes in the amount of waste treatment for (**a**) general waste and (**b**) industrial waste [3]

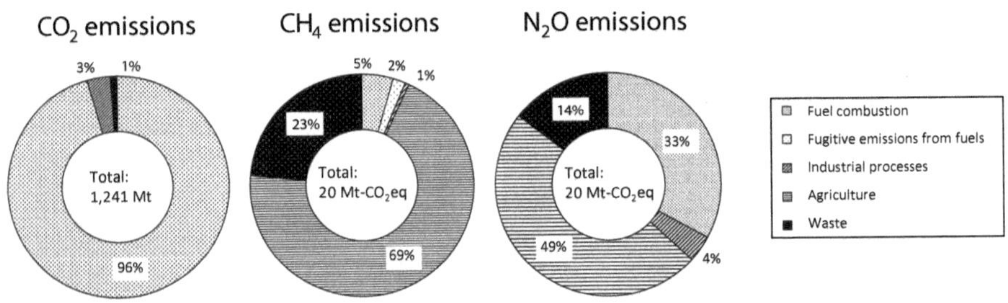

Fig. 6 GHG emissions of Japan and share of waste industry [4]

methane was generated from waste treatment such as landfilling, wastewater treatment, and composting. For nitrous oxide (N_2O), 3.0 Mt-CO_2eq (14 % of national total emissions) was generated from wastewater treatment and incineration [4].

To decrease GHG emissions from waste treatments, incineration and final disposal should be reduced. In addition, recycling industry activities emit GHG in the transportation sector through collection and transportation of waste and energy consumption during the recycling activity.

The current situation and issues in waste treatment such as incineration, recycling, and final disposal are described in the following section.

2.3 *Treatment of General Waste*

2.3.1 Energy Recovery from Incineration Process

Concurrent with the construction of a sound material-cycle society, recovery and effective use of thermal energy from incineration have been promoted.

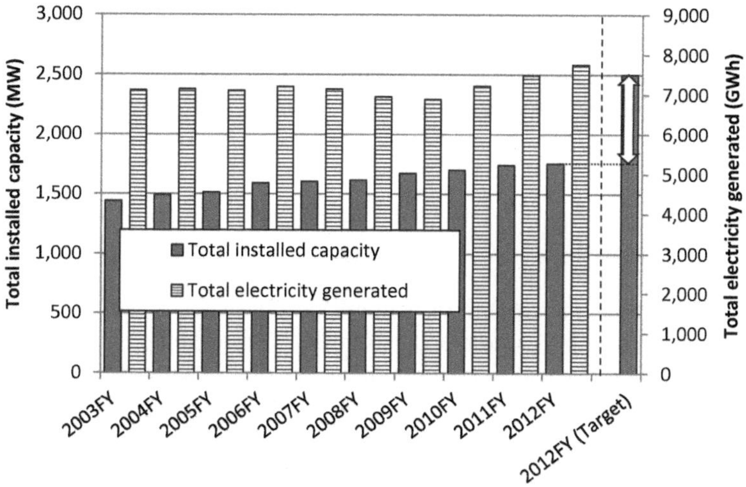

Fig. 7 Changes in total installed capacity and total annual electricity generated for general waste power generation in Japan [3, 5]

Figure 7 shows changes in total installed capacity and total annual electricity generated for general waste power plants.

The Waste Disposal Facility Development Plan [6], approved by the Cabinet in 2008, set an agenda for the promotion of facility development, with consideration of the prevention of global warming and a target for total power generation capacity from waste incineration facilities at 2,500 MW in FY 2012. However, as shown in Fig. 7, the total installed capacity was only 1,750 MW and that target was not reached.

One of the main reasons for this underachievement is aging incineration facilities. According to survey results from the Ministry of the Environment [7], about 50 % and 20 % of incineration facilities are older than 20 and 30 years, respectively. Such aging facilities tend to operate without power generation equipment.

Here, we look at the potential and current state of energy recovery from the incineration process. As shown in Fig. 5a, the reduction by incineration for general waste is 35.2 Mt. Based on data from each general waste incineration facility [3], the average caloric value of incinerated waste is estimated at 7,510 MJ/kg. Energy input to the incineration process is calculated at ~263 PJ, which is 1.2 % of Japanese primary energy supply.

According to data from the facilities [3], 27.6 PJ of thermal energy and 7,750 GWh (27.9 PJ) of electricity are recovered and used in the incineration facilities or supplied externally. The total energy recovery ratio for incineration facilities is calculated at 21 % and is potential for additional energy recovery.

There is a technology called "super waste electric power generation," which enhances power generation efficiency. The technology includes a gas turbine in a conventional waste power generation system and can reheat steam from incineration to realize >25 % power generation efficiency. Owing to problems such as high construction costs and increasing fuel prices, this technology has only been adopted in a few locations.

To increase power generation by promoting energy recovery from incineration, it is necessary to update and improve aging facilities with a long-term equipment implementation plan.

2.3.2 Final Disposal

As mentioned above, the amount of final disposal has been greatly reduced by the development of the legal system and progress of waste treatment efforts, but the residual life of final disposal sites for general waste in Japan was 19.4 years in FY 2011 [3]. It is necessary to continue efforts to diminish final disposal and develop new disposal sites.

2.4 Treatment of Industrial Waste

Illegal dumping has decreased in recent years by means of tightened regulations. Nevertheless, illegal dumping of industrial waste is a serious issue with harmful effects on the environment.

Other issues regarding the treatment of industrial waste are the residual life of final disposal sites and energy recovery from the incineration process. The residual capacity of final disposal sites for industrial waste is more restricted than that for general waste. The residual life of final disposal sites remained in a severe state in FY 2011. Although this life in Japan is 14.9 years, it is only 5.3 years in areas around Tokyo [3].

Power generation in industrial waste incineration facilities is mainly used for self-consumption by paper-manufacturing companies. The principal fuel for this generation was black liquor from the pulping process in the 1990s. This fuel has been in transition to industrial waste such as wood waste, construction waste, old tires, and recycled plastic fuel during the 2000s. The total installed capacity for industrial waste power generation was 1,100–1,200 MW in FY 2012, less than that for general waste power generation [8].

2.5 Material and Chemical Recycling

As mentioned above, recycling of waste is progressing for the building of a sound material-cycle society by means of recycling-related laws. In this section, resources used for recycling are discussed without distinction between general and industrial waste.

The recycled ratio of paper increased from 57 % in 2000 to 63 % in 2011 [9]. For waste plastic, the total recycled ratio including thermal recovery, material, and chemical[3] recycling increased from 50 % in 2000 to 78 % in 2011 [10]. The recycled ratios for other waste such as cans (from 81 to 93 % for aluminum cans), bottles (63–71 % for glass bottles), food waste, and home appliances have also been increasing [11].

It is generally believed to be possible for recycling to eliminate resource and fuel consumption required to make new products. However, it is necessary to establish an effective recycling system that achieves environmental and economic sustainability, as recycling activities also generate GHG and have some costs.

GHG eliminated by recycling efforts are:

1. GHG emitted from landfills or incineration of waste
2. GHG emitted by making new products

GHG emitted by recycling efforts are from:

4. Waste separation, collection, and transportation
5. Making recycled products

It is necessary to have a recycling system satisfying the following equation with the above elements to form a sound material-cycle society:

$$(1) + (2) > (3) + (4)$$

When considering cost, it is also necessary to have a system satisfying the above equation.

It is reported that CO_2 emissions from fossil fuel during paper recycling are greater than those during paper production from wood [12]. Moreover, wood as a raw material of paper is a biological resource that is considered carbon neutral for energy resources.

For plastic recycling, some dirty plastic used for material recycling is costly and energy intensive in the recycling process, which is an issue. Waste plastics are useful energy resources with high caloric value, but they are evaluated only as a source of CO_2 emissions because they are derived from petroleum.

In recent years, waste paper and plastic have been exported to neighboring countries (particularly China), because they are valuable resources as materials and energy. More than 20 % of recovered waste paper was exported in 2013 [9], and ~18 % of generated waste plastic was exported in 2012 [10]. It is a future challenge to construct an Asia-wide recycling system to ensure it is properly used for these exported resources.

[3] Material recycling is a process without chemical reaction that makes plastic products from waste plastic. Chemical recycling is a process with chemical reaction such as pyrolysis and gasification and makes oil, gas, and cokes from waste plastic.

A sound material-cycle society is a concept of resource circulation having a deep relationship with social issues such as energy and the environment.

3 Technology Roadmap

The major issues of waste management described in section "Present status" are summarized as energy recovery from incineration, residual life of final disposal sites, GHG emission from final disposal sites, and recycling optimization. Technical aspects for solving these issues are described in the technology roadmap. For a description of the contents of this roadmap for 2020 and 2030, the target value listed in the national policy is used as a reference, because optimization of waste treatment requires assistance within a policy perspective.

3.1 Waste Power Generation

Figure 8 shows waste treatment amount (facility scale) and superheated steam temperature dependence of power generation efficiency, which was determined from heat balance calculations [14]. This figure clearly demonstrates that a higher steam temperature and larger facility scale yield greater efficiency. The efficiency is extremely low at a facility scale smaller than 300 t/day.

Power generation efficiency for the waste incineration process is considerably lower than that (~40 %) of fossil fuel power plants. The main reasons for such low

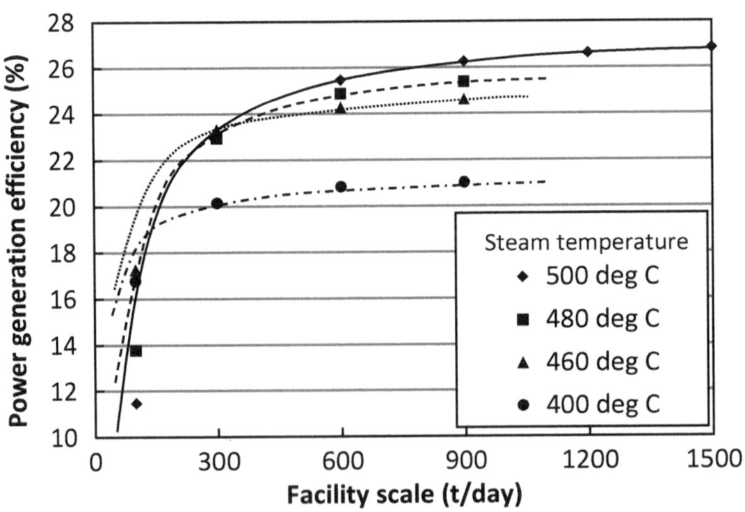

Fig. 8 Estimated results for power generation efficiency [14]

efficiency are considered to be the low superheated steam temperature, recovery loss of heat, and small plant scale.

Superheated steam temperature was controlled to <300 °C until the 1990s, to prevent severe corrosion by hydrogen chloride gas contained in the waste combustion gas. Owing to the development of corrosion-resistant superheater tube material and the ingenuity of superheater design, the steam temperature was raised to 400 °C in the 2000s, which results in a 20 % power generation efficiency. There are operational power plants with maximum steam temperature in the 500 °C class and 30 % power generation efficiency [13].

3.2 Semi-aerobic Landfills

The semi-aerobic landfill structure was developed in a joint study by Fukuoka University and the city of Fukuoka. A leachate collecting pipe is set up at the landfill bottom to remove leachate from the landfill as soon as practicable, so that leachate will not remain where the waste is deposited (Fig. 9).

Natural air is brought in from the open pit of the leachate collection pipe to the landfill layer in the opposite direction of leachate, because heat generated by microbial degradation of waste produces thermal convection from the difference between the internal temperature and ambient air temperature. This enables early stabilization of waste and prevents the generation of methane as GHG, which makes it an effective technology for the prevention of global warming.

Improvement of existing landfills by semi-aerobic landfill technology with the Fukuoka method was authenticated as a new method of the Clean Development Mechanism (CDM) in 2011, which was defined by the United Nations Framework Convention on Climate Change.

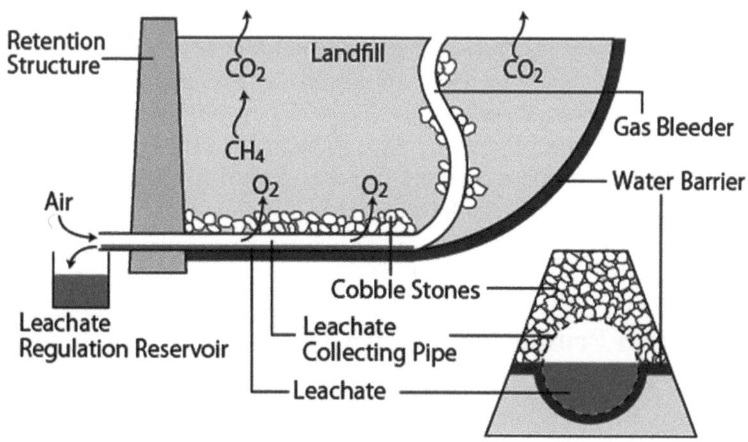

Fig. 9 Schematic diagram of semi-aerobic landfill structure [15]

3.3 National Waste Management Strategies

For a description of contents in the technology roadmap for 2020 and 2030, the target value listed in the national policy is used as a reference.

In the new *Waste Disposal Facility Development Plan* [6] approved by the Cabinet in 2013, the following numerical targets for general waste processing are listed for the period FY 2013 through FY 2017:

- Raising average power generation efficiency of installed facilities by 21 % over the period
- Maintaining the life of final disposal sites at 20 years by increasing the waste recycling rate from 22 to 26 %

In the *Technology Strategy Map 2009* developed by the Ministry of Economy, Trade and Industry and New Energy and Industrial Technology Development Organization in 2009, target values of resource productivity, recycling rate, and final disposal amount for 2015, 2020, and 2030 were established. The target for final disposal amount in 2020 is reduced by more than 25 % compared with that in 2010 and that in 2030 by an additional 25 % or more. To achieve such targets, technology development is needed, such as the suppression of waste generation, waste conversion, and regeneration of final disposal sites including recovery of useful substances [16].

The *Technology Strategy Map 2009* indicates the following technology goals aimed at the creation of a sustainable society, by achieving a balance between the environment and economy. However, specific numerical targets for 2020 and 2030 are not shown:

- Application of eco-design over the entire life cycle of a product
- Establishment of systems and technologies for resource collection and/or recycling
- Establishment of rare resources such as rare metal alternative technologies

Balance between the environment and economy is an important concept. As described in Chap. 2, this is needed to quantitatively evaluate both GHG emission and cost reductions through recycling, relative to new products. For example, the *Containers and Packaging Recycling Law* placed responsibilities for the separate collection of containers and packaging on local governments, and taxes collected from citizens are used for this collection.

4 Benefit and Prospective Future Vision

Conventionally, environmental protection has been considered an inhibitory factor in economic growth. Economic growth in the environmental field has received increased attention in developed countries, because a major challenge for such countries is future sustainable growth. Meanwhile, environmental progress has

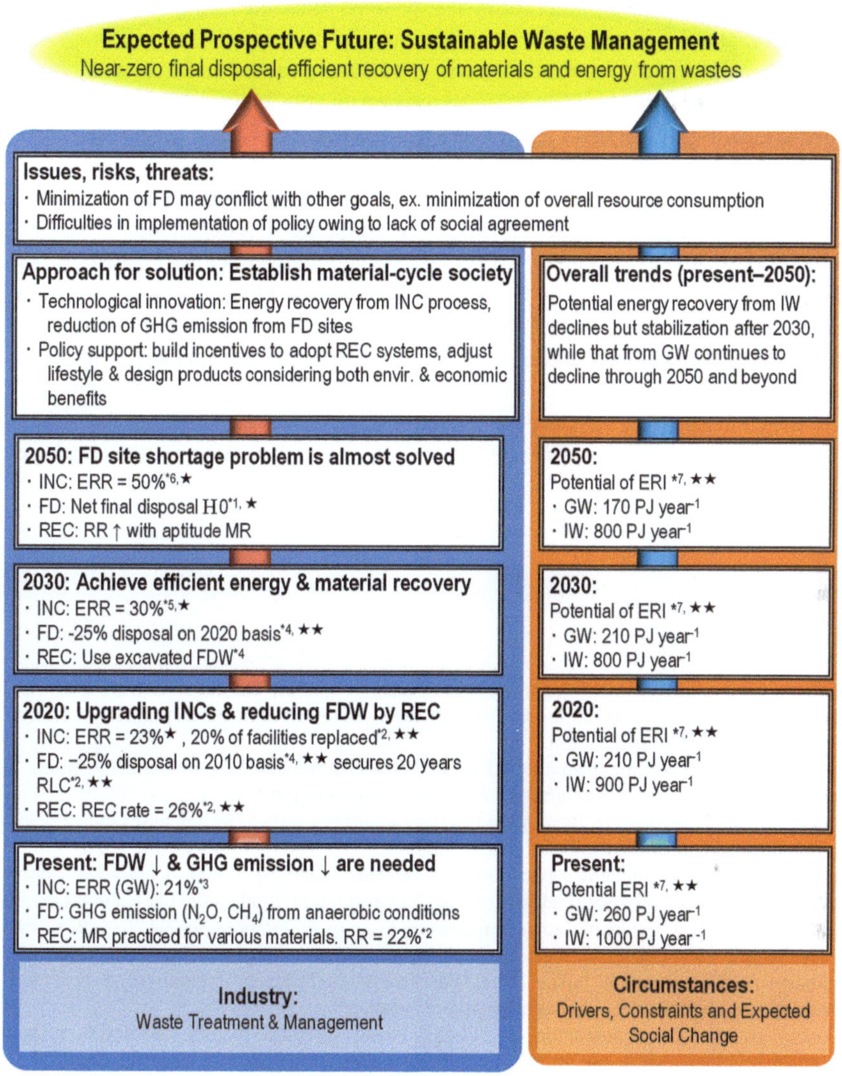

*1 (FDW) – (Excavated FDW recycled), *2 Ministry of Environment, Wastes Facility Development Plan [6], *3 Calculated from waste treatment statistics [3], *4 Ministry of Economy, Trade, and Industries, Technology Strategy Map [16], *5 Equivalent to Netherlands in 2006 [19], *6 Equivalent to Germany in 2006 [19]
*7 Amounts of 1) GSW: estimated assuming proportional to the population, 2) GIW: projected proportional to GDP assuming 1% economic growth rate

Abbreviations: envir.: environmental, ERI: Energy Recovery at INC, ERR: Energy Recovery Rate, FD: Final Disposal, FDW: FD Waste, GHG: GreenHouse Gas, IW: Industrial Waste, non-hazardous waste from industrial activities, GW: General Waste, non-hazardous & non-industrial waste except for night soil, INC: Incinerator, MR: Material (or Mechanical) Recycling, REC: RECycling, RR: Recycling Rate, RLC: Remaining Landfill Capacity

Fig. 10 Technology roadmap for waste-derived energy [3, 6, 16, 19]

become a concern in developing countries, because they have major challenges in poverty eradication and economic development.

In the international discussion of environmental measures and economic activity, the concepts of "green economy" advocated by the United Nations Environment Programme (UNEP) and "green growth" advocated by the Organisation for Economic Co-operation and Development (OECD) are well known.

UNEP defined the green economy as one that results in "improved human well-being and social equity, while significantly reducing environmental risks and ecological scarcities" [17]. The OECD emphasizes the importance of green growth, which achieves economic growth while overcoming resource constraints and reducing environmental impacts [18].

Both green economy and green growth are considered concepts aimed at pursuing sustainability in any aspect of the environment, economy, and society. In particular, green growth reverses the conventional way of thinking, because it regards investment in the environmental field as a factor that promotes economic growth (Fig. 10).

References

1. Ministry of the Environment Japan (2014) Annual report on the environment, the sound material-cycle society, and the biodiversity in Japan 2013 (in Japanese). http://www.env.go.jp/policy/hakusyo/h26/index.html
2. Ministry of the Environment Japan (2014) History and current state of waste management in Japan. https://www.env.go.jp/en/recycle/smcs/attach/hcswm.pdf
3. Ministry of the Environment Japan (2013) Waste treatment statics (in Japanese). http://www.env.go.jp/recycle/waste/wastetoukei_index.html
4. Ministry of the Environment Japan, Greenhouse Gas Inventory Office, Center for Global Environmental Research, National Institute for Environmental Studies (2014) National greenhouse gas inventory report of Japan. http://www-gio.nies.go.jp/aboutghg/nir/2014/NIR-JPN-2014-v3.0.pdf
5. Ministry of the Environment Japan (2014) Waste treatment in Japan (in Japanese). http://www.env.go.jp/recycle/waste_tech/ippan/h24/data/disposal.pdf
6. Ministry of the Environment Japan (2013) Waste disposal facility development plan (in Japanese). http://www.env.go.jp/council/former2013/03haiki/y0320-01/ref03_2.pdf
7. Ministry of the Environment Japan. General waste processing survey results (in Japanese). http://www.env.go.jp/recycle/waste_tech/ippan/index.html
8. Institute for Sustainable Energy Policies (2014) Renewables Japan status report 2014 (in Japanese). http://www.isep.or.jp/images/library/JSR2014All.pdf
9. Paper Recycling Promotion Center (2014) Paper recycling in Japan. http://www.prpc.or.jp/linkfile/english-paperrecycling.pdf
10. Plastic Waste Management Institute (2012) Situation of production, disposal, recycling, and treatment of plastic products (in Japanese). http://www.pwmi.or.jp/flow_pdf/flow2012.pdf
11. Japan Environmental Management Association for Industry (2014) Recycle data book 2014 (in Japanese). http://www.cjc.or.jp/data/pdf/book2014.pdf
12. Nippon paper group (2007) News release (in Japanese) http://www.nipponpapergroup.com/news/news07042401.html

13. IEA bioenergy Task36 (2009) Appendix 2: study tour of Japan http://www.ieabioenergytask36.org/Publications/2007-2009/Appendix_2_Final.pdf
14. Ogawa K, Takigawa T (1999) Study on thermal efficiency improvement of waste to power generation plant. Therm Nucl Power 50(7):819–830 (in Japanese)
15. Ministry of the Environment Japan (2012) Solid waste management and recycling technology of Japan - toward a sustainable society. https://www.env.go.jp/recycle/circul/venous_industry/en/brochure.pdf
16. Ministry of Economy, Trade and Industry Japan (2010) Technology strategy map (in Japanese) http://www.meti.go.jp/policy/economy/gijutsu_kakushin/kenkyu_kaihatu/str-top.html
17. United Nations Environment Programme. http://www.unep.org/greeneconomy/
18. Organisation for Economic Co-operation and Development. http://www.oecd.org/greengrowth/
19. Ministry of the Environment Japan (2011) The waste treatment technology research business in Japan and overseas (in Japanese). https://www.env.go.jp/recycle/report/h24-03.pdf

Topic: CO_2 Breakthrough Program by COURSE50 in Japanese Steel Industry Sector

Yutaka Ujisawa, Shigeaki Tonomura, Natsuo Ishiwata, Yuki Nabeshima, and Koji Saito

Abstract Since FY2008, four Japanese blast furnace steelmakers and one engineering company have been working on the *CO_2 U*ltimate *R*eduction in *S*teelmaking Process by Innovative Technology for Cool *E*arth *50* (COURSE50) project, which is one of the national projects commissioned by the New Energy and Industrial Technology Development Organization of Japan, aimed at developing powerful new CO_2 emission mitigation technologies for steelworks. The goal is to mitigate those emissions in the steelmaking process by approximately 30 % under the precondition of establishment of economic rationality of the process and availability of CCS infrastructure. This is done through a technology that reduces iron ore using hydrogen-amplified coke oven gas to curb CO_2 emissions from blast furnaces and that separates and recovers CO_2 from blast furnace gas using unused exhaust heat from steelworks. In this chapter, an overview of the COURSE50 project is provided.

Keywords COURSE50 • CO_2 emissions mitigation • Hydrogen reduction • CO_2 capture • Recovery of wasted heat

1 Introduction

As an example of technology development through an innovative R&D program, the *CO_2 U*ltimate *R*eduction in *S*teelmaking Process by Innovative Technology for Cool *E*arth *50* (COURSE50) project [1] to mitigate CO_2 emissions from the iron-

Y. Ujisawa (✉) • K. Saito
Nippon Steel & Sumitomo Metal Corporation, Chiba, Japan
e-mail: ujisawa.y6b.yutaka@jp.nssmc.com; saito.5m5.koji@jp.nssmc.com

S. Tonomura
Nippon Steel & Sumitomo Metal Corporation, Tokyo, Japan
e-mail: tonomura.7bf.shigeaki@jp.nssmc.com

N. Ishiwata • Y. Nabeshima
JFE Steel Corporation, Tokyo, Japan
e-mail: n-ishiwata@jfe-steel.co.jp; y-nabeshima@jfe-steel.co.jp

Y. Kato et al. (eds.), *Energy Technology Roadmaps of Japan*,
DOI 10.1007/978-4-431-55951-1_29

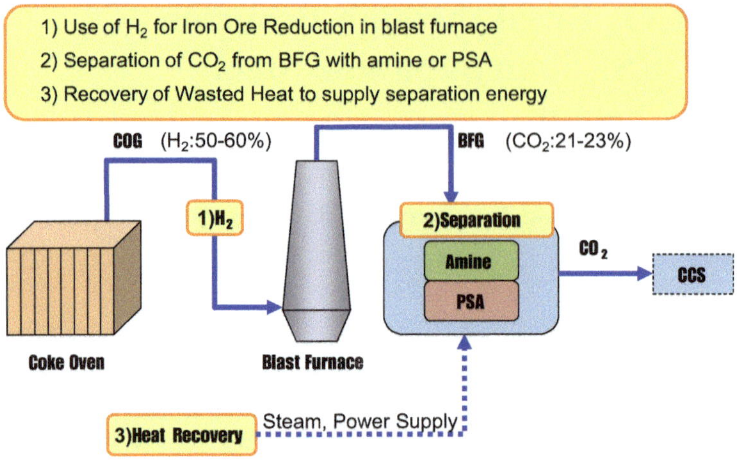

Fig. 1 Outline of COURSE50

making process has been promoted since FY 2008. Figure 1 shows an outline of COURSE50, consisting of two major research activities. The first is development of technology to reduce CO_2 emissions from blast furnaces, involving development of reaction control technology to reduce iron ore using hydrogen, reforming technology of coke oven gas (COG) to increase the hydrogen produced, and technology to manufacture high-strength and high-reactivity coke for hydrogen reduction blast furnaces. The second is development of technology to capture CO_2 from blast furnace gas (BFG) through chemical absorption and physical adsorption methods, using unused waste heat in steelworks. The target of CO_2 emission mitigation is ~30 % in the Japanese iron and steel industry.

2 Technology Details: Future Prospects, Problems, and Risks

The project has completed the development of the basic technologies in STEP1 (2008–2012) as planned and now promotes development of the comprehensive technologies in STEP2 (2013–2017).

2.1 Technology to Reduce CO_2 from Blast Furnaces

Figure 2 shows the concept of CO_2 emission reduction from a blast furnace by increased hydrogen reduction. The project is developing a technology to reduce CO_2 from a blast furnace via iron ore reduction, using hydrogenous reducing agents

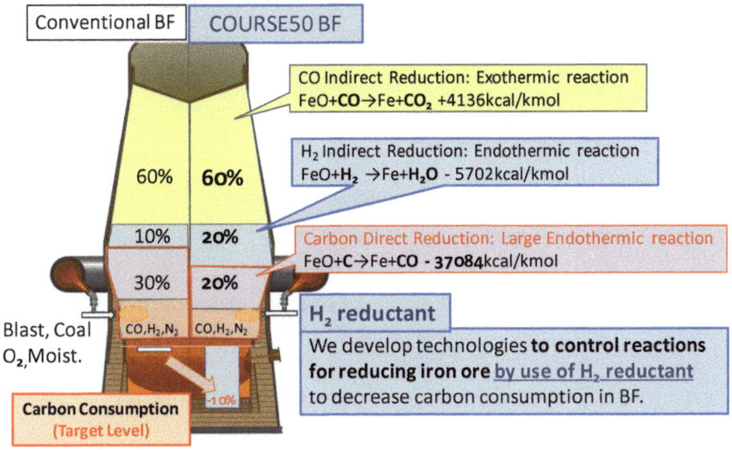

Fig. 2 Concept of CO_2 emission reduction from blast furnace by increased hydrogen reduction

such as coke oven gas (COG) or reformed COG (RCOG). Hydrogen content in COG is enhanced to produce RCOG using a newly developed catalyst and unused waste heat (800 °C) of coke ovens.

A hydrogen reduction promotion trial using a Luossavaara-Kiirunavaara Aktiebolag (LKAB, a Swedish mining company) experimental blast furnace was conducted in STEP1 to investigate and evaluate the potential for replacing coke and coal as reducing agents in the blast furnace with high H_2-containing gas, such as coke oven gas (COG) or reformed COG (RCOG) [2].

Figure 3 shows results of COG and RCOG injection using the LKAB blast furnace. H_2 indirect reduction was increased with both COG injection from a blast tuyere and RCOG injection from a shaft tuyere, because of the fast reaction rate of H_2 indirect reduction. Figure 4 shows results of the experimental blast furnace operation from a mathematical simulation model considering mass and heat transfer, with reactions and flows of gas, solid, and liquid inside the furnace. This confirmed that carbon consumption rates were decreased by about $-3\,\%$ in comparison with the base case of COG blast tuyere injection and RCOG shaft injection, from both experimental and theoretical perspectives.

The characteristics of coke required for hydrogen reduction were also studied. In particular, production of a high-strength coke was attempted to maintain the gas permeability necessary for hydrogen reduction technology under reduced coke-feeding rates relative to conventional blast furnace operation. Figure 5 shows this production procedure, with the use of a high-performance caking (HPC) additive.

HPC begins softening and melting at low temperatures, <300 °C. Therefore, HPC packs coal particles effectively by filling their gaps. As a result, it is possible to produce a high-strength coke without relying on blended coal (Fig. 6).

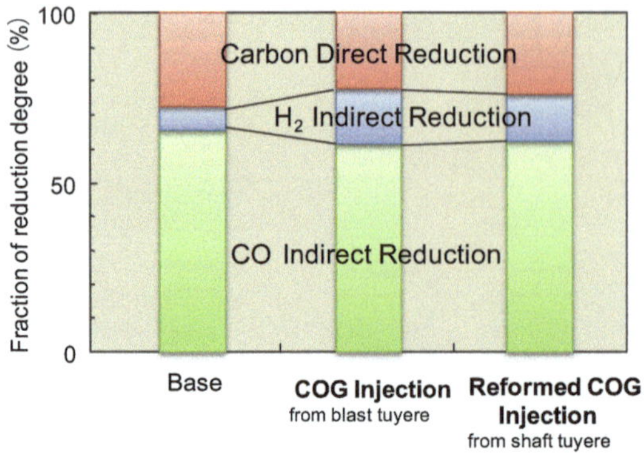

Fig. 3 Comparison of fraction of reduction in hydrogen reduction trial [2]

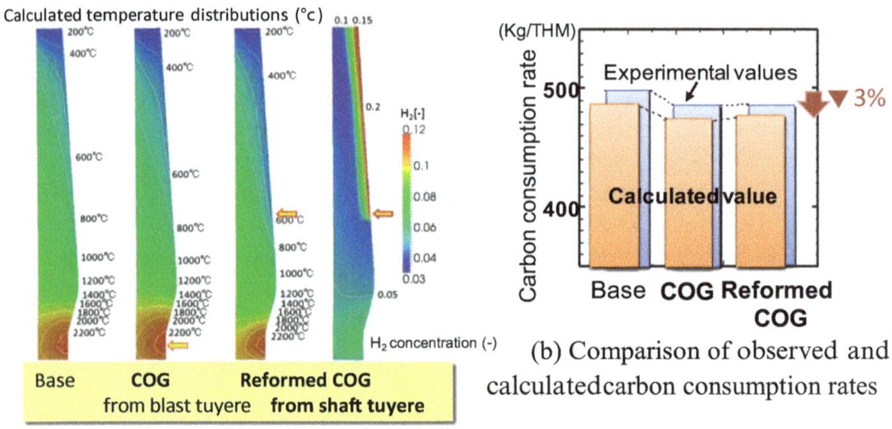

(b) Comparison of observed and calculatedcarbon consumption rates

(a) Temperature profiles calculated by mathematical simulation model

Fig. 4 Calculated results of experimental blast furnace operation. (**a**) Temperature profiles calculated by mathematical simulation model. (**b**) Comparison of observed and calculated carbon consumption rates

Fig. 5 Production of high-strength coke using high-performance caking additive [3]

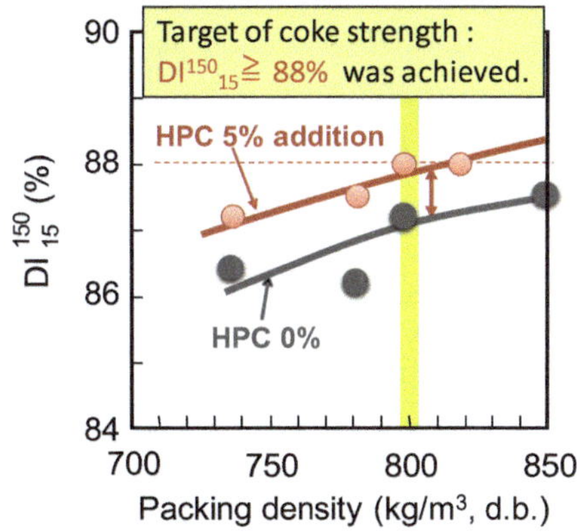

Fig. 6 Experimental results of coke strength enhancement using HPC [3]

2.2 *Technology to Capture CO_2 from Blast Furnace Gas*

The project is also developing technologies to capture CO_2 from BFG through chemical absorption and physical adsorption methods using unused waste heat from steelworks. The typical process flow of chemical absorption is shown in Fig. 7: (1) The absorbent comes in contact with feed gas in the absorber countercurrently and absorbs CO_2 selectively; (2) CO_2-rich absorbent is sent to the stripper and releases CO_2 by heating at ~120 °C; (3) the regenerated absorbent is cooled and sent to the absorber to repeat the cycle.

A chemical absorption method that can reduce thermal energy consumption of CO_2 separation by half (4–2 GJ/t-CO_2) through a high-performance absorbent and improvement of the chemical absorption process was developed (Fig. 8). In STEP1, two pilot test plants were used to develop this process. The first was chemical absorption test 1 (CAT1), whose capacity was 1 ton CO_2/day. It was used for evaluation of the fundamental performance of the absorbent. The second was CAT30, whose capacity was 30 ton CO_2/day. It was used for extensive evaluation of the absorbent selected by CAT1. In addition to its low energy consumption, the new absorbent has another unique and favorable feature. It readily releases CO_2 at a lower temperature than that of the conventional process for regeneration. This reveals the potential to use waste low-temperature heat at low cost.

For separating CO_2 from BFG, we are also developing a physical adsorption method using pressure swing adsorption (PSA) technology. Figure 9 shows the concept of this system. CO is separated in the first PSA unit with CO_2 adsorbent. Off-gas of the first PSA consists mainly of N_2 and CO. This off-gas is introduced to the second PSA unit and CO is adsorbed on the adsorbent and separated from N_2.

Figure 10 shows results of a general test operation using a bench-scale plant called ASCOA-3, whose capacity was 3 ton CO_2/day in STEP1. All targets were achieved during the test operation period. Maximum CO_2 purity was 99.5 % with 3 ton/day CO_2 recovery.

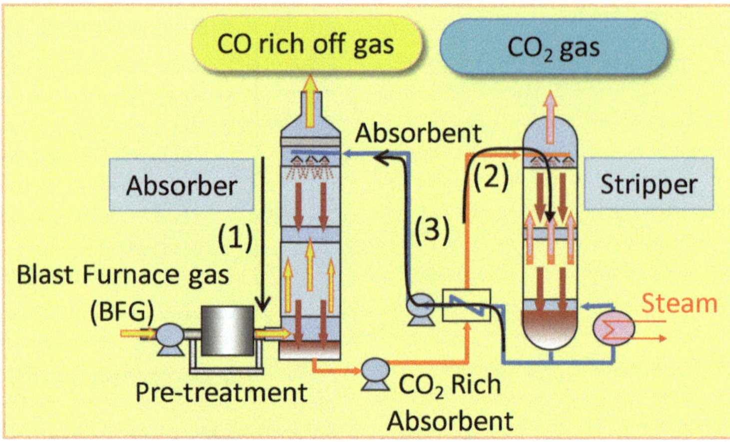

Fig. 7 Chemical absorption process

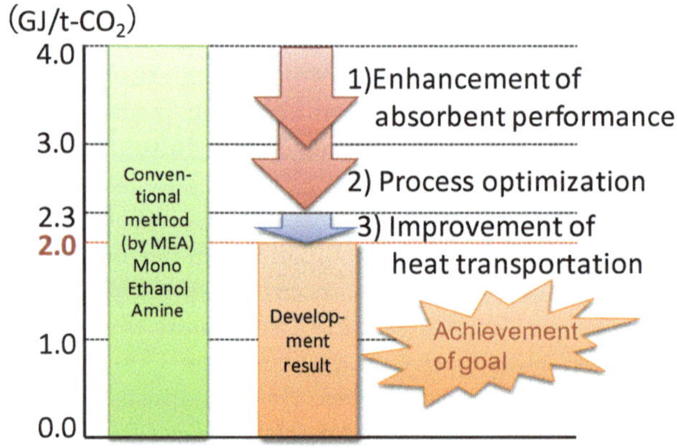

Fig. 8 Thermal energy consumption for CO_2 capture

Steam and electricity are necessary to separate CO_2 from BFG via the aforementioned processes. If energy is supplied from external purchase sources, it increases CO_2 emissions at the production site. Therefore in the project, technology will be developed to use unused waste heat that has not previously been used because of technological and/or economic difficulties in steelworks.

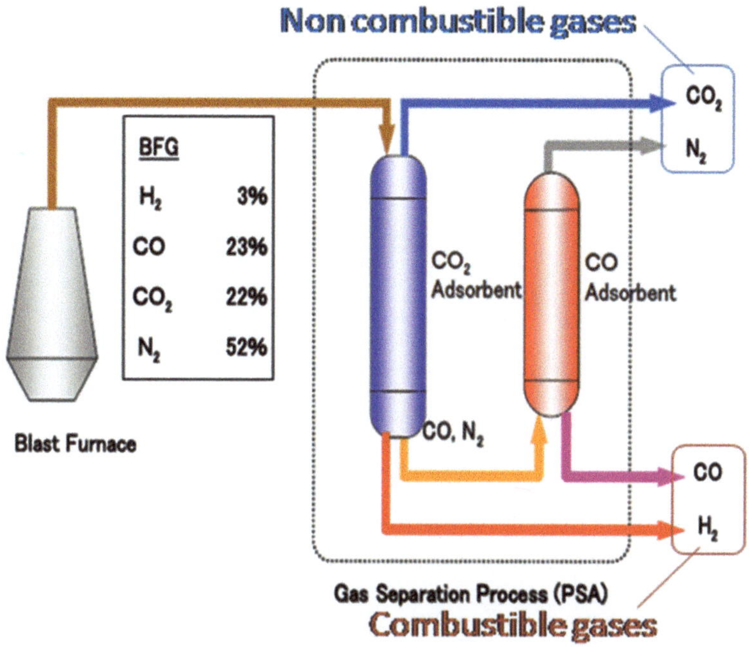

Fig. 9 Concept of two-staged PSA for blast furnace gas [4]

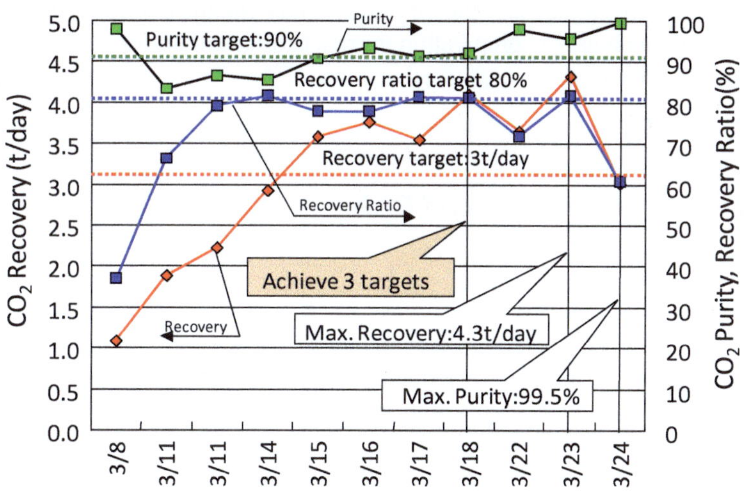

Fig. 10 Test operation results in physical adsorption plant [4]

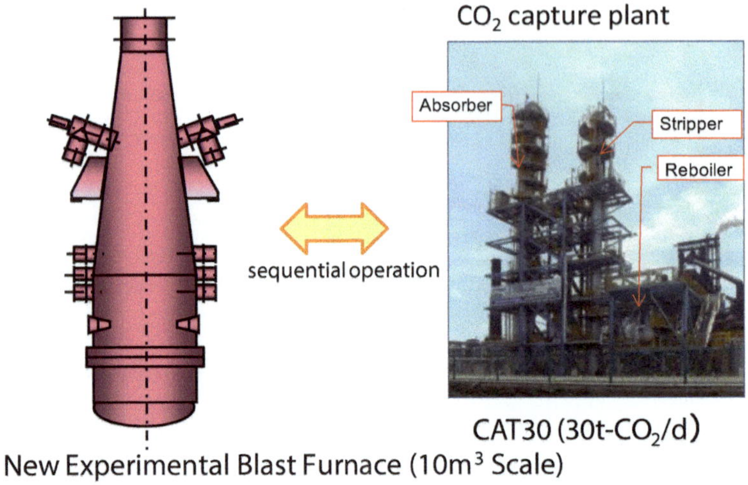

Fig. 11 Development of comprehensive technologies using experimental blast furnace and CO_2 capture plant in STEP2

2.3 *Integrated Technology Development*

In STEP2, integrated verification tests of iron ore hydrogen reduction and CO_2 separation and recovery from BFG are planned, using a newly built 10 m^3-scale experimental blast furnace and the 30 ton CO_2/day chemical absorption test plant (CAT30) simultaneously in preparation for practical development of the next phase in the project (Fig. 11).

2.4 *Technical Problems and Risks*

It is necessary to advance technology development to recover unused waste heat from steelworks economically to realize the CO_2 capture technology. It is also necessary to establish a scale-up technology to enhance the hydrogen reduction reaction in the blast furnace, to reduce CO_2 emission from that furnace. These problems must be solved to achieve the project goal of ~30 % CO_2 emission reduction in steelworks. Furthermore, the environmental benefits of CCS to store CO_2 captured from the BFG must outweigh potential environmental risks.

3 Benefit and Future Vision

The goal of the project is to commercialize the first unit by ~2030 and to generalize the technologies by 2050, in line with the timing of updates to existing blast furnace facilities. This is based on the assumption that CCS infrastructure is available and economic rationality of the process is established.

Iron-making and steelmaking technologies in Japan are already at the highest level in the world, and energy saving by the use of exhaust heat and by-product gas is at a high level within integrated steelworks. Carbon consumption reduction of the blast furnace has reached its limit, and we cannot expect substantial CO_2 emission reduction by extending current technology. Therefore, innovative technology is required to attain further reduction. Japanese steel industry accounts for about 15 % of the national total emissions which amounts to ~39 % of the entire country's industrial sector (about 70 % of the 39 % emissions is derived from blast furnaces). The COURSE50 project will mitigate CO_2 emissions by ~30 % in the Japanese steel industry sector. This value is equivalent to 4.5 % of overall CO_2 emissions in the country.

References

1. Miwa T, Okuda H (2010) J Jpn Inst Energy 89:28
2. Watakabe S et al (2013) ISIJ Int 53:2065
3. Watakabe S (2013) IEAGHG/IETS iron and steel industry CCUS and process integration workshop, Tokyo, 5–7 Nov 2013
4. Saima H (2013) IEAGHG/IETS iron and steel industry CCUS and process integration workshop, Tokyo, 5–7 Nov 2013

Topic: Hybrid Steel Works

Tsuguhiko Nakagawa

Abstract Japanese steelworks achieve the world's highest energy efficiency. However, approaches that seek intra-steelworks optimization alone have limitations when further energy conservation is planned. It is thus necessary to establish a system in which energy can be effectively used across all of society.

Hybrid steelworks have functionality that combines steel manufacturing and energy supply, focusing on improving the energy utilization of by-product gas. Considering the case of steelworks performing energy recuperation for society, two effects can be expected:

1. If, for example, the power generation efficiency rate at the steelworks can be increased from 37 % (current) to 55 % (future), power generation by coal of 4 % would be realized after meeting the power demand of the steelworks. This created power is equivalent in effect to a reduction of coal consumption by 10 % for coal-fired power generation.
2. When the energy recuperated from the steelworks for society is electricity, the power generation cost can be inexpensive relative to coal use for the sole purpose of power generation. This electricity is generated via by-product gas after the coal is used for the reduction of iron ore. In this way, it is possible to reduce electricity costs for society.

In other words, it is possible to achieve simultaneous reduction of CO_2 and societal energy costs. In this manner, a *hybrid steelworks* is defined as an advanced integrated steelworks that combines the functions of energy supply to society and steel manufacturing, by increasing energy utilization of by-product gas (Nakagawa, J Jpn Soc Energy Res 32(4):1, 2011).

Keywords Steelworks • CO_2 emissions • Gas turbine combined cycle • By-product gas • Energy conservation

---

T. Nakagawa (✉)
Okayama Prefectural University, Okayama, Japan
e-mail: nakagawa@cse.oka-pu.ac.jp

Y. Kato et al. (eds.), *Energy Technology Roadmaps of Japan*,
DOI 10.1007/978-4-431-55951-1_30

1 Introduction

The Japanese government lowered its CO_2 emission reduction target in November 2013, announcing a new target of 3.8 % reduction by 2020 (over 2005 levels). In the meantime, the 2013 trade deficit expanded to 13.7 trillion yen, with total fossil fuel imports reaching 28.4 trillion yen, out of 84.6 trillion yen total imports. These values indicate the need as a nation to reduce energy imports, not only from a CO_2 emissions standpoint but also from an economic one [2].

Both the introduction of renewable energy and improvement in energy utilization rate across the country are crucial to reducing energy imports. In fact, Japanese steelworks have maximized energy utilization, achieving the highest energy conservation rates in the world. However, optimization of steelworks alone is a limited approach to planning further energy conservation. It is necessary to construct a system for effective utilization of energy across the entire society [3].

The energy utilization efficiency of a steelworks has reached 46 %, which is 12 % greater than the national average of 34 % [4]. However, the efficiency of coal from power generation to fluid such as water transportation processes is 28 %, 6 % lower than the national average. Moreover, although steel is the final product, coal-fired generation solely has value for the transportation of fluid. Since the energy utilization of steelworks has value that includes energy for utility transport, such as that of cooling water for equipment, the difference between the energy utilizations of steelworks and coal-fired generation is huge. The reason for the high energy utilization in steelworks is that energy used for iron reduction is recovered and reused, e.g., as by-product gas that is used as fuel in the steelworks.

2 Technology Details: Future Prospects, Problems, and Risks

Accordingly, methods for using the aforementioned by-product fuel gas have seen increased attention. Summing the iron ore reduction reaction and by-product gas generation, the conversion rate (=(heat of reduction reaction + calorific value of by-product gas)/calorific value of coal × 100) of energy from coal is very high, 77.5 %. In terms of CO_2 emissions, one-third of the emissions of an entire steelworks is released from the blast furnace, and the other two-thirds is emitted from power generators and heating equipment in the form of combustion exhaust consumed from by-product gas. For the CO_2 emissions of the latter source, the energy utilization rate for processes that improve the performance of steel products, such as rolling, annealing, plating and others, is 20 %. This is 14 % lower than the domestic average energy utilization rate, and there is believed to be room for improvement [1].

To improve by-product gas utilization efficiency, the following technologies will be developed:

1. By-product gas usage is reduced by 20 % with the introduction of a regenerative heat exchange system [5] into the rolling heating furnaces and annealing furnaces.
2. Electric power per unit oxygen is reduced from 0.51 to 0.35 $kWh/O_2-m_N^3$ with the introduction of a high-efficiency oxygen plant [6].
3. Coke ovens are replaced by SCOPE21 (Super Coke Oven for Productivity and Environmental Enhancement toward the twenty-first century) [7, 8]
4. The introduction of the self-heat recuperative [9] amine process [10] to remove CO_2 from blast furnace gas in the COURSE50 project [11, 12].
5. The source of energy required for CO_2 separation is provided by the steam recuperation method from waste heat, such as warm water and exhaust combustion gas of steelworks [13].
6. The gas turbine combined cycle (GTCC) with 52.7 % power generation efficiency [14] is introduced together with conditions 2 and 5.
7. The O_2 concentration of blown air is taken as 35 %, to remove N_2 gas from by-product gas.
8. A high-temperature sensible exhaust heat of the steelworks will be recovered as combustion heat of the by-product gas through the use of endothermic reaction with thermal decomposition of C_nH_m, which is contained in the coke oven gas; the total combustion heat can be 1.3 times greater than that of the original coke oven gas [15].

The lower heating value of blast furnace gas is increased by removing CO_2 and N_2 that are contained more than 70 vol.% of that gas. Therefore, power generation efficiency improves drastically because combustion temperature rises and fuel gas compressor power decreases.

As a result, a power of 570 kWh/t steel is able to be created by a steelworks via energy conservation [1]. Therefore, CO_2 emissions can be reduced by 0.49 t-CO_2/t-steel, which is the evaluated CO_2 emission unit from a base coal-fired power generation of 0.864 kg CO_2/kWh.

3 Benefit and Executable Future Vision

With the application of gas separation in Japanese steelworks, a power of 5.7 billion kWh/year is able to be created by a steelworks with crude steel production of 10 million t-steel/year that uses GTCC for gas separation. By converting all steelworks in Japan (84 million t-steel/year blast furnace crude steel in 2013) [16], it is feasible to produce 48 billion kWh/year of electricity, an amount equivalent to the electricity generated from seven nuclear power plants, without increasing coal consumption. As a result, CO_2 emissions can be reduced by 41 million t/year, which is the evaluated CO_2 emission unit from base coal-fired power generation of 0.864 kg-CO_2/kWh in Japan, by controlling the load on existing coal-fired power stations.

Moreover, 667 billion kWh/year of electricity can be produced without increasing coal consumption using gas separation in steelworks worldwide, which produced 11.7 billion t-steel/year blast furnace crude steel in 2013 [16]. As a result, global CO_2 emissions can be reduced by 570 million t/year.

The development of the new technologies, such as low energy consumption CO_2 separation, high-efficiency by-product gas-fired GTCC and high O_2 operation of blast furnaces, is necessary to realize these effects.

From selling the created electricity, 480 billion yen/year profit will be achieved at the time of 10 yen/kWh electricity unit price. The cost of capital investment is estimated at 2.5 trillion yen in Japan. This represents an IRR (internal rate of return) = 17.1 %. These can be achieved by 2030.

References

1. Nakagawa T (2011) J Jpn Soc Energy Res 32(4):1
2. Ministry of Finance Japan: Trade Statistics of Japan for 2013 (2014) http://www.customs.go.jp/toukei/shinbun/trade-st/2013/2013_214.pdf. Accessed 22 Apr 2014
3. SMART Study Group (2004) Small advanced regional energy technology, Kaibundo, 131
4. Hirata K (2001) Handout for RC185 of the Japan Society of Mechanical Engineers
5. Nakagawa T, Arai N (2000) Kagaku Kogaku Ronbunshu 26:236
6. Haruna K, Miyake M, Sasano H (2005) Sumitomo Kagaku Gijyutsusi (II) 59
7. Nishioka K, Oshima H, Sugiyama I, Fujikawa H (2004) Tetsu Hagane 90:614
8. Kato K (2010) J Jpn Inst Energy 89:2
9. Kishimoto K (2012) Doctoral dissertations of Tokyo University
10. Mimura T, Hayashi M, Hagiu D (2012) Nippon Steel Eng Tech Rev 3:25
11. COURSE50. http://www.jisf.or.jp/course50/outline/. Accessed 5 Apr 2014
12. Miwa T, Okuda H (2010) J Jpn Inst Energy 89:28
13. Nakaso K, Oktariani E, Noda A, Itaya Y, Nakagawa T, Fukai J (2011) J Jpn Soc Energy Res 32(5):9
14. Nakagawa T (2012) Trans JSME Ser B 78:1560
15. Shikata T, Yatabe H, Meguri N, Matsuyama K (2008) Sekitann Kagaku Kaigi Happyou Ronbunshu 45:46
16. The Japan Iron and Steel Federation: World steel Annual Production 2013 (2014) http://www.jisf.or.jp/data/iisi/documents/summary_tables_j2014_02.pdf. Accessed 30 Nov 2014

Topic: Utilization of Heat and Energy by Small- to Medium-Sized Manufacturers: Case of the Molding Industry

Keiko Fujioka

Abstract The need for energy saving by small- and medium-sized manufacturers has been unrecognized because of the difficulty of obtaining actual data on their energy use. However, there is a great deal of potential for reducing energy consumption in this field, which is demonstrated in this article using a case example of molding factories.

Keywords Energy conservation • Small- to medium-sized manufacturer • Plastic molding • Die casting • Thermal energy

1 Introduction

The Japanese government has been promoting energy conservation using support and regulations. Among the regulations, the Law on the Rational Use of Energy (Energy Conservation Law), enacted in 1979, required rational energy usage; that is, it highlighted the need to secure the effective utilization of fossil fuels with improved energy efficiency. After the Great East Japan Earthquake, with realization of the need to ensure power supply, measures were added in response to the need for energy management that included the concept of "time," such as "peak use." The subjects of the regulations were businesses whose annual energy consumption in crude oil equivalent was 1,500 kL and greater. The total energy consumption of unregulated businesses accounted for ~16 % of domestic energy, with ~10 % in the industrial sector and 60 % in the business sector. Unregulated operators are mostly small- to medium-sized enterprises (SMEs) to which energy conservation promotion has become a major challenge.

Conversely, the government has been mainly taking SME support measures, such as the following. Regarding funding, in addition to the conventional energy-saving subsidies (those supporting funds for adopting energy-saving equipment), there have been newly established subsidies. These include those to local factories and to SMEs to acquire such equipment (subsidies for the purchase of latest-model

K. Fujioka (✉)
Functional Fluids Ltd., Osaka, Japan
e-mail: kfujioka@functioanl-fluids.co.jp

Y. Kato et al. (eds.), *Energy Technology Roadmaps of Japan*,
DOI 10.1007/978-4-431-55951-1_31

equipment that can efficiently save energy). Moreover, there are preferential treatment tax and interest subsidy systems for borrowing funds needed for adopting the energy-saving equipment.

According to a survey [1] examining the reasons for the lack of progress in CO_2 reduction at SME factories, many respondents cited the following problems in the introduction of equipment to reduce emissions: (1) It is difficult to obtain financing for capital investment. (2) Information on which equipment to select is lacking. (3) We are unable to quantitatively understand the benefits of such equipment. This shows that in improving production equipment and systems of SME manufacturers toward increasing their energy efficiency, a lack of information and funds as well as engineers is a problem. In response to these problems, environmental and energy measures aimed at SMEs have been taken, primarily from the financial side. These measures include a domestic credit system, low-carbon infrastructure projects through "visualization," funds for environmental and energy measures, a pollution-control tax system, and energy streamlining of company support businesses. For funding addressing these problems, the various subsidy schemes mentioned above have been implemented. Outside of funding, measures have been taken to compensate for the lack of knowledge to promote energy conservation and power-saving diagnoses.

Nevertheless, long-term prospects for the manufacturing industry are unclear, so it is difficult to implement improvements to production processes such as the introduction of new equipment and addition of facilities, which involve capital investment. Therefore, the potential for more effective energy use from initiatives that do not involve capital investment is examined for plants of plastic injection molding and die casting, which consume substantial electricity for heating and cooling.

2 Technology Details: Future Prospects, Problems, and Risks

The processes of plastic injection molding and die casting consist of melting resin and/or metals by heating and pouring them into molds. After cooling, these are made into products. In the plastic molding, the temperature of the heating cylinder is 150–270 °C for resin that melts at low temperature like polyethylene and 350–400 °C for liquid crystal polymer and others whose melting points are high. Therefore, the temperature of the mold during cooling is 20–90 °C for the former and 80 °C for the latter. In die casting, molten metal such as aluminum that has been poured into the molds at >700 °C is cooled to ~350 °C. Although this cooling temperature is above room temperature, many plants have been using refrigerating machines to cool water for use in cooling. This is chiefly because water from cooling towers causes considerable damage, which results from corrosion and

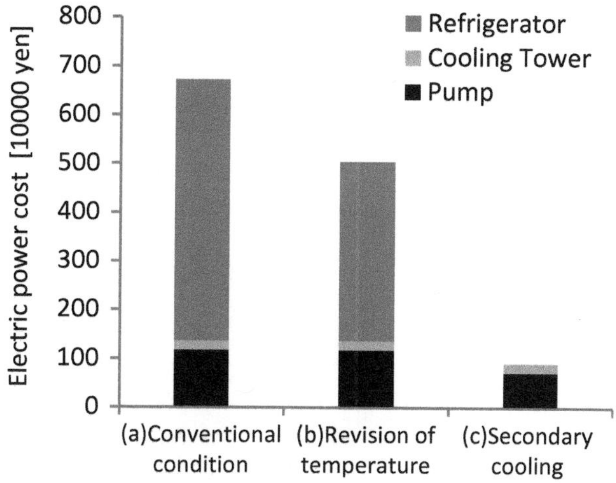

Fig. 1 Comparison of annual electric power cost. (**a**) Conventional condition: refrigerator at 7 °C, (**b**) refrigerator at 15 °C. (**c**) Secondary cooling system making use of cooling tower

scale accumulation and a misunderstanding that the colder the cooling water, the faster it can achieve the desired cooling.

For such processes, the most effective measure to conserve energy is "cooling at suitable temperature." Figure 1 gives a typical case of electric power consumption reduction via several changes in temperature control, which were surveyed at a household appliance company. The situation before those changes was having the production line use two 60-kW refrigerating machines, the temperature of whose cooling water was 7 °C. When that temperature was raised to 15 °C, the electric power consumption declined to 75 %. One of the most important aspects of cooling at suitable temperature is the use of a secondary cooling system that makes it possible to rely on cooling tower water without the problems of corrosion and scale accumulation. When the refrigerating machines were turned off after introduction of the secondary cooling system, power consumption decreased to 14 % [2].

3 Benefit and Attractive Future Vision

SME manufacturing accounted for 15–20 % of CO_2 emissions of the entire manufacturing sector (1990–2010 data; the figure varies annually). Energy-saving measures have already been exhaustively implemented in large corporations, so further improvements are not expected. However, there appears to be room for substantial CO_2 emission reductions among SMEs. For example, annual resin production is 1.4×10^7 ton [3], most of which is processed within Japan. If it is assumed that average resin heat capacity is 1.5 kJ/kg K and the average temperature increase is 200 K, 4.2×10^{15} J of energy is consumed in plastic processing each year. For die casting, annual production of its material (aluminum) is 8.7×10^5 ton [3] and the energy required for temperature increase and melting is 9.6×10^{14} J.

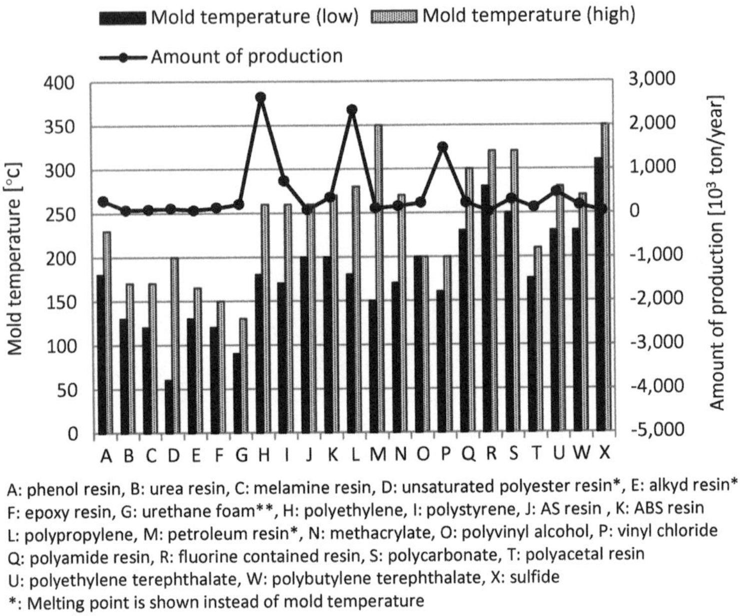

Fig. 2 Mold temperature and annual amount of production of resins in Japan in 2013 [3]

Figure 2 shows the annual production of resins in Japan and mold temperature of each resin [3]. The production amount of resins, whose molding temperature is <150 °C and may need refrigeration in the cooling process, is <10 %. Thus, for resins in excess of 90 % of annual production, there is great potential to reduce energy consumption using cooling water at higher temperatures than in current practice.

Therefore, the first step must be the dissemination of information and knowledge. There are many measures that entail no cost, such as modifying worksite temperature management, which can be immediately implemented. A likely challenge is that different measures must be taken depending on business category and scale. We anticipate significant progress, assuming the development of personnel who can make appropriate judgments in each case (as opposed to providing general guidance) and the construction of databases of effective utilization of thermal energy that incorporate individual cases.

References

1. METI (2008) http://www.meti.go.jp/committee/materials2/downloadfiles/g80908b09j.pdf
2. Fujioka K (2012) Energy saving carried out through cooling process, text for symposium of the society for technology of plasticity, Komatsu
3. Research and Statistics Department Minister's Secretariat METI, Yearbook of Current Production Statistics Paper (2013)

Topic: Regional Utilization of Unused Agricultural Waste

Yutaka Morikawa and Masako Ito

Abstract In Japan's Aichi Prefecture, we promoted cooperation between a farmers' cooperative, firms, and local administrations to advance an enterprise using agricultural waste. We used tomato stems, 30,364 t of which are produced annually in agrarian areas of Aichi, as the principal raw materials. For cost reduction, we studied novel and effective methods for bio-ethanol. With these methods (a heating and high-pressure fluid mill system and novel small water-amount saccharification system), saccharification efficiency became two to four times greater than that of the conventional method, and energy use in the distillation process was decreased to 1/8.

Keywords Biomass • Tomato stems • Bio-ethanol • Saccharification • Nanofiber

1 Introduction

Aichi Prefecture has a warm climate and is near a large consuming area. It is therefore one of the most productive areas for vegetable-growing in Japan (sixth in the country for agricultural production). Japan's highest production of cabbage, Japanese basil, and giant butterbur is found there [1].

As agricultural production increases, so does the amount of agricultural waste (AW). This constitutes 52 % (~300,000 t annually) of total unused (waste) cellulosic biomass in Aichi [2] and mainly consists of tomato stems (30,364 t annually, including ~12 % water). Local systems need to be constructed for AW utilization and novel technologies need developing for conversion of AW into bioenergy or biomaterials.

Here, we introduce a system to efficiently use AW (mainly tomato stems) through cooperation with industry-academia-government and local communities in Aichi.

Y. Morikawa (✉) • M. Ito
Aichi Center for Industry and Science Technology, Aichi, Japan
e-mail: yutaka_morikawa@pref.aichi.lg.jp; masako_5_itou@pref.aichi.lg.jp

Y. Kato et al. (eds.), *Energy Technology Roadmaps of Japan*,
DOI 10.1007/978-4-431-55951-1_32

2 Technology Details: Future Prospects, Problems, and Risks

Figure 1 shows the cooperative system for AW utilization in Aichi. The system consists of agricultural production, AW disposal and collection, conversion of AW to bio-ethanol and fertilizer, and utilization of AW products. To maintain low raw material (stem) prices, we structured the collection system such that each farmer brings the stems to garbage collection points designated by farmers' cooperatives. To enhance business economy, we used by-products of raw materials of plastics, manure, and other items.

At the Aichi Center for Industry and Science Technology, research and development has been conducted to furnish a new method for producing bio-ethanol at a low cost. The production target was set at 1 kL bio-ethanol from 6 tons of AW in 120 days. Costs of AW pretreatment, such as hydrothermal treatment and milling, plus distillation to remove water, were more than 50 % of the thermal energy cost of the entire process [3, 4]. We therefore developed novel technologies to reduce the cost of these processes.

Figure 2 shows a schematic diagram of a novel consecutive heating and high-pressure fluid mill system [5, 6]. Hydrothermal treatment and milling are done separately with batch operation in the conventional method, whereas in the new system, they are performed consecutively in continuous operation. This system significantly improves thermal efficiency and increases AW throughput. Additionally, AW is decomposed to the fiber level with large surface area per unit volume,

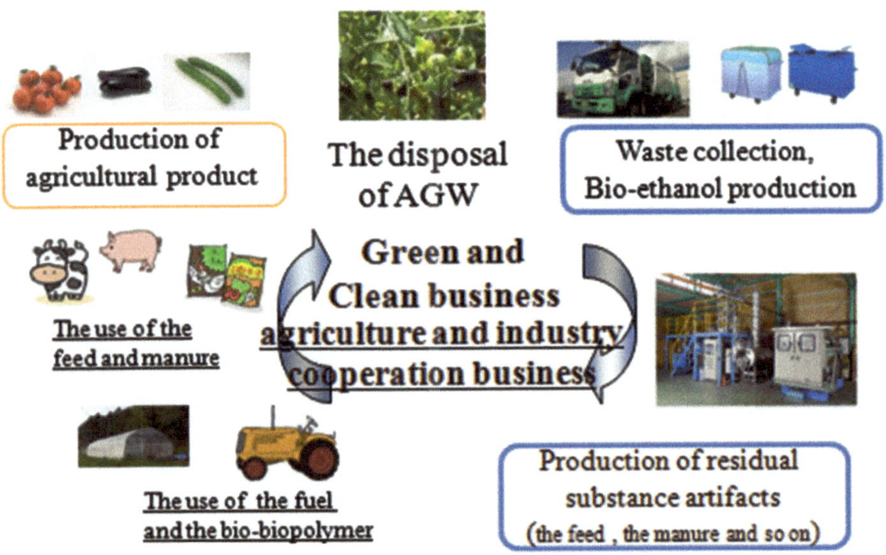

Fig. 1 Cooperative system for AW utilization in Aichi prefecture

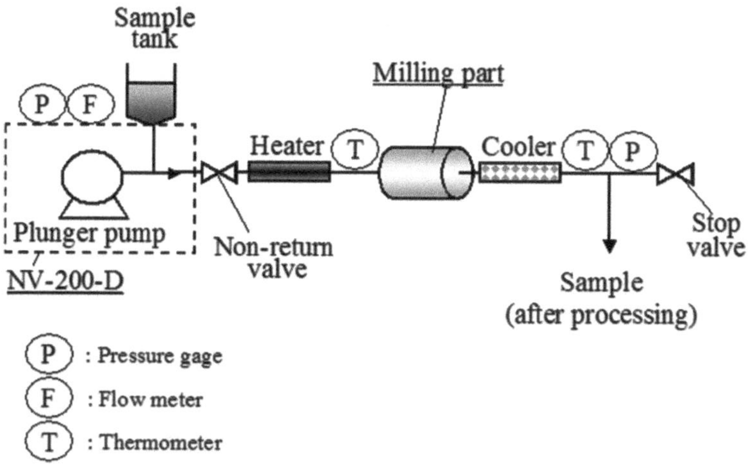

Fig. 2 Schematic diagram of consecutive heating and high-pressure fluid mill system

so the rate of subsequent saccharification becomes two to four times greater than that of the conventional method.

Figure 3 shows a conceptual diagram of a novel small water-amount saccharification system [7]. In the conventional method (Fig. 3a), about ten times the water of the raw materials is used. In the developed method (Fig. 3b), a small amount of water is used for sugar abstraction. We thereby reduce the energy used for the distillation process to 1/8.

3 Benefit and Prospective Future Vision

To ensure an economical business, it is necessary to develop as a priority a value-added, high-profit utilization method and to promote industrialization. Thus, we applied the above milling technology to developing high-value-added cellulose nanofiber [5, 6, 8]. We began to sell a trial product of this nanofiber.

Additionally, to maintain stable AW quantities and quality, the promotion of agricultural industrialization by plant factories and systemized management of unused biomass disposed of in various locations (such as food and building materials, factories and textile mills) will be essential.

In the future, we wish to promote development that will ameliorate problems of domestic energy shortages and environmental destruction through the use of local AW as a green resource.

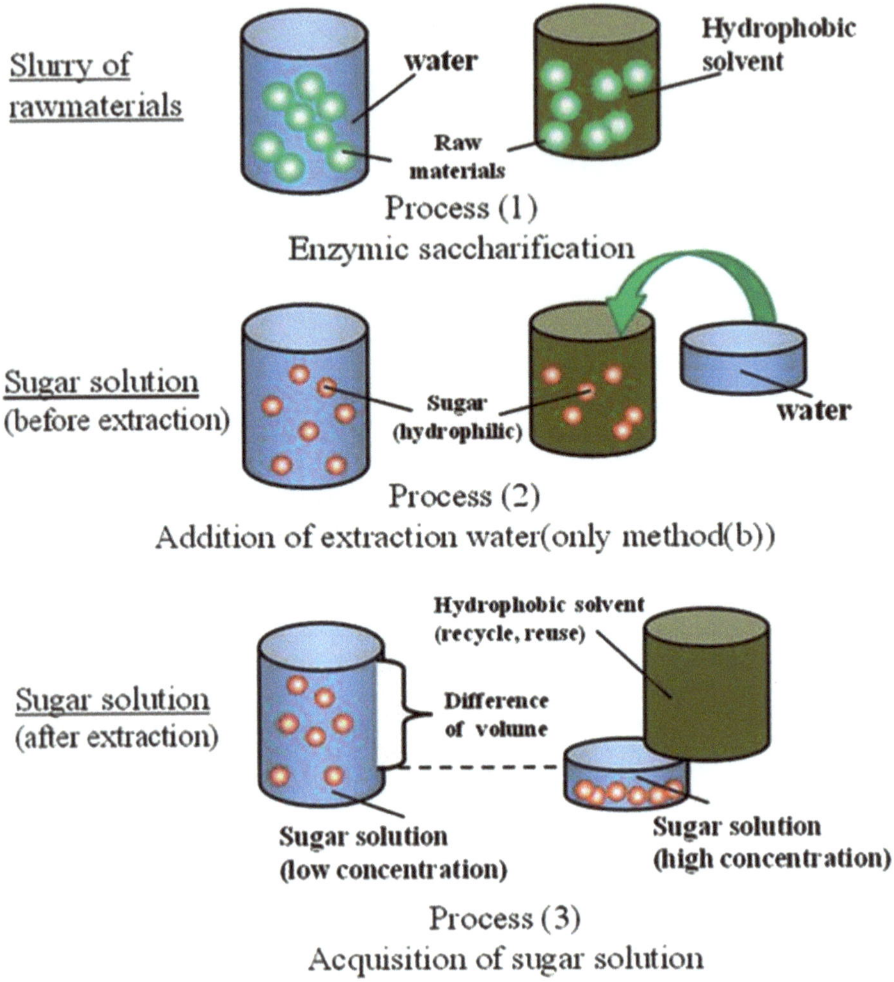

Fig. 3 Acquisition of high-concentration sugar solution (**a**) Conventional methods (**b**) Development methods

References

1. "YokuwakaruAichi no Nougyou 2013 (in jap.)", Foundation Aichi Agricultural Promotion Fund (2013)
2. "Research Report for Biofuel Spread Expansion in Aichi Prefecture, the 19th year of the Heisei era", Aichi Pref. (2008)
3. Saga K, Yokoyama S, Imou K (2008) Net energy analysis of bioethanol production system form rice cropping. J Jpn Soc Energy Res 29(1):31–35

4. Saga K, Fujimoto S, Yanagida T, Tada C, Yuriyivna BL, Elmer B, Minowa T (2009) Comparative evaluation of ethanol production process from cellulosic biomass with different pretreatment and saccharification method. J Jpn Soc Energy Res 30(2):9–14
5. Morikawa Y, Ito M, Umeda S (2010) Continuous decomposition of cellulose biomass by consecutive heating and high-pressure fluid milling. Kagaku Kougaku Ronbunshu 36 (4):259–263
6. Japanese Patent Application No.5232976
7. Japanese Patent Application No.2011-205933
8. http://www.yoshidakikai.co.jp/nano/gijyutu/cellulose_nano.html

Topic: Energy Recovery from Mushroom Culture Waste and the Use of Its Ash as Fertilizer

HeeJoon Kim, Tadaaki Shimizu, Itaru Kourakata, and Yoshihiko Takahashi

Abstract Biomass is one of the most important primary and renewable energy sources. Among agricultural and forestry by-products, cultivation waste from mushroom production is considered to have several advantages, such as low collection cost if used by the same production facility, and it contains many inorganic nutrients, relative to woody biomass resources. Some kinds of mushroom culture waste contain plant nutrients (P, K, Mg, and Si); ash from such biomass waste can be used as fertilizer. Here we suggest new concept for a more economical method of using this waste, namely, cascade utilization of enokidake mushroom culture waste, as fuel, followed by the use of the ash as fertilizer.

Keywords Mushroom culture waste • Fertilizer • Energy recovery • Biomass • Enokidake

1 Introduction

It is well known that biomass is one of the most important primary and renewable energy sources because of its extensive amount [1] and its supply of carbon neutral energy in well-designed systems [2]. Furthermore, biomass use has become increasingly important because of depletion of fossil fuel sources and global warming issues. Among biomass sources, abundant agricultural and forestry by-products are available, but are not used effectively and are only discarded. For examples, many types of mushroom are cultivated and produced for human consumption in Japan. They are produced in indoor "mushroom factories" using cultures consisting of various biomasses such as wood chips, corncobs, and nutrient biomasses. However, after harvest these used mushroom culture biomasses are wasted, and in amounts one or more times the volume of mushroom production.

H. Kim (✉) • T. Shimizu • I. Kourakata • Y. Takahashi
Niigata University, Niigata, Japan
e-mail: kim@eng.niigata-u.ac.jp; tshimizu@eng.niigata-u.ac.jp; kourakata@ccr.niigata-u.ac.jp; takahasi@agr.niigata-u.ac.jp

Y. Kato et al. (eds.), *Energy Technology Roadmaps of Japan*,
DOI 10.1007/978-4-431-55951-1_33

In Japan, total mushroom production in 2012 was 462,000 tons [3]. Among cultivated mushrooms, enokidake (*Flammulina velutipes*) is the most popular, with production reaching 134,000 tons in that year (~30 % of total mushroom production) [3]. The amount of enokidake mushroom culture waste (MCW) produced is nearly the same as that of edible mushrooms, which is estimated ~134,000 tons/year [4, 5]. Its use as biomass fuel is not widespread and most cultivation waste is disposed of as industrial waste. The MCW has some advantages as biomass fuel, as follows:

1. The logistic problem of gathering biomass is minimal, because it is produced in indoor factories.
2. The mushroom cultures are for food production, so the biomass waste is free from harmful materials such as heavy metals.
3. Some mushroom culture wastes contain plant nutrients (P, K, Mg, and Si) and ash from such waste can be used as fertilizer.

In particular, ash in enokidake MCW is rich in P_2O_5 (~25 %) [4]. However, one problem with enokidake MCW is its high moisture content (55–65 %), which makes it disadvantageous for stable combustion in small-scale combustors. In addition, high NO_x emission during combustion is expected owing to the high nitrogen content of enokidake MCW. Therefore, we propose a method of drying MCW by composting or compressing and clean combustion, followed by using the MCW ash as fertilizer for rice production.

2 Technological Details and Future Prospects

Drying enokidake MCW is accomplished using the heat of composting under aerobic conditions. Experimental results using a pilot-scale (3 m^3) composter showed that the moisture content of the MCW can be reduced to 35 % in this way [5]. Combustion experiments using such a combustor (maximum fuel feed 60 kg/h) showed that NO_x and SO_2 can be controlled by injecting urea solution and $Ca(OH)_2$ powder into the combustor, both of which are available as soil fertilizer and pH-controlling reagent, respectively. Ash as fertilizer in paddy fields was applied and the rice crop was found to be equivalent to that grown using ordinary fertilizers.

2.1 Benefits and Attractive Future Vision

These results suggest that MCW used as fuel can contribute to the reduction of greenhouse gas emissions and reuse of natural sources (particularly phosphorus) as fertilizer (Fig. 1).

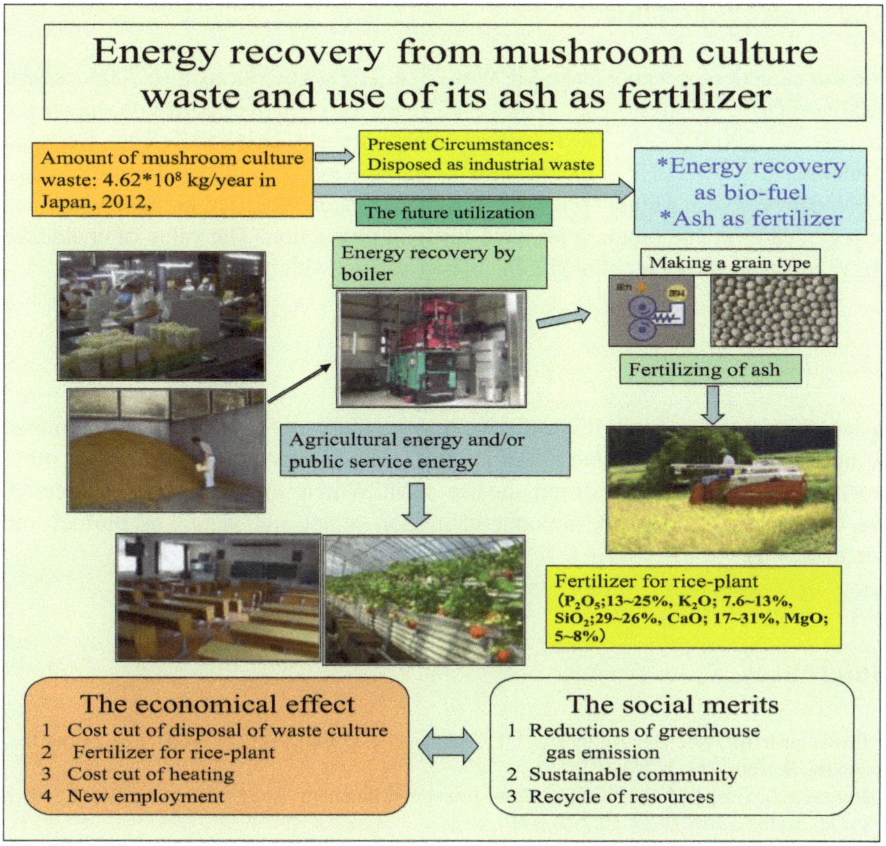

Fig. 1 New concept for more economical use of biomass such as enokidake mushroom culture waste

2.1.1 Value as Biomass Fuel

The moisture content of enokidake MCW is ~55 %. Using this basic figure, the lower heating value (LHV) of wet enokidake MCW (moisture 55 %) is 5.9 MJ/kg by bomb calorimeter experiments. Thus, the energy potential of enokidake MCW after composting is ~790 TJ/year (enokidake production amount was 134,000 tons in 2012).

The average price of crude oil for 3 years from 2012 to 2014 is ~10,000 JPY/bbl = 74,000 JPY/t. Thus, the potential value of enokidake MCW would be 1.3 billion JPY/year.

The total amount of carbon offset would be ~$5.4*10^7$ kg CO_2/year (crude oil base).

2.1.2 Value as Fertilizer

The ash content of wet enokidake MCW is ~5 %. The ash produced is ~6,700 tons/year (equivalent to 1,700 tons P_2O_5). In the past 2 years, the price of imported phosphate ore has been ~20,000 JPY/t [6]. The P_2O_5 content of phosphate ore is ~30 %. Thus, the value of enokidake MCW ash is ~113 million JPY/year. It is expected that ash from enokidake MCW can be sold at a higher price than ore as sustainable fertilizer, because MCW can be regarded as a safe biomass resource for food production. The value of enokidake MCW ash as fertilizer is not negligible in comparison with its value as fuel.

2.1.3 Future Vision

In Japan, the total amount of waste culture from mushroom cultivation is estimated at more than 462,000 tons/year. Although the ash composition of MCW of other mushrooms has not been clarified, the use of MCW in a similar manner is expected. We can easily estimate the amount of carbon offset and values as biofuel and fertilizer.

References

1. Christopher BF, Michael JB, James TR, Falkowski P (1998) Primary production of the biosphere. Science 281:237–240
2. Bernhard S, Gregg M (1996) The role of forest and bioenergy strategies in the global carbon cycle. Biomass Bioenergy 10:275–300
3. e-Stat, Portal site of official statistics of Japan (2014) http://www.e-stat.go.jp/SG1/estat/GL08020101.do?_toGL08020101_&tstatCode=000001021191&requestSender=dsearch
4. Shimizu T et al (2012) Combustion of waste mushroom culture. Proceedings of the 21st International conference on fluidized bed combustion (Naples, Italy), Paper No. TA10-4
5. Report of "Research and development of projects for application in promoting new policy of agriculture, forestry and fisheries", grant number 23054 (2014) https://www.s.affrc.go.jp/docs/sangaku_renkei/kenkyusyokai2014/pdf/p125_23054.pdf
6. Trade Statistics of Japan Ministry of Finance (2012) http://toukei-is.com/h

Topic: Organic Hydride for Hydrogen Energy Carrier

Yasukazu Saito and Yoshimi Okada

Abstract Hydrogen storage and transportation are critically important to realize a low-carbon society. A hydrogen supply chain concept has recently been proposed by Chiyoda Corporation on the successful basis of catalytic methylcyclohexane dehydrogenation, coupled with well-established toluene hydrogenation and existing delivery infrastructures at low cost. The SPERA Hydrogen™ process ensures liquid storage of hydrogen energy for safe large-scale and long-distance transportation (even abroad) under ambient temperature and atmospheric pressure. Overall costs would generate business risks. High-equilibrium temperatures and large enthalpy change (ΔH) are deficit in this reversible reaction couple. Utilization of various types of exhaust heat and development of a compact-size dehydrogenation reactor should be explored.

Keywords Hydrogen supply chain • Methylcyclohexane dehydrogenation • SPERA hydrogen • Compact-size reactor

1 Introduction

Electricity generation from renewable but fluctuating energies (wind, solar, and others) is growing rapidly at relatively large scales. Since chemical energy is superior to electric energy with respect to long-term storage and long-distance transportation, "power-to-gas" technology (electricity conversion to gaseous chemical energy) has emerged. In lieu of gas, "power-to-liquid" technology is anticipated, because easy handling in harmony with existing infrastructure and safety is prerequisite to hydrogen energy carriers. Organic hydrides appear to be the best candidates, as technologies for production, storage, and distribution are already available.

Y. Saito (✉)
New Energy Laboratory, Inc., Tokyo, Japan
e-mail: yassaito@ci.kagu.tus.ac.jp

Y. Okada
Chiyoda Corporation, Yokohama, Japan
e-mail: yookada@ykh.chiyoda.jp

Y. Kato et al. (eds.), *Energy Technology Roadmaps of Japan*,
DOI 10.1007/978-4-431-55951-1_34

2 Technology Details: Future Prospects, Problems, and Risks

Hydrogen storage in liquid organics attracted attention in 1984 for providing onboard hydrogen to vehicles, when the combustion engine of a 16-ton demonstration truck was powered by hydrogen liberated from methylcyclohexane (hydrogen content 6.2 mass%) (Eq. 1). Because of its high dehydrogenation temperatures and strong endothermicity ($\Delta H = 72$ kJ/mol H_2) (Eq. 1), heteroatom candidates such as N-ethylcarbazole have been proposed ($\Delta H = 25$ kJ/mol H_2, 5.7 mass%) (Eq. 3). Another candidate with greater hydrogen content is decalin ($\Delta H = 59$ kJ/mol H_2, 7.3 mass%) (Eq. 2) [1]:

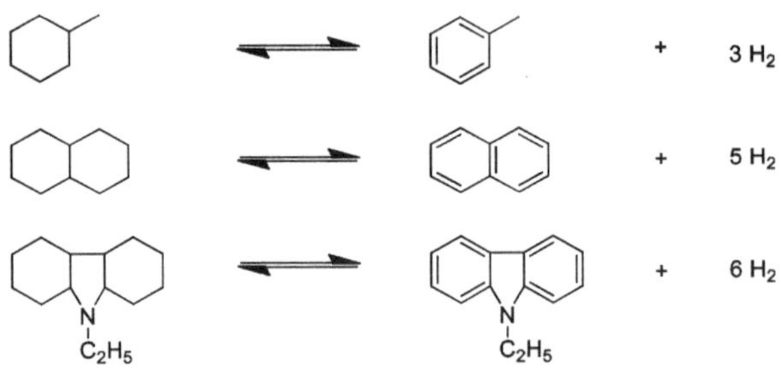

(1)
(2)
(3)

With regard to exergy loss during storage and release processes for standard-state hydrogen, 3.05 MJ/kg H_2 of thermal energy is lost owing to methylcyclohexane (MCH) dehydrogenation, which can be supplied within the temperature ranges of exhaust heats (e.g., 400 °C). In contrast, liquefaction of gaseous hydrogen and compression up to 70 MPa must consume mechanical energies >42 and 22.1 MJ/kg H_2, respectively [2].

For supply capacity of source organic materials, toluene (TOL) is optimum based on the oil refinery and petrochemical industries (annual TOL production in Japan is 1.4 Mton). The use of TOL is limited to industrial solvents and gasoline additives. Moreover, large demands for benzene and xylenes consume TOL by their disproportionation at present. New demand for TOL is therefore anticipated. Recent success of MCH dehydrogenation in the SPERA Hydrogen™ process [3] in Kawasaki, Japan (Fig. 1), will ensure its application to onsite supply at hydrogen stations for fuel cell vehicles (FCVs) and others. The essential innovative aspect of this technology is nano-sized Pt particles supported on alumina, specially modified with sulfur to prolong catalyst life [4].

A new national research laboratory was initiated at Koriyama, Fukushima Prefecture, in April 2014.

According to a future plan presented recently by the Japanese government, candidates for the hydrogen supply chain with large quantities are proposed even from abroad, energy carriers for which include ammonia, liquefied hydrogen, and organic hydride (liquid: −95 to 100 °C). Among these, organic hydride enjoys the greatest confidence, because it is neither toxic nor explosive (albeit flammable) [5].

Fig. 1 Demonstration plant of SPERA Hydrogen™ (2013, Kawasaki). Capacity of 50 Nm^3 H_2/h for TOL hydrogenation and MCH dehydrogenation would be enlarged upon demand requests

Hydrogen demands will increase from the transportation sector (FCV H_2 station as the 2020 target) for energy use in power generation as in the 2030 target. Another compact demand on hydrogen would arise from residential and agricultural sectors. Developments in large-scale fuel cells will be useful to smart eco-cities and greenhouse cultivation, assisted by cogenerated heat and power from solid oxide fuel cell (SOFC).

Technological problems remain for chemical hydride in chemical recuperation of unused low-quality heats at low temperatures and variable magnitudes. The main risks come from high process costs associated with water electrolysis, dehydrogenation, and fuel cells.

Governmental policies on renewable energies, CO_2 mitigation, and nuclear power generation may represent other risks for chemical hydride, since political directions exert a decisive influence on technologies of hydrogen storage and transportation.

3 Benefits and Future Vision

The role of SPERA Hydrogen™ in storing and transporting hydrogen on a large scale would certainly aid CO_2 mitigation by mediating renewable energies with traffic (FCV) and power generation (H_2-added combustion) technologies. The target of halving the 2010 emission (1080 Mt CO_2) by 2050 will be fulfilled with 33.8 Gt H_2 on an equal basis of ΔH_{ox} between C and H_2, which would be supplied from MCH within the framework of annual domestic production 1.40 Mt TOL and its 35-year (2015–2050) accumulation, together with go-and-return transport as frequent as 12 times annually.

Over-season storage and overseas transport of organic hydride enable connection of the Japanese hydrogen economy with large-scale renewable energies abroad, as exemplified in the Aleutians (geothermal), Sakhalin (wind), and Laos (hydroelectric).

Unused heats must be recovered for energy saving. The design and development of compact reactors for highly endothermic MCH dehydrogenation [6] would help recover vehicle exhaust heats. Because SOFC exhaust heats are sufficient to release hydrogen from MCH with respect to temperatures and amounts, electric and thermal cogeneration owing to MCH-driven SOFC would renew greenhouse agriculture in rural (SATOYAMA, i.e., countryside farm near mountain) areas.

With regard to the national tasks of 3E+S (energy security, environment, economics, and safety), the contribution of organic hydride will certainly be substantial.

References

1. Gupta RB (2008) Hydrogen fuel—storage, transport and utilization. CRC Press, New York, Chapt. 1 & 13
2. Saito Y, Hodoshima S, Matsushima Y, Shono A, Otake K (2007) 23rd Conference on energy system, economics & environment, Tokyo (2007.1.25-26), #32-3
3. Okada Y, Yasui M, Shimura M (2013) J Soc Automot Eng Jpn 67(11):63
4. Okada Y, Imagawa K, Shimura M (2014) Fuel Cells Technol 14(1):36
5. Okada Y, Hosono Y, Anzen J (2014) Jpn Soc Saf Eng 53(6):386
6. Saito Y, Aramaki K, Hodoshima S, Saito M, Shono A, Kuwano J, Otake K (2008) Chem Eng Sci 63:4935–4941

Topic: Liquid Biofuel Production

Naomi Shibasaki-Kitakawa

Abstract Liquid biofuels, bioethanol, and biodiesel (fatty acid methyl ester) are practical and offer attractive advantages over other forms of biomass energy resources in terms of high energy density and ease of storage and transport. However, there has been no significant increase in the production of these fuels in the past few years. The reason is that there is no profitability for producers owing to high production cost and no benefits for consumers because of unstable product quality and lower fuel economy and power. Moreover, emergence of next-generation automobiles (e.g., battery-powered and fuel cell electric vehicles) will lower the demand for gasoline. From this perspective, it seems as if there is no role left for biofuel production in Japan's future energy system. However, from another perspective, biofuels may have a vital function under some circumstances, if appropriate and profitable technologies are developed. Using the example of a novel biodiesel production process using resin catalysts, this article discusses those technologies and circumstances in which biofuels can be a part of future sustainable energy systems.

Keywords Biodiesel • Bioethanol • Liquid biofuel • Practical production technology

1 Introduction

Liquid biofuels offer attractive advantages in terms of high energy density and storage and transport. Bioethanol is useful as a gasoline additive (up to 10 %), with production worldwide in 2013 of 88 million kL. Biodiesel, or fatty acid methyl ester (FAME), is also used as a fuel alternative (100 %) to petroleum diesel, with worldwide production in 2013 at 26 million kL. However, there has been no significant increase in the production of these fuels in the past few years. One reason is that the production costs for bioethanol and biodiesel are much higher than those for petroleum fuels; thus there is no profitability for producers. In addition,

N. Shibasaki-Kitakawa (✉)
Tohoku University, Sendai, Japan
e-mail: naomi@rpel.che.tohoku.ac.jp

Y. Kato et al. (eds.), *Energy Technology Roadmaps of Japan*,
DOI 10.1007/978-4-431-55951-1_35

Table 1 Petroleum fuel consumption of automobiles and allowable maximum amount and actual production amount of liquid biofuels in Japan (2013)

Petroleum fuels	Gasoline consumption [million kL]	Petroleum diesel consumption [million kL]
Commercial vehicles	0.83[a]	16.04[a]
Private vehicle	55.97[a]	8.31[a]
Total	56.80[a]	24.35[a]
Biomass fuels	Bioethanol	Biodiesel
Allowable maximum amount [million kL]	5.68	24.35
Actual production amount [million kL]	0.022	0.0086

[a]Reported values [1]

the product quality is unstable and the fuel economy and power are low; thus there are few benefits of using these fuels for consumers.

Table 1 shows the petroleum fuel consumption of automobiles in 2013 in Japan and the allowable maximum amount and actual production amount of liquid biofuels. Gasoline consumption is three times higher than that of petroleum diesel consumption; however, the allowable maximum amount of biodiesel (FAME) is three times higher than that of bioethanol. As next-generation automobiles, electric-powered vehicles, and fuel cell vehicle are developed, gasoline consumption will be reduced. Thus, production of anhydrous bioethanol as a gasoline additive, which requires significant energy input, is likely to stop. For automobiles using an internal combustion engine, high promotion of engine efficiency and effective utilization of waste energy are still important features of the next generation. For these types of vehicles, petroleum diesel consumption will not change for a while. Therefore, it remains justifiable to develop systems for efficient and profitable production of high-quality biodiesel.

2 Technology Details: Future Prospects, Problems, and Risks

Industrial biodiesel production is still mostly performed by transesterification of triglyceride with methanol using homogeneous alkaline catalyst such as NaOH or KOH. However, the catalyst reacts with the triglyceride and its hydrolysate, free fatty acid (FFA) to form soap, which contaminates the product by dispersing the micelles of the by-product glycerol. Complex upstream and downstream processes are necessary to prevent soap formation and to remove impurities (soap, glycerol, and catalyst). Selection of the homogeneous alkaline catalyst causes serious problems: (1) cheaper oils with FFA content of more than 1 wt.%, which are not in competition with food, cannot be used for biodiesel feedstock, and (2) waste, such as the wash water used for removing impurities and the by-product glycerol contaminated with catalyst, must be

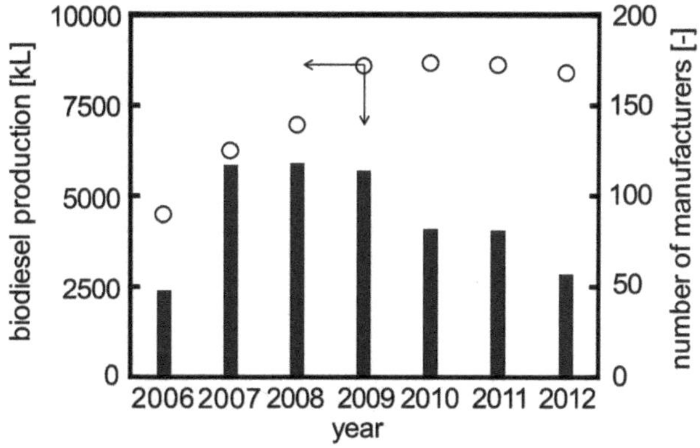

Fig. 1 Changes in biodiesel production and number of manufacturers in Japan

treated. These issues have a large impact on the production cost, and the entire process from upstream to downstream is not profitable.

Changes in biodiesel production and the number of manufacturers in Japan are shown in Fig. 1, as reported by the Council for Biodiesel Fuel Development of Japan Organics Recycling Association [2]. Annual production did not increase from 2009 (8,500 kL), and the number of manufacturers decreased from over 110 in 2008 to almost 50 in 2012. In Japan, only waste edible oil is used as feedstock, which causes further difficulties in producing high-quality biodiesel. The waste edible oil with high FFA content (acid value >2 mg KOH/g) must be discarded to improve product quality, and the discarded amount often reaches half of the total amount of oil. An appropriate recipe for reaction conditions (methanol and catalyst amounts, temperature, time, and so on) must be chosen for each batch depending on the properties of the oil used. In the process of product purification, washing with hot water containing some chemicals must be done five times. Thus, the actual production cost is 167 JPY/L (average values with and without depreciation and amortization and personnel costs, reported by each manufacturer) in 2012, which is much higher than the sales price of petroleum diesel (115–145 JPY/L).

For the continuation and further expansion of biodiesel production, it is indispensable to construct entire systems consisting of a sustainable supply of materials and simple production processes with the lowest environmental impact and cost and to stabilize and improve product quality. Biomass oil mainly consists of triglyceride and its hydrolysate, FFA. FFA is toxic to biomembranes and must be removed from food. The FFA content in the oil increases with the storage period after harvesting biomass, and the cost of oil with higher FFA content is lower. In edible oil refining, there are large amounts of waste FFA, 10–20 wt.% of the edible oil output, 2.57 million tons in 2013 in Japan. Thus, technology to completely convert waste FFA to biodiesel is strongly needed. Heterogeneous catalysts with no soap formation are also needed for more efficient continuous manufacturing and to prevent the generation of waste, such as the wash water and the by-product glycerol contaminated

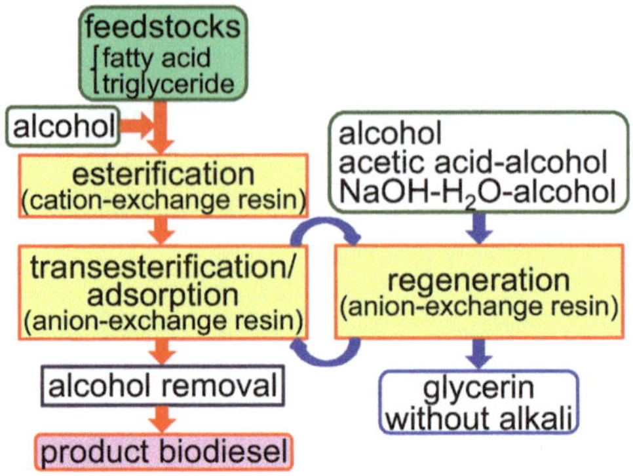

Fig. 2 Flow chart of biodiesel production process with ion-exchange resin catalysts

with catalyst. However, commercially available catalysts are simple, cheap, affordable, and reusable without loss of catalytic activity. To reduce production costs, scale-up of the apparatus is ineffective because production will not increase owing to a shortage of feedstock. It is necessary to construct a small-scale distributed manufacturing system that is profitable, by taking advantage of regional strengths, such as using waste materials and unused heat.

Recently, the use of ion-exchange resin catalysts has been reported to be a useful technology for continuous biodiesel production [3–7]. The process (Fig. 2) permits practical production of high-quality biodiesel from various oils with FFA content up to 100 wt.%, such as waste edible oil and waste acid oil. In addition, all by-products (glycerol and water) are adsorbed on the anion-exchange resin, so that the effluent from the process is free from these and fully meets the international standard specification, with no need for complex downstream processes. Periodic regeneration of the anion-exchange resin is needed, but high-quality glycerol and bioactive compounds can be recovered during regeneration, to increase profitability. A fully automated pilot plant [8] (Fig. 3) has been constructed and has functioned for more than 3 years with no loss of catalytic activity. In addition, the product quality has been confirmed by inspection organizations. Production costs using these resin catalysts are shown in Table 2 and compared with those using conventional alkaline catalyst. In the resin method, the complete conversion of both triglyceride and FFA are attained under stoichiometric condition, so that the feedstock oil cost is decreased owing to high product yield. As the FFA content of the feedstock increases, FFA does not participate in resin deactivation and the impact of the regeneration solution is drastically reduced. The cheapest cost (43.5 JPY/L) is attained using waste acid oil, and when the costs of equipment and personnel are taken into account, the total cost will be lower than that of fossil fuel.

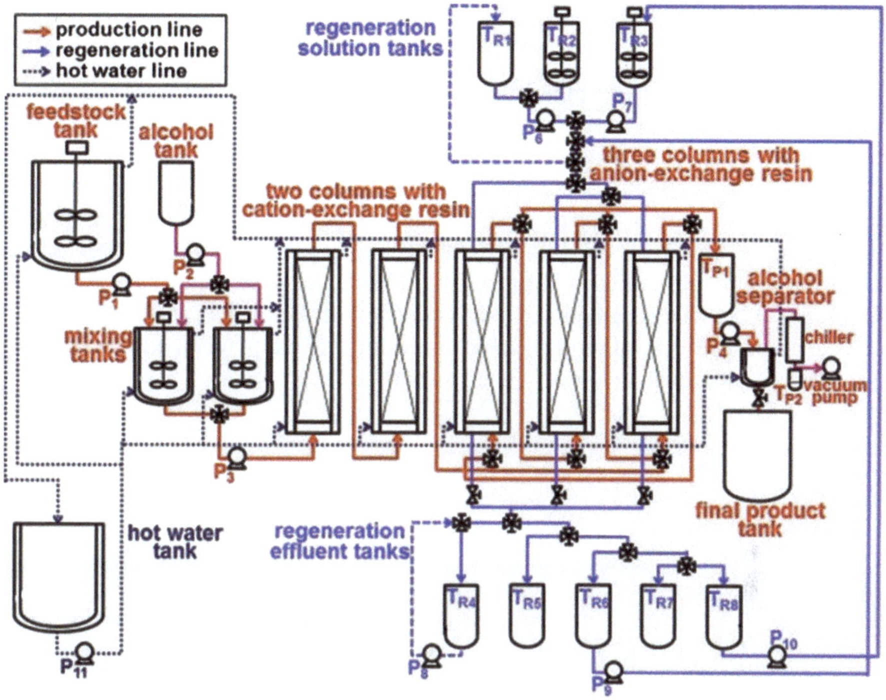

Fig. 3 Process flow diagram of automated pilot plant for continuous biodiesel production with ion-exchange resin catalysts

Table 2 Comparison of production cost using ion-exchange resin catalysts and homogeneous alkali catalyst

	Ion-exchange resin method with solution recycle		Homogeneous alkali method
	Waste edible oil (FFA <1 wt.%)	Waste acid oil (FFA >95 wt.%)	Waste edible oil (FFA <1 wt.%)
1. Feedstock oil cost	**41.1**	**26.0**	**43.0**[b]
2. Running cost	**43.4**	**17.52**	**55.0**[b]
Methanol, alkali	–	–	7.0[b]
Regeneration solution	40.1	12.9	–
Resin	1.6	3.0	–
Others	–	–	9.0[b]
Waste treatment	0.1	0.02	37.0[b]
Electricity	1.6	1.6	2.0[b]
Total 1 + 2	**84.5**	**43.52**	**98.0**[b]

Note: Waste edible oil, 43JPY/L[9]; waste acid oil, 34 JPY/L (= heavy oil A); MeOH, 75.1 JPY/L; acetic acid, 138 JPY/L; NaOH,116 JPY/L; and waste treatment, 1.7 JPY/L
[b]Reported values [9]

3 Benefits and Future Vision

It is possible to develop an entire biodiesel production system that is also profitable for producers. In addition, when bioethanol is used as alcohol to produce fatty acid ethyl ester (FAEE) as biodiesel, this fuel has higher fuel economy and power than FAME. Thus, the benefits for consumers will be improved. However, the amount of available biomass oil in Japan is significantly limited, such that the maximum production of biodiesel is estimated at 3.0 million kL, which is 10 % of the allowable maximum amount. Furthermore, to increase production, it is necessary to discuss whether to extend the cultivation of oil-bearing crops such as rice, soybean, rapeseed, and sunflower, which are conventional plants whose cultivation and harvest methods have already been established.

References

1. Annual report 2013 on fuel consumption of automobiles in Japan (Ministry of Land, Infrastructure and Transport)
2. Survey Results Report by Council for Biodiesel Fuel Development of Japan Organics Recycling Association. http://www.jora.jp/biodz/index.html
3. Shibasaki-Kitakawa N, Honda H, Kuribayashi H, Toda T, Fukumura T, Yonemoto T (2007) Biodiesel production using anionic ion-exchange resin as heterogeneous catalyst. Bioresour Technol 98:416–421
4. Shibasaki-Kitakawa N, Tsuji T, Chida K, Kubo M, Yonemoto T (2010) Simple continuous production process of biodiesel fuel form oil with high content of free fatty acid using ion-exchange resin catalysts. Energy Fuel 24:3634–3638
5. Ren YB, He BQ, Yan F, Wang H, Cheng Y, Lin LG, Feng YH, Li JX (2012) Continuous biodiesel production in a fixed bed reactor packed with anion-exchange resin as heterogeneous catalyst. Bioresour Technol 113:19–22
6. Singh B, Guldhe A, Rawat I, Bux F (2014) Towards a sustainable approach for development of biodiesel from plant and microalgae. Renew Sustain Energy Rev 29:216–245
7. Shibasaki-Kitakawa N, Hiromori K, Ihara T, Nakashima K, Yonemoto T (2015) Production of high quality biodiesel from waste acid oil obtained during edible oil refining using ion-exchange resin catalysts. Fuel 139:11–17
8. https://www.youtube.com/watch?v=kZAT7zoraGs&feature=youtu.be
9. 2010 Report by Kyoto Municipal Waste Edible Oil Fuel Production Facility.

Part VII
Commercial and Residential Energy Utilization

Mitsuhiro Kubota

Commercial and residential sectors account for ~1/3 of final energy consumption in Japan. Energy consumption of commercial and residential buildings should be reduced based on an accurate and reliable evaluation of building energy performance. Moreover, in both sectors, nearly half of consumed energy is supplied as electricity, and the remaining half is used as heat. Therefore, it is important to use electricity and heat efficiently and reduce losses of both energies. In the residential sector, ~6 % of electric usage is dissipated by standby power, indicating that this energy loss should be reduced by a novel device using nanoelectronics technology. Fuel cell combined heat and power (FC CHP) systems were introduced to the residential market in 2009. After the Great East Japan Earthquake in 2011, distributed power generation such as gas-fired CHP and FC CHP systems have been gaining attention for reinforcing energy resiliency and improving energy utilization efficiency. These systems can supply electricity through islanded operation, even during outages. Therefore, the introduction of CHP systems will be promoted in Japan in the future. In addition to the improvement of energy efficiency for each facility, smart energy systems, in which distributed generators are connected to form a network with user equipment and power storage, have also attracted attention in the country.

M. Kubota (✉)
Nagoya University, Nagoya, Japan
e-mail: kubota@nuce.nagoya-u.ac.jp

Commercial and Residential Buildings

Takao Sawachi

Abstract In developed countries including Japan, the development of regulations and evaluation methods for the energy performance of buildings is ongoing and should be near completion by 2020. The major index is primary energy use by the building. Strong correlation between predicted and real energy use is sought, although this is not easy mainly because it requires accord between the energy calculation method and standards of products used as building components. The development of innovative technologies must include product standards for energy performance, which makes product characteristics transparent and the product easily integrated in buildings. By 2050, a 64 % reduction of greenhouse gases over current conditions will be realized globally and an 80 % reduction in currently developed countries including Japan.

Keywords Building • Energy calculation • Standard • Equipment • 2°C scenario

1 Introduction

Energy use in buildings is directly connected to greenhouse gas (GHG) emissions from the building sector, which are a major cause of global warming. It is meaningful to review the footprint of technological ideas and subsequent research and development, as well as results from the types of concepts that have been realized as products, which have spread and contributed to energy conservation since the oil crises of the 1970s. Lessons from success and failure must be fully learned and the resulting experience should be applied through 2050.

This chapter addresses the target for reducing GHG emissions, which is proposed by the International Energy Agency (IEA) and agreed upon by many countries including Japan. The plan of the Japanese government is referenced for the strategy to reach the target in the building sector.

T. Sawachi (✉)
National Institute for Land and Infrastructure Management (NILIM), Ministry of Land, Infrastructure, Transport and Tourism (MLIT), Ibaraki, Japan
e-mail: sawachi-t92ta@nilim.go.jp

Y. Kato et al. (eds.), *Energy Technology Roadmaps of Japan*,
DOI 10.1007/978-4-431-55951-1_36

2 Present Status

In the building sector, policies regarding the spread of low-carbon buildings have already begun, as in other sectors such as industry and transportation. According to Energy Technology Perspectives 2012 [1] of the IEA, global energy-related CO_2 emissions from the building sector in 2009 were 8.06 $GtCO_2$, which represents 26 % of total worldwide energy-related CO_2 emissions, including both OECD and non-OECD countries.

Since the oil crises of the 1970s, policies for improving the energy efficiency of buildings have been implemented in most developed countries. However, there has been a continual increase of energy consumption in buildings, partly because of increases in the number of households and floor areas of service (commercial) buildings. In Japan, CO_2 emissions from the building sector, which is the sum of emissions from the "commercial and other sector" and "residential," exceeded that from "industries" in 2011 (Fig. 1). The first step to save energy, especially for space heating, was insulation in northern and central Europe and North America. Solar shading for reducing energy needs for space cooling has also been considered as part of the thermal performance of the building envelope. As a second step, since around 2000 in most developed countries, enhanced integrated energy performance of buildings has been targeted by including mechanical and electrical systems.

The current worldwide major index in evaluating the energy performance of buildings is annual primary energy consumption.

The evaluation of building energy performance should be as accurate as possible to reflect reality. However, there are many barriers that should be overcome before obtaining the ideal method for this evaluation. Therefore, it should be emphasized

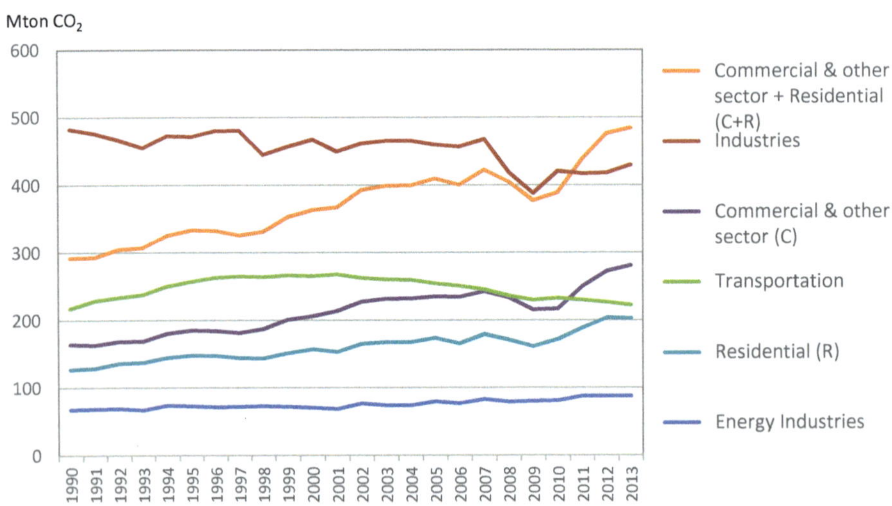

Fig. 1 Change of CO_2 emissions by sectors in Japan [2] (with allocation of CO_2 emissions from power generation and steam generation to each final demand sector)

that improved evaluation methods of building energy performance is one of the most critical technological issues in the building sector and environmentally related building engineering. Once such methods are realized, comparison between design alternatives will be much easier, with consideration of building use and climatic conditions.

Unfortunately, in reality, improvement of methods for evaluating building energy performance cannot be achieved rapidly, because of the following barriers: [3].

- Diversity of buildings and their functions
- Fragmented sector, involving various business and industries
- Lack of coordinated standards and codes for energy performance of buildings
- Scarce standards and guidance on renovation of existing buildings

To attain an accurate and reliable method for evaluating building energy performance and efficient coordination between the building sector and chemical engineering, the exchange of transparent knowledge is key. Standardization of products whose core technologies are based on chemical engineering should be coordinated with energy-related standards for buildings, which are still under development. For example, the contribution of fuel cells to energy saving in residential buildings cannot be evaluated solely by electricity generation and heat recovery efficiencies. This contribution is strongly dependent on the demand for electricity and hot water. In the latest Japanese building energy standard [4] considering time-series electricity and hot water use, packaged products of fuel cells (0.7–1.0 kW rated electricity output) including hot water tanks and auxiliary boilers must be tested for winter, summer, and intermediate season conditions. Background knowledge is reflected in Japanese industrial standards of stationary fuel cell power systems and in international standards, which have been led by Japanese industries.

Figure 2 shows how CO_2 emissions shall be reduced toward 2050 [1]. In the most influential scenario (2DS), which limits average global temperature increase

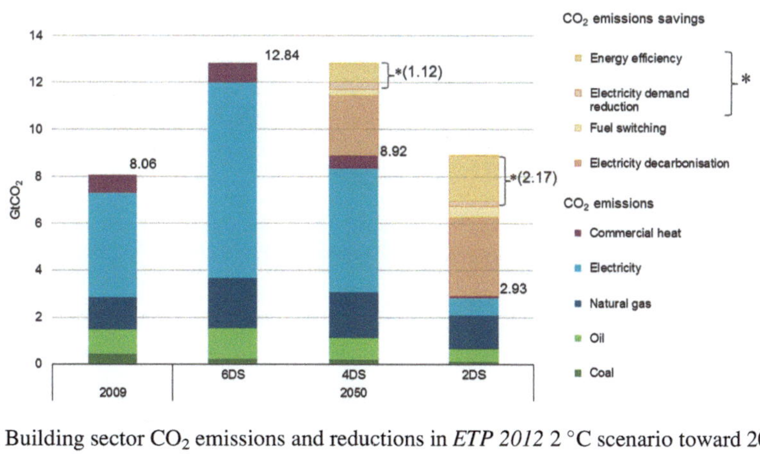

Fig. 2 Building sector CO_2 emissions and reductions in *ETP 2012* 2 °C scenario toward 2050 [1]

in 2050 to 2 °C, a 64 % reduction of CO_2 emissions in the global building sector is assumed (8.06–2.93 $GtCO_2$). This scenario considers the 2050 population and economic growth in developing countries, such as 1.5 times the current population, seven times the annual income, and four times the floor area of commercial buildings in India. This means that much greater reduction of CO_2 emissions in developed countries is necessary to realize the 2 °C scenario. Means to reduce this emission in the building sector involve the demand and supply sides of energy flow. It is noteworthy that a third of this reduction is on the demand side, i.e., by design of buildings and systems, and two thirds on the supply side, i.e., by electricity decarbonization and fuel switching.

Japan has set its target for reducing energy consumption and GHG emissions in accord with the policy of the G8, which has taken IEA's perspectives into consideration. That target is a worldwide 50 % reduction of GHGs by 2050 and an 80 % reduction in Japan and other well-developed countries [5]. Following the Japanese target, the new low-carbon technology plan was made by the Council for Science and Technology Policy in September 2013 [6]. In this plan, various promising technologies for energy production (supply), consumption (demand), distribution, and other purposes have been addressed, and their potentials for reducing GHGs are estimated. Among these technologies, "energy-efficient houses/buildings" are addressed and their technology roadmap is shown in Fig. 3.

To assure the spread of energy-efficient houses and buildings in Japan, the government has implemented measures to improve the environment with a view toward phased introduction of a requirement for new houses and buildings by 2020 to comply with energy-saving standards. Included are revisions of such standards and promotion of their more widespread use [7].

Among 37 promising environmental and energy technologies listed and evaluated in the new low-carbon technology plan [6], there are ten technology categories including energy-efficient houses/buildings. Those categories are shown in Table 1 with potential and maturity assessment.

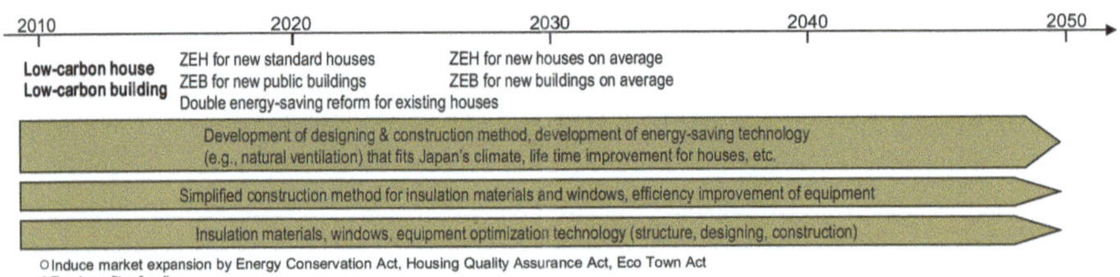

Fig. 3 Technology roadmap for "energy-efficient houses/buildings" in the new low-carbon technology plan of the Council for Science and Technology Policy [6]

Table 1 Evaluation of environmental and energy technologies [6]

Subcategory (only building sector-relevant categories are extracted from the original table)	Global GHG reduction effect in 2050[a]	Maturity phase
4. Solar energy (Light)	A	Basic research diffusion
5. Solar energy (Heat)	A	Basic research diffusion
8. Biomass	A	Basic research diffusion
18. Innovative devices (information system, lighting, display)	A	Applied research diffusion
22. Energy management system	A	Applied research diffusion
23. Energy-efficient houses/buildings	A	Applied research diffusion
25. High-efficiency heat pumps	B	Applied research diffusion
30. Fuel cells	B	Demonstration diffusion
31. High-performance electricity storage	[b]	Applied research diffusion
32. Heat storage/insulation technology	C	Applied research diffusion

[a]A = 1 billion ton or more, B = 0.3–1 billion ton, C = Less than 0.3 billion ton
[b]No evaluation made

3 Technology Roadmap

Figure 4 shows a roadmap toward realization of high-energy performance buildings in 2050. Energy use and relevant GHG emissions from the building sector have aroused public attention. This sector shares a major part of global GHG emissions with the industry and transportation sectors. Upon comparing these three sectors, the problem of the building sector can be highlighted as shown in Fig. 4 ("Issues, risks, threats"), based on reference [3].

There are various types of buildings (especially commercial buildings), and it is usually difficult even to find buildings with the same pattern of usage and function. Energy use is typically recorded for each type of energy medium, and it is not easy to determine this use for each building purpose and function. This causes difficulty in evaluating the effectiveness of countermeasures to save energy in buildings.

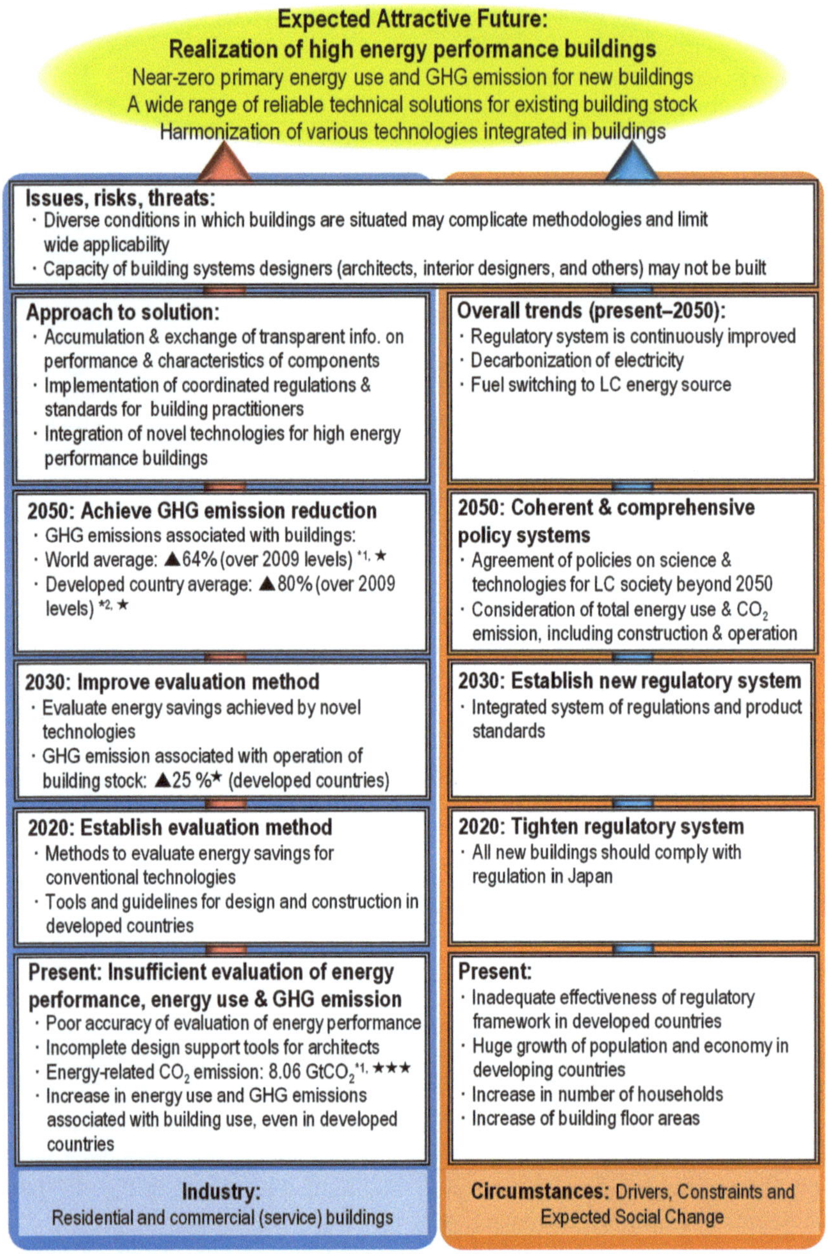

*1 Energy Technology Perspectives 2012 [1]
*2 Proactive Diplomatic Strategy for Countering Global Warming [4]

Abbreviations:
CO_2: Carbon Dioxide, GHG: GreenHouse Gas, LC: Low Carbon

Fig. 4 Roadmap toward realization of high-energy performance buildings in 2050

In contrast, drivers of motor vehicles can easily estimate the energy efficiency of their vehicles by dividing mileage by gasoline fill volume.

The envelope and energy systems of buildings contain various kinds of mechanical, electrical, and chemical products and materials, whose characteristics influence energy usage. A building is a complicated system integrating these various components. The energy-saving effectiveness of the system components depends on the building pattern of use and function and on outdoor climatic conditions. There are also interactions among various components. This is why the availability of transparent information on performance and characteristics of the components and their background technologies is important. Without such transparency, no single innovative technology or component can be satisfactorily integrated in the building with strong energy performance. Without such integration, it is very difficult to expect full-scale procurement and spread of the technology. This has been a lesson since energy conservation became a target of building engineering in the 1970s. The product standard is essential to define a product and make its characteristics transparent by prescribing relevant parameters and methods to measure them. Once the parameters are defined, they can be referenced in the design of buildings and energy systems such as air conditioning, lighting, and others.

One umbrella standard is necessary for calculating energy use (building energy performance) by referring to product characteristics, building configuration, building use, and climatic conditions. This standard is usually defined in national regulations, with criteria on how much energy use is permitted. In addition to the umbrella standard, there must be product standards, in which indices representing product characteristics are referenced in energy calculation at the building level.

Technological goals and solutions can be grouped into two categories, "energy demand side" and "energy supply side." This chapter mainly deals with the demand side, but national and international targets for CO_2 reduction have been presented as a whole.

According to the scenario by the IEA [1] mentioned above, CO_2 reduction in the building sector toward 2050 is expected more on the supply side, such as through solar/wind power, biomass, and carbon capture storage. The 26 % CO_2 reduction in developed countries by 2050 relative to the present is anticipated on the demand side, mainly by improving building energy efficiency. Overall, in the building sector, between the present and 2050, a 64 % reduction of CO_2 emissions is a target in the 2DS scenario, in which temperature increase in 2050 is limited to <2 °C. It appears that a nearly 80 % reduction is the target for 2050 only for currently developed areas such as Japan, North America, and the EU. This is the same as the Japanese target in that year, which has been announced by the Prime Minister [7].

As of January 2015, Japan has completely revised its national regulations for building energy performance and has planned to make operation of this regulation more stringent by 2020 [8]. According to that plan, by 2020, all new building construction including detached houses shall not be permitted without complying with the requirement in the regulation. Such more stringent regulation operation shall proceed incrementally from large buildings (except residential ones) ultimately to relatively small buildings like detached houses. The principal index of evaluation is total primary energy use of the building, and calculation methods and programs have been supplied by a group led by national research institutes [4]. The calculation methods are prescribed in detail in relevant materials like the umbrella standard for calculating energy use, and some portion shall be shared as international standards [9] in the business plan of a Japanese team for the ISOs. Although there is still no sign that energy use in Japanese buildings has begun to decrease, preparation for full-scale application of evaluation tools and their regulation is underway. The period before 2020 is considered the preparation phase.

Between 2020 and 2050, it is expected that energy use and relevant CO_2 emissions will be gradually reduced with the aid of energy calculation methods, which enables practitioners to select better countermeasures for energy saving. These methods will guide the development of innovative technologies at the component level as to how they should be evaluated for energy-saving effectiveness. In that process, innovative technologies being readied for full-scale application shall be given their own product standards, consonant with the umbrella calculation methods and standards.

In the roadmap, a contrast between conventional and innovative technologies is highlighted. The conventional technologies include insulation, other building materials, motors, pumps, boilers, heat pumps, ducts, pipes, refrigerants, and sensors (temperature, humidity, illumination, CO_2, CO, air flow rate, liquid flow rate, and others). Their control and design methods for improved combination and cost-effectiveness are also included. In particular, standardization of control methods is strongly prioritized, as its influence on building energy performance cannot be evaluated without definition of the technology.

4 Benefit and Future Vision

Near-zero primary energy use and carbon dioxide emission solutions have been adopted in new buildings and communities, and a wide range of reliable technical solutions have been made available for the existing building stock. Both developed and currently developing countries can enjoy energy to maintain healthy, comfortable, and productive indoor spaces in buildings.

References

1. International Energy Agency (2012) Energy technology perspectives 2012. International Energy Agency, Paris. ISBN: 978-92-64-17488-7
2. Green House Gas Inventory Office of Japan, http://www-gio.nies.go.jp/aboutghg/nir/nir-e.html
3. IEA Energy in Buildings and Communities program Strategic Plan 2014–2019, October 2013, http://www.iea-ebc.org/fileadmin/user_upload/images/Pictures/EBC_Strategic_Plan_2014_19.pdf

4. Relevant Materials for 2013 Energy Standard (Promulgated in January 2013). Manual of program for primary energy consumption in buildings, Technical Note of National Institute for Land and Infrastructure Management, MLIT (Ministry of Land, Infrastructure, Transportation and Tourism), Japan, No. 762, Nov 2013 (in Japanese)
5. Ministry of Foreign Affairs of Japan: Proactive Diplomatic Strategy for Countering Global Warming, November 2013. http://www.mofa.go.jp/press/release/press3e_000007.html
6. Council for Science and Technology Policy: new low carbon technology plan, Sept 2013. http://www8.cao.go.jp/cstp/english/doc/new_low_carbon_tec_plan/index.html
7. Prime Minister of Japan and his Cabinet: Japan revitalization strategy (Revised in 2014), Japan's challenge for the future, http://www.kantei.go.jp/jp/singi/keizaisaisei/pdf/honbunEN.pdf
8. Panel on Infrastructure Development, Ministry of Land, Infrastructure, Transportation and tourism: the fist report on future countermeasures for energy conservation in residential and non-residential buildings, Jan 2015, http://www.mlit.go.jp/report/press/house04_hh_000571.html
9. ISO DIS18523-1 Energy performance of buildings – Schedule and condition of building, zone and room usage for energy calculation – Part 1: non-residential buildings, ISO CD18523-2 energy performance of buildings – schedule and condition of building, zone and room usage for energy calculation – Part 2: residential buildings

Smart Community

Takao Shinji

Abstract In addition to energy conservation and cost savings, the introduction of distributed power generators is expected to bring new benefits such as users' resilient access to electric energy. To enhance the capacity of distributed power generation, co-introduction of combined heat and power coupled with solar panels and wind turbines is a promising approach. The Smart Community is a state of energy use in the residential and commercial sectors where distributed power generators are connected to form a network with user equipment and power storages as its nodes. Here, connection refers to transmission and utilization of power and information on the network state. To build a new energy system that contributes to energy conservation and the spread of renewable energy, the following benefits are pursued in the Smart Community: (1) energy conservation by "networked area energy use"; (2) contribution to stable electric power supply by total energy management, including demand-response measures and suppression of frequency fluctuation caused by intermittent renewable energy; and (3) Secure energy in designated areas.

Keywords Networked area energy use • Energy management • Renewable energy output leveling • Resilient power supply

1 Introduction

Combined heat and power (CHP) systems have been installed in buildings or factories in Japan. Because some CHP systems supplied electricity even during outages after the Great East Japan Earthquake (GEJE), expected merits from CHP systems are not any more merely confined to energy conservation and cost saving. Now, reinforcement of energy resiliency and stabilization of frequency has grown to be an important reason for investment.

Fossil fuel becomes more and more scarce. To allow CHP systems that consume precious fossil fuel to realize aforementioned benefits, we must construct the

T. Shinji (✉)
Tokyo Gas Co., Ltd., Tokyo, Japan
e-mail: tshinji@tokyo-gas.co.jp

Y. Kato et al. (eds.), *Energy Technology Roadmaps of Japan*,
DOI 10.1007/978-4-431-55951-1_37

infrastructure for networked area energy use. In this chapter, the present status is briefly summarized, and future visions for the Smart Community are illustrated.

2 Present Status

The supply-demand balance became tight after the GEJE due to halting of nuclear power plants. This situation made it more urgent to increase power generation capacity while reinforcing energy resiliency for disasters. To achieve this with CHP systems, merely increasing installed capacity in buildings and factories cannot unleash their maximum potential – a "networked area energy use" (Fig. 1) is needed.

Energy efficiency improves by several percent when CHP systems are integrated over the network. Energy efficiency improves because the ratio of peak demand to total demand is reduced by the accumulation of loads when their patterns vary. A more balanced demand over the 24 h with less peaks enables a generator to operate at large load factor and high efficiency.

For example, the power load peak of an office building is around noon and that of a residence is in the morning and evening (Fig. 2). Thus, the power load curve of the sum of an office building and residence has a smaller peak and larger minimum, which allows a generator to operate at high load factor and efficiency (Fig. 3).

The same applies in the case of "heat demand." To operate CHP systems more efficiently, heat recovery is crucial. To recover additional heat, a higher load factor achieved by networked area energy is necessary. Integration of CHP systems over a network can realize a better heat recovery.

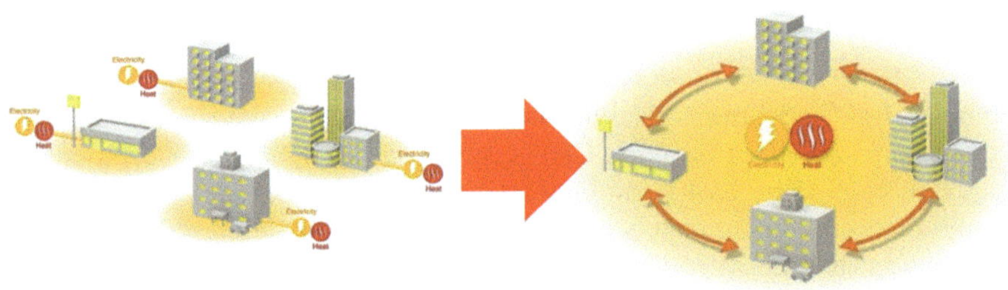

Fig. 1 Networked area energy use

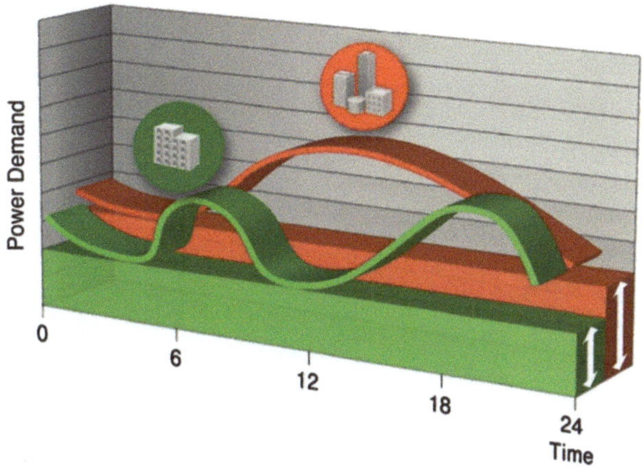

Fig. 2 Load curve of office and residence

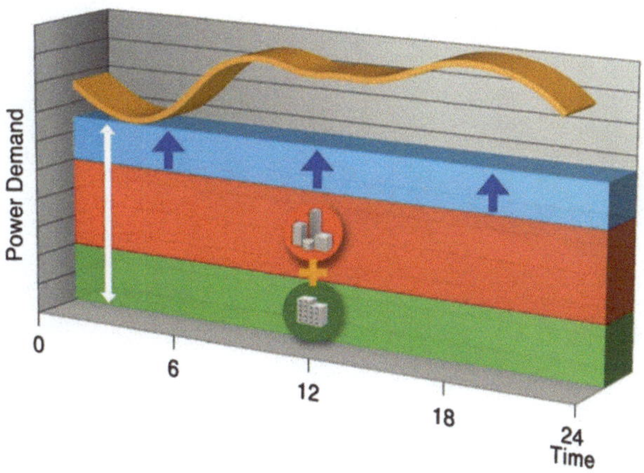

Fig. 3 Load curve of sum of office and residence

2.1 Improvement of Energy Resiliency (Electric Power Supply During Outage)

Normally, CHP systems are connected to the power grid. In case of outage, many of CHP systems merely supply heat and power to the building or factory but not to the neighborhood. This can be amended by allowing the interconnected CHP systems to change the mode to an islanded operation, supplying electricity to a limited area such as a Smart Community (Figs. 4 and 5). Installation of gas-fired CHP systems and islanded CHP operation during outages facilitates compensation for the damaged electric networks with gas networks (Fig. 6).

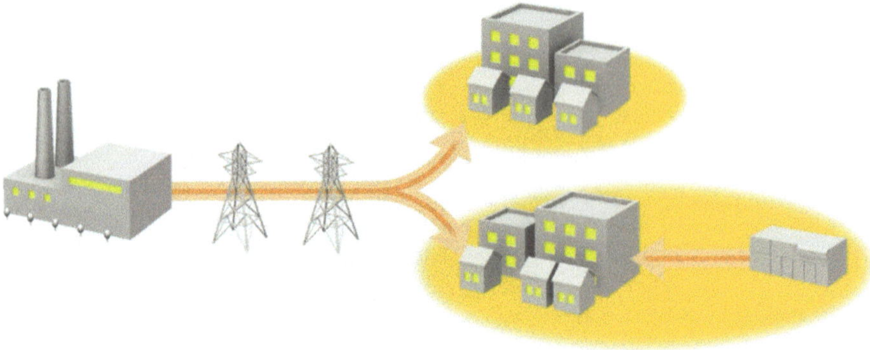

Fig. 4 Normal operation

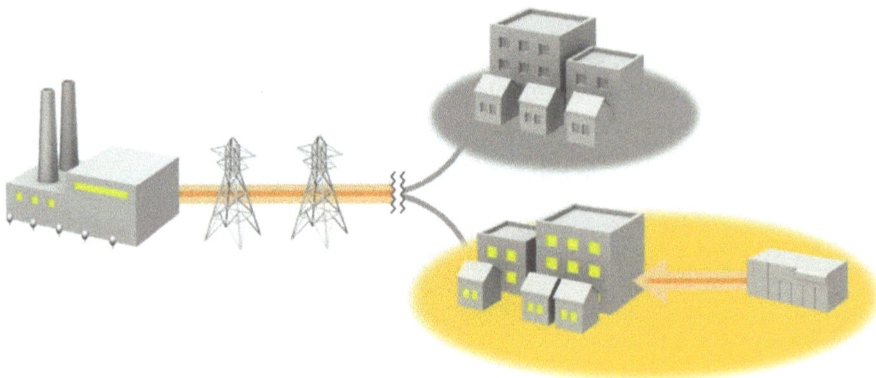

Fig. 5 Islanded operation during outage

2.2 Suppression of Fluctuation Caused by Intermittent Renewable Energy

Renewable power sources (i.e., solar cells and wind turbines) have increased its importance after the GEJE to compensate for the loss of electricity supply from nuclear power plants. However, this development has influenced the stability of the electric power grid (Fig. 7).

Fossil fuel-consuming generators such as gas engines, gas turbines, and diesel engines, the components of CHP systems, can suppress the fluctuation caused by intermittent renewable energy because of the rapid controllability (Fig. 8).

Aforementioned "networked area energy use" can maximize the benefit of CHP systems in stabilization of power supply [1, 2]. The amount of compensation realized by a CHP system is very small compared with large-scale gas or oil-fired power plants.

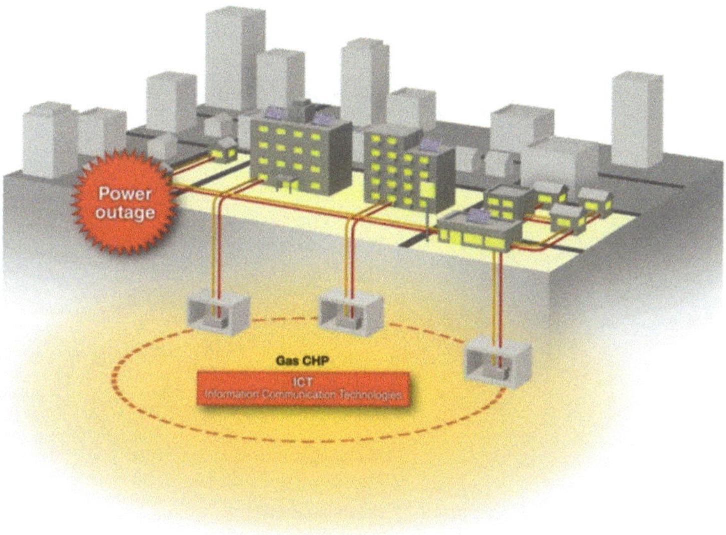

Fig. 6 Islanded operation in a Smart Community (*ICT* Information Communication Technology)

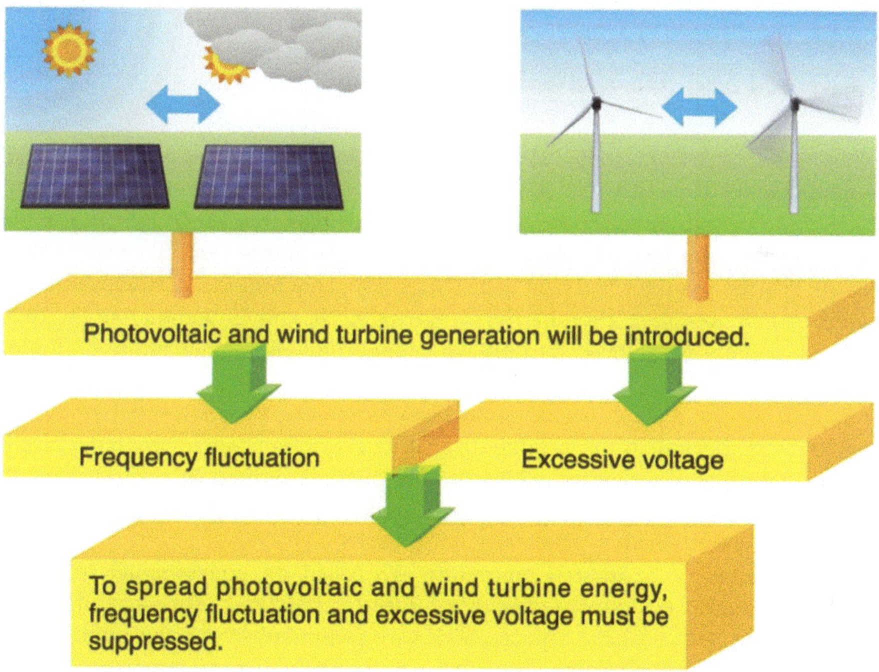

Fig. 7 Penetration of renewable energy

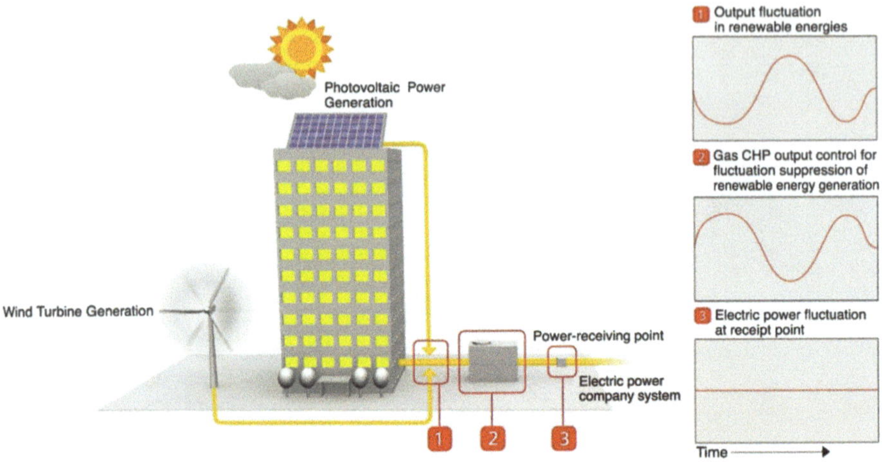

Fig. 8 Suppression of fluctuation caused by renewable energy

However, if CHP systems are interconnected, the total amount will be greater and they would be able to enhance grid stability.

3 Benefit and Attractive Future Vision

In the future, all CHP systems will be connected to each other by information and communication technology, which enables (1) operations at a high efficiency, (2) islanded operation under the disasters in Smart Communities, and (3) a stable power supply with lots of solar and wind power is introduced. To realize this, electric distribution lines and heat distribution pipes are important. However, these face economic and regulation barriers. Figures 9, 10, and 11 summarize the road map for coping with these barriers to enhance the implementation of Smart Communities.

The Smart Community has economic benefits and energy resiliency, but it is not easy to realize because of its installation cost. Energy efficiency improvement from flexible operation via electric cables or heat pipes is limited to several percent. Thus, it is not easy to recover initial cost (including construction cost) from energy charges that are retained by efficiency improvement.

The facilities of a Smart Community are based on conventional technical standards such as *Technical Standards for Electric Facilities*. This standard often does not consider the new aspects of Smart Community and thus hinders its

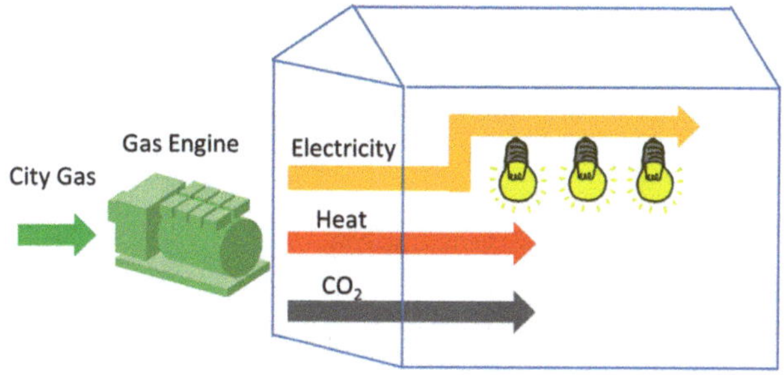

Fig. 9 Trigeneration for greenhouse

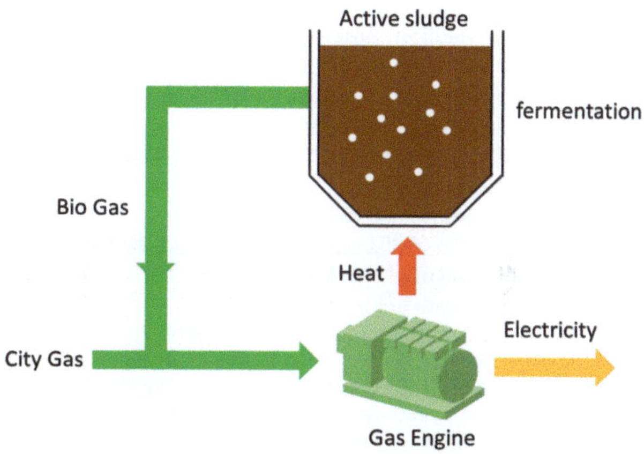

Fig. 10 Heat for fermentation

introduction. There is another regulation that is hindering the implementation of Smart community. To install electric cables or heat pipes under roads, one needs the permission of the road owner. Improvement of current regulation (including deregulation) is required. In the road map summarized in Fig. 11, this is realized step by step, in tandem with development of management technology and increase in social recognition.

To accelerate the social recognition that drives reregulation, it is important to better emphasize that there are greater benefits other than what is currently internalized in the economy. In addition to economic benefits and energy resiliency with islanded operation, the Smart Community will provide benefits such as distributed CO_2 generation. The CO_2, together with heat and electric power generated, can be used to grow plants in a greenhouse (Fig. 9). Another use of generated heat is to aid fermentation in the production of biogas from wastewater (Fig. 10). In this way, Smart Communities can assist promotion of rural areas.

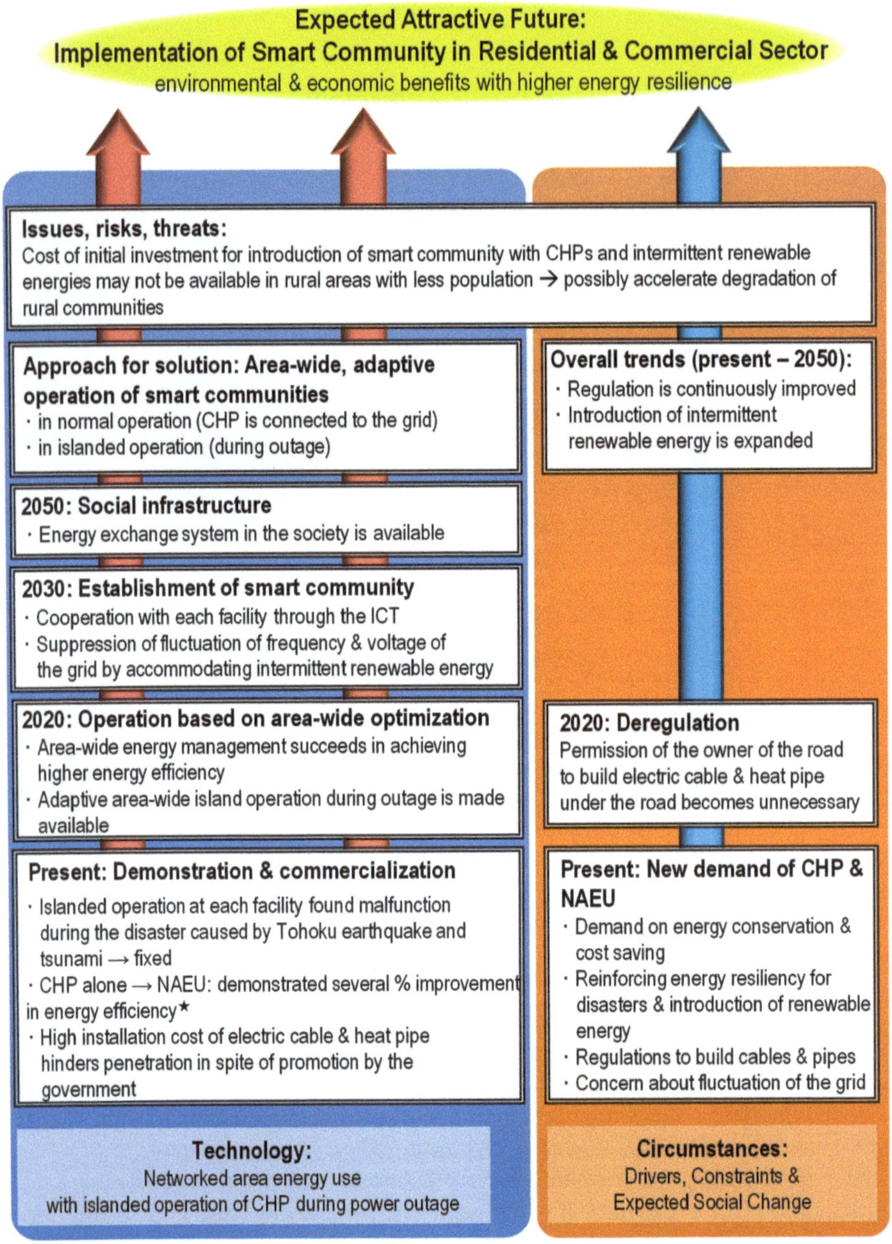

Abbreviations:
CHP: Combined Heat and Power, NAEU: Networked Area Energy Use, ICT: Information and Communication Technology

Fig. 11 Road map for development of Smart Community

References

1. Shinji et al (2009) Evaluation of micro grid including power storage device. J Jpn Inst Energy 88 (3):199–204. doi:10.3775/jie.88.199
2. Shinji T, Yokoyama A, Hayashi Y (2009) Distributed generation in Japan. Power Energy Soc Gen Meet (PES'09 IEEE), pp. 1–5. doi:10.1109/PES.2009.5275263

Fuel Cell Combined Heat and Power Systems in Residential Sector

Junichiro Otomo

Abstract To realize future widespread use of residential fuel cell systems, a technology road map of 1-kW-class fuel cell combined heat and power (FC CHP) systems is proposed based on technology assessment and cost analysis. In this chapter, residential applications of polymer electrolyte fuel cells (PEFCs) and solid oxide fuel cells (SOFCs) are discussed in terms of performance and cost. Currently, the cumulative number of shipments of PEFC CHP systems is the largest among several FC types in Japan and worldwide. The share of SOFC CHP systems with high electrical efficiency, i.e., small heat-to-power ratio (H/E), will increase gradually in the future, because a relatively small H/E is desirable in the Japanese residential sector. The cost reduction target for FC CHP systems is addressed in terms of economic and energy efficiencies. The cost analysis shows the break-even point of the FC CHP system price for residential heat and power demands. The result suggests that both the PEFC CHP and SOFC CHP systems can be applied to suitable demands of heat-to-power ratios. Considering the cost payback time, the final goal of FC CHP system price is discussed. The proposed technology road map suggests a technological pathway to achieve the final target of 1-kW-class FC CHP system price with relevant technological improvements and cost reduction. Improvements of electrical efficiency and durability (i.e., system lifetime) and material development for state-of-the-art fuel cell technology are also treated.

Keywords Polymer electrolyte fuel cell • Solid oxide fuel cell • Residential fuel cell system • Combined heat and power • Cost analysis

1 Introduction

The electricity mix in Japan greatly changed after the Great East Japan Earthquake, and currently thermal power generation makes up ~90 % of this mix. Considering economic and environmental impacts, the need for technology development of

J. Otomo (✉)
The University of Tokyo, Chiba, Japan
e-mail: otomo@k.u-tokyo.ac.jp

Y. Kato et al. (eds.), *Energy Technology Roadmaps of Japan*,
DOI 10.1007/978-4-431-55951-1_38

power generation is increasing further. Fuel cells (FCs) are promising alternative power generation technologies. Regardless of the magnitude of power generation, FCs can directly convert chemical energy to electricity, resulting in very efficient power generation with even small systems such as a residential 1-kW-class system. In fact, a recent 1-kW-class solid oxide FC combined heat and power (SOFC CHP) system developed as a commercial product has achieved an electrical efficiency (LHV) of 46.5 % (total efficiency = electrical efficiency + heat recovery efficiency, LHV = 90 % in 2012) [1]. Although electrical efficiency of state-of-the-art LNG combined cycles has reached ~60 % (LHV), average electrical efficiency of thermal power plants including coal fuel is approximately the same as that of the SOFC CHP system (46.5 %) [2, 3]. Furthermore, FC CHP systems can use waste heat as well as generate electricity. Thus, FC CHP systems can play a role in alternative power sources of thermal power plants and as a decentralized power generation system. In 2009, a residential polymer electrolyte fuel cell (PEFC) CHP system was developed into a commercial product for residential use in Japan. In 2011, an SOFC CHP system was also commercialized. Furthermore, a fuel cell vehicle (FCV, "Mirai") was launched in the Japanese market by Toyota Motor Co. at the end of 2014. Thus, the wide diffusion of FC technologies has only just begun in recent years, especially residential FC CHP systems (PEFC CHP and SOFC CHP). However, to realize large-scale diffusion of FC systems in the near future, several technological challenges remain for FC CHP systems, such as improvements of electrical efficiency, power density, lifetime, cell and system design, and reduction of system cost. In this chapter, a technology road map of 1-kW-class FC CHP systems for stand-alone houses and condominiums is addressed in terms of cost, performance, and relevant material development, to provide a research strategy for FC technologies.

2 Present Status

2.1 Stationary Applications of Fuel Cells

In the USA and Japan, stationary systems of phosphoric acid fuel cells (PAFCs) for industrial use were commercialized in the 1990s. The number of units shipped gradually increased, but the cumulative number became saturated in the 2000s. The cumulative number of PAFC systems shipped in Japan was 230 units in 2009, and the global number in 2006 was 205 units [4]. In contrast to PAFCs, in the 1990s, there was extensive development of polymer electrolyte fuel cells (PEFCs). Specific advantages of PEFCs are (1) high power density, (2) compact structure, (3) rapid start-up, and (4) excellent dynamic response [5]. In 2009, 700-W residential PEFC CHP systems were commercialized in Japan. The cumulative number of PEFC units shipped was ~77,000 from 2009 to 2013. The electrical efficiency of PEFC CHP systems with city gas is 39 % (LHV) and heat recovery efficiency 56 %

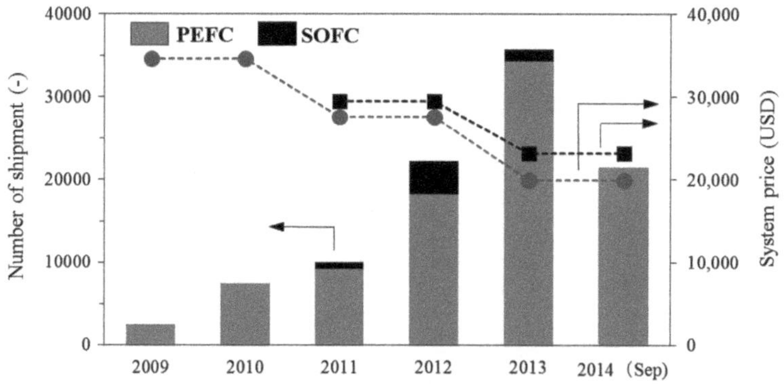

Fig. 1 Number of units shipped for PEFC CHP and SOFC CHP systems and their system price in Japan (system price does not include subsidy by Japanese government)

(LHV) (total efficiency 95 % in 2013) [6]. System price was ~3,500,000 JPY (~35,000 USD) in 2009, which declined to 2,000,000 JPY (~20,000 USD) by 2013 [7–9]. Globally, annual shipments of FC systems in 2013 reached 66,800 units, growing by 46 % compared with 2012 [10]. PEFCs still led unit shipments in 2013, accounting for 88 % of the total. Regionally, Asia dominated unit shipments in that year, with a 76 % share of total units [10]. Therefore, the current leader among stationary fuel cells is PEFCs, and the largest share of PEFC shipment is in Japan. The number of shipments of PEFC CHP systems and system price in the country are shown in Fig. 1 [11, 12]. The system price decreased continuously with increasing shipments from 2008 to 2014. Lifetime has been improved, from 8,000 h in 2005 to 40,000 h in 2008, which can also contribute to reducing system cost as well as a scale-up effect of annual production. Thus, the status of PEFC CHP systems has progressed steadily in terms of performance, cost, and diffusion in society.

In 2011, an SOFC CHP system with a flat tubular cell was developed into a commercial product for residential use in Japan. As stated above, the power generation efficiency of a residential SOFC CHP system is already superior to that of a residential PEFC CHP (SOFC, 46.5 % [1, 13–17]; PEFC, 39 % for city gas [18–23]). Although rated power generation efficiency of the SOFC CHP system is 46.5 %, the largest reported value has been 60 % [24]. Therefore, the SOFC has great potential as an alternative decentralized power generation system. Because the system price is being continually reduced as shown in Fig. 1, the number of shipments of SOFC CHP systems will increase in the near future. The improvement in power generation efficiency of SOFC and PEFC systems is summarized in Fig. 2 [14–23]. In field testing of SOFC CHP systems, a 40,000-h lifetime has been achieved [25].

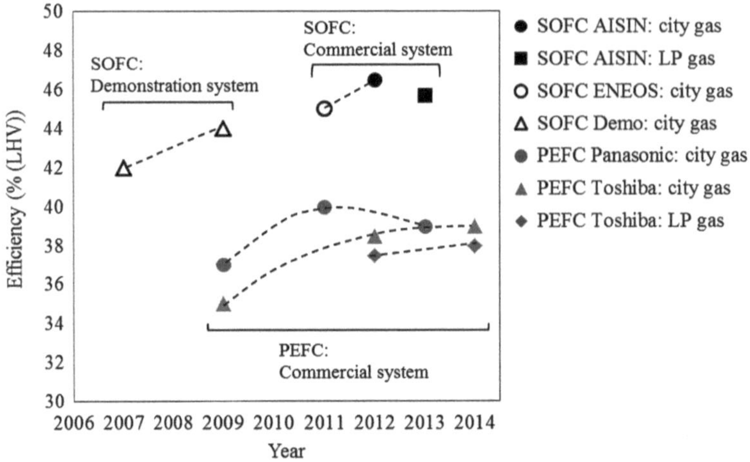

Fig. 2 Electrical efficiencies (*LHV*) of PEFC CHP and SOFC CHP systems

Table 1 Deregulations

For electricity utility industry laws:
1. No requirement for full-time monitoring
2. No requirement for exchanging fuel gas with inert gas
3. No need for apparatus to prevent pressurization
4. No requirement for appointing electrical engineers
5. No requirement for submitting safety regulations
For fire service laws:
1. No requirement for submitting setup
2. Deregulation of retention of the interval of FC systems
3. No need for backfire flame arrester

2.2 Deregulations

There have been several deregulations to promote widespread use of stationary FC systems in Japan [26]. These deregulations are summarized in Table 1.

These deregulations can help spread FC CHP systems and technology developments.

3 Technology Road Map

In Japan, the New Energy and Industrial Technology Development Organization (NEDO) has promoted the development of FCs by manufacturers [27, 28]. In the road map proposed by NEDO, 2030 targets for PEFC CHP and SOFC CHP energy

Table 2 Future target for FC CHP systems proposed by NEDO [27, 28]

	Present	2030
700-W PEFC CHP		
Power generation efficiency, LHV (%)	39	>40
Durability (hr)	>40,000	>90,000
System price (USD/W)	29	<4
700-W SOFC CHP		
Power generation efficiency, LHV (%)	46.5	>55
Durability (hr)	>40,000	>90,000
Power density (kW/L)	0.1–0.2	0.4–1
System price (USD/W)	33	<4

conversion efficiencies exceed 40 % and 55 %, respectively. System price should be <400 JPY/W (4 USD/W) for both systems (Table 2).

Recently, NEDO also published a road map toward hydrogen economy, separated into three phases [29]: phase 1 (present–2020), scaling up installation of FC CHP systems and FCVs; phase II (2020–2030), large-scale installation of hydrogen power generation systems and hydrogen supply systems; and phase III (2040), establishment of CO_2-free hydrogen supply systems with carbon capture and storage and/or renewable energies. Regarding this scenario, targets for residential FC CHP systems are as follows:

1. The target value for shipment of residential FC CHP systems in Japan is 1.4 M units in 2020 and 530 M in 2030.
2. Cost payback time is 7–8 years in 2020 and 5 years in 2030.

Relevant projects are promoted by NEDO [30, 31].

Regarding the targets, cost analysis and a related technology road map from the present (2015) to 2050 are discussed in the next section.

3.1 Evaluation of Cost Target for FC CHP Systems

To realize widespread use of FC CHP systems, required system price P_{FC} was evaluated by comparison with a conventional system that combines grid electricity and hot water supply with a city gas boiler. The impact of FC CHP system price on power and heat generation cost (i.e., electricity and waste heat) with changing electrical efficiency was investigated using Eqs. 1 and 2. Power and heat generation cost C_p (JPY/kWh) was estimated by

$$C_p = C_{system} + C_{fuel}, \tag{1}$$

where C_{system} and C_{fuel} are FC system and fuel costs (i.e., city gas cost), respectively. The FC CHP system cost includes stack, auxiliaries (heat exchanger, power converter, air blower, gas burner), fuel processing system, control system, casing, and a hot water tank. C_{system} is calculated by $(P_{FC} \cdot \alpha)/E_{an}$, where P_{FC}, α, and E_{an} are FC CHP system price, annual expenditure rate, and electricity generated annually, respectively. Key assumptions are operating rate of FC system $= 0.8$, lifetime $=$ 10 years, and $\alpha = 0.1$. The functional unit is 1 kWh of electricity generated by the FC CHP system. C_{cost} was calculated by

$$C_{fuel} = \frac{UC_{CG}}{\eta_{elec}} \tag{2}$$

where UC_{CG} is the unit cost of city gas (CG) and η_{elec} is net AC electrical efficiency of the FC CHP system. For that system, $C_{p,FC}$ includes electricity and waste heat; it was calculated by changing η_{elec} from 0.35 to 0.6, corresponding to variation of heat-to-power ratio (H/E) from 1.57 to 0.5 under the assumption of a fixed total energy conversion efficiency ($\eta_{elec} + \eta_{heat} = 0.9$). Power generation cost $C_{p,conv}$ of the conventional system (grid electricity and hot water supply with city gas boiler) was calculated for comparison. $C_{system,conv}$ of the conventional system was calculated similar to the FC CHP system. $C_{p,conv}$ was calculated by summation of grid electricity cost (23 JPY/kWh) and hot water supply cost with change of H/E from 1.57 to 0.5. Fuel cost of the hot water supply was from $C_{fuel,conv} = UC_{CG}/\eta_{heat}$. The values of the parameters are summarized in the Appendix.

Figure 3 shows the ratio $C_{p,FC}/C_{p,conv}$ vs. P_{FC} with variable power generation efficiency, assuming typical city gas and electricity prices for home use in Japan,

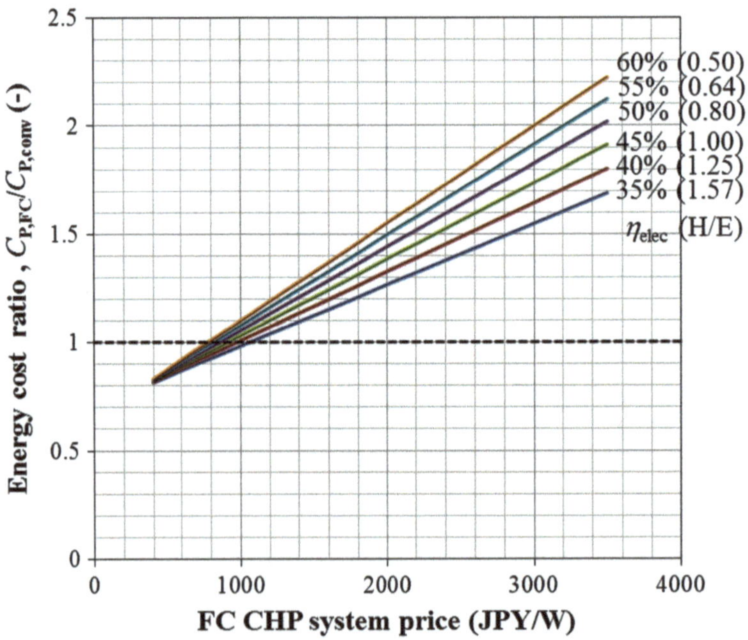

Fig. 3 Energy cost ratio $C_{p,FC}/C_{p,conv}$ vs. FC CHP system price (P_{FC}) with various power generation efficiencies (LHV) for 700-W residential FC CHP systems. The *broken line* represents the break-even point

i.e., 140 JPY/Nm^3 and 23 JPY/kWh, respectively. The result shows that the ratio can reach the break-even point ($C_{p,FC}/C_{p,conv} < 1$) if the FC CHP system price is less than 800–1,000 JPY/W (8–10 USD/W). This suggests that both PEFC CHP and SOFC CHP systems can be applied to suitable demands of H/E = 0.5–1.57. In fact, the ratio of total load (H/E) depends on location in Japan, and PEFC CHP systems save energy in cold regions [32]. As described above, the current share of PEFC CHP system shipments is largest among several FC types in Japan and the world. In the near future, however, the share of SOFC CHP system will increase gradually, because the average H/E of residential use in Japan is closer to that of the SOFC CHP system, given its high power generation efficiency. The break-even point was also evaluated for use of a CO_2 heat pump as a competitive technology of hot water supply. This point is achieved if the FC CHP system price is <600–800 JPY/W (6–8 USD/W).

The analysis suggests that the break-even point for the FC CHP system price is 800–1,000 JPY/W, compared with the conventional system. As mentioned above, however, target values of cost payback time are 7–8 years in 2020 and 5 years in 2030 [29]. Japanese gas companies claim that annual cost benefits are 50,000–60,000 JPY for a 700-W PEFC CHP system [33] and ~90,000 for a 700-W SOFC CHP system [34]. Therefore, FC CHP system prices are calculated as follows: PEFC CHP 350,000–480,000 JPY in 2020 (5–7 USD/W) and 250,000–300,000 JPY in 2030 (4 USD/W) and SOFC 630,000–720,000 JPY in 2020 (9–10 USD/W) and 450,000 JPY in 2030 (6 USD/W). These values largely correspond to the target values in Table 2.

3.2 Technology Road Map

Quantitative technological requirements are addressed relative to the achievement of targets in Table 2 and the results of Fig. 3. As mentioned above, targets for shipments of residential FC CHP systems in Japan are 1.4 M units in 2020 and 530 M units in 2030 [29]. The number of houses in the country in 2008 was 57,590,000, among which ~7,600,000 were empty. Therefore, the total number of households was ~50,000,000 (stand-alone houses 27,460,000 and apartment complexes 20,690,000; the remainder are 1,330,000 row houses) [35]. Considering the values in reference [35], 10,000,000 units of 1–10-kW-class FC CHP systems was assumed as the highest number of shipments (20 % of total household number in 2050). Current cost structures of PEFC CHP and SOFC systems are shown in Table 3 [29]. The results suggest that the cost ratio of balance of system (BOS: fuel processing system, auxiliaries, controller, case, and hot water tank) to module is 2.3–5.7. This means that BOS cost reduction is very important, as is reduction of module production cost.

Considering the above, the proposed technology road map from the present to 2050 is summarized in Table 4, suggesting that the required system price in Table 2, i.e., <400 JPY/W (4 USD/W) in 2030, can be fully achieved for both PEFC and

Table 3 Cost structures of FC CHP systems [29]

	PEFC CHP	SOFC CHP
Module	0.15	0.3
BOS	0.85	0.7
Fuel processing system	0.15	0.2
Auxiliaries	0.15	0.1
Controller	0.1	0.1
Case	0.15	0.1
Hot water tank	0.3	0.2

SOFC CHP systems. In fact, detailed cost analysis of SOFC module production revealed that achievement of the required module production cost is possible [36]. In this analysis, cost functions including relevant scale factors (i.e., learning curves) were used for all apparatus in every stack production process. Thus, although rigorous evaluation of FC production cost remains difficult, the proposed values of module production costs in Table 4 are mostly consistent with the previous study [36]. The BOS cost of a PEFC CHP system was also evaluated in the auxiliaries cost analysis project [37]. Based on the results, the BOS cost can be reduced to <200,000 JPY/system (290 JPY/W). In conclusion, if several key parameters such as production scale, power generation efficiency, and production costs are achieved, the targeted system price will be reached. For production scale, the share of SOFC CHP systems will increase gradually as discussed above, because of its high power generation efficiency. However, considering the target for total shipments, a tremendous scale-up of module production is essential. For example, as shown in Table 2, a total shipment of 1.4 M units in 2020 will require a huge and rapid production scale-up, such as 400,000 unit/year. Therefore, consistency between the target value of total shipment and scale-up feasibility of module production should be carefully considered.

Considering the targets of total shipments of FC CHP systems and the development of a hydrogen economy as proposed by NEDO [29], phase II (2020–2030, large-scale installation of hydrogen power generation systems) is considered a turning point at which the production scale of SOFC CHP systems is greater than that of PEFC CHP systems. Technological innovation of SOFC module production must promote large-scale installation of SOFC CHP systems. However, if the share of FCVs increases steadily, costs of relevant accessories of PEFC systems such as stack, auxiliaries, and fuel processing systems will decline. This may contribute to widespread use of PEFC CHP systems. Development of the hydrogen economy will also foster widespread use of these systems, because it can reduce the cost of hydrogen production and allow PEFC CHP systems to simplify relevant accessories, thereby decreasing system cost. Therefore, related technological developments and market status will affect the diffusion of PEFC and SOFC CHP systems in society.

The technology road map clearly shows the innovation pathway to reach the cost target. The required conditions of system cost and power generation efficiency can

Table 4 Proposed technology road map for stationary FC CHP systems

	Present (2015)	2020	2030	2050	Remark
Total shipment	100,000 (100 k)	1,400,000 (1.4 M)*	5,300,000 (5.3 M)*	10,000,000 (10 M)	*Ref. [29]
Production scale (stack/year): PEFC	30,000	300,000	300,000	300,000	
SOFC	1,000	100,000	500,000	500,000	
Power generation efficiency[a, b] (%): PEFC	39	40	40	40–45	Ref. [18–23]
SOFC	45	50	55	60	Ref. [14–17]
Lifetime (year): PEFC	>5	10	15	20	
SOFC	5	10	15	20	
Module cost[c] (JPY/W): PEFC	–	50	40	30	
SOFC	–	80	50	35	Ref. [27–29, 36]
BOS cost[c] (JPY/W): PEFC	–	250	200	150	Ref. [37]
SOFC	–	300	250	150	Ref. [36, 37]
System cost[c] (JPY/W): PEFC	–	300	240	180	
SOFC	–	380	300	185	
System price[d] (JPY/W (USD/W)): PEFC	2,900 (29)	400–500 (4–5)	<400 (<4)	<300 (<3)	Ref. [7–9, 27–29]
SOFC	3,300 (33)	500–1,000 (5–10)	<400 (<4)	<300 (<3)	Ref. [13, 27–29, 36]

[a]Efficiency calculated on LHV base
[b]Power loss due to module, DC-AC power inverter, and auxiliaries
[c]Module, BOS, and system costs correspond to production cost
[d]System prices correspond to market prices (system price < system cost)

also be realized by both FC CHP systems. The following issues should be resolved for the development of those systems in the near future:

(A) PEFC CHP system

For PEFC systems, three subjects have been investigated, i.e., fundamental, technological development for practical applications, and future technology [30]. In the current phase, studies of practical applications are important for PEFC CHP systems to attain widespread use:

A1. Fundamental: degradation mechanism of membrane-electrode assembly (MEA), new material development for electrolyte and electrode catalyst, new methods for analysis, and measurement of MEA and materials
A2. Technological development for practical applications: development of production technologies for reducing production cost and amount of Pt catalyst used, safety technology, and improvement of durability of MEA and system
A3. Future technology: new technological research that can realize significant cost reduction and high durability of PEFC systems

The current phase of stationary PEFC system development is practical use and popularization. Thus, the subjects of A1 and A2 are important for PEFC CHP systems. The subjects of A3 focus on the development of FC vehicles rather than stationary PEFCs. Therefore, improvement of durability and the development of production technologies are indispensable to fulfill the proposed road map.

(B) SOFC CHP system

In comparison with PEFCs, the development of SOFCs requires more fundamental research [31]. The requirement for power generation efficiency is also much greater than for PEFCs, but achievement of such stringent requirements for this efficiency has been demonstrated by an SOFC system [24]. This suggests that SOFCs have great potential as power generation systems. The following tasks should be completed:

B1. Fundamental (1), for improvement of power generation efficiency to >50 %:

1.1. Evaluation and optimization of microstructures of electrode-electrolyte interface
1.2. Mechanism of transport phenomena at interfaces
1.3. New material development

B2. Fundamental (2), for improvement of lifetime to >10–15 years:

2.1. Clarification of degradation mechanism
2.2. Investigation of cation diffusion and oxygen potential at interfaces
2.3. Development of robust cell and stack

B3. Fundamental (3), for improvement of cell design:

3.1. Thin film, miniaturization

B4. Technological development of practical applications for development of production technologies:

4.1. Improvement of process yield, scale-up
4.2. Optimization of sintering process

Considering material development and microstructural improvement, requirements for the technology road map in Table 2 will be achieved. For other aspects of the development of FC CHP systems, spatially constraining factors such as allocation of FC units and hot water tank, tank size, and stand-alone or collective systems are also interesting subjects.

4 Benefit and Attractive Future Vision

Finally, CO_2 emission reduction potential and household electricity and city gas cost savings for the installation of residential FC CHP systems are addressed. Based on the energy-flow diagram of an FC CHP system in Fig. 4, CO_2 emission reduction was estimated as a function of power generation efficiency. The household electricity and city gas cost savings for FC system installation were also evaluated according to that diagram. To simplify the discussion, data of 700-W FC CHP

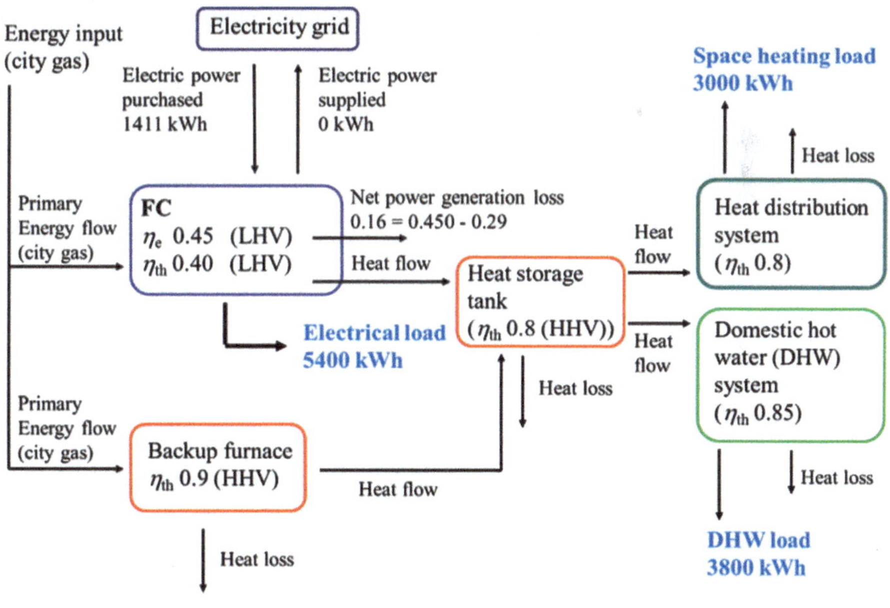

Fig. 4 Energy-flow diagram of a residential FC CHP system. Electrical efficiency was presumed to be 45 % (*LHV*)

systems were used for the calculations, assuming electrical efficiencies of 40–60 % (LHV). Also, a stand-alone house with a family of four was used as a typical case to calculate the total impact in Japan. Conditions for the estimates are summarized in the Appendix. In the calculation, heat-first-and-electricity-second-type cogeneration was applied to the FC CHP operation. In fact, this type of cogeneration is a general operational mode for PEFC CHP systems. Reverse power flow, i.e., power generating operation exporting power back to the grid, was forbidden. Therefore, under present conditions, CO_2 emission reduction and cost benefit became saturated for power generation efficiency >55 %.

The comparison between the primary energy input for a home with FC CHP systems and that with a gas combination room heater and water heater revealed the CO_2 emission reduction and overall cost savings of household electricity and city gas (Table 5). In present conditions, the potential CO_2 emission reduction per year can reach ~2 ton/year. The overall cost saving of household electricity and city gas is ~40,000 JPY/year. This means a cost payback time of 5–10 years in the future, which would be a strong driving force for widespread use of residential FC CHP systems. In the calculation, heat-first-and-electricity-second-type cogeneration was presumed for SOFC CHP operation. Reverse power flow was again forbidden. Therefore, under current conditions, the CO_2 emission reduction and overall cost saving of household electricity and city gas were saturated for power generation efficiency >55 %. As already stated, cost-benefit values reported by gas companies (50,000–90,000 JPY/year) [33, 34] are twice those in Table 5. The main reason is discounts on city gas prices by gas companies. Although the discount on city gas prices is not assumed in this chapter, the overall cost saving of household electricity and city gas is sufficiently large, as stated above. The contribution of FC systems to the CO_2 emission reduction in Japan is discussed in Part 1 of this book. Since FC systems are applicable to industrial sectors and automobiles as well as residential use, greater significant impacts on this reduction are expected. The FC technologies will be further diffused by the growth of smart grid systems and hydrogen production and transport technologies (Fig. 5).

Table 5 CO_2 emission reduction and cost benefit for residential FC CHP systems

Power generation efficiency[a] (%)	40	45	50	55	60
CO_2 reduction (ton/year)	1.5	1.6	1.8	1.9	1.9
Overall cost saving of household electricity and city gas (JPY/year)	19,000	30,000	37,000	43,000	47,000

[a]Efficiency calculated on LHV base

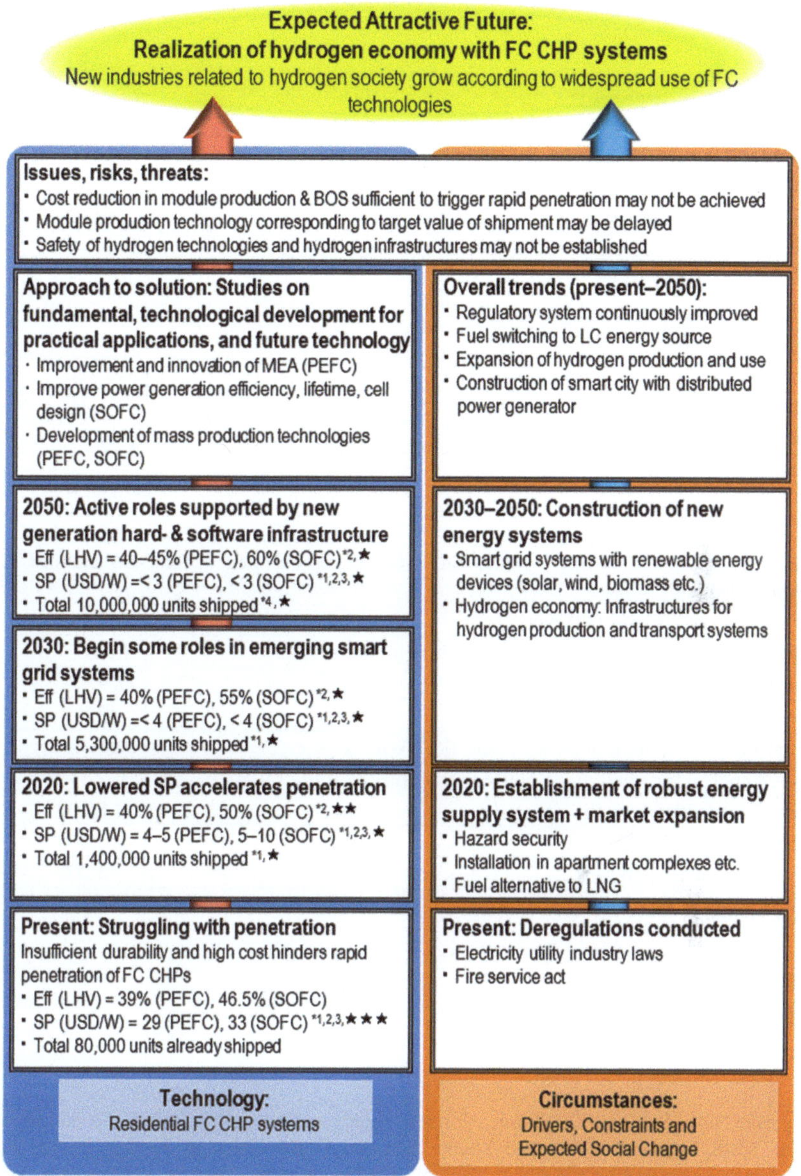

*1 Hydrogen/fuel cell/strategy roadmap in 2014 [29]
*2 NEDO report: Roadmap on fuel cells and hydrogen technology development 2010 [27]
*3 Websites of manufacturers [7-9, 13], *4 Target value chosen based on total number of households in Japan. Housing and land Survey 2008 by the statistics bureau, MIC [35]

Abbreviations: FC: Fuel Cell, CHP: Combined Heat and Power, Eff: Electrical Efficiency, LHV: Lower Heating Value, PEFC: Polymer Electrolyte Fuel Cell, SOFC: Solid Oxide Fuel Cell, SP: System Price, LNG: Liquefied Natural Gas, MEA: Membrane-Electrode Assembly, LC: Low Carbon
BOS: Balance Of System (fuel processing system, auxiliaries, controller, case and hot water tank)

Fig. 5 Technology road map of residential FC CHP systems

Appendix

Values for systems competitive with FC CHP systems for Fig. 3	
Hot water supply	
Price (JPY)	150,000
Lifetime (year)	10
Annual heat generation (kWh)	4,000
Annual expenditure rate	0.1
System cost (JPY/kWh)	3.8
Efficiency (–)	0.9
CO_2 heat pump system	
Price (JPY)	300,000
Lifetime (year)	10
Annual heat generation (kWh)	4,000
Annual expenditure rate	0.1
System cost (JPY/kWh)	7.5

Parameters for calculation of CO_2 emission reduction and cost benefit for 1-kW-class FC CHP systems in Table 5	
Heating value of city gas (MJ Nm^{-3})	45
City gas price (JPY Nm^{-3}) (typical value used in this chapter)	140
CO_2 emission for city gas (kg Nm^{-3})	2.29
Thermal power plant efficiency[a] (%)	42
CO_2 emission from thermal power plant[a] (kg kWh^{-1})	0.61
Required energy for stand-alone house with family of four (150 m^2):	
Electrical load (kWh)	5,400
Hot water supply (kWh)	3,800
Living room heater (kWh)	3,000

[a]Values estimated based on Refs. [38, 39]

References

1. Kyocera News Releases (2012) http://global.kyocera.com/news/2012/0305_woec.html, Osaka Gas, Aisin, Kyocera, Chofu and Toyota Announce Completion of World-Class Efficiency Residential-Use Solid Oxide Fuel Cell (SOFC) Cogeneration System Co-Development and Commercialization of "ENE-FARM Type S"
2. TEPCO HP, Thermal power generation (2014) http://www.tepco.co.jp/solution/power_equipment/thermal_power/index-j.html
3. Koyama M, Kimura S, Kikuchi Y, Nakagaki T, Itaoka K (2014) Present status and points of discussion for future energy systems in Japan from the aspects of technology options. J Chem Eng Jpn 47:499–513
4. Yoshioka H (2010) Prospect of PAFC. Hydrogen Energy Syst 35:15–18
5. Sharaf OZ, Orhan MF (2014) An overview of fuel cell technology: fundamentals and applications. Renew Sustain Energy Rev 32:810–853

6. Panasonic Headquarter News (2013) Launch of new 'Ene-Farm' home fuel cell product more affordable and easier to install, http://panasonic.co.jp/corp/news/official.data/data.dir/2013/01/ en130117-5/en130117-5.html#about_e
7. PEFC CHP system price1. http://panasonic.co.jp/ap/FC/en_doc03_00.html
8. PEFC CHP system price2. http://www.osakagas.co.jp/company/press/pr_2013/1206874_7831.html
9. PEFC CHP system price3. http://www.osakagas.co.jp/company/press/pr_2011/1195483_4332.html
10. The Fuel Cell Today Industry Review 2013. http://www.fuelcelltoday.com/analysis/industry-review/2013
11. JEMA (2014) Topics: report on the number of units shipped for stationary fuel cell systems in 2013, Denki 770:62–65. https://jema-net.or.jp/Japanese/res/fuel/images/chousakekka.pdf
12. ACEJ HP (2014) Report on the number of units shipped for ENE-FARM. http://www.ace.or.jp/web/works/works_0090.html
13. SOFC CHP system price. http://www.osakagas.co.jp/company/press/pr_2014/1209206_10899.html
14. SOFC CHP system performance1. http://www.noe.jx-group.co.jp/newsrelease/2011/20110915_01_0950261.html
15. SOFC CHP system performance2. http://www.osakagas.co.jp/company/press/pr_2012/1196121_5712.html
16. SOFC CHP system performance3. http://www.osakagas.co.jp/company/press/pr_2014/1209206_10899.html
17. Suzuki M, Iwata S, Higaki K, Inoue S, Shigehisa T, Miyachi I, Nakabayashi H, Shimazu K (2009) Development and field test results of residential SOFC CHP system. ECS Trans 25 (2):143–147
18. PEFC CHP system performance1. http://www.tokyo-gas.co.jp/Press/20110209-01.html
19. PEFC CHP system performance2. http://www.osakagas.co.jp/company/press/pr_2009/1174857_1256.html
20. PEFC CHP system performance3. http://news.panasonic.com/press/news/official.data/data.dir/jn110209-1/jn110209-1.html
21. PEFC CHP system performance4. http://www.toshiba.co.jp/about/press/2011_12/pr_j2004.htm
22. PEFC CHP system performance5. http://news.panasonic.com/press/news/official.data/data.dir/2013/01/jn130117-2/jn130117-2.html
23. PEFC CHP system performance6. http://www.toshiba.co.jp/product/fc/products/04.htm
24. Payne R, Love J, Kah M (2009) Generating electricity at 60 % electrical efficiency from 1–2 kWe SOFC products. ECS Trans 25:231–239
25. Hosoi K, Ito M, Fukae M (2011) Status national project for SOFC development in Japan. ECS Trans 35:11–18
26. NEDO report (2010) Ex-post valuation: maintenance project of common infrastructure for establishment of hydrogen society. http://www.nedo.go.jp/content/100159111.pdf
27. NEDO report: roadmap on fuel cells and hydrogen technology development 2010. http://www.nedo.go.jp/news/other/FF_00059.html
28. NEDO report: fuel cell & hydrogen 2009–2010. http://www.nedo.go.jp/content/100079666.pdf
29. Hydrogen/fuel cell strategy roadmap in 2014. http://www.meti.go.jp/press/2014/06/20140624004/20140624004-2.pdf
30. PEFC project. http://www.nedo.go.jp/activities/FF_00142.html
31. SOFC project. http://www.nedo.go.jp/activities/ZZJP_100060.html
32. Yoshida Y, Kohno T, Enokibori M, Matsuhashi R (2008) CO_2 emission characteristics of residential energy supply systems in consideration of the variation of personal pattern of life. J Jpn Inst Energy 87:1010–1018
33. Tokyo gas HP (2014) http://home.tokyo-gas.co.jp/ryokin/enefarm/index.html

34. Osakagas HP (2014) http://www.osakagas.co.jp/company/press/pr_2014/1209206_10899.html
35. Housing and land Survey 2008 by the statistics bureau, Ministry of Internal Affairs and Communications, MIC. http://www.stat.go.jp/english/data/jyutaku/index.htm
36. Otomo J, Oishi J, Mitsumori T, Iwasaki H, Yamada K (2013) Evaluation of cost reduction potential for 1kW class SOFC stack production: implications for SOFC technology scenario. Int J Hydrogen Energy 38:14337–14347
37. NEDO report, Common list of specifications for auxiliaries of residential fuel cell systems (2010) http://www.nedo.go.jp/content/100079684.pdf
38. Imamura E, Nagano K (2010) Evaluation of life cycle CO_2 emissions of power generation technologies – update for state-of-the-art plants- The Central Research Institute of Electric Power Industry (CRIEPI) report, Y09027. http://criepi.denken.or.jp/jp/kenkikaku/report/detail/Y09027.html
39. The federation of electric power companies of Japan, Electricity mix in 2012. http://www.fepc.or.jp/about_us/pr/sonota/1227416_1511.html

Nanoelectronics with Low Power Consumption

Takashi Kimura

Abstract Conventional electronic devices have been based on semiconductor transistor technology, in which electron charge is controlled by electrical means. Since the emergence of the integrated circuit concept for semiconductor devices, device performance has significantly advanced via large-scale integration with the miniaturization of transistors. However, these devices are not energy efficient because of their substantial Joule heat generation and volatile characteristics. In addition, owing to recent development of top-down nanofabrication technology, device dimensions are close to the intrinsic physical scalability limits within a few-nanometer range. To overcome these serious obstacles, innovative materials, device structures, and operational principles have been recently demonstrated. In this chapter, future prospects of next-generation nanoelectronics with low power consumption are discussed with consideration of the aforementioned proposals.

Keywords Nanoelectronics • Information storage • Nonvolatile memory • Beyond CMOS • Large-scale integration

1 Introduction

Recent development of nanofabrication technology has significantly advanced integrated circuits of semiconductor devices based on the silicon metal-oxide semiconductor (MOS) transistor. This has brought immense progress to electrical devices and information and communication technology (ICT). These have become a major industry and essential technologies for daily life. Thus far, the performance of semiconductor devices has been developed by increasing the number of transistors in an integrated circuit and reducing individual transistor size. Such reduction was believed to decrease power consumption and increase operating speed. This is known as Moore's law in large-scale integration (LSI). According to this law, the power dissipation of an individual transistor is scaled by a factor of S^2, where S is a number used as a multiplier in scaling, i.e., the scaling factor, and is <1. The

T. Kimura (✉)
Kyushu University, Fukuoka, Japan
e-mail: t-kimu@phys.kyushu-u.ac.jp

Y. Kato et al. (eds.), *Energy Technology Roadmaps of Japan*,
DOI 10.1007/978-4-431-55951-1_39

operational speed of the transistor is increased by a factor of $1/S$. Thus, performance of semiconductor devices was increased by scaling the MOS field effect transistor (FET). However, because of rapid progress in miniaturizing silicon MOS transistors, the device dimension is close to the intrinsic physical scalability limits within a few-nanometer range, and it becomes impossible to apply similar scaling rules in this regime [1]. Moreover, power density of advanced semiconductor devices such as a CPU is approaching 100 mW/cm^2, which is much greater than a hot plate power density. Because continuous growth of device performance is at the core of the electronics industry, further development of nanoelectronic devices is indispensable. Therefore, the physical limitation and large power consumption are serious obstacles.

2 Present Status

With the rapid increase and popularization of information technology (IT), power consumption has greatly increased. By taking advantage of IT technology, increased use of IT apparatus is facilitated, such as the remote operation of household electrical equipment or industrial equipment and exploitation of smart grids to minimize energy consumption. In 2010, the energy consumption of IT equipment became ~10 TWh, and it is believed that this figure will grow continuously in Japan in the future. Therefore, reduction of energy consumption of IT equipment is directly connected with a reduction of all electric energy consumption in society.

In a conventional computer device, typical power consumption for each component is summarized in Fig. 1. The CPU is a major power consumer because of fast operation for processing. As additional processing speed is required in the near

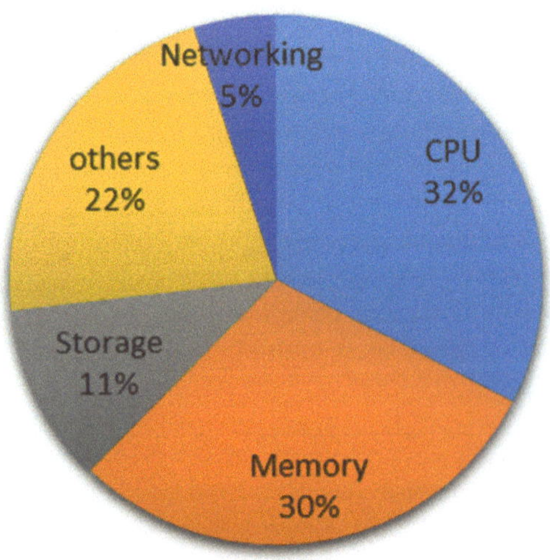

Fig. 1 Power consumption for each component in a typical computer

future, power consumption will increase significantly. The main reason for the substantial power consumption of a CPU is the large on-resistance of nano-sized MOS transistors. The resistance significantly increases with the reduction of size. The motherboard, graphics board, and other processing devices also dissipate significant power. These are also attributable to the large on-resistance of the MOS transistor. Therefore, a paradigm shift for planar MOS transistor and/or the development of innovative devices are indispensable in future nanoelectronics.

Power consumption by dynamic random access memory (RAM) is not small, mainly because of its volatility. Preserving information without any processing, dynamic RAM dissipates finite power. Therefore, the development of nonvolatile memory is key to reducing power consumption. The representative device for storage is the hard disk drive (HDD). Although the HDD is nonvolatile, power dissipation during its operation is large. This is because present HDDs are mechanical. Instead of an HDD, a solid state disk based on a floating MOS transistor (described below) has lately been rapidly developing. Currently, storage capacity for the solid state disk is much less than that of an HDD. However, the situation can be improved by the development of new architecture such as three-dimensional integration.

3 Technology Roadmap

As mentioned in the previous section, there are many milestones for reducing power consumption while increasing device performance. To tackle these serious issues, several approaches have been proposed. The first is to develop an innovative material instead of silicon. By changing the material, MOS transistor performance can be improved, even with the same size. This approach is known as "more Moore." The second approach is an innovative structure for the MOS transistor, e.g., vertical and other device structures with performance superior to the conventional planar configuration. This is known as "more than Moore." The third method is a device operated by an innovative operation principle. This is known as "beyond CMOS." Only some of these devices have been demonstrated. Moreover, their potential for realistic nanoelectronic devices, involving compatibility with conventional devices and fabrication and running costs, should be considered. In this chapter, advantages and disadvantages of these novel devices are introduced, and their feasibility for replacing the current semiconductor transistor with low power consumption is discussed.

3.1 Candidate Materials for Replacing Silicon-Based Semiconductor Devices

Silicon(Si) has been the most used material in semiconductor devices because it has the second largest Clark's number. More importantly, the well-controlled insulating layer of SiO_2 strongly enhances the potential of Si-based devices. As the electron and hole in germanium (Ge) and III-V semiconductors such as GaAs and InAs are known to have higher mobilities [2, 3], improved performance of the semiconductor device is expected by developing the FET with such materials. However, Ge FET is known to have two major issues, difficulties in growth of the high-quality insulating layer and n-channel formation. Gate insulation is also a problem in III-V semiconductors, as is the high cost of the substrate. Carbon-based materials such as carbon nanotube and graphene are candidates for nano-sized transistors because of their excellent properties such as high mobility and ballistic transport. However, design integration and reliable contact resistance are unsolved issues. Therefore, at present, Si remains a promising material for future nanoelectronics. Other materials require further improvements to surpass Si technology.

3.2 Candidate Structures for Replacing Conventional FET

The constant scaling technology of the conventional planar FET (Fig. 2a) enables continuous improvement of IC performance and cost, leading to innovation in recent electronics and ICT. However, the insulating layer thickness is close to the intrinsic physical limit less than within 1 nm, increasing leakage current from the gate. The increase of leakage current between source and drain because of a short-channel effect is another serious obstacle. Therefore, it is believed that fabrication of an FET with gate length as little as 20 nm is difficult. To overcome these serious obstacles, a fin-type FET (Fig. 2b) has been demonstrated [4]. This is known as fin FET, consisting of a thin silicon body wrapped in gate electrodes. This innovative structure can reduce the aforementioned leakage currents and facilitates scaling down of the gate width to <10 nm. Fin FET LSIs have already been manufactured by several semiconductor companies and will be a standard for future semiconductor technology. However, even using fin FET, if the feature scale is <5 nm, it will be difficult to continue the scale-down. As an alternative device structure, an FET based on vertical nanowire (Fig. 2c) has been considered, using Si and InAs nanowires [5]. Although the physical size limitation of the nanowire FET can be reduced to as small as 5 nm, integration and cost reduction are important milestones for these technologies.

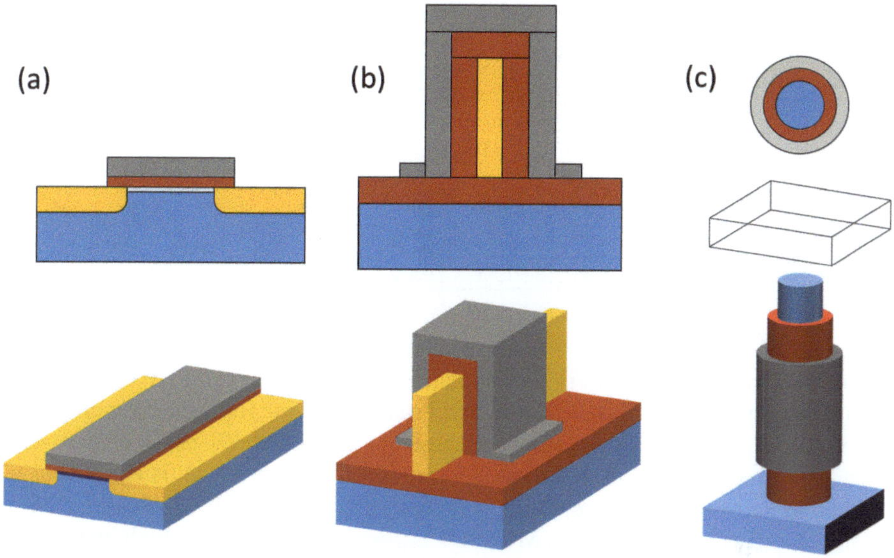

Fig. 2 **a** Conventional planar-type field effect transistor (*FET*), **b** fin-type FET, and **c** nanowire-type FET

3.3 Candidate Device Architectures for Replacing Conventional Dynamic Random-Access Memory

RAM is important in digital ICs. Normally, RAM circuits are composed by semiconductor devices such as dynamic and static RAMs. However, they have volatile properties, meaning that stored information disappears when power is no longer supplied to the RAM. Therefore, in most digital ICs, a finite power is dissipated only for preserving information, without any operation processing. This is known as standby power dissipation. As mentioned above, because of the ultimate reduction of FET size, transistor resistance will become extremely high, causing substantial standby power dissipation in information storage. To solve this problem, considerable attention has been given to nonvolatile memory. There are numerous approaches to achieve nonvolatile RAM. Table 1 compares various memories. Spin RAM is based on ferromagnetic metals of small size. The discovery of spin-dependent transport such as giant magnetoresistance and tunnel magnetoresistance enables the use of spin information by electrical means. Spin RAM has fast writing and reading speed as well as high endurance [6]. In particular, feature size can be reduced to 20 nm. Moreover, the memory cell can be formed by a relatively small unit cell area $6F^2$ (F is feature size). These advantages greatly enhance the potential of spin RAM. However, the profile distribution for each device is a crucial issue for present spin RAM technology. Fe-RAM is based on ferroelectricity and has strong compatibility with dynamic RAM [7]. However, slow writing speed is a serious issue for Fe-RAM. Moreover, the minimum size of

Table 1 Performance comparison of random access memories for different operation principles

	SRAM	DRAM	NAND	FeRAM	STT-MRAM	PRAM	ReRAM
Sell size (F^2)	~150	~8	≤2	~2	~6	≤2	≤2
Non-volatility	×	×	○	○	○	○	○
Read time	~ns	~30 ns	~50 ms	~30 ns	~1 ns	~30 ns	30–1 ns
Write time	~ns	~30 ns	~1 ms	~30 ns	~3 ns	~300 ns	30–1 ns
Endurance	$>10^{15}$	$>10^{15}$	10^5	10^{12}	$>10^{15}$	$>10^6$	$>10^6$
Random access	○	○	×	○	○	○	○
Highest voltage	~1 V	~3 V	~20 V	~3 V	1–3 V	~5 V	~5 V

the Fe-RAM unit cell is at the micron or submicron scale. Nevertheless, stability of device performance is a powerful advantage in practical application. Exploration of innovative ferroelectric materials is an important milestone for further development of Fe-RAM. Resistive RAM [8] and phase-change RAM [9] use the resistance change of phase transition. Because the unit cell for these RAMs is equivalent to one diode and one resistance, there is a small unit cell area F^2, indicating that a highly seamless integration with MOSFET is possible. Therefore, a high integration density is expected. However, the detailed mechanism is still controversial and retention and endurance require improvement. Therefore, it is still difficult to use these RAMs in practical applications.

Nonvolatile memory based on semiconductor devices was also demonstrated using an FET with floating gate. This is known as flash memory [10]. The device can be fabricated by extending conventional semiconductor technology. Therefore, seamless device integration with CMOS technology and cost-effective devices are great advantages of flash memory. There are two major device types based on NAND and NOR logics. In particular, as can be seen in Fig. 3, an extremely small cell size can be achieved in NAND flash memory, owing to its simplicity. In addition, three-dimensional device integration based on a vertical FET significantly increases the integration density. However, because of its series-connecting structure, NAND flash is not suitable for random access operation. This slows reading and writing operations. Moreover, writing and standby power are much greater than other nonvolatile memory, so they are mainly used in storage devices.

3.3.1 Spin FET and Spin MOSFET

As one of the beyond complementary MOS technologies, which are different from the aforementioned technology, devices called spin FET [11] and spin MOSFET [12] have been proposed. Tremendous innovation is anticipated for devices made with such processing and memory, based on non-volatility. A schematic of the spin MOSFET is shown in Fig. 4. Information is stored using the relative angle of the source and drain of the semiconductor, which are composed of ferromagnetic materials. Capacitance and magnetic tunnel junction MTJ are not required for this device. This revolutionary device has not been demonstrated even after 10 years,

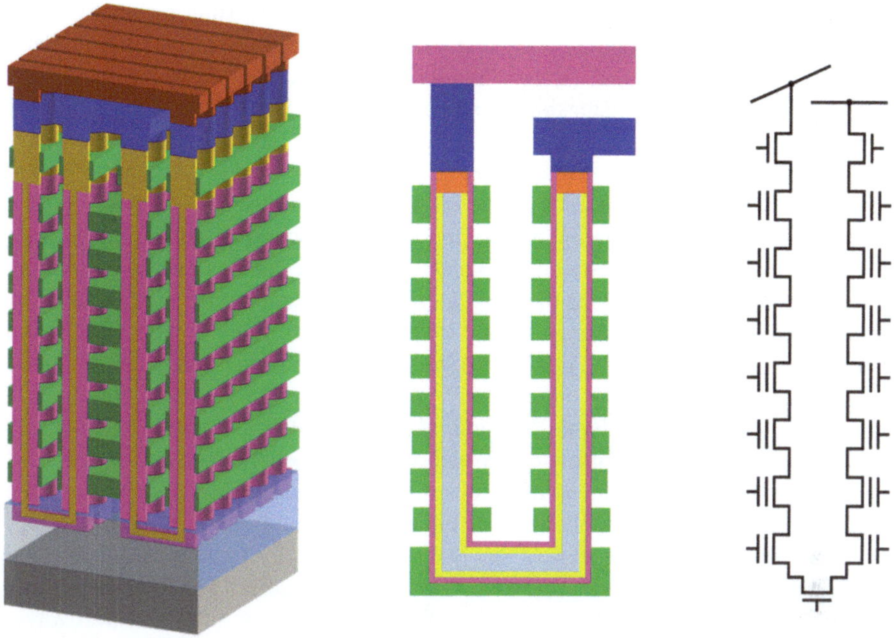

Fig. 3 NAND flash memory with three-dimensional architecture

Fig. 4 Spin-based processing device with memory function (*left*: Spin FET; *middle*: Spin MOSFET; *right*: expected I–V characteristics)

because of significant technical challenges such as fabrication techniques for a high-quality ferromagnetic metal/semiconductor interface and efficient spin injection technology.

3.3.2 Storage

In the operation of electronic devices, information storage is indispensable technology for increasing process speed, reducing power consumption, and enhancing operating efficiency. Magnetic recording is considered the most common means to store information for a long period, and it remains the dominant method at present. This recording stores information by manipulating the orientation of a magnet, in which the magnetization direction is represented by 0 or 1, using bistable properties. This recording technology was widely used in early products such as magnetic tapes and floppy disks and is now mainly used in hard disk drives. Since the permeability is always greater than 1 in principle, the magnetic field can penetrate any medium (solid, liquid, vacuum, or others). These advantages facilitate easy manipulation and detection of the magnetization, and the cost of production is very low. However, magnet size for a single bit becomes a physical limitation with the explosion of digital information and improved nanoprocessing technology, similar to the development of transistors. Magnet size for one bit has been significantly reduced, however to a few tens of nanometers. Further reduction is accompanied by the loss of magnetic ordering, owing to thermal disturbance. The application method of the magnetic field and limitation of the spatial distribution of magnetic field strength are other major milestones. In the nanometer regime, to maintain high thermal stability, research based on materials with substantial magnetic anisotropy has advanced. However, for materials with great magnetic anisotropy, a powerful magnetic field is required for writing information, because of the tradeoff between magnetic anisotropy and application of the magnetic field. Thus, further development of magnetic storage cannot be realized simply by extending the present technique. Accordingly, to achieve further miniaturization and high recording density, a stage of exploitation of new fusion techniques such as microwave-assisted and thermally assisted magnetic storage has arrived [13].

To deal with the physical limitation of magnetic recording, the rapidly growing aforementioned flash memory technology with floating gate has been developed. Owing to their simple structure, floating-gate transistors could easily realize three-dimensional fabrication, with a dramatic increase of bits per unit area and cost reduction. Moreover, another advantage of the transistor is high processing speed. In principle, one system possesses one head, used for reading and writing information in magnetic storage. Therefore, it is necessary to move the magnet to the head position, which greatly reduces the processing speed. However, a floating-gate transistor structure enables immediate reading and writing of the electronic information of each bit and does not require a drive mechanism. Further, processing speed is noticeably improved. As a consequence, this has been rapidly popularized with the name solid state drive (SSD) in devices with relatively little information quantity, such as personal computers. However, this system has many problems such as failure of long-term data storage, limitation of writing time, and increased cost consumption under large capacity, which restricts the use of devices inferior to the personal computer.

3.3.3 Wiring

Wiring between devices and/or chips is also important for nano-integration and fast electrical processing and transportation. Thus far, metallic Cu electrodes have been widely used. However, for wire width <100 nm, electrical conductivity and critical current density for electrical migration are known to decrease with that width. This is a serious obstacle in nano-sized integration. Integrating the large-scale circuit and transporting the electrical signal over long distances are also important milestones.

From these points of view, remarkable progress has recently been made with techniques using optical wiring in lieu of electrical wiring. High efficiency can be achieved using optical fiber, especially for long-distance wiring. This is an important advantage in recently developed cloud computing, because of the high speed and low energy loss during information transmission. An important issue for optical wiring is the development of a conversion technique from electricity into light and vice versa. A novel technique using self-assembled quantum dots has been promoted. Individual performance has considerably improved. As a consequence, establishment of mass production technology with reduced variation is considered one of the most critical research topics in the future.

4 Technology Roadmap

Nanoelectronic devices, which are critical for future IT, can be classified into four main elemental components, processor, memory, storage, and the wiring that connects them (Fig. 5). In each element, current techniques are approaching technological and/or physical limits and are about to enter a transitional period. Therefore, innovative technology must be developed for further improvement of device performance. This situation has arisen from the tremendous development of nanotechnology. Once a specific technology has been established, its miniaturization has rapidly advanced. As stated earlier, each type of new technology has unique problems and shortcomings. Material and/or structural innovations are indispensable to deal with the current competitive situation. Given present knowledge, it is difficult to determine which technology will survive.

Nevertheless, exploring the aforementioned techniques is ultimately feasible by combining the spin-based techniques. During ideal operation, it is possible to reduce power consumption by 10 % or more with introduction of spin MOSFET into the processor and memory devices, instead of the current MOSFET-based semiconductor devices. Moreover, operating power can be expected to be effectively zero. Regarding storage, establishment of new architecture on the basis of HDD or SSD technology is essential. Using cloud computing, power consumption for a single device may be notably reduced and minimized using optical wiring in the information transmission technology. Power consumption by wiring is estimated to decrease by 10–10 % of current levels, as optical wire is related to the aforementioned storage. A reduction by ~10 % of total power consumption is also anticipated.

Devices using new principles are usually based on phenomena unique to the nanoscale regime. However, the microscopic structure or origin of these phenomena remains controversial. Realization of practical application would be difficult without the origin elucidated. To do so, the use of observation and measurement technology in nanostructure is essential. Given these viewpoints, semiconductor

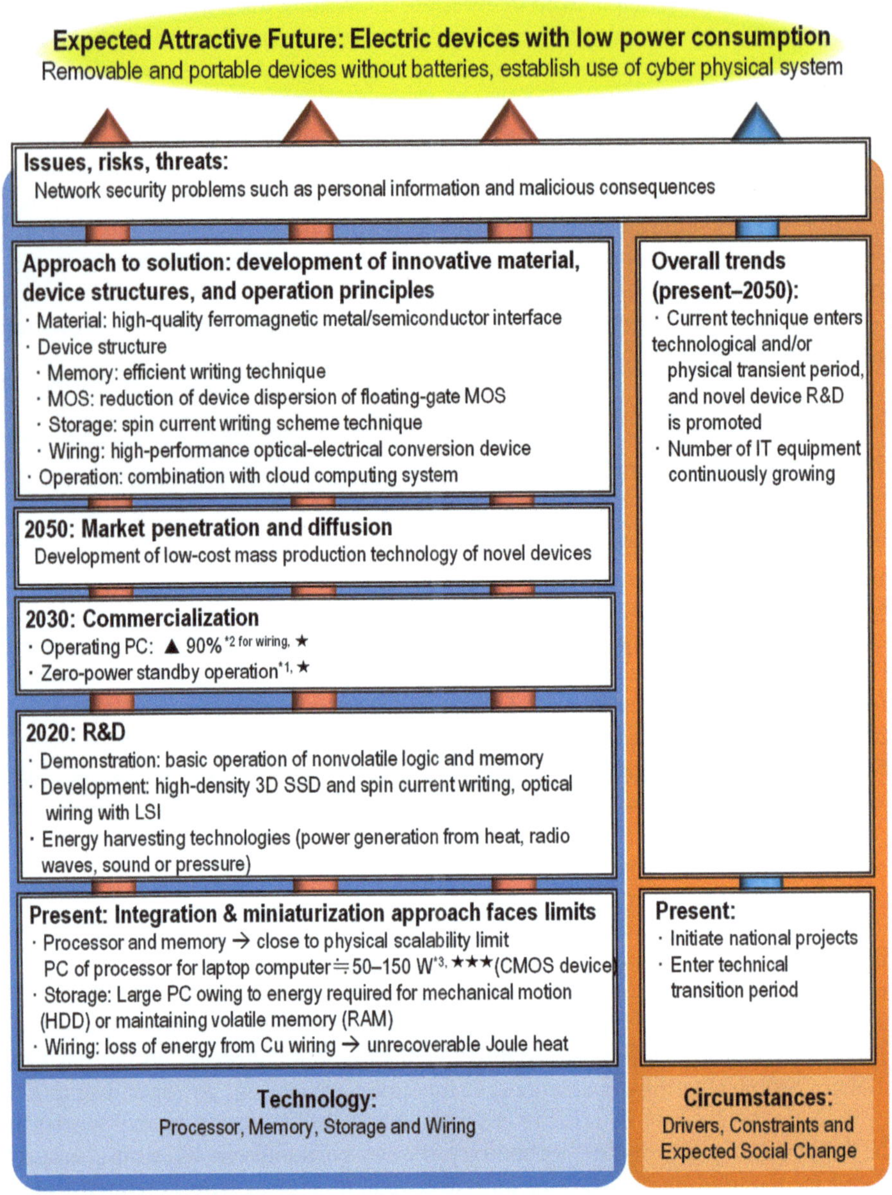

*1 ImPACT (Impulsing Paradigm Change through Disruptive Technologies):
Achieving ultimate green IT Devices with long usage times without charging program

*2 PETRA (Photonics Electronics Technology Research Association) project

*3 https://en.wikipedia.org/wiki/List_of_CPU_power_dissipation_figures#Intel_Core_i5

Abbreviations:

CMOS: Complementary Metal Oxide Semiconductor; HDD: Hard Disk Drive
SSD: Solid State Drive; LSI: Large-Scale Integration; PC: Power Consumption

Fig. 5 Technology roadmap for future nanoelectronics

devices developed through material and/or structural improvement without altering the basic operation principle are still reasonable and straightforward means of enhancing device performance.

Examining the trend of science and engineering research in recent years, university researchers have focused on new phenomena as basic research. However, research and development aimed at performance improvement receive little attention from those researchers, because of its weak academic impact. To resolve this dilemma, it is necessary to stimulate industry-academia collaboration, establish evaluation systems of related study, and enhance the attractiveness of this endeavor as an object for researchers.

5 Benefit and Future Vision

The rapid development of information technology has brought enormous wealth in society. Electronic devices, which are the main players in this innovation, are continuing to increase in performance, and it appears that the development of these devices will not cease. However, the idea that miniaturization equals high performance has come to an end, because advanced elemental technology for most parts has arrived at a transitional period.

Energy consumption has greatly diversified with the development of society. There is more unused (such as thermal) energy released into the environment, owing to waste heat and electromagnetic waves in wireless communication. One of the attractive features of recent nanoelectronic devices introduced in this section is extremely low power consumption. This indicates that electronic devices without power drive (zero-power devices) can be achieved using discarded waste energy. To realize this type of device, it is very important to decrease power consumption and establish new energy-harvesting technology for converting heat, radio waves, sound, or pressure into electrical energy [14]. These technologies constitute a cyber physical network system, which is one of the collaborating computational elements controlling physical entities.

In brief, realization of the aforementioned electronic devices without power drive requires combination of the advantages of various materials, precise manipulation of the system, and training of researchers and technicians with a wide range of knowledge in specialized fields.

References

1. Proc IEEE 89(3), March 2001 Limit Si technology
2. Takagi S, Maeda T, Taoka N, Nishizawa M, Morita Y, Ikeda K, Yamashita Y, Nishikawa M, Kumagai H, Nakane R, Sugahara S, Sugiyama N (2007) Gate dielectric formation and MIS interface characterization on Ge. Microelectron Eng 84(9–10):2314–2319

3. Mimura T (2002) The early history of the high electron mobility transistor (HEMT). IEEE Trans Microw Theory Tech 50(3):780–782
4. Bin Yu, Leland Chang, Ahmed S, Haihong Wang, Bell S, Chih-Yuh Yang, Tabery C, Chau Ho, Qi Xiang, Tsu-Jae King, Bokor J, Chenming Hu, Ming-Ren Lin, Kyser D (2002) FinFET scaling to 10 nm gate length. IEDM technical digest:251–254
5. Xiang J, Lu W, Hu Y, Wu Y, Yan H, Lieber CM (2006) Ge/Si nanowire heterostructures as high-performance field-effect transistors. Nature 441:489–493
6. Yoda H, Fujita S, Shimomura N, Kitagawa E, Abe K, Nomura K, Noguchi H, Ito J (2012) Progress of STT-MRAM technology and the effect on normally-off computing systems. 2012 I.E. international electron devices meeting (IEDM) technical digest 11(3)
7. Scott JF, Paz de Araujo CA (1989) Ferroelectric memories. Science 246:1400–1405
8. Wei Z et al (2008) Highly reliable TaOx ReRAM and direct evidence of redox 997 reaction mechanism. In: Proceedings of the IEDM. pp 293–296
9. Lai S (2003) Current status of the phase change memory and its future. Electron devices meeting
10. Masuoka F, Momodomi M, Iwata Y, Shirota R (1987) New ultra high density EPROM and flash EEPROM with NAND structure cell. Electron devices meeting
11. Datta S, Das B (1990) Electronic analog of the electrooptic modulator. Appl Phys Lett 56:665–667
12. Sugahara S, Tanaka M (2006) Spin MOSFETs as a basis for spintronics. ACM Trans Storage 2 (2):197–219
13. Li W, Buford B, Jander A, Dhagat P (2014) Acoustically assisted magnetic recording: a new paradigm in magnetic data storage. IEEE Trans Magn 50(3)
14. Priya S, Inman DJ (eds) (2009) Energy harvesting technologies. Springer

Topic: Thermally Driven Heat Pumps

Mitsuhiro Kubota

Abstract Thermally driven heat pumps represent technology not only for recovering thermal energy but also for supplying it to processes after upgrading the quality of recovered heat. The pumps are classified into three types: absorption, adsorption, and chemical reaction types. The absorption and adsorption heat pumps have already been commercialized to generate cooling energy for air conditioning by recovering thermal energy at temperatures below 363 K and around 453 K. Reducing the initial apparatus cost and finding a suitable market for each heat pump system are important for promoting their introduction into the market. These pumps are expected to save energy and improve total energy efficiency of heat utilization systems by recovering and upgrading thermal energy discharged from various processes.

Keywords Heat pump • Absorption • Adsorption • Chemical reaction • Thermal energy

1 Introduction

Thermally driven heat pumps are mainly classified into three types based on the absorption mechanism of the working medium, i.e., absorption, adsorption, and chemical reaction types. An absorption heat pump (AbHP) commercialized in Japan uses the water absorption/desorption nature of lithium bromide solution. An adsorption heat pump (AdHP) uses an adsorption/desorption phenomena of solid adsorbent. Silica gel/H_2O and functional adsorbent material (FAM)/H_2O systems have also been commercialized in Japan. A chemical heat pump (CHP) uses chemical reactions between a reactant and working medium. Many reaction systems have been proposed and developed, but none have been commercialized. These heat pumps can be operated without electricity but use thermal energy as a driving energy source. They have the greatest advantage in that they can not only recover thermal energy but also transform the temperature of that energy and supply

M. Kubota (✉)
Nagoya University, Nagoya, Japan
e-mail: kubota@nuce.nagoya-u.ac.jp

Y. Kato et al. (eds.), *Energy Technology Roadmaps of Japan*,
DOI 10.1007/978-4-431-55951-1_40

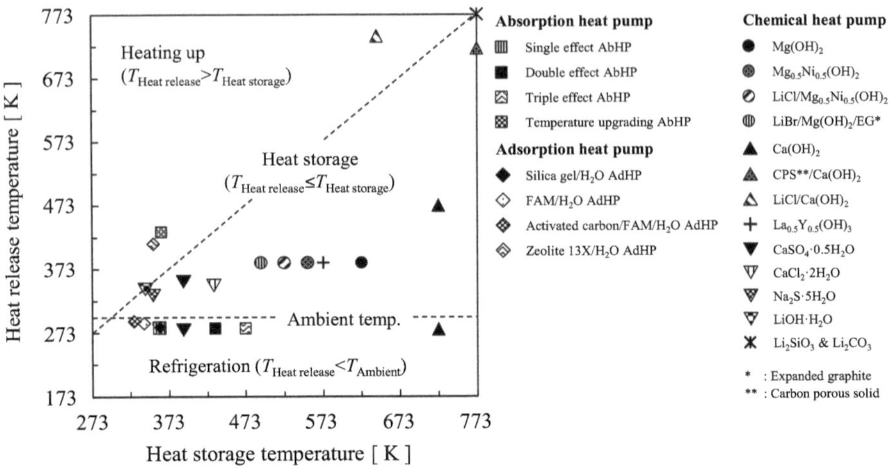

Fig. 1 Relationship between heat storage and heat release temperatures of the heat pumps in Japan

the upgraded heat to its demand. There are four modes of thermally driven heat pumps for increasing the quality of thermal energy, i.e., heat storage, heat upgrade, heat enhancement, and refrigeration. Figure 1 shows the relationship between heat storage and heat release temperatures of the heat pumps researched and commercialized in Japan.

Among these modes, AbHPs and AdHPs with refrigeration mode have been commercialized in Japan. In both systems, thermal energy at <363 K (or ~453 K for the double-effect AbHP) is used as regeneration energy to desorb water from an absorbent/adsorbent. Meanwhile, cooling energy at ~280 K can be produced for air conditioning by water evaporation, accompanied by absorption/adsorption. The refrigeration mode allows thermal energy recovered at <363 K and ~453 K to be transformed into it at ~280 K for air conditioning. In Japan, an AbHP with cooling capacity 17.6–17,600 kW has been already commercialized, meaning that it can cover a wide range of cooling capacity. The AdHP in the country has a maximum cooling capacity of 499 kW, owing to its complicated gas-solid adsorber. The coefficient of performance (COP), defined as the ratio of cooling capacity to thermal energy supplied for regeneration, is reported to be around 0.6 for single-effect AbHPs and AdHPs and 1.2–1.4 for a double-effect AbHP. About 56,000 AbHPs (34.5 GW) have been introduced in Japan for waste heat recovery, cogeneration, and large-scale district heating and cooling (DHC) systems over the last 20 years [1]. Meanwhile, 300 or more AdHPs have been introduced [2]. For the chemical heat pump, fundamental studies have been done on both preparation of a novel reactant and development of the reactor with high heat output performance.

2 Technology Details: Future Prospects, Problems, and Risks

An AbHP is technically well developed and has been widely used in the aforementioned applications [3]. An AdHP has also been commercialized as mentioned above. However, the cooling power per unit apparatus volume of the AdHP (=8.5 kW/m^3) is about 25 % that of the AbHP (32.4 kW/m^3). The AbHP is a competing technology with the AdHP when cooling energy is generated using thermal energy at ~363 K. To enhance AdHP competitiveness, there has been much research in Japan toward reducing apparatus volume by improving heat and mass transfer in the adsorber, which occupies a major part of the AdHP system. Negishi et al. prepared a thick coating of mesoporous silica by the electrophoretic deposition process for improvement of AdHP cooling power [4]. To take advantage of AdHPs over AbHPs, lowering the regeneration temperature has been attempted by changing an adsorbent from silica gel to FAM. The AdHP with FAM was commercialized to generate cooling energy even at regeneration temperature of 338 K. Rahman et al. proposed a four-bed, three-stage AdHP cycle, indicating cooling energy generation numerically at regeneration temperature of 313 K [5]. For large-scale penetration of AdHPs, reducing the initial apparatus cost is nearly as important as developing technologies.

For marketing CHPs, improvement of both (i) reaction material and (ii) system performance is still required, that is, (i) improvement of reaction rate and repeatability of the reaction material and adjustment of heat storage temperature and (ii) heat and mass transfer enhancement of the reactor and system design of the CHP. Research trends of chemical heat storage material are described in detail in the next topic. For improvement of system performance, Zamengo et al. attempted to enhance heat transfer in a packed bed reactor by adding expanded graphite to $Mg(OH)_2$ [6]. Shimazu et al. proposed a porous composite structure sintered with $Ca(OH)_2$ and fine copper particles to improve heat transfer and durability of the composite material [7]. Ogura et al. evaluated the energy efficiency of a $CaSO_4$ CHP in heat enhancement mode with other types of heat pump systems, finding that the $CaSO_4$ CHP could achieve a larger COP than the others [8].

Finally, in the thermally driven heat pumps, heat supply and demand of the market must match those of the heat pump in terms of temperature and amount of heat. For large-scale penetration of these heat pumps, market research is required to find a market suitable for its application.

3 Benefit and Attractive Future Vision

The thermally driven heat pumps are the technology not only for recovering thermal energy but also for supplying it to processes after raising the quality of recovered heat. AbHPs and AdHPs are promising technologies for generating cooling energy for air conditioning by recovering thermal energy at temperatures below 363 K and around 453 K. These heat pumps can be used in solar water heaters, fuel cells, boilers, gas heat pumps, geothermal heat systems, and other systems that discharge such low-temperature heat. Meanwhile, CHPs have the potential to recover heat in the wide temperature ranges of the heat source. Reducing the initial apparatus cost and finding a suitable market for these heat pump systems are important to promote their introduction into the markets. These heat pumps are expected to save energy and improve total energy efficiency of heat utilization systems by recovering and boosting thermal energy.

References

1. The Japan Refrigeration and Air Conditioning Industry Association HP. http://www.jraia.or.jp/download/excel/shipments_products.xls
2. Union Industry Co., Ltd. HP. http://www.union-reitouki.com/news/cat4/index.html#12
3. Best R, Rivera W (2015) A review of thermal cooling systems. Appl Therm Eng 75:1162–1175
4. Negishi H, Miyamoto A, Endo A (2013) Preparation of thick mesoporous silica coating by electrophoretic deposition with binder addition and its water vapor adsorption-desorption properties. Microporous Mesoporous Mater 180:250–256
5. Rahman AFMM, Ueda Y, Akisawa A, Miyazaki T, Saha BB (2013) Design and performance of an innovative four-bed, three-stage adsorption cycle. Energies 6(3):1365–1384
6. Zamengo M, Ryu J, Kato Y (2014) Composite block of magnesium hydroxide – expanded graphite for chemical heat storage and heat pump. Appl Therm Eng 69:29–38
7. Shimazu T, Mitsui H, Fukushima Y, Wakayama H, Fuse T, Saegusa H (2011) Japan Patent P2011-196661A
8. Shimizu T, Ogura H (2013) Experimental studies on energy efficiencies of $CaSO_4$ chemical heat pump for waste heat recovery and cold/hot heat production. Trans Jpn Soc Mech Eng 79 (808):2597–2601

Topic: Materials for Thermochemical Energy Storage

Junichi Ryu

Abstract Materials for thermochemical energy storage based on gas-solid reaction have higher energy storage densities than conventional materials for latent and sensible heat storage. These materials furnish the potential to use industrial waste heat as a heat source. Here, research and development of materials for thermochemical energy storage are introduced.

Keywords Thermochemical energy storage • Gas-solid reaction • Industrial waste heat • Mixed hydroxide • Reaction enhancement

1 Introduction

Thermal energy storage systems with sensible, latent, and reaction heats are widely investigated to improve the efficiency of energy systems and reduce fossil fuel consumption.

Figure 1 compares the energy density of these energy systems. Conventional energy storage systems with sensible and latent heats are based on physical phenomena. These systems have good reproducibility and durability, but their performance depends on the physical property of materials. It is therefore difficult to improve this performance. Conversely, thermochemical energy storage systems based on the reaction heat of reversible chemical reactions have greater energy densities (~1000 kJ/kg) than latent heat (~500 kJ/kg) or sensible heat (~300 kJ/kg) systems. Here, materials for a thermochemical energy storage system based on a gas-solid reaction are introduced.

J. Ryu (✉)
Chiba University, Chiba, Japan
e-mail: jryu@chiba-u.jp

Y. Kato et al. (eds.), *Energy Technology Roadmaps of Japan*,
DOI 10.1007/978-4-431-55951-1_41

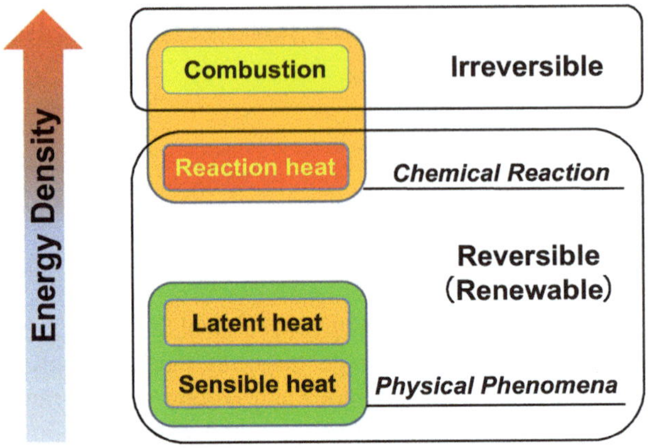

Fig. 1 Comparison of energy density for thermal energy storage system

2 Technology Details: Future Prospects, Problems, and Risks

In a thermochemical energy storage system, mixing of reactants and separation of products should be simple. In addition, a reversible reaction system is required for repeatable use of the system. Therefore, several gas-solid reaction systems are investigated for this system. Examples of the reaction system, temperature for typical heat storage operation ($T_{storage}$), and thermal energy storage density are expressed as follows [1–3]:

$$MgO + H_2O(g) \rightleftarrows Mg(OH)_2, T_{storage} = 350°C, \;\; 1.4 \;\; MJ/kg - Mg(OH)_2 \quad (1)$$

$$CaO + H_2O(g) \rightleftarrows Ca(OH)_2, T_{storage} = 500°C, \;\; 1.5 \;\; MJ/kg - Ca(OH)_2 \quad (2)$$

$$CaO + CO_2(g) \rightleftarrows CaCO_3, T_{storage} = 800°C, \;\; 1.7 \;\; MJ/kg - CaCO_3 \quad (3)$$

Here, forward reactions can be used for heat output operation and backward reactions for heat storage operation. The temperature for heat storage operation and value of thermal energy density can be predicted from the enthalpy change (ΔH) and entropy change (ΔS) of these reactions and molecular weight of the materials.

To use unused thermal energy such as industrial waste heat, it is necessary to develop a new material for chemical heat storage, which can function at temperatures less than the heat source. The temperature for industrial waste heat is generally <250 °C, so a reduction of temperature for heat storage operation is expected.

Per Eq. (1), thermal decomposition of magnesium hydroxide ($Mg(OH)_2$) can be used for heat storage operation at 350 °C. However, this reaction does not proceed at <250 °C because of thermodynamic constraints. Thermal decomposition of calcium hydroxide and calcium carbonate also do not proceed under this condition. In recent years, chemically modified magnesium hydroxides, such as Mg-based mixed hydroxide and metal salt-added $Mg(OH)_2$, and composite reactant and porous materials, such as $Mg(OH)_2$/carbon and $Mg(OH)_2$/vermiculite, have been developed [4–6]. Magnesium-based mixed hydroxide can be decomposed to magnesium-based mixed oxide at 250–300 °C, owing to the change of reaction heat for decomposition. In addition, the thermal decomposition temperature of magnesium hydroxide decreases by metal salt addition, owing to reduction of activation energy for that decomposition [4]. Thermal decomposition of magnesium hydroxide is enhanced by the formation of $Mg(OH)_2$/carbon and $Mg(OH)_2$/vermiculite composite, because the presence of carbon and vermiculite enhances thermal conductivity and provides sufficient water diffusivity [5, 6]. Even though these new materials could decrease the temperature of heat storage operation, a severe hydration condition with water vapor at 85 °C is still required for hydration reaction in the heat output operation. Consequently, another heat source for heat output operation is required, and this is a disadvantage for practical use. Therefore, new materials with greater hydration reactivity in mild conditions with water vapor ~30 °C should be developed. It is also necessary to consider the safety and durability of the material for thermochemical energy storage.

3 Benefit and Prospective Future Vision

Materials for thermochemical energy storage systems can facilitate practical use of waste heat from chemical processes, nuclear reactors, cogeneration systems, waste incineration facilities, the steel industry, and others. The thermochemical energy storage system will reduce fossil fuel consumption and CO_2 emission.

References

1. Pardo P, Deydier A, Anxionnaz-Minvielle Z, Rouge S, Cabassud M, Cognet P (2014) A review on high temperature thermochemical heat energy storage. Renew Sustain Energy Rev 32:591–610. doi:10.1016/j.rser.2013.12.014
2. Cabeza LF, Gutierrez A, Barreneche C, Ushak S, Fernandez AG, Fernandez AI, Grageda M (2015) Lithium in thermal energy storage: a state-of-the-art review. Renew Sustain Energy Rev 42:1106–1112. doi:10.1016/j.rser.2014.10.096
3. Yan T, Wang RZ, Li TX, Wang LW, Fred IT (2015) A review of promising candidate reactions for chemical heat storage. Renew Sustain Energy Rev 43:13–31. doi:10.1016/j.rser.2014.11.015
4. Ryu J, Mizuno T, Ishitobi H, Kato Y (2014) Dehydration and hydration behavior of Mg–Co mixed hydroxide as a material for chemical heat storage. J Chem Eng Jpn 47:579–586. doi:10.1252/jcej.13we303
5. Myagmarjav O, Ryu J, Kato Y (2014) Kinetic analysis of the effects of mixing mole ratios of LiBr-to-Mg(OH)2 on dehydration and hydration. J Chem Eng Jpn 47:595–601. doi:10.1252/jcej.13we301
6. Shkatulov A, Ryu J, Kato Y, Aristov Y (2012) Composite material "$Mg(OH)_2$/vermiculite": a promising new candidate for storage of middle temperature heat. Energy 44:1028–1034. doi:10.1016/j.energy.2012.04.045

Part VIII Transportation

Yukitaka Kato

Battery electric vehicles (BEVs) and fuel-cell electric vehicles have been developed as next-generation automobiles, because these cars have high energy efficiency and low CO_2 emission. However, as alternatives to gasoline cars, they have major problems such as mileage, price and infrastructure.

Thermal management of automobiles improves their total energy efficiency, not only for internal combustion engine vehicles (ICEVs) but also for BEVs and hybrid electric vehicles (HEVs). Chemical heat pump systems can save on energy usage of expensive batteries for heating and cooling.

Power electronics that enables DC-AC or AC-DC conversion of electric power is a critical technology toward energy saving. Wide-bandgap semiconductors such as SiC or GaN, which can be driven at high frequency under high temperature with downsized converters, are important in future power electronics.

ICEVs including HEVs are expected to remain dominant up to 2050. Thermal efficiency of HEVs with next-generation ICEVs will reach 50 % in the future through ICEV technology developments of high compression/expansion ratio, super lean combustion and higher boost pressure.

Y. Kato (✉)
Tokyo Institute of Technology, Tokyo, Japan
e-mail: yukitaka@nr.titech.ac.jp

Automotive Internal Combustion Engines

Hiroshi Kawanabe

Abstract To achieve further reductions in fossil fuel consumption and carbon dioxide (CO_2) emissions, the transportation sector will need to assume a key role. Based on the BLUE Map scenarios for 50 % reduction in CO_2 emissions by 2050 in the Energy Technology Protocol 2010 of the International Energy Agency, conventional engine vehicles and hybrid electric vehicles (HEVs)—including plug-in hybrid electric vehicles (PHEVs)—are expected to account for 90 % of the market and will still account for 50 % of the market by 2050. This means that it is important that thermal efficiency of the internal combustion engine (ICE) be increased. Research and development into high-efficiency ICEs has been progressing in Japan, Europe, and the United States. In this paper, current research status and future prospects of ICEs and HEVs (PHEVs) are described.

Keywords Powertrain of automotive • Internal combustion engine • Thermal efficiency • CO_2 reduction • Hybrid electric vehicle

1 Introduction

This paper examines progress toward future reductions in fossil fuel consumption and carbon dioxide (CO_2) emissions by the transportation sector. In Japan, ~98 % of the energy source for this sector depends on imported oil. In addition, the sector accounts for 18 % of total CO_2 emissions, as shown in Fig. 1 [1]. To reduce fossil fuel consumption and CO_2 emissions, the transportation sector must assume a key role. The worldwide reduction scenario for these two items, based on the two-degrees scenario (2DS) of the Energy Technology Protocol (ETP) 2014 [2] of the International Energy Agency (IEA), indicates that consumption of primary energy by 2050 will have increased that in 2011. However, fossil fuel consumption should decrease to 50 % by 2050, as shown in Fig. 2.

To envision a reduction strategy for energy consumption and CO_2 emissions from automobiles, it is important to visualize energy flow. Figure 3 shows a

H. Kawanabe (✉)
Kyoto University, Kyoto, Japan
e-mail: kawanabe@energy.kyoto-u.ac.jp

Y. Kato et al. (eds.), *Energy Technology Roadmaps of Japan*,
DOI 10.1007/978-4-431-55951-1_42

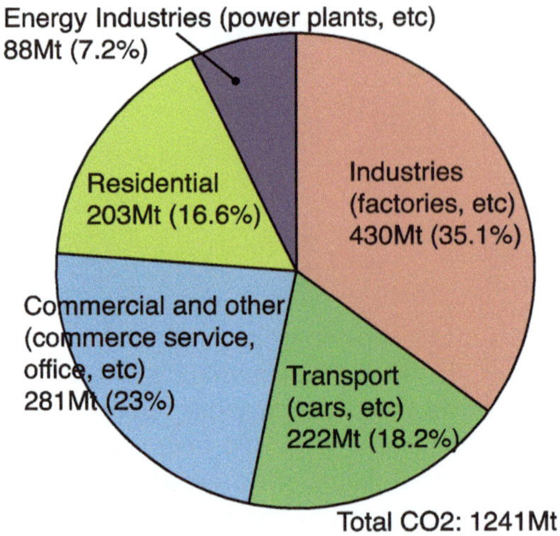

Fig. 1 Energy-origin CO_2 emissions from each sector in Japan (2012) [1]

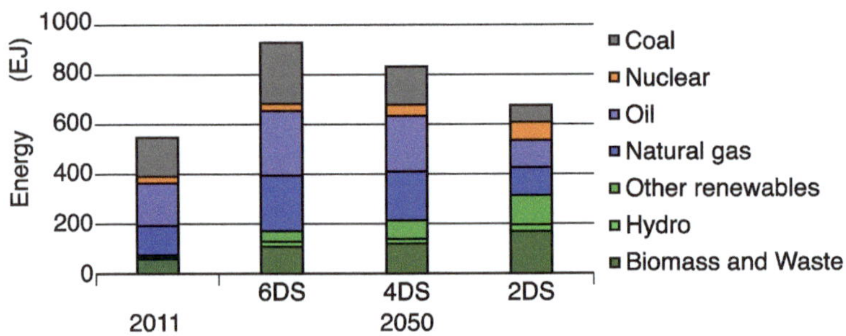

Fig. 2 Global primary energy supply [2]

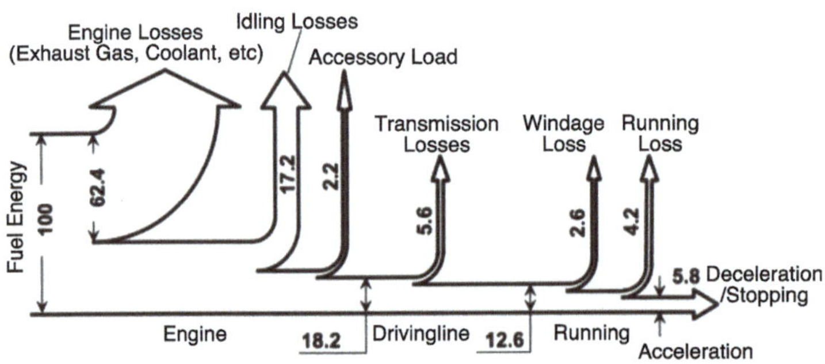

Fig. 3 Energy flow of passenger car for urban-mode driving [3]

schematic diagram of the energy flow of a conventional passenger car. The chemical energy of fuel is converted into kinetic energy of the automobile by the engine; however, the conversion efficiency is only 12.6 % [3]. The remainder of the produced energy is wasted as thermal energy emitted from the car. In addition, the kinetic energy changes into thermal energy during vehicle deceleration and stopping. To increase total energy efficiency, it is vitally important to recover this energy loss and improve the thermal efficiency of engines. However, for automotive engines, a wider speed range and greater power are required. It is difficult to improve efficiency to obtain broader engine-operation conditions. Furthermore, to achieve the long-term target of reductions in energy use and emissions, feasible technologies must be rapidly introduced into the market because the duration of on-road use for passenger vehicles is more than 10 years.

2 Present Technologies and Possibilities

Recent key technologies in automotive development are downsizing and hybridization of the powertrain. In particular, the hybrid electric vehicle (HEV) system is suitable for recovery of lost energy in passenger vehicles. Figure 4 shows a schematic diagram of a serial/parallel hybrid system [3]. This type of powertrain consists of an engine and electric motor/generator. Electric energy is accumulated during deceleration for use during acceleration. Generally, the thermal efficiency of a partial load on the engine is relatively low; therefore, total efficiency is significantly improved by providing assistance via an electric motor during starting and accelerating. Recently, the plug-in HEV (PHEV) system has been developed. The basic concept of this system is a battery electric vehicle (BEV) that is assisted by a relatively small engine, used only for generating electricity. The current problem with the BEV is the low energy and power density of an electric battery compared with liquid fuels, so the BEV travel range is relatively short. To supply electricity during driving, a small engine system is used.

Here, we describe prospects for future powertrain system strategies. Figure 5 shows the evolution of light-duty passenger vehicle sales by technology type in the baseline and BLUE Map scenarios from the ETP 2010 issued by the IEA [2]. Until 2030, conventional engine vehicles and HEVs (including PHEVs) will account for 90 % of the market, and will still account for 50 % of the market by 2050. In addition, difficulty in market penetration by BEVs has recently been reported because of problems related to production, infrastructure, and the market environment. Based on this situation, the internal combustion engine (ICE) will remain in use as the main power source of light-duty vehicles (LDVs). Therefore, improvement in thermal

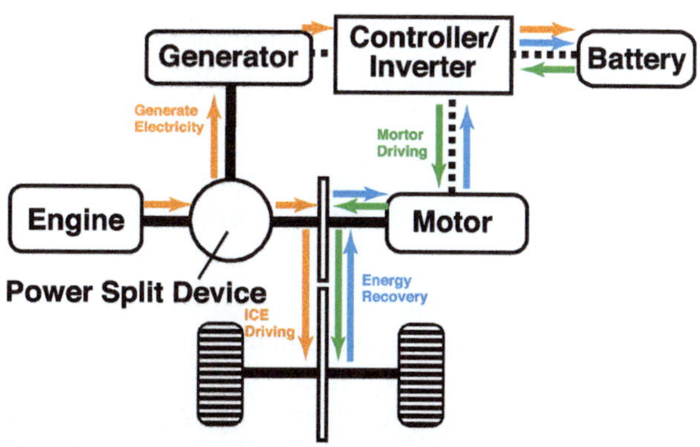

Fig. 4 Schematic diagram of serial/parallel hybrid system [3]

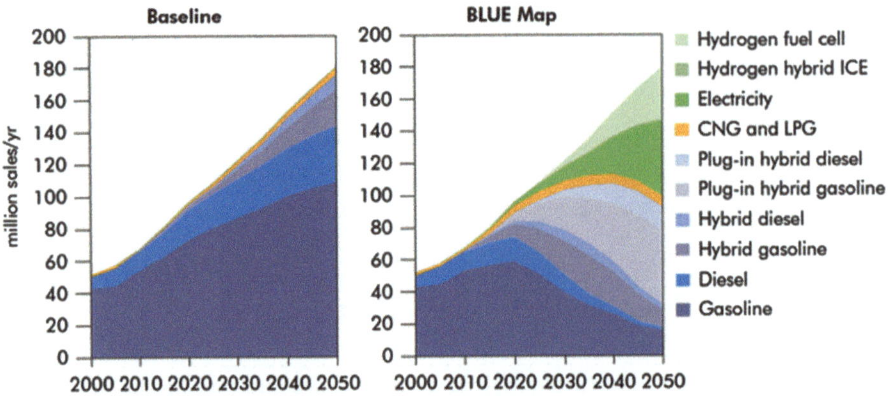

Fig. 5 Evolution of passenger car sales for baseline and BLUE Map scenario [2]

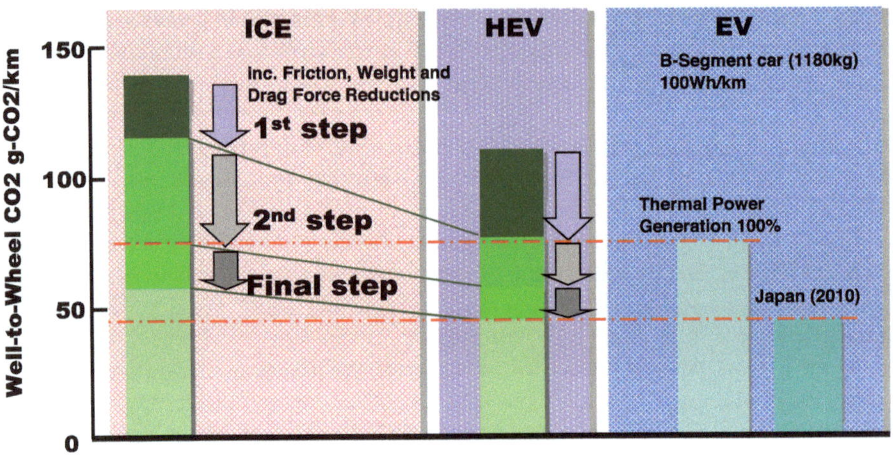

Fig. 6 Reduction strategy of well-to-wheel CO_2 emission [4]

efficiency of the ICE is essential. CO_2 emissions from a vehicle can be reduced by 50 % via increasing engine thermal efficiency. Figure 6 shows a reduction strategy of well-to-wheel CO_2 emissions for the ICE, HEV, and BEV [4]. Here, the first through third (final) steps represent improvement stages of engine efficiency. The final step will be realized by improvement in engine efficiency by as much as 50 %, a reduction of running loss by 20–30 %, and reducing vehicle weight. Finally, CO_2 emissions of an HEV with ICE in the final stage can be reduced to that of a BEV.

3 Research Trends for Future ICEs

Research trends for ICEs worldwide are presented here. Figure 7 shows the interannual trend for thermal efficiency of gasoline and diesel engines for LDVs [5]. The efficiency of diesel engines is generally greater than that of gasoline engines, but diesel engine efficiency has reached a plateau over the past 5 years. Emission regulations have become stricter each year, such that fuel consumption is

Fig. 7 Interannual trend of thermal efficiency [5]

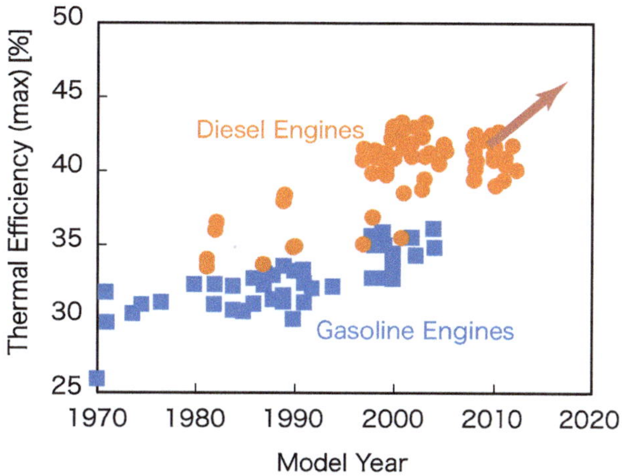

being sacrificed to achieve emission reductions. To achieve the 2DS scenario of the IEA, thermal efficiency of the ICE must be increased to 50 %.

The US Department of Energy has received funding to not only improve the battery systems of HEVs and BEVs but also for research and development of the advanced combustion engine (ACE). The objective of ACE research and development was to increase maximum thermal efficiency by as much as 45 % for LDVs and 50 % for heavy-duty vehicles (HDVs) (2010). Furthermore, fuel consumption of LDVs will be improved by 25 % (gasoline engine) and 40 % (diesel engine) (by 2015), and the maximum efficiency of HDVs will be increased to 55 %. As an example of this project, the Advanced Combustion Concepts—Enabling Systems and Solutions (ACCESS) for high-efficiency light-duty vehicles calls for improvement in fuel consumption of more than 25 %. In addition, some Reactivity Controlled Compression Ignition projects are advancing and studies are progressing toward an HDV thermal efficiency of 55 % by reducing heat loss from combustion chamber walls. In addition, the European Council for Automotive R&D (EUCAR) and the Forschungsvereinigung Verbrennungskraftmaschinen eingetragener Verein (FVV, Research Association for ICE) in European countries are studying a wide variety of means to improve thermal efficiency, without declaring a final target value. Partially premixed combustion (PPC) R&D is progressing at Lund University in Sweden. The final target of this project is a thermal efficiency of 60 %.

Automotive companies in Japan have been researching the ICE combustion process. For example, the Toyota Motor Company has been working to improve thermal efficiency with a super-long stroke and lean boost system using ethanol fuel, using a large compression ratio and high research octane number (RON) fuel. Figure 8 shows the experimental results. Here, an indicated thermal efficiency of ~44 % is achieved, much higher than that of conventional engines or the Atkinson cycle engine [6]. In addition, to improve thermal efficiency under partial load, homogeneous charge compression ignition (HCCI) combustion has been widely studied for both gasoline and diesel engines. For example, Chiba University has

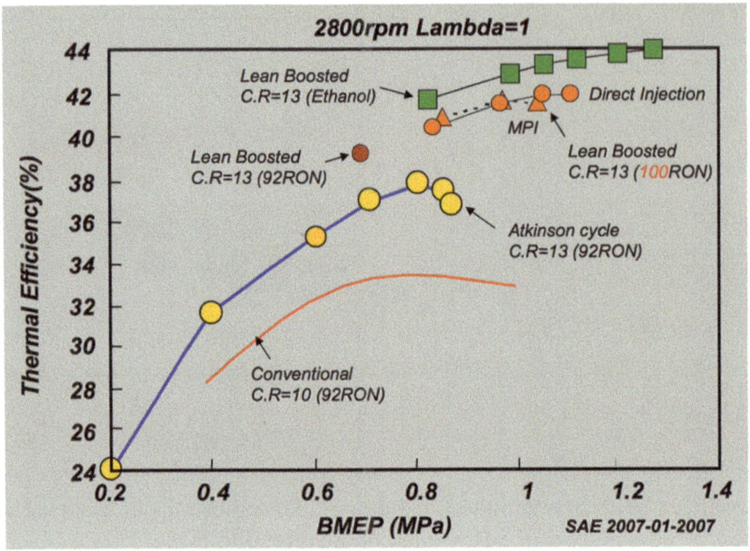

Fig. 8 High efficiency with lean boost system [6]

developed a blow-down supercharge system, which uses the pressure wave from the exhaust process to control intake gas and improve the stability of HCCI combustion under low-load conditions in a gasoline engine [7].

Recently, the Japanese government launched a comprehensive research project called the Cross-ministerial Strategic Innovation Promotion Program (SIP), which includes energy-related tasks. Innovative combustion technology for the ICE of passenger vehicles is one of the SIP tasks. The main target of this task is an increase of maximum brake thermal efficiency of gasoline and diesel engines for passenger vehicles and LDVs to 50 %. This target will be achieved by lean boost technologies using high RON fuel for gasoline engines and by low heat-loss combustion for diesel engines. The New Advanced Combustion Engine Research Center has developed the super clean diesel engine for HDVs. Higher performance, thermal efficiency, and lower emissions have been simultaneously realized using a higher boost pressure system.

4 Benefits and Future Vision

Based on the information in the previous section, maximum thermal efficiency will reach 50 % in the near future. Common key technologies for this include a high compression/expansion ratio, super lean combustion, and higher boost pressure. A possible 50 % reduction in fuel consumption is presented in Fig. 6, with the ICE achieving a brake thermal efficiency of 50 %. This scenario is schematically described in Fig. 9. Each target will be achieved while meeting emission regulations.

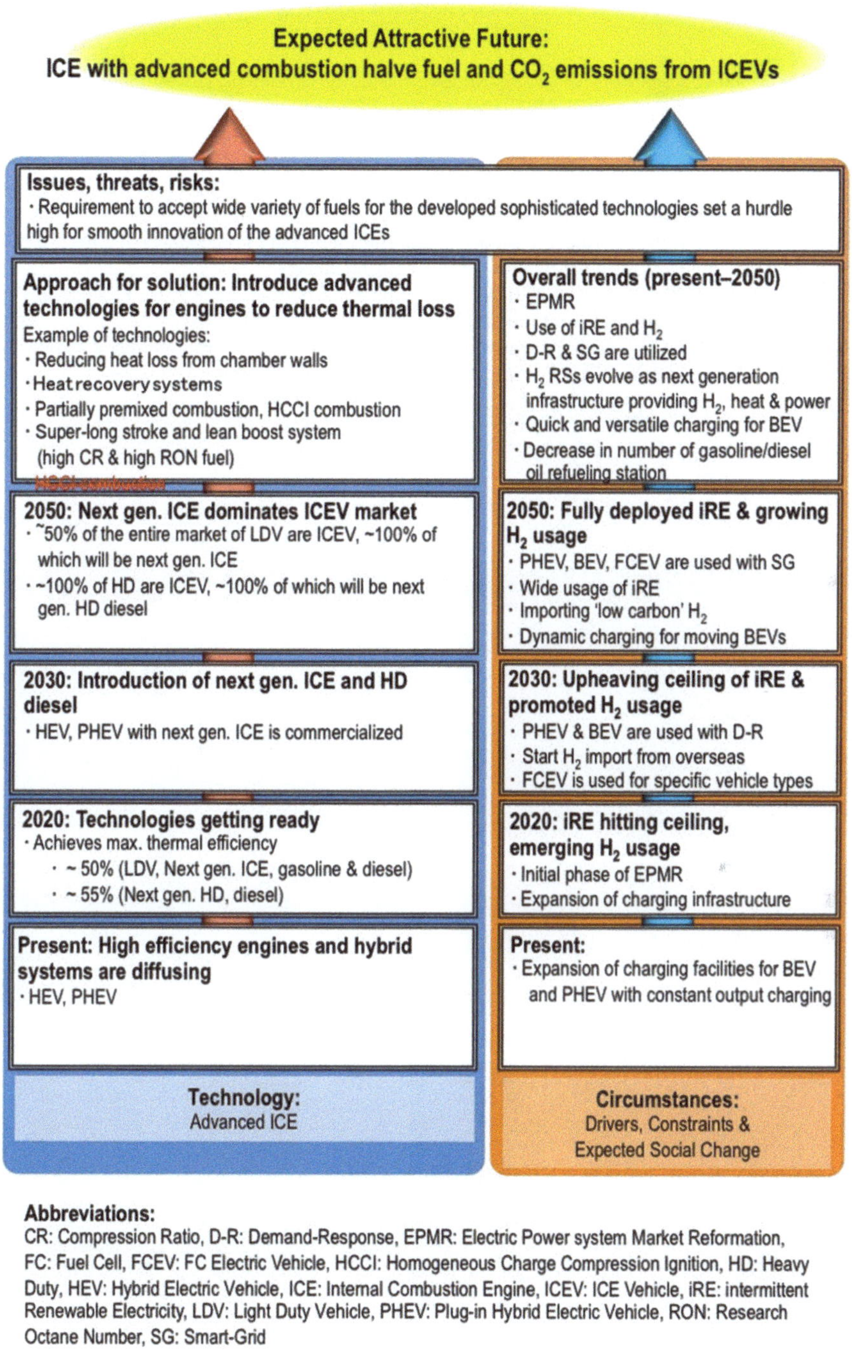

Fig. 9 Future scenario

References

1. Japan's Climate Change Policies (2014) Ministry of the Environment, Government of Japan
2. Energy Technology Perspectives (2010) International Energy Agency
3. Shimizu K (2004) Review of components and sub-system for HEV. Engine Technol 6(3):22–27, 32 (in Japanese)
4. Hitomi M (2010) Future prospect of internal combustion engine. In: Proceedings of 21st Internal combustion engine symposium, keynote speech 1 (in Japanese)
5. Furuno S (2012) History and future prospect of gasoline engine. In: Proceedings of 23rd Internal combustion engine symposium, keynote speech 1 (in Japanese)
6. Nakata K et al (2007) The impact of RON on SI engine thermal efficiency. In: Proceedings of JSAE/SAE 2007 fuel and lubricant meeting, Paper No. 2007-01-2007
7. Kuboyama T et al (2009) An experimental study of a gasoline HCCI engine using the blow-down super charge system. SAE Technical paper, No. 2009-01-0496

Secondary Batteries and Fuel Cell Systems for Next-Generation Vehicles

Gen Inoue

Abstract As next-generation automobiles, battery electric vehicles (BEVs) and fuel cell electric vehicles (FCEVs) have been developed because of their high-energy efficiency and low CO_2 emission. These technologies are extremely important for the next-generation energy society. This is because electric energy has greater efficiency than heat and kinetic energy of internal combustion engines and because of the diversification of primary energy sources and collaboration with the smart community as mobile energy storage. However, to present an alternative to gasoline automobiles, they must overcome major problems such as mileage, price, and infrastructure. In this chapter, an overview of BEV and FCEV is given, and a technology road map of secondary battery and fuel cell systems for the next-generation automotive is introduced.

Keywords BEV • FCEV • Tank-to-wheel • PEFC • LiB

1 Introduction

The energy consumption of vehicles in the transportation sector is ~3000 PJ/year, or ~20 % of all energy consumption in Japan [1]. Thus, the improvement of vehicle energy efficiency is extremely important to achieve a low-energy society in the country. There are related topics such as worldwide environmental regulations, reduction of greenhouse gas (GHG), price increases, exhaustion of petroleum, and the introduction of renewable energy resources. Automotive is a key industry for economic growth in Japan. As next-generation automobiles, battery electric vehicles (BEVs) and fuel cell electric vehicles (FCEVs) have been developed. An FCEV has a polymer electrolyte fuel cell (PEFC), which is a low-temperature type, <100 °C. The full reaction is "$H_2 + 1/2O_2 \rightarrow H_2O$ (liquid)." Because chemical energy is converted to electrical energy directly, without the path of heat and kinetic energy, theoretical power generation efficiency ($\Delta G°/\Delta H°$) is 83 %. It is difficult

G. Inoue (✉)
Kyoto University, Kyoto, Japan
e-mail: ginoue@cheme.kyoto-u.ac.jp

Y. Kato et al. (eds.), *Energy Technology Roadmaps of Japan*,
DOI 10.1007/978-4-431-55951-1_43

to increase effective heat efficiency in a single automotive vehicle because the optimum range of energy efficiency for load variation is very small. This is a problem unique to internal conversion engines. The PEFC is suitable for automotive power generation because of short start and stop times, a compact system, and strong stability to vibration. The BEV has a lithium-ion battery (LiB). LiB is an energy storage system that uses intercalation and deintercalation reactions in anode and cathode active material. Because a nonaqueous electrolyte is used, a higher voltage than that of water electrolysis can be attained. Thus, the LiB energy density is greater than that of nickel metal hydride and other batteries.

Table 1 shows the tank-to-wheel efficiencies (MJ/km) of various automotive types, which are defined as consumption energy of primary energy for 1-km operations [2]. The energy efficiency of BEVs and FCEVs strongly depends on energy resources, transportation, and infrastructure, so they cannot be estimated as unequivocal values. Nevertheless, these energy efficiencies are much higher than those of ICE vehicles, given current feasible resources and the composition of electrical sources. The table shows shares in Japan [3, 4], where most vehicles have internal combustion engines (ICEs) using gasoline, whose share is 96 %. The hybrid electric vehicle (HEV) share is 3.8 %. The Japanese government has established a target of 50 % share of next-generation vehicles (HEVs, BEVs, and FCEVs) by 2030. From the *Energy Technology Perspectives 2012* (ETP 2012) of the International Energy Agency (IEA) [5], significant diffusion of BEVs and FCEVs will begin in 2020 and 2030, respectively. In 2050, the share of these cars will be ~40 % and that of electric energy cars including BEVs, FCEVs, and HEVs >90 %.

In 2009, FCEVs of the Toyota Motor Corp., Honda Motor Co., Ltd., and Nissan Motor Co., Ltd., were operated over 1100 km from Tokyo to Fukuoka in 2 days, with two hydrogen charges under a Japan Hydrogen and Fuel Cell Demonstration

Table 1 Tank-to-wheel efficiencies (MJ/km) of various automotive types reported by the Japan Hydrogen and Fuel Cell Demonstration Project (JHFC) in 2011 [2] and shares in 2012 and 2030 target [3, 4]

Type of automotive vehicle		Tank to wheel [2] [MJ/km] (JC08)	Share in Japan [3, 4] 2012	2030 target
Internal combustion engine (gasoline)	ICEV	1.69	96.00 %	30~50 %
Internal combustion engine (diesel)	DICEV	1.63	0.10 %	5~10 %
Compressed natural gas vehicle	CNGV	1.69	0.05 %	–
Hybrid electric vehicle	HEV	1.09	3.80 % (HEV, PHEV)	30~40 %
Plug-in HEV (BEV mode)	PHEV	0.36		20~30 % (PHEV, BEV)
Plug-in HEV (HEV mode)	PHEV	1.17		
Battery electric vehicle	BEV	0.36	0.05 %	
Fuel cell electric vehicle	FCEV	0.73	0 %	~3 %

Tank-to-wheel efficiency (T to W) is calculated for each technology in 2010

Project (JHFC). These cars can achieve a mileage similar to that of an ICE [6]. Problems of cold start and performance under high-temperature and low-humidity conditions were solved by test driving in a desert and a cold area at less than −30 °C. Finally, Toyota Motor Corp. began selling the world's first commercial FCEV (called the Mirai, which means "future" in Japanese) in Japan on 15 December 2014 [7]. The vehicle price is ¥7.23 million (including consumption tax). The Japanese government gives the customer a subsidy of ¥2 million per one car. The vehicle contains two hydrogen tanks and can travel ~650 km on one charge. Honda Motor Co. announced an FCEV to be sold in Japan on 1 April 2015 [8]. Nissan Motor Co. aims to launch its own model in 2017 [9]. Japan will be the first nation in the world to introduce FCEV technology in the transportation society.

Mitsubishi Motors Corp. and Nissan Motor Co. began BEV sales in 2009 and 2010, respectively. These cars have a LiB. Recently, there have been many rapid charging stands for BEVs in towns, such as large shopping centers. The cars are expected to be mobile storage systems of the smart community (V2X, vehicle to home, vehicle to community, and others). For example, the fuel cell bus of Toyota can output 3 kW of power for 100 h as backup storage in an emergency [10]. For such a case, the effect of introduction must be estimated across all energy systems in the society.

As mentioned above, although the performances of both BEVs and FCEVs are increasing steadily, many issues remain for wide diffusion per scenarios of the IEA and Japanese government. In this chapter, an overview of BEVs and FCEVs is given, and a technology road map of secondary battery and fuel cell systems for the next-generation automotive is introduced. HEVs and plug-in HEVs (PHEVs) are not addressed as next-generation automotives in this chapter. However, because the battery is a common technology in HEVs, PHEVs, BEVs, and FCEVs, the outlook for HEVs and PHEVs is mentioned in the BEV technology road map.

2 Present Status

FCEV well-to-wheel efficiency is greater than that of an ICE; the mileage is ~700 km and the hydrogen refueling time is ~3 min. Given these facts, the performance of FCEVs has already reached the performance of the general car. The most important subject for FCEV diffusion is cost except for hydrogen infrastructure (information on this hydrogen infrastructure and stationary PEFC systems is given in other chapters). The initial FCEV price upon market introduction is determined not only by vehicle cost but also various other factors, such as production volume, diffusion scenario, and R&D expenses. The FCEV target price for wide diffusion will fluctuate with the hydrogen price, if the FCEV were to be a perfect alternative to ICEs, that price would have to be one-third to one-fourth of its present value. For this, the fuel cell stack cost would have to be decreased to $30/kW.

In BEVs, battery capacity is limited because of vehicle space and weight. Mileage on one charge is ~200 km, about a third that of an ICE. Charging time is ~8 h via domestic power (200 V). The BEV sales price was ~¥3 million as of November 2014, so cost reduction is needed. Current performances of BEV battery systems are energy density 70 Wh/kg, power densities 1400–2000 W/kg, and battery cost ¥100,000/kWh. For BEVs to be perfect alternatives to ICE vehicles, performances must be ¥5000/kWh and 700 Wh/kg.

Understandably, goal setting for FCEV and BEV performance based on ICE vehicles is not realistic. It is important to discuss instead the ideal situation for BEVs and FCEVs in future societal systems. For example, BEVs are for short drives in cities, and FCEVs for long-distance excursions. In this chapter, the technology and development of battery and fuel cell systems, which are the most important components of the two vehicle types, are addressed from the standpoint of present technology and vision without considering them as alternatives to ICE cars. There is the possibility for improvement of energy density and costs of the battery and fuel cell systems. By summarizing present technologies and approaches, the author hopes that this will be helpful to battery and fuel cell researchers toward collecting information and gaining perspective on those technologies. Further, the battery and fuel cell system road map is summarized for BEVs and FCEVs.

3 Technology Road Map

3.1 Technology Road Map for BEVs

To reduce cost, the development of new, active materials, the separator and electrolyte, is needed. High-energy-density materials can reduce the amount of material used. To solve the problem of charge-discharge rate, a structural design based on kinetic theory is a must. Given these factors, the following four approaches are examined: (1) development of innovative active materials and electrolytes, (2) development of post-lithium-ion batteries, (3) optimization of fabrication processes and quality control, and (4) new structural designs from the perspective of reaction kinetics and mass transport dynamics. Details are given below.

3.1.1 Development of Innovative Active Materials and Electrolytes

Weight and volume energy density increase with high-energy-density material. Innovative materials are screened and designed through both experimentation and computational science [11]. However, advanced design of porous electrodes is necessary because of large expansion and contraction by intercalation effects such as Si. Given this, the unique Si particle structure has been examined [12]. Incombustible and durable electrolytes have been developed. The molecular design of innovative electrolytes for high conductivity has been researched by computational technology [13]. In a separator, composite and advanced uniform membranes such as nano-porous membrane have been developed to reduce ion transfer resistance [14]. In addition, high-efficiency solid-solid and solid-liquid interface-forming technology has been studied.

3.1.2 Development of Post-lithium-Ion Batteries

The theoretical limit of LiB energy density is 300 Wh/kg, so an ordinal LiB cannot achieve 700 Wh/kg. Given this, new high-capacity and high-output batteries have been developed, such as all solid state, metal air, and Mg ion [15, 16]. Sulfide and oxide materials have been investigated as electrolyte materials. The design of the interface between solid and solid is particularly important [17].

3.1.3 Optimization of Fabrication Processes and Quality Control

The electrode layer is made by various fabrication processes, namely, granulating, mixing, coating, and drying. Among these processes, drying must be simplified because heating cost is included in the battery system cost. However, porous electrode structure depends on the drying rate. Figure 1 shows a cross section of an LiB electrode from focused ion beam-scanning electron microscopy (FIB-SEM)

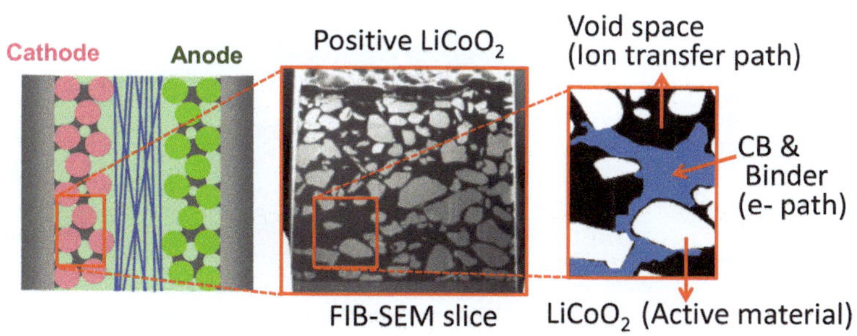

Fig. 1 Cross section of LiB

[18]. This heterogeneous structure is strongly related to effective ion and electron conductivity and the diffusion coefficient. Thus, the relationship between process and electrode structure must be well understood from the standpoint of improving battery performance and quality control [19].

3.1.4 New Structural Designs (Reaction Kinetics and Mass Transport Dynamics)

To achieve a charging time equal to the refueling time of an ICE, a high rate of charging must be realized. To do so, both the reduction of mass transport resistance in active material through the development of nanoscale particles and of ion transport in void space of the porous electrode layer through uniform distribution and optimal arrangement of binder and conductive materials are crucial. To decrease ion transfer resistance between electrodes, three-dimensional electrode structures of interdigitated shape have been developed [20].

In addition, thermal storage systems and motors as BEV technology are currently under investigation. Advanced observation and visualization techniques have been followed to understand internal phenomena and electrochemical reaction mechanisms.

3.2 *Technology Road Map for FCEV*

To reduce FCEV cost, a low-cost PEFC stack is needed. Figure 2 shows a schematic image of a PEFC and reaction mechanism of the cathode catalyst layer. Figure 3 shows an 80-kW PEFC system cost and a cost reduction scenario, using data reported by the US Department of Energy in 2012 [21]. The major cost of the

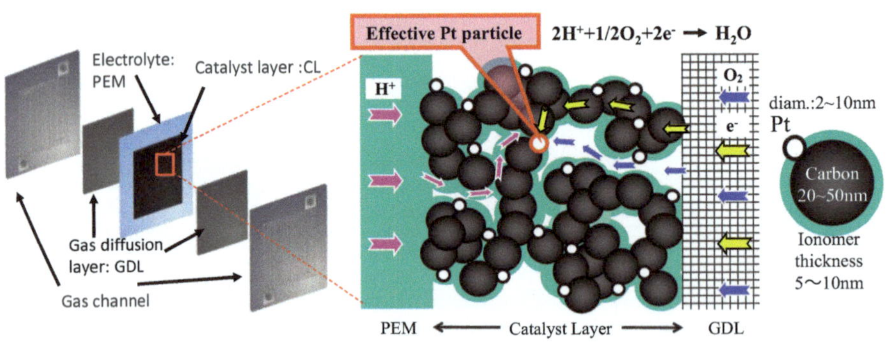

Fig. 2 Schematic of PEFC and reaction mechanism of cathode catalyst layer

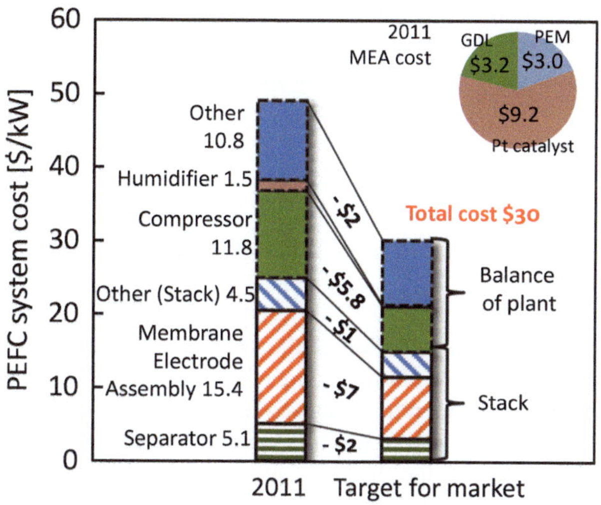

Fig. 3 80-kW PEFC system cost and cost reduction scenario (500,000 cars/year) (DOE Hydrogen and Fuel Cell Program 2012)

PEFC is from platinum particles for the electrode catalyst. However, to reach the target cost that can promote wide diffusion, both reduction in the amount of Pt and a comprehensive cost reduction with the balance of plant (BOP) are necessary. In particular, the following approaches are examined: (1) cost reduction with non-Pt catalyst and a small amount of Pt catalyst, (2) new cell design and improvement of flexibility of cell components, and (3) development of high-current-density cells by improving mass transfer performance. Details are given below.

3.2.1 Cost Reduction with Non-Pt Catalyst and a Small Amount of Pt Catalyst

In a PEFC, overvoltage of the oxygen reduction reaction (ORR) at the cathode is the dominant phenomenon, and so a reduction of this overvoltage, i.e., improvement of ORR activity, is required. For improving the activity of the Pt catalyst, development of monodispersed Pt particles, nanoscale particles, and surface-modified particles has been attempted. In particular, to prevent Pt dissolution and agglomeration, a Pt catalyst covered by a SiO_2 nano-layer has been developed [22]. Further, to increase the effectiveness of Pt in ORR, a core-shell catalyst with a thin Pt layer has been developed [23]. Also, carbon-alloy [24] and oxide-based [25] non-Pt catalysts have been investigated. Oxide-based support material has been developed by improvement of carbon corrosion [26]. The catalyst layer consists of Pt particles, carbon support, void, and ionomer. Reactant gas, protons, and electrons transfer this porous catalyst layer. Therefore, ORR effectiveness depends on these mass transport phenomena. Accordingly, to increase Pt utilization, the reaction mechanism at the Pt surface and mass transport mechanism in the catalyst layer have been examined via experimental and computational approaches [27].

3.2.2 New Cell Design and Improvement of Flexibility of Cell Components

In a typical PEFC, a Pt catalyst with high durability for electrochemical reactions is used, because the catalyst layer has a strong acid condition. With an anion exchange membrane as electrolyte, various catalyst metals that are of low durability and cheaper than Pt can be used. Given this, the high ion-conductive and high-durability anion exchange membrane has been developed for this new type of FC [28]. Furthermore, the optimal catalyst layer structure has been reconsidered with carbon nano-fiber and mid-temperature-type electrolyte of polybenzimidazole, instead of carbon black and low-temperature-type fluorine electrolyte. With this cell, a one-eighth reduction of Pt amount has been reported [29]. A gas-diffusion layer-free cell has been developed to reduce mass transfer resistance and increase limiting current density [30].

3.2.3 Development of High-Current-Density Cells by Improving Mass Transfer Performance

To increase output density per cell by high-current density, electrode area and cell number can be reduced. This can in turn reduce the amount of Pt. At present, the limiting current is controlled by oxygen transfer performance. Optimization of cell structure and material properties of pore size, porosity, and wettability is needed to increase that performance from the gas channel to catalyst surface [31]. In connection with this approach, direct visualization of liquid water with soft X-ray CT [32] and measurement of temperature and humidity distributions [33, 34] have been investigated.

In this section, technical aspects for LiB and PEFC are discussed. Although the quantitative impacts and goals of this technology must be demonstrated to improve the performance of the overall system, it is difficult to show its effects quantitatively, because the novel materials have the potential to drastically alter specifications of the battery and cell. Furthermore, because these factors will be affected by future society, the composition of the energy resources, hydrogen supply situation, and diffusion rate of these technologies cannot be estimated by current information.

4 Benefit and Prospective Future Vision

Most LiB and PEFC technical aspects in Sect. 3 are related to cost reduction and improving energy density. Lessening internal resistance is one approach to cost reduction, such as the increase of current density and decrease of mass transport resistance in the case of the thick electrode layer. The author does not believe that tank-to-wheel efficiency can be improved by solving the aforementioned technical aspects. It is therefore essential that mileage depends on the improvement of energy capacity. To increase mileage, improving the energy density of the LiB and the design of the hydrogen tank for the PEFC is vital. From the perspective of mileage and infrastructure, both BEVs and FCEVs need not be perfect alternatives to the gasoline car. In an environment in which societal systems are diversifying, the appropriate next-generation vehicles should be introduced in the right locations. Moreover, the significance of these technologies as backup sources and mobile storage and power generation systems in smart grids must be considered. Advanced relationships to the charging infrastructure network, V2X, and intelligent transportation systems (ITS) can be expected. The vision of a future mobile society must be addressed from standpoints of next-generation vehicle performance, infrastructure for these motilities, decreasing birthrate, aging populations (the rate of aging in Japan will be ~40 % by 2050), bipolarization of populated and depopulated areas, potentially low mobility, and new mobility systems such as car sharing. The effects of the diffusion of next-generation vehicles (HEVs, BEVs, and FCEVs) are extremely powerful from the perspectives of reduced energy consumption and CO_2 emissions, increasing energy self-sufficiency, diversification of energy resources, cooperation with smart grids, use as backup and mobile power sources, and enhancement of competitiveness in the automotive industry. For mass introduction, the risk of material resources must be considered. Finally, environmental burdens from production processes and material recycling must be addressed in the context of a sustainable society.

In conclusion, road maps of the LiB and PEFC are shown in Fig. 4. The comprehensive approach to the various technical aspects in Sect. 3 is critical. It is difficult to meet the specified requirements of BEVs and PEFCs in a sustainable future society using current technology, so breakthroughs are needed with various approaches. In this figure, cost targets are taken from other reports [4, 21]. Accordingly, relationships between technical aspects and the cost targets are not shown.

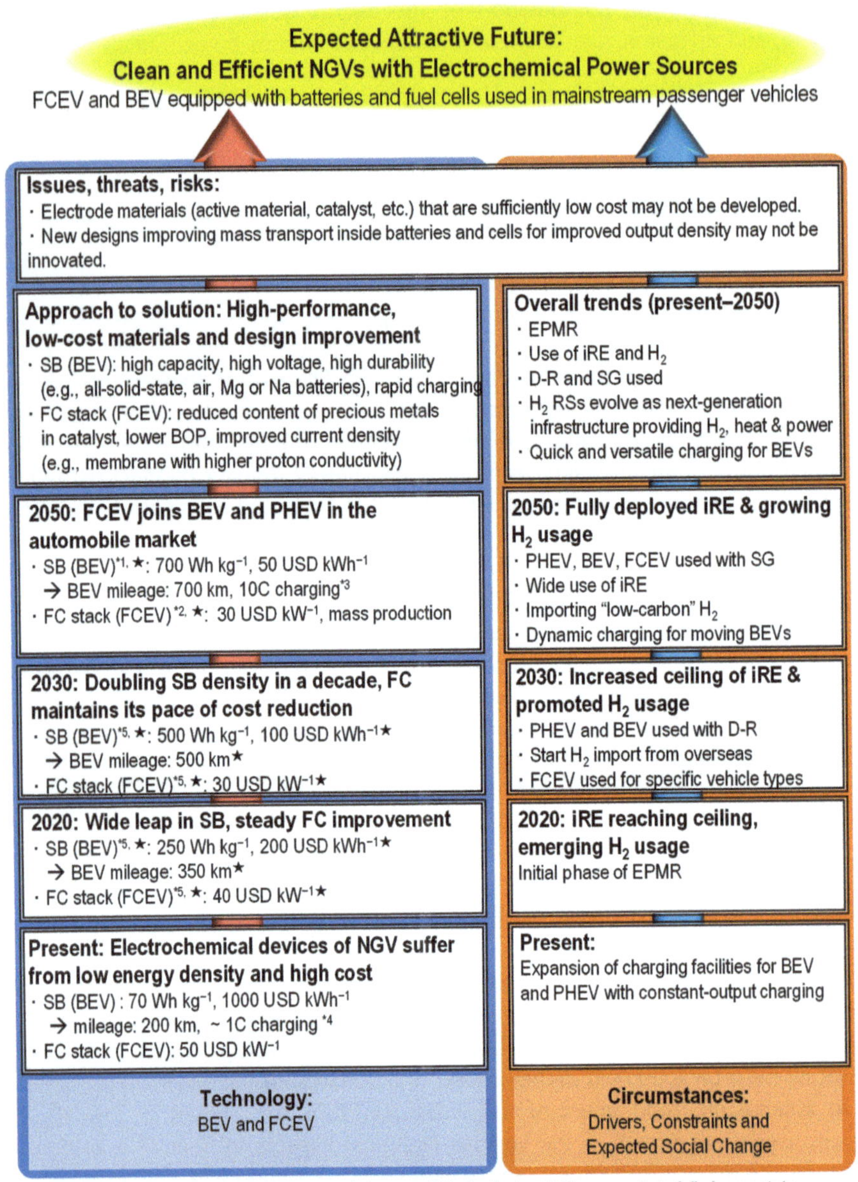

*1 [4], *2 [21] *3 From empty to full state of charge within 6 minutes *4 From empty to full charge state within 1 hour, *5 interpolation using present and 2050 values from author's perspective

Abbreviations:

BEV: Battery Electric Vehicle, BOP: Balance Of Plant, D-R: Demand-Response, EPMR: Electric Power system Market Reformation, FC: Fuel Cell, FCEV: FC Electric Vehicle, HEV: Hybrid Electric Vehicle, iRE: intermittent Renewable Electricity, NGV: Next Generation Vehicles, PHEV: Plug-in Hybrid Electric Vehicle, RS: Refueling Station, SB: Secondary Battery, SG: Smart Grid, USD: U.S. Dollars

Fig. 4 Road map of secondary batteries and fuel cell systems for next-generation vehicles

References

1. Ministry of Economy, Trade and Industry (2014) The 2013 annual report on energy (Japan's Energy White Paper 2013). http://www.meti.go.jp/english/press/2013/0614_02.html. Accessed 25 Dec 2014
2. JHFC report (2014) Objectives and activities of the JHFC project. http://www.jari.or.jp/Portals/0/jhfc/e/data/pdf/stake_d_e.pdf. Accessed 25 Dec 2014
3. Japan Automotive Manufactures Association, Inc. (2014) 2013 report on environmental protection efforts. http://www.jama-english.jp/publications/env_prot_report_2013.pdf. Accessed 25 Dec 2014
4. Ministry of Economy, Trade and Industry (2014) Announcement of the next-generation vehicle strategy 2010. http://www.meti.go.jp/english/press/data/20100412_02.html. Accessed 25 Dec 2014
5. IEA (2014) Energy technology perspectives 2012. http://www.iea.org/textbase/npsum/ETP2012SUM.pdf. Accessed 25 Dec 2014
6. JHFC (2014) JHFC news March 2010. http://www.jari.or.jp/portals/0/jhfc/data/pdf/jhfcnews_vol4.pdf (in Japanese). Accessed 25 Dec 2014
7. Toyota FCEV (2014) http://www.toyota.com/fuelcell/fcv.html. Accessed 25 Dec 2014
8. Honda FCEV (2014) http://world.honda.com/news/2014/4141117All-New-Fuel-Cell-Vehicle-FCEV-CONCEPT/index.html. Accessed 25 Dec 2014
9. InsideEVs (2014) http://insideevs.com/nissan-coo-says-2017-fuel-cell-vehicle-wont-compete-with-nissan-leaf/. Accessed 25 Dec 2014
10. Toyota (2014) TMC develops fuel-cell-bus power supply system http://www2.toyota.co.jp/en/news/12/08/0831.html. Accessed 25 Dec 2014
11. Jalem R et al (2014) J Mater Chem A 2:720
12. Wada T et al (2014) Nano Lett 14(8):4505–4510
13. Sodeyama K et al (2014) J Phys Chem C 118(26):14091–14097
14. Munakata H et al (2010) Abstract #180, The 15th international meeting on lithium batteries – IMLB 2010
15. Ikezawa A et al (2014) Abstract #32, A1: batteries and energy technology joint general session, 2014 ECS and SMEQ
16. Zhang R et al (2012) Electrochem Commun 23:110–113
17. Hayashi A et al (2001) J Am Ceram Soc 84(2):477–479
18. Inoue G et al (2014) Abstract #374, A5: lithium-ion batteries, 2014 ECS and SMEQ
19. Hatakeyama N et al (2014) Proceedings of the 55th battery symposium in Japan, 2A03, p 58 (in Japanese)
20. Kanamura H et al (2011) IOP conf. series: materials science and engineering 18, 122007
21. U.S. Department of Energy (2014) 2012 annual merit review and peer evaluation meeting, fuel cell sub-program. http://www.hydrogen.energy.gov/pdfs/review12/fc000_papageorgopoulos_2012_o.pdf. Accessed 25 Dec 2014
22. Takenaka S et al (2007) J Phys Chem C 111(42):15133–15136
23. Inaba M et al (2010) ECS Trans 33(1):231–238
24. Nabae Y et al (2009) ECS Trans 25(1):463–467
25. Suito K et al (2012) ECS Trans 50(2):1777–1783
26. Masao A et al (2009) Electrochem Solid-State Lett 12(9):B119–B122
27. Inoue G et al (2009) ECS Trans 25(1):1519–1527
28. Tanaka M et al (2011) Polym Chem 2:99–106
29. Fujigaya T et al (2013) Polym J 45:326–330
30. Kotaka T et al (2013) ECS Trans 58(1):1033–1046
31. Yokoyama K et al (2013) State-of-the-art fuel cells and hydrogen technology in Japan (FC-DIC Transactions)
32. Sasabe T et al (2010) Int J Hydrogen Energy 35:11119–11128
33. Kawase M et al (2009) 216th ECS meeting, Abstract #1073
34. Araki T et al (2014) ECS Trans 64(3):619–628

Power Electronics for Vehicles and Energy Systems

Takaji Umeno

Abstract Power electronics that enable DC-AC, AC-DC, or DC-DC conversion of electric power are one of the key technologies for the spread of electric-drive vehicles and renewable energy systems such as PV and wind turbine generators that address the global warming issue and contribute to energy saving. In particular, electrification of vehicles that eliminate about 20 % of total CO_2 emissions contributes to overall reduction of those emissions, and power electronics become an important technology in vehicle electrification. Wide-bandgap semiconductors such as SiC or GaN are important to future technical development of power electronics. Compared with Si, SiC has superior properties in that it has three times the bandgap, twice the saturation velocity, ten times greater breakdown field, and three times greater thermal conductivity. Given these characteristics, wide-bandgap power devices can be driven at high frequency and high temperature, and drastic downsizing of the converter and loss reduction are enabled. By replacing conventional Si power devices with SiC, significant energy saving is anticipated.

Keywords Power device • SiC • GaN • Hybrid electric vehicle • Battery electric vehicle • Power conversion

1 Introduction

In recent years, the reduction of CO_2 emission has been regarded as important for countermeasures against global warming, and the influence of rising oil prices has become a significant economic issue. Power electronics that enable DC-AC, AC-DC, or DC-DC conversion of electric power are one of the key technologies for the spread of electric-drive vehicles and renewable energy systems such as PV and wind turbine generators that address the global warming issue and contribute to energy saving. The core technology of power electronics is a power device, power conversion circuit, and their control. These have been applied in various areas,

T. Umeno (✉)
Toyota Central R & D Labs. Inc., Aichi, Japan
e-mail: umeno@mosk.tytlabs.co.jp

Y. Kato et al. (eds.), *Energy Technology Roadmaps of Japan*,
DOI 10.1007/978-4-431-55951-1_44

including electric power systems, railroads, household electrical appliances (air-conditioners, washing machines, refrigerators, and others), and automotives such as hybrid and electric vehicles.

In particular, electrification of vehicles that eliminate about 20 % of total CO_2 emissions contributes to overall reduction of those emissions. In this chapter, the present status of power electronics is reviewed, mainly regarding their application to electric-drive vehicles. A technical roadmap for the power device and its application are described along with its CO_2 emission reduction for society.

2 Present Status

Some scenarios estimate that the automobile market share of electric-drive vehicles represented by hybrid electric vehicles (HEVs) will reach 15 % in 2020, growing to 40 % in 2030 and 90 % in 2050 [1]. In the background, certain CO_2 emission vehicle regulations become more stringent each year. For example, in Europe, car makers are obliged to meet a 95-g/km CO_2 emission regulation by 2021. Automakers need to promote development and diffusion of electric-drive vehicles to meet such regulations.

There are three types of HEV, depending on the capacity of the electric system [2], namely, strong, mild, and micro. A strong HEV is designed to drive like a battery electric vehicle (BEV). A mild HEV has a motor with small capacity, which is connected to the engine in parallel and adds power during acceleration. A micro HEV uses an alternator to produce energy regeneration of 5–10 kW during deceleration [3]. The strong HEV has a mileage performance about twice that of a conventional vehicle, and a mileage improvement ~10 % is reported for the micro HEV.

Figure 1 shows an example electric system of the strong HEV. In a compact car, the voltage of the main battery is ~200 V and is controlled as much as 650–700 V by the boost converter [4]. A three-phase DC-AC inverter drives a motor for traction. During deceleration, kinematic energy is regenerated through the motor to charge the battery. The boost converter controls the voltage so the inverter and motor can operate under optimal conditions. Power from the engine is converted to electric AC power by the generator and converted to DC power by the inverter. DC power is used for motor drive or battery charge. The boost converter and inverter control output voltage by pulse-width modulation (PWM). PWM is realized by an on-off switching sequence with power devices like a Si-insulated gate bipolar transistor (Si-IGBT) and diode. The main battery is connected to an isolated DC-DC converter and supplies power to auxiliaries such as steering and the air conditioner and charges an auxiliary battery. In that converter, Si metal-oxide-semiconductor field-effect transistors (Si-MOSFET) are mainly used. In this way, the semiconductor power conversion equipment is important in electric-drive vehicles, and the significant component is a power device such as IGBT and MOSFET.

The maximum power of the motor and inverter of the compact car is ~60 kW. The hybrid system is used in medium and large vehicles and the inverter power exceeds 160 kW. When the inverter has a large capacity, it is necessary to reduce

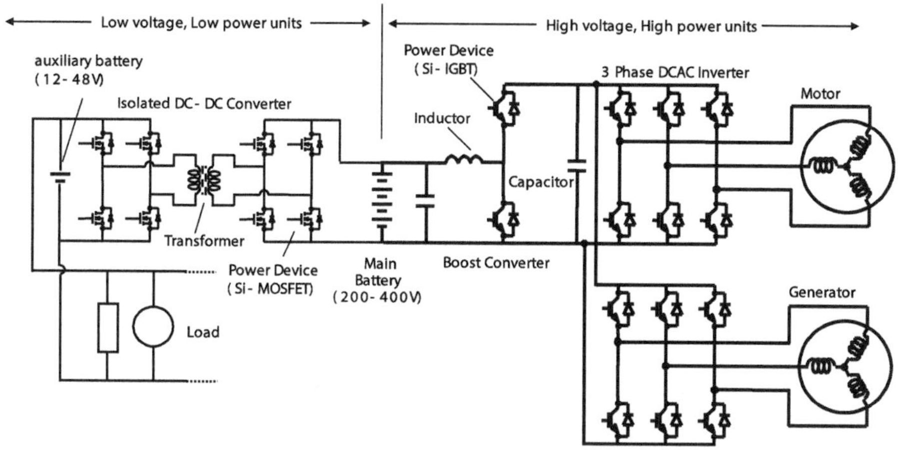

Fig. 1 Example of power electronics in strong hybrid electric vehicle

the volume and mass of the power control unit (PCU), which consists of inverters and a converter because of the requirement of mountability.

It is effective for PCU downsizing to use wide-bandgap semiconductors such as silicon carbide (SiC) or gallium nitride (GaN) for switching devices. Compared with Si, SiC has superior properties of three times the bandgap, twice the saturation velocity, ten times greater breakdown field, and three times greater thermal conductivity [5]. Given these characteristics, wide-bandgap power devices can be driven at high frequency and high temperature, and extreme downsizing of the PCU and loss reduction are facilitated. Figure 2 shows characteristics of the power devices. In this figure, dashed lines indicate performance limits derived from material properties, and ovals with solid lines show the area in which the reported devices currently exist. As shown in Fig. 2, wide-bandgap semiconductors have lower on-resistance than Si. Miniaturization of passive elements such as the inductor and capacitor by high-frequency switching and cooling system simplification will significantly reduce PCU size.

Because SiC makes high output possible, as opposed to GaN, SiC power devices will be used in the boost converter and inverter, and GaN power devices will be used in the isolated DC-DC converter or low-voltage converter of auxiliary systems [6]. By using SiC power devices in the PCU of HEVs, mileage improvement of about 5–10 % is expected [7]. In addition, by replacing power devices of the isolated DC-DC converter with GaN from Si, further high-frequency drive will be possible to reduce the size of passive components.

However, the market size of the wide-bandgap power device is extremely small, because a wafer is expensive compared with Si. The Czochralski method enables growth of large Si crystals, up to 450 mm in diameter and several meters in length, at a high growth rate of several centimeters per hour. Therefore, cost reduction of Si wafers has progressed rapidly with mass production. However, SiC and GaN are greatly inferior in diameter, length, and growth rate of bulk crystals used as the basis of a wafer, so the cost reduction is not substantially progressing for these

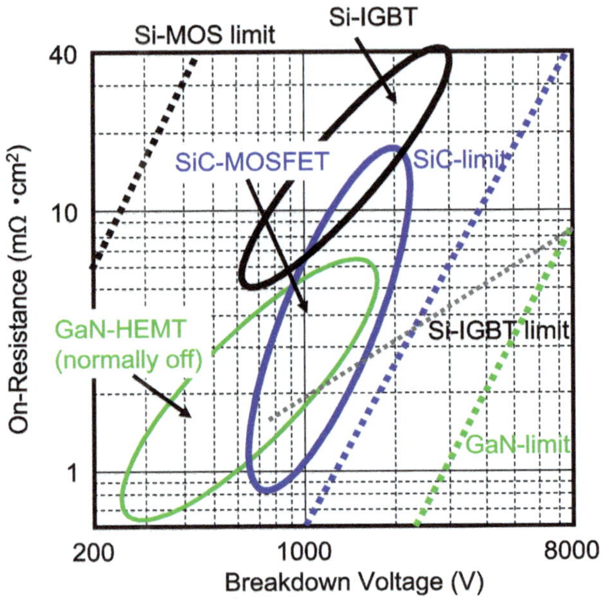

Fig. 2 Characteristics of power devices

materials. At present, the Schottky barrier diode (SBD) is largely commercialized, and it has begun to be used in industrial instruments and household electrical appliances.

Performance enhancement of the conventional Si power device has been ongoing. For example, a reverse-conducting IGBT (RC-IGBT), in which the IGBT and diode are unified on one chip, is being developed [8]. The RC-IGBT will become one of the solutions for reducing the size and loss of the PCU before the diffusion of SiC power devices and will be used in various power electronics apparatuses.

3 Technology Roadmap

At present, the global market for the SiC power device is limited to SiC-SBD and is ~1600 million yen. Future market size is anticipated to increase through cost reduction of the device, commercialization of the switching device, and SiC used in the power electronics apparatus. The market size of the power device is ~1100 billion yen, and that of the power electronics ~6 trillion yen [9]. Through expansion of the markets for renewable energy and electric-drive vehicles, SiC power devices will be widely used in these markets. Consequently, it is expected that the global market size of the power device will be ~10 trillion yen by 2050 [10].

Table 1 shows wide-bandgap power electronics technology development as surveyed by the National Institute of Advanced Industrial Science and Technology (AIST) [11]. The current wafer diameter of SiC is 4 in., which is for the first-generation wafer, and is a minimum for mass production. Recently, 6-in.-diameter

Table 1 Generation of wide-bandgap power electronics technology [11]

	First generation	Second generation	Third generation
Water stage	SiC	SiC	SiC
	Sublimation growth for 4 in.	For 6 in. sublimation growth	Sublimation growth for 8 in.
	GaN	High-speed and thick epitaxy	Liquid phase growth, gas method
	Hetero-epitaxy	GaN, diamond, etc.	GaN, diamond, etc.
		Bulk	Large-diameter bulk
Device stage	1 kV class	5 kV class (3–6 kV)	10 kV class (over 10 kV)
	High current	High reliability, high temperature	Bipolar device
		High functionality, integration	
System stage	10 W/cm^3	25 W/cm^3	50 W/cm^3
Application	Consumer, general use	Automobile, transportation infrastructure	Electric power system infrastructure

SiC wafer in the second generation has been obtained using the physical vapor transport (PVT) technique [12]. In addition, alternative SiC growth methods such as solution growth have enabled high-quality SiC crystals [13]. Such technology would reduce the cost and improve the quality of SiC wafers. A 6-in. wafer will be mass-produced by 2020, and stable supply will be accomplished with improvement of the yield rate by 2030. By 2050, it is anticipated that the cost of the SiC power device would be competitive with Si-IGBT by using a third-generation 8-in. wafer. Then, replacement of Si-IGBT by SiC-MOSFET will be achieved in many power electronics apparatuses, including electric-drive vehicles.

As on-resistance of the SiC power device is less than Si-IGBT, the chip area of that device can be reduced. Therefore, it is believed that practical use of PCUs replacing Si-IGBT with SiC-MOSFET would begin with small-lot production after 2020 [14]. By 2030, a high-frequency drive based on the high-speed switching characteristic of SiC will be mass-produced, and PCU volume will become ~20 % that of conventional PCUs with Si-IGBT, through downsizing of passive elements such as the inductor and capacitor. The SiC advantage of high-temperature operation (>200 °C) will achieve unification of the cooling system and engine [15], which will further reduce PCU volume and cost.

In most developed countries, government measures for the diffusion of electric-drive vehicles have been implemented. Furthermore, wireless power transmission technology to make charging convenient is evolving, and this system will be standardized. By 2020, charging infrastructure, which is key to the diffusion of electric-drive vehicles, will be expanded. This fact removes psychological barriers of consumers for BEV purchase. It is presumed that the 95-g/km CO_2 emission regulation will be met by widespread use of electric-drive vehicles. Then, many of those vehicles will be connected to houses and/or the electric power grid and will constitute a business model of the smart grid system. The battery of the electric-drive vehicle will be a buffer of electricity and will smooth its flow from fluctuating renewable energy [16]. By 2050, a dynamic charging system for moving BEVs [17] could enter the practical-use phase. By automated driving on the highway, precise control of the power transmission is enabled. Because high-speed driving on the highway consumes substantial electrical energy, power transmission of a BEV significantly extends its range, and this promotes the diffusion of electric-drive vehicles.

4 Benefit and Prospective Future Vision

Figure 3 shows an application map of the several types of power devices. In most applications of Si-IGBT and Si-MOSFET power devices, the replacement of SiC and GaN power devices is gradual. Energy saving is assumed when wide-bandgap devices are used in a power electronics apparatus, including inverters for household appliances, electric-drive vehicles, industrial inverters, computer power supplies, UPS, power conditioners for PV power generation, and fuel cells [18].

Generally, the efficiency of power electronics apparatus is 85–97 %, and about half the loss is from power devices. A SiC power device can reduce loss by ~70 %. Therefore, efficiency can be improved by 2–5 % using SiC power devices. According to this estimation, by 2030, substantial energy savings of 53 million kL per year of crude oil equivalent, or 83 million tons CO_2 equivalent, are expected worldwide [9]. In this estimation, the contribution of vehicular SiC power devices comprises about half the reduced emission. With the expansion of electric-drive vehicles, this effect is anticipated to increase significantly (Fig. 4).

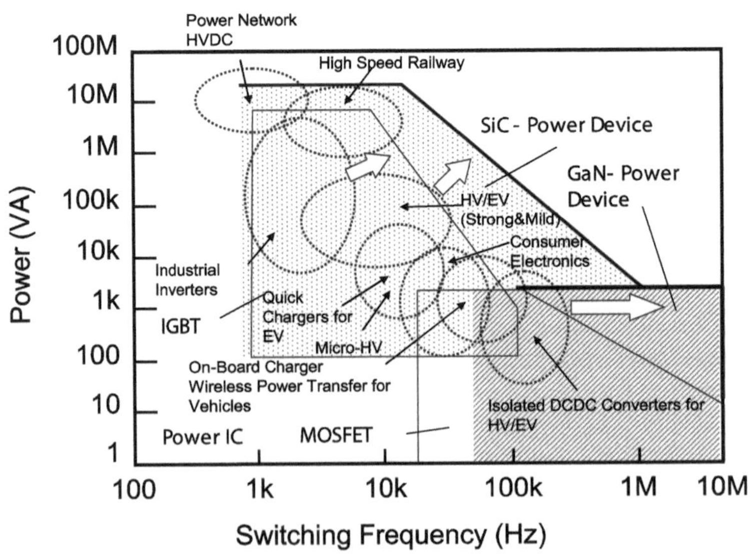

Fig. 3 Application area of power devices

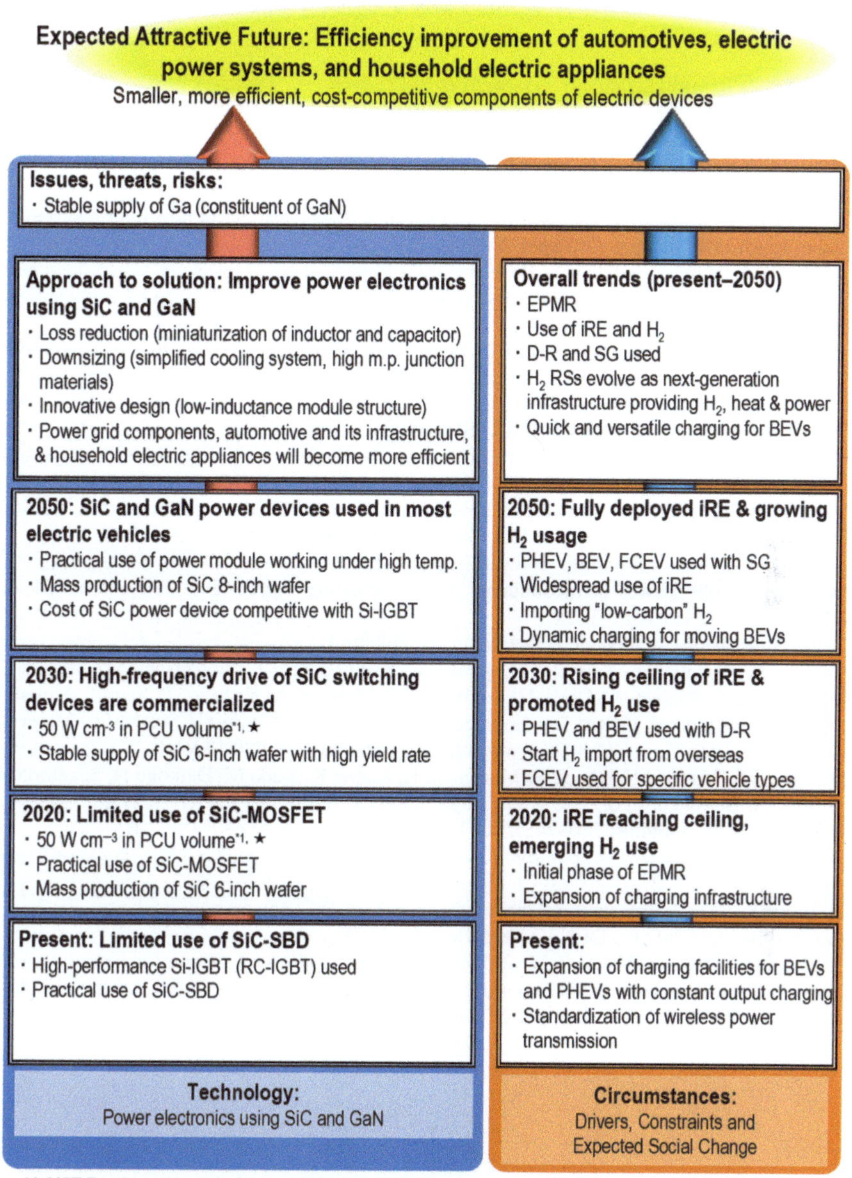

*1 AIST Roadmap: https://unit.aist.go.jp/adperc/cie/research/research8.html

Abbreviations:
D-R: Demand-Response, EPMR: Electric Power system Market Reformation, FC: Fuel Cell, FCEV: FC Electric Vehicle, HEV: Hybrid Electric Vehicle, IGBT: Insulated Gate Bipolar Transistor, iRE: intermittent Renewable Electricity, m.p.: melting point, PCU: Power Control Unit, PHEV: Plug-in Hybrid Electric Vehicle, RC: Reverse-Conducting SG: Smart-Grid, MOSFET: Metal-Oxide-Semiconductor Field-Effect Transistors, SBD: Schottky Barrier Diode

Fig. 4 Technology roadmap

References

1. IEA (2012) Energy technology perspectives 2012
2. Teratani T (2014) Impact of DC48 for automotive power supply systems. Proc. EVTeC & APE2014, No. 20144061
3. Kume A, Kotani K (2014) Comparison of low-voltage regenerative braking system in WLTP. Proc. EVTeC & APE2014, No. 20144083
4. Matsubara T, Yaguchi H, Takaoka T, Jinno K (2009) Development of new hybrid system for compact class vehicles, SAE Paper 2009-01-1332
5. Tolbert L, Ozpineci B, Islam S, Peng F (2002) Impact of SiC power electronic devices for hybrid electric vehicles, SAE 2002 transactions. J Passenger Cars: Electron Electr Syst, 2003, ISBN 0-7680-1291-0, pp 765–771
6. Kachi T, Kikuta D, Uesugi T (2012) GaN power device and reliability for automotive applications, reliability physics symposium (IRPS), 2012, pp 3D.1.1–3D.1.4. doi: 10.1109/IRPS.2012.6241815
7. http://newsroom.toyota.co.jp/en/detail/2656842/
8. Takahashi K, Yoshida S, Noguchi S, Kuribayashi H, Nashida N, Kobayashi Y, Kobayashi H, Mochizuki K, Ikeda Y, Ikawa O (2014) New reverse-conducting IGBT (1200V) with revolutionary compact package, Power Electronics Conference (IPEC-Hiroshima 2014 – ECCE-ASIA), 2014, pp 2569–2574 doi: 10.1109/IPEC.2014.6869951
9. http://cocn.jp/common/pdf/greenpowerele.pdf
10. Mori M (2008) Power semiconductor devices creating comfortable low-carbon society. Hitachi Rev 90(12):1022–1023
11. https://unit.aist.go.jp/adperc/cie/research/research8.html
12. Kondo H, Takaba H, Yamada M, Urakami Y, Okamoto T, Kobayashi M, Masuda T, Gunjishima I, Shigeto K, Ooya N, Sugiyama N, Matsuse A, Kozawa T, Sato T, Hirose F, Yamauchi S, Onda S (2014) Development of RAF quality 150mm 4H-SiC wafer. Mater Sci Forum 778–780:17. doi:10.4028/www.scientific.net/MSF.778-780.17A
13. Kusunoki K, Yashiro N, Okada N, Moriguchi K, Kamei K, Kado M, Daikoku H, Sakamoto H, Suzuki H, Bessho T (n.d.) Growth of large diameter 4H-SiC by TSSG Technique Materials Science Forum (volumes 740–742). doi:10.4028/www.scientific.net/MSF.740–742.65
14. http://corporatenews.pressroom.toyota.com/releases/toyota+develops+wafer+boost+mileage.htm
15. Tsuruta K (2011) Prospects of the practical use of SiC power semiconductor devices in automotive applications. DENSO Tech Rev 16:90–95
16. Su W, Eichi H, Zeng W, Chow M (2012) A survey on the electrification of transportation in a smart grid environment. IEEE Trans Ind Informa 8(1):1–10. doi:10.1109/TII.2011.2172454
17. Thomson A (2014) The future of road transportation is electric and wireless. Proc. EVTeC & APE2014, No. 20144021
18. http://www.jeita.or.jp/japanese/exhibit/06MrTakami.pdf

Effective Thermal Energy Utilization for Automobiles

Hironao Ogura

Abstract Automobiles with internal combustion engines (ICEV) can use only around 30 % of fuel energy for operation, and under current conditions, around 70 % of the unused energy is lost as waste heat. Therefore, thermal management of automobiles greatly improves the total energy efficiency of vehicles with such engines. Even for an electric vehicle (EV) and hybrid electric vehicle (HEV), when compression-type heat pumps (HPs) with an electric compressor are used as the air conditioner, it is not effective if the vehicle uses energy a lot from an expensive battery.

In this chapter, a thermal energy utilization roadmap for vehicles is addressed for present systems such as thermoelectric conversion, Rankine cycle, sensible and latent heat storage, and next-generation systems such as chemical HPs for waste heat recycling that recover waste heat energy and convert it into cold and high-temperature heats.

Until around 2050, vigorous promotion of the engine efficiency of a conventional automobile and effective utilization of engine waste energy as shown in the technology roadmap are important, in addition to next-generation-type automobiles. In 2050, hybrid application of a thermally driven HP and high-performance thermoelectric conversion units should be implemented for automobiles in practical use. For electric power, exhaust heat recovery by thermoelectric conversion is not very efficient, but useful if a vehicle needs more electricity. Hot/cold heat use by recycling exhaust heat called “heat–heat use” has superior efficiency. In particular, automobile systems of chemical heat storage/chemical HP recovery of waste heat that reform it to cold/hot heat are anticipated. The chemical HP systems can be driven not only by engine waste heat but also various waste heats in HEVs/PHEVs or electric power in charging equipment and save energy of expensive batteries for heating and cooling.

Keywords Fuel efficiency • Waste heat • Heat recovery • Chemical heat pump • Heat storage

H. Ogura (✉)
Chiba University, Chiba, Japan
e-mail: hiro_ogura@faculty.chiba-u.jp

Y. Kato et al. (eds.), *Energy Technology Roadmaps of Japan*,
DOI 10.1007/978-4-431-55951-1_45

1 Introduction

Over the more than 100-year history of practical automobiles, almost all of them have had internal combustion engines (ICEV). However, in recent years, the hybrid vehicle (HV) and hybrid electric vehicle (HEV) have been rapidly put into use. Automobiles with internal combustion engines can use only ~30 % of the fuel energy for operation, and under present conditions, ~70 % of the unused energy is lost as waste heat. Therefore, thermal management of automobiles greatly improves their total energy efficiency in operation with such engines. For an EV and HEV, when compression-type HPs with electric compressors are used as the air conditioner, it is not effective if energy from an expensive battery is used for long periods.

The introduction of energy reuse and recycling systems for the substantial waste energy is anticipated, not only for conventional vehicles with internal combustion engines but also for EVs and HEVs with electric power. In this chapter, a thermal energy utilization roadmap for vehicles is addressed for present systems such as thermoelectric conversion, Rankine cycle, sensible and latent heat storage, and next-generation systems such as chemical HPs of waste heat recycling that recover waste heat energy and convert it into cold and high-temperature heats.

2 Present Status

2.1 Unused Energy of Automobiles

Because improvement of automobile engines is described in other chapters, I describe here effective utilization of unused energy from automobiles. Figure 1 shows an energy flow example [1] of a conventional automobile. Around 70 % of fuel energy is wasted heat from cooling and exhaust for a gasoline engine. Around 60 % is wasted in the case of a diesel engine. These waste heats vary by location and operating characteristics (Fig. 2) [2]. However, there is potential vehicle energy-use efficiency improvement of engine performance if the otherwise lost thermal energy can be recovered and recycled.

2.2 Waste Heat Utilization Systems of Automobiles

To make effective use of waste heat from automobiles, systems are proposed such as "heat–heat use," which reuses and recycles heat by storing and reforming it, "heat–motive power use," which converts heat to motive power, and "heat–electric power use," which converts heat to electric power. Various unused heat utilization

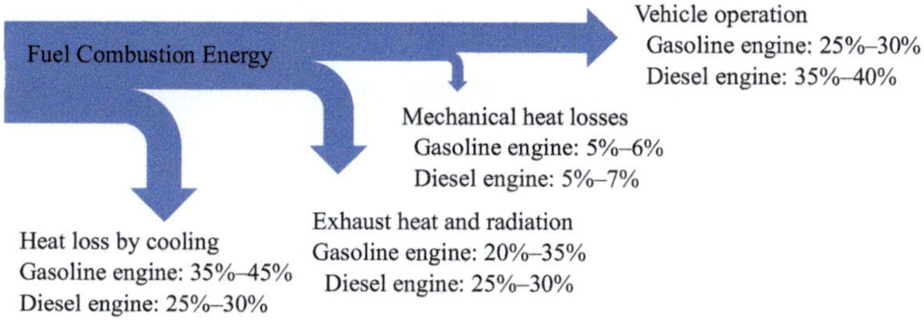

Fig. 1 Energy flow example of conventional vehicle [1]

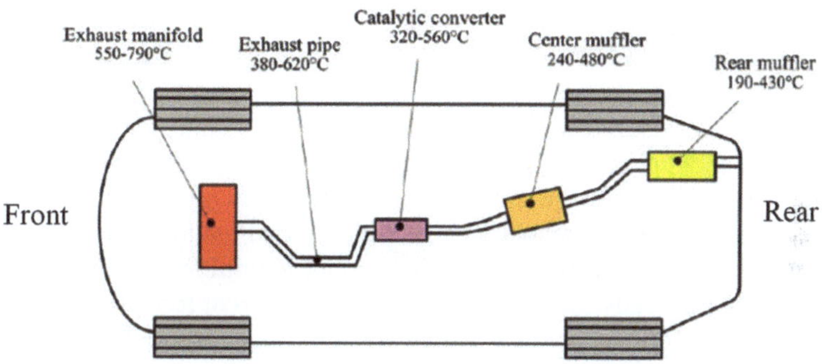

Fig. 2 Temperature distribution example of exhaust system [2]

systems have been researched and developed for automotive waste heat management technology and are described as follows.

2.2.1 Heat–Electric Power Use

For automotive waste heat management, investigation of thermoelectric element material and systems for automobiles have been studied in various countries. Examinations of automobiles are supported by NEDO and JST in Japan. If the quantity of thermoelectric generation reaches 0.5–1 kW, the thermoelectric element system could substitute for the alternator.

There are thermoelectric conversions that use the Seebeck effect. Because the system efficiency including that of a heat exchanger in actual conditions is low (5–7 %) [3], development of high-efficiency material is required. However, the systems are relatively small and are an effective technology in the current electrification of automobiles.

2.2.2 Heat–Motive Power Use

A heat engine, including automobile engines, is a technology to convert thermal energy into power. To use waste heat from an automobile engine, an exhaust heat recovery system using the Rankine cycle has been developed. The Rankine system with an engine-exhaust heat source generates electricity and charges the battery of an HEV to assist in driving the motor [4]. From energetic analysis data for travel at 100 km/h, exhaust heat recovery efficiency is improved by 3.8 %.

However, this type of external combustion engine tends to be heavy for attaining high power and may affect fuel efficiency.

2.2.3 Heat–Heat Use

As effective waste heat reuse and recycling technologies, there are heat storage and HPs driven by thermal energy. Heat storage technologies to reuse heat are of three types: the sensible heat storage using heat capacity, latent heat storage using phase-change heat, and chemical heat storage using chemical reaction energy. As an HP technology to recycle heat for automobiles, an air conditioner compression HP is mainly used, but it needs external power like engine drive force. Heat pumps that can function with only a waste heat source are absorption, adsorption, and chemical types.

Because details on HP principles are given in other chapters, I introduce only R&D examples for automotive technology. Efficiencies of the heat–heat use technology are much higher than those of heat–electric power use (Sect. 2.2.1) and heat–motive power use (Sect. 2.2.2.). The heat storage technology facilitates the reuse of dozens of percent of the waste heat source. The chemical HP technology permits recycling of the waste heat source and releases several hundreds of percent of hot/cold heat.

2.2.3.1 Sensible Heat Storage Technology

For this system that recovers waste heat by sensible heat storage, an R&D example shows cooling water warmed by engine heat in a thermal storage tank and warming of the engine before its next start [5]. The system is equipped with a commercialized automobile, reduces hydrocarbon release by 14 %, and shortens warm-up time by about half. However, this type of system may make the automobile larger/heavier, because the heat storage density is low and heat loss is substantial.

2.2.3.2 Latent Heat Storage Technology

Latent heat storage uses phase-change heat of a material, such as ice, inorganic hydrated salt, paraffin, or organic substance. It is better to use a higher storage density at nearly constant temperature in the phase-change scenario in comparison with sensible heat storage. However, selection of the temperature level is required, as is an examination of price and durability.

There are still few examples of practical use, but a latent heat storage material at 3–30 °C of Eco-Joule has been developed in Japan and used to improve the performance of a car component canister [6].

2.2.3.3 Chemical Heat Storage Technology

Chemical heat storage uses the chemical reaction energy of a material and has some distinctive features, such as advantages of high heat storage density, small heat loss, and constant reaction temperature. Furthermore, it can be developed in a chemical HP that can alter the heat storage/release temperature.

Chemical heat storage is already in practical use as with a portable body warmer, which can release heat for a long period. However, an upgraded chemical HP function controlling the temperature of released heat is desirable for introduction into next-generation automobile heat management, as follows.

2.2.3.4 Heat Pump Technology

Figure 3 shows a schematic diagram of heat engine and HP operation. A conventional mechanical heat pump (MHP) generates high-temperature heat $Q_{H,R}$ at temperature $T_{H,R}$ or low-temperature heat $Q_{L,R}$ at $T_{L,R}$ by work W using engine/electric power, as shown on the right side of the figure. A chemical heat pump (CHP) does not need W. Instead of W, a CHP creates energy for W by chemically storing high-temperature heat $T_{H,S}$ in the heat storage step as in heat engine operation (left side of figure).

The coefficient of performance (COP) based on input and output enthalpy is defined for CHP by Eq. 1. In our studies, the COP often exceeded 1 [7].

$$COP_h = \frac{Q_{L,R} + Q_{H,R}}{Q_{H,S}} \tag{1}$$

Next are shown some next-generation automobile waste heat recovery technologies with CHP, developed for practical use.

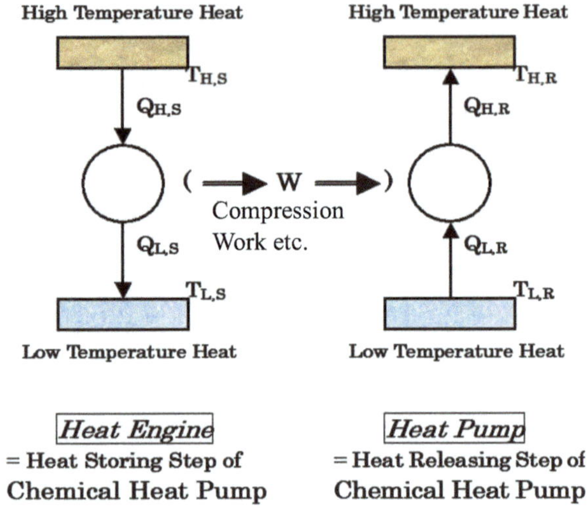

Fig. 3 Operation principle of heat engine and heat pump

2.2.3.5 Chemical Heat Pump System for Engine Waste Storage and Cold-Start Heating

While an automobile is running (heat storage step of CHP), waste heat from the engine is stored by the CHP using calcium oxide/water/calcium hydroxide ($CaO/H_2O/Ca(OH)_2$) as shown in Fig. 4. The heat is stored in the form of thermo-chemical energy by decomposition of $Ca(OH)_2$ (s) in the high-temperature side reactor. Released water vapor H_2O (g) flows into the low-temperature side reactor (condenser) because of a pressure difference between the two reactors. When the car starts (engine off: heat-release step of CHP), hot/cold heat is generated rapidly by CHP for warming the engine and cooling the passenger cabin [8]. The water vapor H_2O (g) flows from the low-temperature side reactor (evaporator) to the high-temperature side reactor by opening a valve, owing to the pressure difference between reactors. The exothermic hydration reaction of the reactant CaO (s) releases high-temperature heat $Q_{H,R}$ in the high-temperature side reactor. The evaporator is cooled by releasing its latent heat $Q_{L,R}$ for the air conditioner (A/C in Fig. 4).

It was found in our lab-CHP experiments that chemical heat storage in the reactor proceeded by actual 400-cc engine-exhaust gas at ~773 K. The stored heat was calculated at 4.5 MJ in a 5.3-MJ-scale heat exchanger. In the heat-release step, the stored heat in the 2.5 kg CaO reactor density ~1.8 MJ/kg CaO was released for heating the heat-exchange media for a radiator. This was heated to 373 K in 10 min and began boiling. These results show the practical potential for a $CaO/H_2O/Ca(OH)_2$ CHP to recycle automobile waste heat. However, the reactor of a CHP should be designed for individual vehicles to meet the waste heat condition.

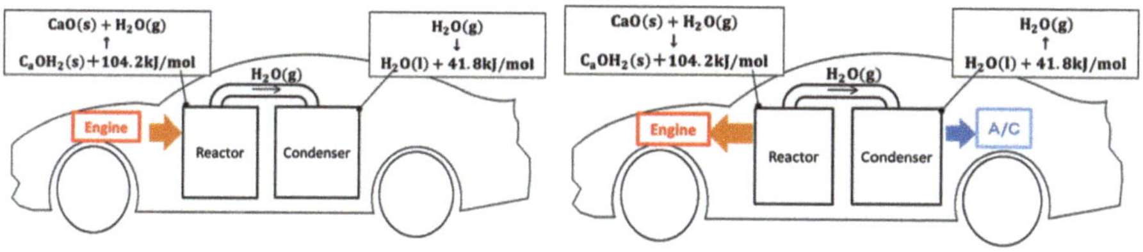

Fig. 4 CHP R&D example for engine waste storage and cold-start heating. *Left*, running (heat storage step); *right*, starting (heat-release step)

3 Technology Roadmap

For future automotive diffusion forecasting, the Ministry of the Environment hosts an environmental response automobile diffusion study meeting. An environmental response vehicle diffusion strategy [9] was published in 2010, which included a forecast of future environmental response vehicle diffusion numbers. The holding number of environmental response vehicles such as HEVs, PHEVs (plug-in hybrid electric vehicle), and EVs was only 2 % in 2010, but reaches nearly 50 % in 2030 and 90 % in 2050. However, HEVs and PHEVs are equipped with an engine, so internal combustion engine deployment vehicles were still ~80 % in 2030 and 50 % in 2030. Therefore, until around the year 2050, strong promotion of engine efficiency of conventional automobiles and effective utilization of engine waste energy as shown in the technology roadmap Fig. 5 are important, as are next-generation automobiles.

Before considering the technical roadmap for effective thermal energy utilization of automobiles, the political roadmap should be explored, especially for regulation of environmental response vehicles. At present, the government is simulating transportation systems with those vehicles in the future. Until around 2020, feasibility studies will be done for tax control and energy efficiency under various scenarios. Until around 2030, tax control and financial support for the ICEV, HEV, and PEHV should be nearly complete and applied. Until around 2050, those for FCEVs (fuel cell electric vehicle) should also be nearly fixed and applied. In this way, social systems including regulation for tax control and financial support are implemented for environmental impacts based on vehicle energy efficiencies.

For the aforementioned political roadmap, the technical roadmap of effective thermal energy utilization for automobiles will be mainly manifested by two technologies, as described in the following.

The first is the waste heat to heat utilization technology. At present, the heat storage and HP technologies are used in automobiles, but are limited as stated in Sect. 2. Until around 2020, thermally driven HPs such as CHPs should be used in automobiles, even in feasibility studies. Until 2030, these HPs should be put to practical use.

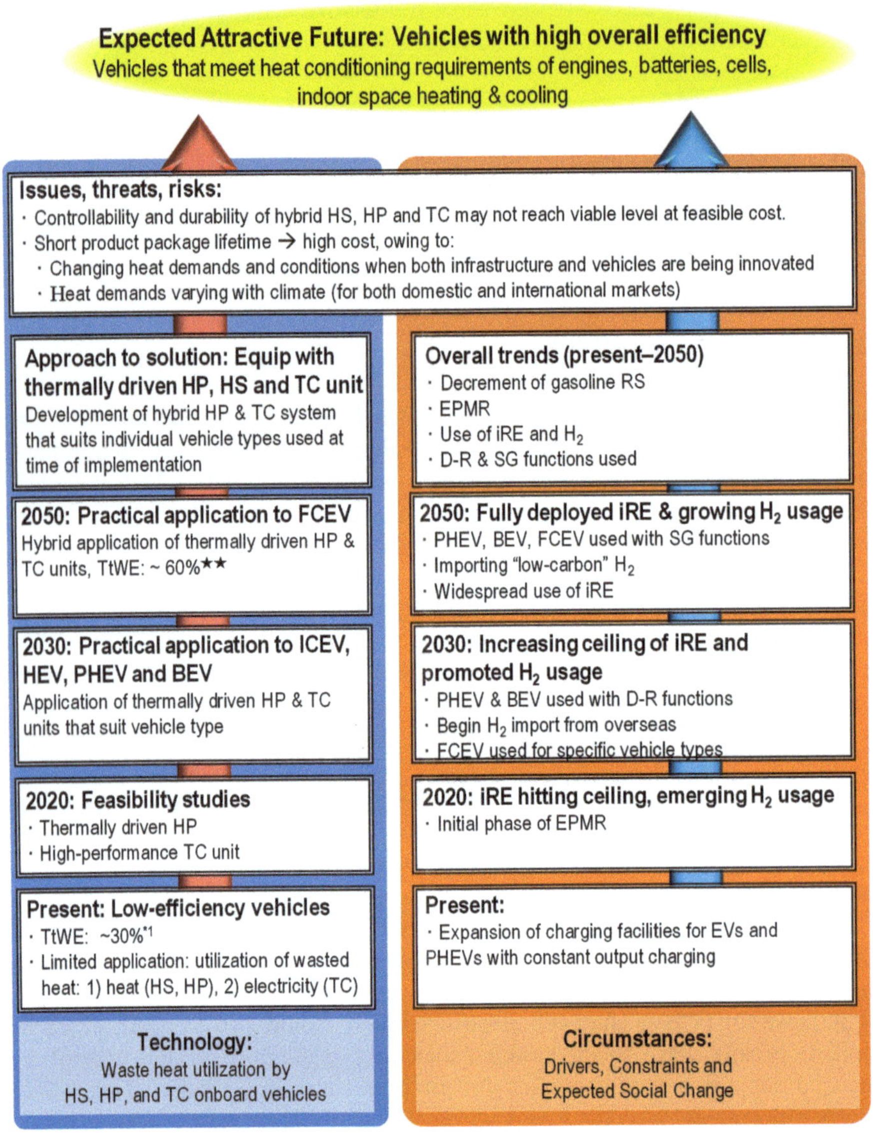

*1 [1]

Abbreviations:
BEV: Battery Electric Vehicle, D-R: Demand-Response, EPMR: Electric Power system Market Reformation, FCEV: Fuel Cell Electric Vehicle, HP: Heat Pump, HS: Heat Storage, HEV: Hybrid Electric Vehicle, ICEV: Internal Combustion Engine Vehicle, iRE: intermittent Renewable Electricity, NC: Normal Charger (200 V), TC: Thermoelectric Conversion, PHEV: Plug-in Hybrid Electric Vehicle, QC: Quick Charger, RS: Refueling Station, SG: Smart-Grid, stns.: stations, TtWE: Tank-to-Wheel Efficiency

Fig. 5 Technology roadmap of effective thermal energy utilization for automobiles

Second, the waste heat to electric power utilization technology will be developed for automobiles. Current thermoelectric conversion technologies are also used in automobiles, but are limited as mentioned in Sect. 2. Until around 2020, high-performance thermoelectric conversion units should be used in automobiles, even in feasibility studies. Until 2030, high-performance thermoelectric conversion units should be in practical use.

In 2050, hybrid application of the above thermally driven HP and high-performance thermoelectric conversion unit should be realized for automobiles in practical use.

4 Benefit and Prospective Future Vision

The goal of the technology roadmap is the reduction of environmental impacts of automobiles by raising their total energy efficiency. The energy efficiency of tank-to-wheel is presently ~30 % (Fig. 1) [1]. This should be ~60 % using the aforementioned hybrid application of thermally driven HP and high-performance thermoelectric conversion unit. This will be achieved by recycling about half the energy of the waste heat of current vehicles.

The viability will depend not only on the performance of the thermally driven HP and high-performance thermoelectric conversion unit but also on waste heat quality. The amount of waste heat will decrease by vehicle improvements and temperatures will decline. Therefore, the thermally driven HP and high-performance thermoelectric conversion unit must match waste heat qualities. In particular, controllability and durability of the systems to match the heat source must be developed for individual vehicles.

As a simulated example of fuel consumption [10], we compared model A, a popular conventional ICEV, weight 1500 kg, 2000-cc engine equipped with MHP for air conditioning, and model B, equipped with hot/cold heat generation CHP for warming up and air conditioning of a model A vehicle. These simulations were performed with consideration of MHP performance, CHP weights, fuel efficiencies at engine temperature, and others. As a result, although the model A fuel efficiency of 13.5 km/l became 11.9 km/l with the air conditioner on in a JC08 mode, the fuel efficiency of model B was 13.9 km/l even with the air conditioner on. For a cold-start condition with the air conditioner on, the fuel efficiency of model A became 9.9 km/l, and that of model B was 12.9 km/l. In this case, model A consumed 880 l/year and released 2.04 t CO_2/year, whereas model B consumed 680 l/year and released 1.58 t CO_2/year.

Until around 2050, vigorous promotion of engine efficiency of conventional automobiles and effective utilization of engine waste energy as shown in the technology roadmap are important, as are next-generation automobiles. By 2050, hybrid application of the thermally driven HP and high-performance thermoelectric conversion unit should be realized for automobiles in practical use. For electric power, exhaust heat recovery by thermoelectric conversion is not very efficient, but would be valuable if a vehicle needs more electricity. However, the hot/cold heat use by recycling exhaust heat called "heat–heat use" has superior efficiency. In particular, automobile systems of chemical heat storage/chemical HP recovery of waste heat and reformation as cold/hot heat are expected. CHP systems can be driven not only by engine waste heat of ICEV but also by various waste heats from HEVs/PHEVs or electric power in charging equipment and can save on energy used by expensive batteries for heating and cooling.

These types of next-generation technologies with highly effective energy recycling systems of automobile waste energy have little environmental impact, have low cost, and are anticipated to be in practical use in the near future.

References

1. Society of Automotive Engineers of Japan (JSAE) (2008) Automotive technology handbook, basics/theory, revised edition. Seikosya
2. Chau KT, Chan CC (2007) Emerging energy-efficient technologies for hybrid electric vehicles. Proc IEEE 95(4):821–835
3. Matsubara K (2011) Prospects of the thermoelectric conversion technologies. J Plasma Fusion Res 87(12):818–824
4. Ibaraki S, Endo T, Kojima Y, Takahashi K, Baba T, Kawajiri S (2007) Study of efficient on-board waste heat recovery system using ranking cycle. Trans Soc Automot Eng Jpn 38 (4):73–78
5. Miyata Y, Morikawa T, Ito S, Toyoshma T (2005) Development of heat storage system for warming up engine. Denso Tech Rev 10(1):88–95
6. Petrochemical Part of Japan Energy (2010) Application of paraffin latent heat storage materials to electric vehicle, The second management strategy lecture in Chugoku local industry-university co-operation official cooperation consortium
7. Ogura H, Yamamoto T, Kage H (2003) Efficiencies of $CaO/H_2O/Ca(OH)_2$ chemical heat pump for heat storing and heating/cooling. Energy 28(14):1479–1493
8. Ogura H, Enomoto J, Hanatsuka D, Hori R, Kobayashi H (2009) Experimental study on chemical heat pump for waste heat utilization of automobiles. Proc. of Spring Meeting of Society of Automotive Engineers of Japan (JSAE), 20–22 May, Yokohama, Japan, 236
9. Ministry of the Environment (2010) Environmental response vehicle diffusion strategy. https://www.env.go.jp/air/report/h22-02/index.html
10. Ogura H, Fujita H (2015) Possibility of waste heat recycle utilization of automobiles by chemical heat pump. In: Proceedings of Autumn Meeting of Society of Automotive Engineers of Japan (JSAE), 14–16 October, Kitakyushu, Japan

Index

Y. Kato et al. (eds.), *Energy Technology Roadmaps of Japan*,
DOI 10.1007/978-4-431-55951-1

T

U